Principles of Soil Chemistry

Principles of Soil Chemistry

Editor: Rufus Mitchell

Callisto Reference
www.callistoreference.com

Callisto Reference,
118-35 Queens Blvd., Suite 400,
Forest Hills, NY 11375, USA

Visit us on the World Wide Web at:
www.callistoreference.com

ISBN: 978-1-63239-913-7 (Hardback)

Printed in China.

Cataloging-in-Publication Data

Principles of soil chemistry / edited by Rufus Mitchell.
p. cm.
Includes bibliographical references and index.
ISBN 978-1-63239-913-7
1. Soil chemistry. 2. Soil science. I. Mitchell, Rufus.
S592.5 .P75 2018
631.41--dc23

Table of Contents

Preface

This book aims to highlight the current researches and provides a platform to further the scope of innovations in this area. This book is a product of the combined efforts of many researchers and scientists, after going through thorough studies and analysis from different parts of the world. The objective of this book is to provide the readers with the latest information of the field.

The branch of science that studies the chemical characteristics of soil is known as soil chemistry. It is affected by mineral composition and environmental factors. Soil chemistry also helps in understanding the contaminants that are present in soil. As this field is emerging at a rapid pace, the contents of this book will help the readers understand the modern concepts and applications of the subject. This book discusses the fundamentals as well as modern approaches of soil chemistry. Coherent flow of topics, student-friendly language and extensive use of examples make this book an invaluable source of knowledge.

I would like to express my sincere thanks to the authors for their dedicated efforts in the completion of this book. I acknowledge the efforts of the publisher for providing constant support. Lastly, I would like to thank my family for their support in all academic endeavors.

Editor

1

β-Diversity of Functional Groups of Woody Plants in a Tropical Dry Forest in Yucatan

Jorge Omar López-Martínez[1], Lucía Sanaphre-Villanueva[1], Juan Manuel Dupuy[1]*, José Luis Hernández-Stefanoni[1], Jorge Arturo Meave[2], José Alberto Gallardo-Cruz[2]

1 Centro de Investigación Científica de Yucatán, Mérida, México, 2 Departamento de Ecología y Recursos Naturales, Facultad de Ciencias, Universidad Nacional Autónoma de México, México, México

Abstract

Two main theories have attempted to explain variation in plant species composition (β-diversity). Niche theory proposes that most of the variation is related to environment (environmental filtering), whereas neutral theory posits that dispersal limitation is the main driver of β-diversity. In this study, we first explored how α- and β-diversity of plant functional groups defined by growth form (trees, shrubs and lianas, which represent different strategies of resource partitioning), and dispersal syndrome (autochory, anemochory and zoochory, which represent differences in dispersal limitation) vary with successional age and topographic position in a tropical dry forest. Second, we examined the effects of environmental, spatial, and spatially-structured environmental factors on β-diversity of functional groups; we used the spatial structure of sampling sites as a proxy for dispersal limitation, and elevation, soil properties and forest stand age as indicators of environmental filtering. We recorded 200 species and 22,245 individuals in 276 plots; 120 species were trees, 41 shrubs and 39 lianas. We found that β-diversity was highest for shrubs, intermediate for lianas and lowest for trees, and was slightly higher for zoochorous than for autochorous and anemochorous species. All three dispersal syndromes, trees and shrubs varied in composition among vegetation classes (successional age and topographic position), whilst lianas did not. β-diversity was influenced mostly by proxies of environmental filtering, except for shrubs, for which the influence of dispersal limitation was more important. Stand age and topography significantly influenced α-diversity across functional groups, but showed a low influence on β-diversity –possibly due to the counterbalancing effect of resprouting on plant distribution and composition. Our results show that considering different plant functional groups reveals important differences in both α- and β-diversity patterns and correlates that are not apparent when focusing on overall woody plant diversity, and that have important implications for ecological theory and biodiversity conservation.

Editor: Jerome Chave, Centre National de la Recherche Scientifique, France

Funding: Fondo Mixto CONACYT-Gobierno del Estado de Yucatán (FOMIX-Yucatán) project YUC-2008-C06-108863. Centro de Investigación Científica de Yucatán, A.C. (CICY). The funders had no role in study design, data collection and analysis, decision to publish, or preparation of the manuscript.

Competing Interests: The authors have declared that no competing interests exist.

* E-mail: jmdupuy@cicy.mx

Introduction

Uncovering the mechanisms underlying variation in species composition (β-diversity) is a central goal of community ecology and biogeography [1], [2], [3]. Different theories have been proposed to explain variation in species composition, and many of our current ideas are based on the observation that species differ in their environmental requirements [4], [5]. The niche theory proposes that most of this variation is related to particular adaptations of species to the environment [6], [7], [8]. Therefore, species abundance may be an indicator of how suitable environmental conditions are for species to grow and reproduce. In this sense, functional attributes such as growth form would be important to understand the mechanisms assembling plant communities, because they represent different strategies of resource partitioning [4].

In contrast, the neutral theory [9], [10], [11] suggests that all species are competitively equal and able to grow equally well under a wide range of environmental conditions. According to the neutral theory, the main factor determining plant community composition is dispersal limitation or recruitment limitation, that is, the limited capacity of plants to effectively disperse their seeds and establish seedlings away from parent individuals, which causes the spatial location of an individual to be constrained by the location of its parent. As a result, sites located closer together should be more similar in composition than those lying far apart [10], [12] regardless of their environmental conditions [13]. Therefore, dispersal syndrome appears to be an important functional trait that should be considered to understand the mechanisms affecting β-diversity.

Other factors may also influence β-diversity patterns. For example, previous studies have shown that the magnitude of β-diversity is influenced by secondary succession [14], soil properties [1], [2], [15], [16], fragmentation [17] and topography [18]. These factors produce changes in community patterns by modifying the local environment and resource availability [19], hence creating opportunities for the establishment of new species and reducing populations of those that are already established (see reviews by [20], [21], [22]).

Additionally, species with specific traits may be favored over others at different successional stages, and different ecological strategies are dominant at different moments of vegetation development [23], [24], [25]. For instance, it has been found that plants of early successional stages in Neotropical lowland forests have primarily anemochorous (i.e. wind-dispersed) or autochorous (explosive) dispersal, while plants with zoochorous (i.e. animal-dispersed) seeds are more common in old-growth forests [26,27]. Similarly, different life forms alternate dominance along secondary succession. In tropical wet forests, herbs, shrubs and vines dominate during the initial successional stages, and are subsequently replaced by short- and long-lived pioneer trees – which in turn are replaced by shade-tolerant trees [28]. In Bolivian dry forests the abundance of shrubs, lianas and bromeliads increases with successional age, whereas grasses become increasingly rarer [29]. Moreover, it has been proposed that species with long-distance dispersal, such as pioneer species, should be able to locate most of the appropriate habitats in a landscape, and such dispersal may compensate for low persistence rates at any given site [30], [31]. Accordingly, spatial variation in species composition (i.e. β-diversity) is expected to be lowest for species with the longest dispersal distance.

β-diversity also varies with plant community type, even within a single landscape [32]. For instance, β-diversity tended to increase when moving from a tree-dominated community to a shrub- or herb-dominated one [33]. However, β-diversity patterns among growth forms are not always consistent. For example, in Appalachian spruce-fir forests β-diversity values were twice as large for herb assemblages compared to woody plants [30], while in a Chinese subtropical forest, trees showed higher β-diversity than shrubs [34]. Thus, dispersal syndrome and growth form are important plant life-history traits that could shed light on the processes and mechanisms influencing plant community assembly [4], [35], [36].

The objectives of this study were twofold. First, we wanted to explore how α- and β-diversity of plant functional groups defined by dispersal syndrome and growth form vary with successional age and topographic position. Second, we sought to assess and compare the relative influence of environmental filtering (niche differentiation) vs. dispersal limitation on β-diversity of these functional groups. We used elevation, soil properties and forest stand age as putative environmental filters, whereas dispersal limitation was assessed in terms of geographical distance among sampling sites.

We made three predictions based on the findings of previous studies and on the following arguments. Since β-diversity is expected to be inversely related to dispersal distance [30], [31], our first prediction was that β-diversity would be highest for autochorous species and for shrubs, and lowest for anemochorus species and for trees (P1). Second, since plant diversity and structural complexity generally increase during tropical forest succession [20], [28], [37], [38], [39], we expected an increase in overall species richness (α-diversity) across growth forms and dispersal syndromes during succession, with a prevalence of shrubs and of anemochorous and autochorous species in early successional ages, and of zoochory and tree species in more advanced successional ages (P2). Finally, traits that are relevant for plants' ecological strategy could have evolved in response to environmental filters [4], [40], [41], [42]. Since a greater dispersal distance may increase the chances of finding appropriate environmental conditions for establishment, we predicted that the association between environmental filters and β-diversity would be strongest for trees and for anemochorous species, and weakest for shrubs and for autochorous species (P3).

Materials and Methods

Study area

The study area (ca. 22×16 km) is centrally located in the Yucatán Peninsula, Mexico (20°01′–20°16′N, 89°39′–89°60′W) (Fig. 1). Climate is warm sub-humid, with summer rain (May – October) and a marked dry season (November – April). Mean annual temperature is ca. 26°C, and mean annual precipitation ranges between 1,000 and 1,200 mm [43]. The landscape consists of Cenozoic limestone hills with moderate slopes (10–25°), which alternate with flat areas; elevation ranges from 60 to 180 m a.s.l. Soil heterogeneity is strongly associated with topography; with shallow, rocky Lithosols and Rendzines dominating the hills, while deeper, clayey Luvisols and Cambisols prevail in flat areas [44]. Regional vegetation is classified as seasonally dry tropical forest (50–75% of the species shed their leaves during the dry season) and it is composed of forest stands of different successional ages since abandonment after traditional slash-and-burn agriculture, which has been practiced for over 2000 years [45], and permanent agricultural fields concentrated around the villages (Fig. 1). Land tenure is mostly communal and approvals for field work were obtained from each of the "ejidos" whose lands fell within the study area: Xkobenhaltún, Xul and Yaxhachén.

Vegetation and environmental data

A land cover map of the studied area was obtained from a SPOT 5 satellite image (10 m spatial resolution) acquired in January 2005, after conducting a supervised classification of previously geo-referenced and radiometrically corrected bands 2 (red), 3 (near infrared), and 4 (mid infrared). Six land-cover types were defined: (1) 3–8 yr-old secondary forest (vegetation class 1 or: VC1); (2) 9–15 yr-old secondary forest (VC2); (3) >15 yr-old secondary forest on flat areas (VC3); (4) >15 yr-old secondary forest on hills (VC4) (5) agricultural fields; (6) urban areas and roads. The vegetation classes correspond to different ages of forest succession and different topographic positions. The ranges of successional ages were selected according to the differences in vegetation reflectance using the SPOT 5 imagery, while topographic position represented the hills and flat areas present in the landscape. Both attributes have been shown to influence vegetation structure, diversity and composition, with canopy height, basal area and α-diversity increasing with forest stand age, and most attributes, including species composition, differing between hills and flat areas [39]. The accuracy of the classified map was assessed using the overall accuracy and Cohen's Kappa statistic (see [46] for details on the classification procedure).

Field data were collected from June to September of 2008 and 2009 through a hierarchical design. First, we randomly selected 23 sampling landscapes (SL) of 1 km^2 each, encompassing the whole range of forest fragmentation in the study area. We then randomly located three plots for each of the four vegetation classes (VC1-VC4) within each SL (276 sampling sites in total) using a GPS unit. Each plot consisted of a circular 200 m^2 area where all woody plants ≥5 cm in DBH were recorded, plus a concentric 50 m^2-subplot where all woody plants ≥1 cm in DBH were recorded. Stand age of each site was determined through interviews with local residents who had lived in the area for >40 yr. Fifty three sampling sites belonged to VC1, 75 to VC2, 86 to VC3, and 62 to VC4, corresponding to the approximate proportional areas covered by each vegetation class in each SL.

To characterize local soil conditions, three 10 cm-deep soil samples (ca. 400 g each) were taken at the centre, northern and southern limits of each sampling plot and subsequently pooled to obtain a single compound sample. The following

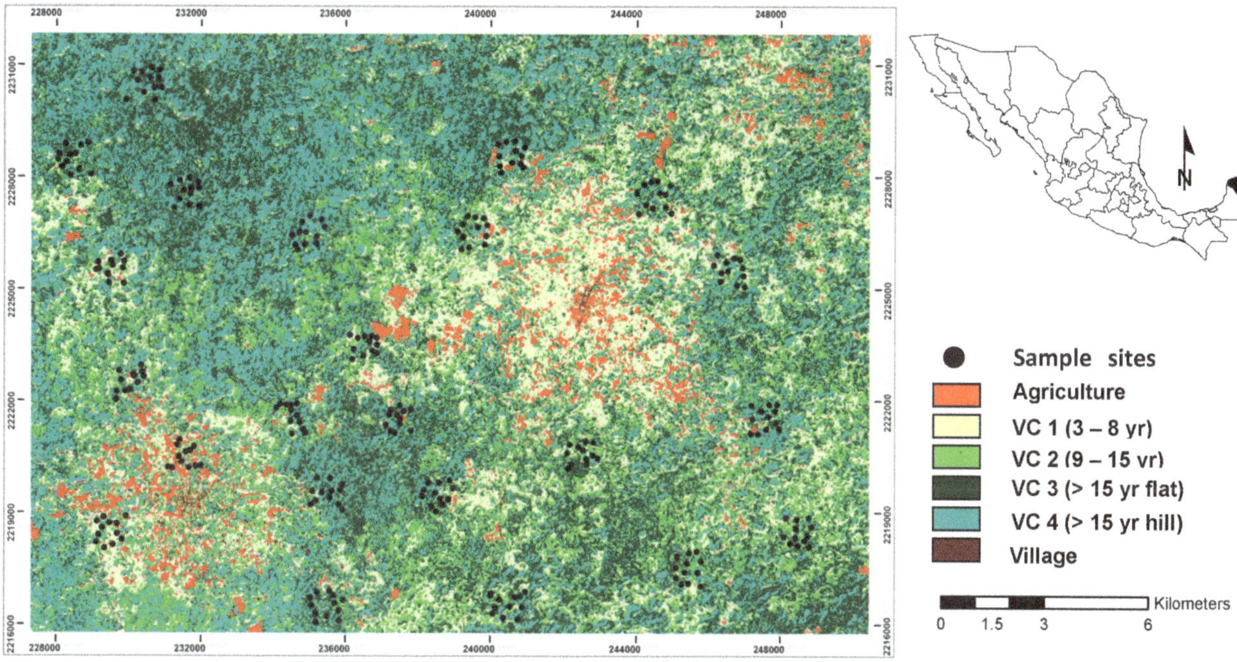

Figure 1. Location and land-cover thematic map of the study area showing the location of sampling sites.

physical-chemical properties were analyzed: pH in water (1:2 ratio), electric conductivity (EC; 1:5 extracts), cation exchange capacity (CEC; Olsen), soil organic matter (SOM, combustion or Walkley-Black), total nitrogen (N, Kjeldahl), available phosphorous (P, Bray), interchangeable potassium (K; with ammonium acetate), and soil texture (percent lime, clay and sand content, Bouyoucos) [39]. Finally, we recorded elevation and slope for each plot using a Digital Elevation Model (DEM, spatial resolution: 50 m) from the Instituto Nacional de Estadística, Geografía e Informática (INEGI).

Plant species were identified in the field by expert regional parataxonomists. Plant specimens of those species that could not be identified in the field were collected and identified by experts from the herbarium at the Center of Scientific Research of Yucatan. We categorized species according to two functional attributes: growth form (shrubs, trees and lianas) and dispersal syndrome (authochory, anemochory and zoochory) based on information obtained from the herbarium at the Center of Scientific Research of Yucatan (http://www.cicy.mx/ofer-tec-herbario/herbario), descriptions provided in the Flora of Veracruz (http://www1.inecol.edu.mx/floraver/inicio.htm), the Flora Mesoamericana (http://www.tropicos.org/Project/FM), the Flora of Barro Colorado Island (http://biogeodb.stri.si.edu/bioinformatics/croat/home), the Flora of Guatemala (http://archive.org/search.php?query=flora%20of%20guatemala), The Missouri Botanical Garden (http://www.mobot.org/mobot/fm/), and based on the expert opinion of regional parataxonomists.

Spatial data

We used a principal coordinate of neighbor matrices analysis (PCNM, [47]) using the R software with the 'spacemakeR' library (version 2.10.1, R Development Core Team) to generate a set of explanatory spatial variables that represent the spatial structure of sampling sites. This set of variables (called PCNM vectors) represents a spectral decomposition of the spatial relationships among the sampling sites [48]. PCNM vectors are also uncorrelated variables that can be used as predictors in regression analysis to describe spatial relationships in community data, because they are not subject to multicollinearity [47]. This analysis was performed separately for each functional group. In each analysis we selected those PCNM vectors having positive, significant autocorrelation values (Moran's I; $P>0.001$) [48].

Data analysis

We compared the relative magnitude of variation in woody species composition among dispersal syndromes and growth forms using Detrented Correspondence Analysis (DCA). This method indicates the variation in species composition in standard deviation units, which are directly comparable among datasets [49]. To assess the differences in species composition of each dispersal syndrome and growth form among vegetation classes we used an analysis of similarity (ANOSIM) in PRIMER-E 6.1.12 [50].

We performed a variation partitioning analysis [51] to evaluate the relative contribution of the abiotic environment (elevation and soil properties), stand age, and spatial structure of sampling sites on β-diversity for each dispersal syndrome and growth form. The total variation of each assemblage was dissected into five components: (1) a pure abiotic environmental effect, (2) a pure effect of spatial structure of sampling sites, (3) a stand age effect, (4) the combined variation explained by all components, and (5) the unexplained variation. To this end, we transformed species abundances using Hellinger's transformation and conducted a redundancy analysis (RDA) [52] in CANOCO 4.0 [53], down-weighing the most abundant species to reduce their influence on the analysis [54]. We recorded the proportion of the variation (R^2) explained by all significant ($P<0.05$) variables of each component (i.e. soil properties, stand age and PCNM vectors) using forward selection with unrestricted Monte-Carlo permutations.

Additionally, we evaluated differences in species richness (α-diversity) among functional groups and vegetation classes using a generalized linear model (GLM) with a Poisson error distribution and a logarithmic link [55], [56], in which dispersal syndrome and growth form were nested within vegetation class. The best explanatory model was selected using backward stepwise selection starting from a global model [55]. *Post hoc* comparisons among vegetation classes, growth forms and dispersal syndromes were performed using contrast tests. These analyses were carried out in R v. 2.13.0 [57].

Results

Β diversity by functional group and vegetation class

A total of 22,245 individuals belonging to 200 species and 50 families were recorded in the 276 plots (Table S1). Trees were by far the dominant growth form, both in terms of species richness and abundance, with shrubs and lianas lagging far behind. Zoochory was the most specious (but least abundant) dispersal syndrome, followed by autochocory and anemochory. Lianas were predominantly anemochorous, whilst shrubs were predominantly zoochorous and trees were more evenly distributed among and dominated all dispersal syndromes (Table 1).

Spatial variation in floristic composition (β-diversity) was highest for shrubs (DCA ordination axes represented 19.13 SD), lowest for trees (6.97 SD), and intermediate for lianas (12.83 SD). With respect to dispersal syndrome, β-diversity was only slightly higher for the zoochorous guild (9.44 SD), than for the autochorous and anemochorous guilds (7.59 and 7.55 SD, respectively). The analysis of similarity (ANOSIM) revealed that tree species composition was similar between the youngest vegetation classes VC1 and VC2 ($P = 0.15$), but differed among all other classes (Table 2). Shrubs differed in species composition among all vegetation classes ($P < 0.011$), except between VC3 and VC4, which were within the same range of successional age (16–70 years; $P = 0.12$). In contrast, liana species composition did not differ among vegetation classes, except between VC1 and VC4 ($P = 0.02$). Dispersal syndromes generally differed significantly among vegetation classes, except between the two youngest ones (Table 2).

Species richness by functional group across secondary succession

Species richness (α-diversity) differed significantly among vegetation classes, as well as among functional groups, and showed an increase with successional age (VC1<VC2<VC3) with a peak in VC4 (>15 yr-old forests on hills) (Fig. 2). Zoochory was the dominant dispersal syndrome in all vegetation classes, except for VC1 (3–8 yr-old forests), and followed the same general pattern shown by overall species richness (VC4>VC3>VC2>VC1). Autochory dominated in VC1, and showed higher richness than anemochory in all vegetation classes ($P < 0.001$), except for VC3 (>15 yr-old forests on flat areas), where these syndromes were not significantly different. Trees were the dominant growth form, and showed the same general pattern as overall species richness (VC4>VC3>VC2>VC1). Shrubs and lianas had similar richness in all vegetation classes except for VC3, where there was a significant difference between them ($P < 0.001$) (Fig. 2).

Variation partitioning

Abiotic environmental variables (elevation and soil properties), successional age and spatial structure of sampling sites all influenced β-diversity, but accounted for less than one third of its total variation; furthermore, their relative influence differed among functional groups (Table 3). In terms of growth form, overall variation in species composition explained by all independent variables was greater for trees (29.2 % of total variation), compared to lianas (12.9%) and shrubs (10.3%). Moreover, the relative influence of explanatory variables varied considerably among growth forms. For trees, abiotic environmental variables accounted for most of the variation in composition that was explained by the model (51%), while shrubs were more strongly influenced by the spatial structure of sampling sites (46.8%), and lianas were similarly influenced by environmental variables and spatial structure (42.1% and 39.5%, respectively). Stand age accounted for around 12% of β-diversity of trees and shrubs, but had no significant influence on lianas. The single most important environmental variable for trees was soil organic matter (SOM; 74.5% of the variation due to abiotic environment), for shrubs it was cation exchange capacity (CEC; 76.5%) whereas for lianas it was pH (44.0%; Table 3).

Regarding dispersal syndromes, explanatory variables were slightly more strongly associated with the autochorous and anemochorous guilds (31.2% and 27.4%, respectively), than with the zoochorous guild (20.4%). Across dispersal syndromes, abiotic environmental variables (elevation and soil properties) consistently accounted for most of the variation in species composition (47.6–59%), followed by spatial structure of sampling sites (22.4–40.2%) –although, in the case of zoochory, the percentage of variation explained by abiotic variables and spatial structure was similar (47.6% and 40.2%, respectively). Forest stand age had a stronger influence on anemochorous species (20%), compared to autochorous and zoochorous ones (9.3% and 6.9%, respectively). Soil organic matter (SOM) was consistently the single most important environmental variable, accounting for 61.9–82.1% of the variation in composition due to abiotic environment across dispersal syndromes (Table 3).

Discussion

We found marked differences in the magnitude and correlates of β-diversity among growth forms and smaller differences among

Table 1. Species richness and abundance of woody plants by growth form and dispersal syndrome.

Dispersal syndrome	Growth form						Total abundance	Total richness
	Trees		Shrubs		Lianas			
	Abundance	Richness	Abundance	Richness	Abundance	Richness		
Anemochory	6431	17	8	4	1143	30	7582	51
Autochory	7594	42	324	14	19	3	7937	59
Zoochory	6013	61	575	23	138	6	6726	90
Total	20038	120	907	41	1300	39	22245	200

Table 2. Results of ANOSIM comparisons of community composition among vegetation classes by growth form and dispersal syndrome.

Type of comparison		**Growth form**						**Dispersal syndrome**					
		Trees		**Shrubs**		**Lianas**		**Anemochory**		**Autochory**		**Zoochory**	
		r	*P*	*r*	*P*	*r*	*P*	*r*	*P*	*r*	*P*	*r*	*P*
All vegetation classes		0.28	<.01	0.11	<.01	0.014	0.17	0.18	<.01	0.24	<.01	0.18	<.01
Pairwise comparision	1–2	0.021	0.15	0.071	<.01	0.003	0.51	0.004	0.39	0.028	0.1	0.053	0.03
	1–3	0.29	<.01	0.21	<.01	0.031	0.24	0.241	<.01	0.128	<.01	0.212	<.01
	1–4	0.71	<.01	0.37	<.01	0.078	0.02	0.488	<.01	0.65	<.01	0.514	<.01
	2–3	0.16	<.01	0.05	0.011	0.009	0.18	0.129	<.01	0.113	<.01	0.055	<.01
	2–4	0.43	<.01	0.122	<.01	0.001	0.44	0.299	<.01	0.423	<.01	0.22	<.01
	3–4	0.24	<.01	0.016	0.12	0.006	0.34	0.053	<.01	0.25	<.01	0.229	<.01

dispersal syndromes. Our results suggest that both dispersal limitation and environmental filtering contribute to the assembly of woody plant species, but their relative influence varies among functional groups – especially growth forms.

Β-diversity by dispersal syndrome and growth form

In agreement with our first prediction (P1), variation in floristic composition was greatest for shrubs and lowest for trees. Other studies have shown a positive influence of plant height on dispersal distance [58], [59]. This is logical for wind-dispersed species, but could also hold for zoochorous species, possibly reflecting systematic variation in the animal species that forage at different canopy heights; as well as for autochorous species, since taller plants tend to have broader crowns [59]. Thus one may reasonably expect plant height to be directly related to dispersal distance and inversely related to β-diversity – although this does not explain why β-diversity was almost twice as high for lianas as for trees.

The magnitude of β-diversity differed only slightly among dispersal syndromes, possibly reflecting wide variation in seed dispersal distance within each dispersal syndrome [59]. The greater magnitude of β-diversity for zoochorous species than for autochorous and anemochorous species provides only partial support to our first prediction (P1) based on species dispersal distance potential. As expected, the lowest value of β-diversity corresponded to anemochory. However, autochory had a similarly low value of β-diversity, which contradicts our prediction. This latter result may be related to factors other than dispersal capacity, such as the ability to resprout after frequent human disturbance, such as extraction of wood and palm leaves for construction, and slash and burn agriculture, which has been practiced in our study area for over 2000 years [60]. Resprouting allows plants to survive after disturbance [61], [62] and uncouples species composition from seed availability [63], this could help explain why species with low dispersal capacity can dominate throughout the landscape and across succession [39], thereby reducing variation in species composition (i.e. β-diversity). Some examples of

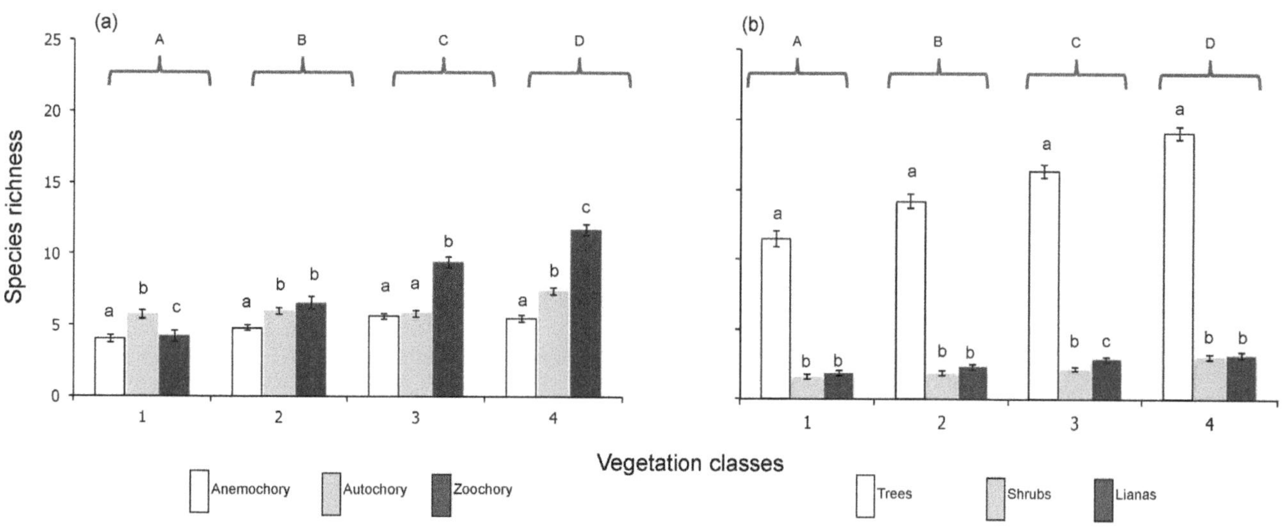

Figure 2. Differences in species richness by functional group among vegetation classes (upper-case letters) and among functional groups (lower-case letters). Different letters represent significant differences (*P*<0.001).

Table 3. Variation partitioning (percentage) of β-diversity by plant growth form and dispersal syndrome through partial Redundancy analysis (RDA). Only significant variables ($P<0.05$) are included.

	Growth form			Dispersal syndrome		
% Variation	*Trees*	*Shrubs*	*Lianas*	*Anemochory*	*Autochory*	*Zoochory*
Total explained variation	29.2	10.3	12.9	27.4	31.2	20.4
Total unexplained variation	70.8	89.7	87.1	72.6	68.8	79.6
Sets of variables						
Environmental variables	51.0	36.2	42.1	48.8	59.0	47.6
Stand age	12.0	11.7		20.0	9.3	6.9
Spatial structure	27.7	46.8	39.5	24.2	22.4	40.2
Shared variation	9.3	5.3	18.5	7.0	9.3	5.4
Environmental variables only						
SOM[a]	74.5		18.0	61.9	82.1	81.4
pH	8.7		44.0	14.3	7.1	11.3
K	6.7	23.5		12.4	3.3	7.2
P					2.7	
% Sand content	6.0		24.0	11.4		
% Clay content	4.0					
CEC[b]		76.5			2.7	
Elevation			14.0		2.2	

[a]SOM: soil organic matter; [b]CEC: cation exchange capacity.

autochorous species that have a strong resprouting capacity and are among the most abundant species across successional ages and topographic positions are *Caesalpinia gaumeri* Greenm., *Bauhinia ungulata* L., *Croton reflexifolius* Kunth, and *Leucaena leucocephala* (Lam.) de Wit.

Contrary to our first prediction (P1), zoochory showed the highest value of β-diversity. This is likely the result of a highly variable dispersal range among zoochorous plant species depending on dispersers' motility, body size, diet and home range, as well as on fruit traits and the availability of alternate food sources [59]. Furthermore, it has been found that few animal-dispersed seeds are brought into recently abandoned agricultural areas by frugivores, even when forest fragments are nearby, probably because of fruit scarcity, higher predation risk or the unfamiliar habitat that an open area represents for frugivores [64], [65], [66]. Therefore, frequent disturbance, such as that produced by traditional shifting cultivation (as practiced in our study area) could reduce the dispersal of zoochorous species, thereby increasing this guild's β-diversity. Moreover, clumping of seed deposition was reported to be significantly higher for zoochorous species compared to anemochorous and autochorours species [66], which could also help explain the higher β-diversity of zoochorous plant species found in this study.

α- and β-Diversity by functional group across secondary succession

As predicted (P2), overall species richness increased across functional groups along secondary forest succession (Fig. 2). Moreover, zoochory dominated in late successional ages and it also showed the highest α-diversity overall. This result agrees with our second prediction and with previous findings in other tropical dry forests in Brazil [67] and Ecuador [68], as well as in tropical wet forests in general [69], [70]. Zoochory also showed a noticeable increase in richness during secondary forest succession, as we initially predicted and as has been previously reported in tropical wet and dry forests [69], [71].

Contrary to our second prediction, however, anemochory had the lowest species richness along forest succession and showed a slight tendency to increase with successional age. Our study site is a relatively short-stature (8–12 m) semi-deciduous forest, which could help explain in part this result, since a considerable percentage of woody plants (25–50%) retain their leaves throughout the year. Both factors could have a negative effect on the efficiency of anemochorous seed dispersal, since the advantage of this dispersal syndrome in dry forests is associated with higher wind circulation during the dry season both through the upper canopy and lower vegetation layers [67]. The increase in anemochoruos species richness with successional age may result from the incorporation of new species of long-lived pioneers, such as *Albizia tomentosa* (Micheli) Standl., as well as shade-tolerant species such as *Gyrocarpus americanus* Jacq.

Regarding growth form, we also found only partial support for our second prediction. Trees were the most species-rich growth form not only in late successional ages (as predicted), but throughout succession (Fig. 2b). Tree species richness also increased with successional age, as expected and as previously reported in tropical forests [69], [71], [72]. Contrary to our prediction, however, shrubs did not dominate in the early phases of succession, possibly because some shrub species, such as *Bunchosia swartziana* Griseb., *Aphelandra scabra* (Vahl) Sm., and *Chiococca alba* (L.) Hitchc. are shade-tolerant. Overall relative richness of lianas (19.4%) was similar to that reported in other Neotropical forests ~20% [73]. Absolute and relative liana species richness increased slightly from early to more advanced successional age classes, as also reported for the Mata Seca in Brazil [74]. The increase in liana richness with successional age may be related to the increasing availability of large trees that can provide structural support [75], [76], combined with the relatively high light availability that characterizes tropical dry forests in general [77], [78].

Species composition of trees was similar between the youngest vegetation classes (VC1 and VC2), although species richness differed. Since most of our VC3 and VC4 sites were >30 y-old, this finding may be attributed to the dominance of a group of pioneer species during the first 15–20 years of succession, which are subsequently replaced by shade-tolerant species, as has been reported in several studies both in wet and in dry tropical forests [24], [28], [37], [79]. All other pairwise comparisons among vegetation classes differed in tree species composition, including VC3 and VC4 (>15 yr-old forests on flat areas and on hills, respectively), which indicates that topographic position also contributes to β- diversity of trees, as well as to α- diversity (Fig. 2b).

In contrast, species composition of shrubs differed among all vegetation classes except for VC3 and VC4, indicating that β-diversity of plants with this growth form varies with forest succession, but not with topographic position. Liana species composition did not differ significantly among vegetation classes (except between VC1 and VC4), suggesting that their β-diversity is relatively unaffected by forest succession and topographic position. This may be the consequence of the dominance across vegetation classes of three species of lianas that represented more than 50% of their total abundance. Besides, lianas typically produce deep roots and a highly efficient water-transport system, and they generally require well-lit environments for their successful establishment [80], [81]. These characteristics could help explain why lianas seem to be able to establish, grow and persist equally well under different topographic positions and successional age classes, since they may be able to cope with water stress and may not be limited by low light availability.

Variation partitioning

The proportion of the total variation in species composition accounted for by the explanatory variables, as well as the relative importance of explanatory variables, varied markedly among growth forms and slightly among dispersal syndromes (Table 3). These results clearly indicate that β-diversity along with the factors that influence it depend on key functional attributes, such as growth form and, to a lesser extent, dispersal syndrome. As predicted (P3), environmental filters explained the largest percentage of β-diversity for trees and the lowest for shrubs. In fact, abiotic environmental variables contributed most to β-diversity across growth forms and dispersal syndromes, except for shrubs. The dominant role of environmental factors suggests that environmental filtering has a strong influence on β-diversity, as has been reported in previous studies [2], [15], [16], [82], [83], thus providing support to the niche partitioning theory.

Shrubs were the only guild for which the spatial structure of sampling sites explained a larger percentage of variation in species composition than environmental filters, suggesting that dispersal limitation plays a greater role on shrub β-diversity than environmental filters. This may be partly attributed to their relatively short stature, which may impose critical limits on their dispersal capacity, resulting in a strong spatial autocorrelation in species composition. Using the same data set and (local) spatial grain of this study, [14] found that both environmental variables and the spatial structure of sampling sites had similar influences on overall β-diversity of woody species. Our results clearly show that this general pattern masks major differences in the relative importance of environmental filters vs. dispersal limitation among growth forms. However, it is also possible that the greater influence of spatial structure of sampling sites relative to environmental filters on shrubs is partly related to key environmental variables, such as light and soil moisture availability, that were not measured, that are spatially structured, and that may have a greater influence on shrubs compared to trees and lianas – given the former shorter stature and generally shallower root system [84]. Further studies are required to evaluate this alternative explanation.

Dispersal syndromes showed fairly consistent patterns of variation partitioning of β-diversity, with a clear predominance of environmental filters, except for zoochory (Table 3). Contrary to our prediction (P3), the autochorous guild showed a slightly stronger association with environmental variables compared to the other dispersal syndromes. This prediction was based on the assumption that dispersal distance is generally greater for anemochorous species compared to autochorous ones, and that greater dispersal ability should allow species to colonize sites with the appropriate environmental conditions. Alternatively, autochorous species may indeed have the most limited dispersal distances, and assuming that environmental conditions are spatially structured, it is logical that short-distance dispersing, spatially aggregated species should show higher association with spatially autocorrelated environmental factors. This would require that the spatial pattern of species composition of autochorous plants closely matches the spatial structure of sampled environmental conditions, resulting in a large percentage of variation explained by the joint effect of environmental factors and spatial structure of sampling sites (*i.e.* a spatially structured environment). Although the variation explained jointly by environment and spatial structure was indeed highest for the autochorous guild, it was still small relative to the variation explained by pure environmental or pure spatial components (Table 3). Moreover, if autochorous species were indeed strongly limited by dispersal limitation, one would expect that the spatial structure of sampling sites would explain a large percentage of variation in species composition of this syndrome. Yet, this syndrome showed the lowest percentage of β-diversity explained by spatial structure of sampling sites. As previously discussed, this unexpected result could be partly attributed to the dominance of tree species with strong resprouting capacity, displaying different dispersal syndromes, predominantly autochory. On the other hand, the fact that the largest percentage of variation in species composition explained by spatial structure of sampling sites corresponded to the zoochorous guild could be partly attributed to the greater clumping of seed deposition that has been reported for this syndrome compared to the autochorous and anemochorous syndromes [59].

The effect of forest succession on β-diversity varied among functional groups. Stand age explained almost 12% of the total variation in species composition of trees and shrubs, but was unrelated to liana β-diversity. This result may be due to the capacity of lianas to establish, grow and persist equally well under different topographic positions and successional age classes, as discussed previously. Similarly, stand age explained less than 10% of variation in species composition of zoochorous and autochorous species, and 20% of the anemochorous β-diversity. In this same study area, [14] found that stand age explained only 3% of the total variation in species composition of all woody species. These contrasting results clearly illustrate the importance of considering different functional groups when evaluating the ecological factors that influence β-diversity.

Soil variables explained over half of the variation in tree species composition. Previous findings in the same study area showed that hills have a thinner soil layer with high rock and nutrient contents, whereas flat areas have deeper soils with high clay content [39]. These characteristics have a major influence on woody species composition, which is dominated by trees [39]. The most important environmental variable explaining β-diversity differed among growth forms: soil organic matter (SOM) for trees, cation exchange capacity (CEC) for shrubs, and pH for lianas. This result suggests that growth forms are limited by different soil variables, which is in agreement with niche partitioning theory. SOM was consistently the single most important environmental factor influencing β-diversity of all dispersal syndromes and of trees (Table 2). This result may be due to the positive influence of SOM on soil fertility and on the capacity of soils to hold water and incorporate nutrients [20], [60], [85]. Our results agree with findings from many tropical forests worldwide [15], [16], [86], suggesting that soil properties can be regarded as major drivers of variation in plant species composition (β-diversity).

To conclude, our results clearly show that the magnitude of β-diversity varied markedly among growth forms, but much less among dispersal syndromes. They also suggest that environmental filtering is an important factor explaining β-diversity of different functional groups, except for shrubs, for which dispersal limitation appears to be more important, possibly due to their low stature that may limit their dispersal capacity. Additionally, stand age clearly influenced α-diversity across functional groups, but showed a relatively low influence on β-diversity, possibly due to the prevalence of tree species with strong resprouting capacity. Finally, our results clearly show that considering different functional groups may reveal important differences in β-diversity patterns and correlates that are not apparent when one considers overall woody plant diversity. Such differences have important implications for ecological theory, and can inform conservation, restoration and management strategies needed to stem the relentless deforestation and degradation of tropical dry forests.

Acknowledgments

We thank the ejidos of Xkobenhaltún, Xul and Yaxhachén for allowing us to work in their lands and for their assistance with fieldwork and determination of stand age. James Callaghan and Kaxil Kiuic A.C. provided logistic support. We are grateful to Filigonio May-Pat and Fernando Tun-Dzul for their help with species identification, fieldwork and GIS processing, to Rodrigo Hernández-Juárez and Lizette Cicero-Jurado for their assistance with data analysis, and to Jorge Carlos Trejo and Carmen Salazar for their help with the determination of plant functional groups.

Author Contributions

Conceived and designed the experiments: JOLM JMD JLHS. Performed the experiments: JOLM JMD JLHS. Analyzed the data: JOLM LSV JLHS JAGC. Contributed reagents/materials/analysis tools: JMD JLHS. Wrote the paper: JOLM LSV JMD JLHS JAM JAGC.

References

1. Tuomisto H, Ruokolainen K, Kalliola R, Linna A, Danjoy W, et al. (1995) Dissecting Amazonian biodiversity. Science 269: 63–66.
2. Tuomisto H, Ruokolainen K, Yli-Halla M (2003) Dispersal, environment, and floristic variation of western Amazonian forests. Science 241: 241–244.
3. Tuomisto H (2010) A diversity of beta diversities: straightening up a concept gone awry. Part 1. Defining beta diversity as a function of alpha and gamma diversity. Ecography 33: 2–22.
4. Grubb PJ (1977) The maintenance of species-richness in plant communities: the importance of the regeneration niche. Biol Rev 52: 107–145.
5. Poorter L (2007) Are species adapted to their regeneration niche, adult niche, or both? Am Nat 169: 433–442.
6. Hutchinson GE (1957) Concluding remarks. Cold Spring Harb Symp Quant Biol 22: 415–427.
7. Leibold MA (1995) The niche concept revisited: mechanistic models and community context. Ecology 76: 1371–1382.
8. Leibold MA (2008) Ecology: return of the niche. Nature 454: 39–41.
9. Hubbell SP (1979) Tree dispersion, abundance, and diversity in a tropical dry forest. Science 203: 1299–1309.
10. Hubbell SP (2001) The unified neutral theory of biodiversity and biogeography. Princeton: Princeton University Press. 448 p.
11. Hubbell S (2005) Neutral theory in community ecology and the hypothesis of functional equivalence. Funct Ecol 19: 166–172.
12. Condit R, Pitman N, Leight EJ, Chave J, Terborgh J, et al. (2002) Beta-diversity in tropical forest trees. Science 295: 666–669.
13. Primack RB, Miao SL (1992) Dispersal can limit local plant distribution. Conserv Biol 6: 513–519.
14. López-Martínez JO, Hernández-Stefanoni JL, Dupuy JM, Meave JA (2013) Partitioning the variation of woody plant β-diversity in a landscape of secondary tropical dry forests across spatial scales. J Veg Sci 24: 33–45.
15. Tuomisto H, Poulsen DA, Ruokolainen K, Moran RC, Quintana et al. (2003) Linking floristic patterns with soil heterogeneity and satellite imagery in Ecuadorian Amazonia. Ecol Appl 13: 352–371.
16. Paoli GD, Curran LM, Zak DR (2006) Soil nutrients and beta diversity in the Bornean Dipterocarpaceae: evidence for niche partitioning by tropical rain forest trees. J Ecol 94: 157–170.
17. Fahrig L (2003) Effects of habitat fragmentation on biodiversity. Ann Rev Ecol Evol Syst 34: 487–515.
18. Gallardo-Cruz A, Pérez-García EA, Meave JA (2009) β-diversity and vegetation structure as influenced by slope and altitude in a seasonally dry tropical landscape. Landsc Ecol 24: 473–482.
19. Motzkin G, Wilson P, Foster DR, Allen A (1999) Vegetation patterns in heterogeneous landscapes: the importance of history and environment. J Veg Sci 10: 903–920.
20. Guariguata MR, Ostertag R (2001) Neotropical secondary forest succession: changes in structural and functional characteristics. For Ecol Manage 148: 185–206.
21. Chazdon RL, Letcher SG, van Breugel M, Martínez-Ramos M, Bongers F, et al. (2007) Rates of change in tree communities of secondary Neotropical forests following major disturbances. Philos Trans R Soc Lond Ser B 362: 273–289.
22. Quesada M, Sanchez-Azofeifa GA, Alvarez-Añorve M, Stoner KE, Avila-Cabadilla L, et al. (2009) Succession and management of tropical dry forests in the Americas: review and new perspectives. For Ecol Manage 258: 1014–1024.
23. Chazdon RL, Finegan B, Capers RS, Salgado-Negret B, Casanoves F, et al. (2010) Composition and dynamics of functional groups of trees during tropical forest succession in Northeastern Costa Rica. Biotropica 42: 31–40.
24. Lebrija-Trejos E, Pérez-García EA, Meave JA, Bongers F, Poorter L (2010) Functional traits and environmental filtering drive community assembly in a species-rich tropical system. Ecology 91: 386–398.
25. Álvarez-Añorve MY, Quesada M, Sánchez-Azofeifa GA, Avila-Cabadilla LD, Gamon JA (2012) Functional regeneration and spectral reflectance of trees during succession in a highly diverse tropical dry forest ecosystem. Am J Bot 99: 816–826.
26. Opler PA, Baker HG, Gordon WF (1980) Plant reproductive characteristics during secondary succession in neotropical lowland forest ecosystems. Biotropica 12: 40–46.
27. Chazdon RL, Careaga S, Webb C, Vargas O (2003) Community and phylogenetic structure of reproductive traits of woody species in wet tropical forests. Ecol Monogr 73: 331–348.
28. Chazdon RL (2008) Chance and determinism in tropical forest succession. In: Carson WP, Schnitzer SA, editors. Tropical forest community ecology. Oxford: Blackwell Publishing. 384–408.
29. Kennard DK (2002) Secondary forest succession in a tropical dry forest: patterns of development across a 50-year chronosequence in lowland Bolivia. J Trop Ecol 18: 53–66.
30. Nekola JC, White PS (1999) The distance decay of similarity in biogeography and ecology. J Biogeogr 26: 867–878.
31. Qian H, Ricklefs RE, White S (2005) Beta diversity of angiosperms in temperate floras of eastern Asia and eastern North America. Ecol Lett 8: 15–22.
32. Gallardo-Cruz J, Meave J, Pérez-García E, Hernández-Stefanoni J (2010) Spatial structure of plant communities in a complex tropical landscape: implications for β-diversity. Community Ecol 11: 202–210.
33. Whittaker RH (1960) Vegetation of the Siskiyou Mountains, Oregon and California. Ecol Monogr 30: 279–338.
34. Bin Y, Wang Z, Wang Z, Ye W, Cao H, et al. (2010) The effects of dispersal limitation and topographic heterogeneity on beta diversity and phylobetadiversity in a subtropical forest. Plant Ecol 209: 237–256.
35. Amarasekare P (2003) Competitive coexistence in spatially structured environments: a synthesis. Ecol Lett 6: 1109–1122.

36. Ben-Hur E, Fragman-Sapir O, Hadas R, Singer A, Kadmon R (2012) Functional trade-offs increase species diversity in experimental plant communities. Ecol Lett 15: 1276–1282.
37. Finegan B (1996) Pattern and process in neotropical secondary rain forests: the first 100 years of succession. Trends Ecol Evol 11: 119–124.
38. Lebrija-Trejos E, Bongers F, Pérez-García EA, Meave JA (2008) Successional change and resilience of a very dry tropical deciduous forest following shifting agriculture. Biotropica 40: 422–431.
39. Dupuy JM, Hernández-Stefanoni JL, Hernández-Juárez RA, Tetetla-Rangel E, López-Martínez JO, et al. (2012) Patterns and correlates of tropical dry forest structure and composition in a highly replicated chronosequence in Yucatan, Mexico. Biotropica 44: 151–162.
40. Howe HF, Smallwood J (1982) Ecology of seed dispersal. Annu Rev Ecol Syst 13: 201–228.
41. McGill BJ, Maurer BA, Weiser MD (2006) Empirical evaluation of neutral theory. Ecology 87:1411–1423.
42. Webb CT, Hoeting JA, Ames GM, Pyne MI, LeRoy Poff N (2010) A structured and dynamic framework to advance traits-based theory and prediction in ecology. Ecol Lett 13: 267–283.
43. Flores JS, Espejel I (1994) Tipos de vegetación de la Península de Yucatán. Etnoflora Yucatanense, fascículo 3. Mérida: Universidad Autónoma de Yucatán. 135 p.
44. Bautista-Zuñiga F, Batllori-Sampedro E, Ortiz-Pérez MA, Palacio-Aponte G, Castillo-González M. (2003) Geoformas, agua y suelo en la Península de Yucatán. In: Colunga-GarcíaMarín P, Larqué SA, editors. Naturaleza y sociedad en el área maya: pasado, presente y futuro. Mérida: Academia Mexicana de Ciencias, Centro de Investigación Científica de Yucatán. pp. 21–35.
45. Rico-Gray V, García-Franco G (1991) The Maya and the vegetation of the Yucatan Peninsula. J Ethnobiol 11: 135–142.
46. Hernández-Stefanoni JL, Dupuy J, Tun-Dzul F (2011) Influence of landscape structure and stand ages on species density and biomass of a tropical dry forest across spatial scales. Landsc Ecol 26: 355–370.
47. Borcard D, Legendre P (2002) All-scale spatial analysis of ecological data by means of principal coordinates of neighbour matrices. Ecol Model 153: 51–68.
48. Borcard D, Legendre P, Avois-Jacquet C, Tuomisto H (2004) Dissecting the spatial structure of ecological data at multiples scales. Ecology 85: 1826–1832.
49. Jongman RHG, ter Braak CJF, van Tongeren OFR (1987) Data analysis in community and landscape ecology. Cambridge: Cambridge University Press. p. 324.
50. Clarke KR, Warwick RM (1994) Changes in marine communities: an approach to statistical analysis and interpretation. Plymouth: PRIMER-E. 144 p.
51. Borcard D, Legendre P, Drapeau P (1992) Partialling out the spatial component of ecological variation. Ecology 73: 1045–1055.
52. Legendre P, Gallagher ED (2001) Ecologically meaningful transformations for ordination of species data. Oecologia 129: 271–280.
53. ter Braak CJF, Smilauer P (1998) CANOCO Reference Manual and User's Guide to CANOCO for Windows: Software for Canonical Community Ordination (version 4). New York: Microcomputer Power. 351 p.
54. Jones MM, Tuomisto H, Borcard D, Legendre P, Clark DB, Olivas PC (2008) Explaining variation in tropical plant community composition: influence of environmental and spatial data quality. Oecologia 155: 593–604.
55. Crawley MJ (2007) The R Book. West Sussex: John Wiley and Sons. 950 p.
56. Zuur AF, Leno EN, Walker NJ, Saveliev AA, Smith GM (2009) Mixed effects models and extensions in ecology with R. New York: Springer. 596 p.
57. R Development Core Team (2012) A language and environment for statistical computing. R Foundation for Statistical Computing, Vienna. Available: http://www.Rproject.org/. Accessed 15 August 2012.
58. Westoby M, Rice B, Howell J (1990) Seed size and plant growth form as factors in dispersal spectra. Ecology 71: 1307–1315.
59. Muller-Landau HC, Wright SJ, Calderón O, Condit R, Hubbell SP (2008) Interspecific variation in primary seed dispersal in a tropical forest. J Ecol 96: 653–667.
60. Durán R, Méndez M. (2010) Biodiversidad y Desarrollo Humano en Yucatán. Mérida: CICY, PPD-FMAM, CONABIO, SEDUMA. 496 p.
61. Swaine MD (1992) Characteristics of dry forest in West Africa and the influence of fire. J Veg Sci 3: 365–374.
62. Miller PM, Kauffman JB (1998) Seedling and sprout response to slash-and-burn agriculture in a tropical deciduous forest. Biotropica 30: 538–546.
63. Vieira DLM, Scariot A (2006) Principles of natural regeneration of tropical dry forests for restoration. Restor Ecol 14: 11–20.
64. Duncan RS, Chapman CA (1999) Seed dispersal and potential forest succession in abandoned agriculture in tropical Africa. Ecol Appl 9: 998–1008.
65. van Breugel M, Martínez-Ramos M, Bongers F (2006) Community dynamics during early secondary succession in Mexican tropical rain forests. J Trop Ecol 22: 663–674.
66. van Breugel M, Bongers F, Martínez-Ramos M (2007) Species dynamics during early secondary forest succession: recruitment, mortality and species turnover. Biotropica 39: 610–619.
67. Sobral LM, Machado ICS (2001) Fruiting phenology and seed dispersal syndromes in Caatinga, a tropical dry forest in the northeast of Brazil. J Trop Ecol 17: 303–321.
68. Jara-Guerrero A, De la Cruz M, Méndez M (2011) Seed dispersal spectrum of woody species in south Ecuadorian dry forests: environmental correlates and the effect of considering species abundance. Biotropica 43: 722–730.
69. Armesto JJ, Rozzi R (1989) Seed dispersal syndromes in the rain forest of Chiloé: evidence for the importance of biotic dispersal in a temperate rain forest. J Biogeogr 16: 219–226.
70. Du Y, Mi X, Liu X, Chen L, Ma K (2009) Seed dispersal phenology and dispersal syndromes in a subtropical broad-leaved forest of China. For Ecol Manage 258: 1147–1152.
71. Medina E (1995) Diversity of life forms of higher plants in neotropical dry forests. In: Bullock S, Mooney HA, Medina E, editors. Seasonally Dry Tropical Forests. Cambridge: Cambridge University Press. 221–242.
72. Butler DW, Green RJ, Lamb D, McDonald WJF, Forster PI (2007) Biogeography of seed-dispersal syndromes, life-forms and seed sizes among woody rain-forest plants in Australia's subtropics. J Biogeogr 34: 1736–1750.
73. Burnham RJ (2004) Alpha and beta diversity of lianas in Yasuní, Ecuador. For Ecol Manage 190: 43–55.
74. Madeira B, Espírito-Santo M, Neto S, Nunes Y, Sánchez-Azofeifa GA, et al. (2009) Changes in tree and liana communities along a successional gradient in a tropical dry forest in south-eastern Brazil. Plant Ecol 201: 291–301.
75. Clark DB, Clark DA (1990) Distribution and effects on tree growth of lianas and woody hemiepiphytes in a Costa Rican tropical wet forest. J Trop Ecol 6: 321–331.
76. Nabe-Nielsen J (2001) Diversity and distribution of lianas in a neotropical rain forest, Yasuní National Park, Ecuador. J Trop Ecol 17: 1–19.
77. Ceccon E, Huante P, Rincón E (2006) Abiotic factors influencing tropical dry forest regeneration. Braz Arch Biol Techn 49: 305–312.
78. Lebrija-Trejos E, Pérez-García EA, Meave JA, Poorter L, Bongers F (2011) Environmental changes during secondary succession in a tropical dry forest in Mexico. J Trop Ecol 27: 477–489.
79. Swaine MD, Hall JB (1983) Early succession on cleared forest land in Ghana. J Ecol 71: 601–627.
80. Schnitzer SA, Bongers F (2002) The ecology of lianas and their role in forests.Trends Ecol Evol 17: 223–230.
81. Schnitzer SA (2005) A mechanistic explanation for global patterns of liana abundance and distribution.Am Nat 166: 262–276.
82. Tuomisto H, Poulsen AD (1996) Influence of edaphic specialization on Pteridophyte distribution in neotropical rain forests. J Biogeogr 23: 283–293.
83. Ruokolainen K, Tuomisto H (2002) Beta-diversity in tropical forests. Science 297: 1439.
84. Dodd MB, Lauenroth WK, Welker JM (1998) Differential water resource use by herbaceous and woody plant life-forms in a shortgrass steppe community. Oecologia 117: 504–512.
85. Swaine MD (1996) Rainfall and soil fertility as factors limiting forest species distributions in Ghana. J Ecol 84: 419–428.
86. Costa FRC, Magnusson WE, Luizao RC (2005) Mesoscale distribution patterns of Amazonian understorey herbs in relation to topography, soil and watersheds. J Ecol 93: 863–878.

2

Dissolved Organic Carbon in Headwater Streams and Riparian Soil Organic Carbon along an Altitudinal Gradient in the Wuyi Mountains, China

Wei Huang[1], William H. McDowell[2], Xiaoming Zou[1,3], Honghua Ruan[1]*, Jiashe Wang[4], Liguang Li[1,5]

1 Faculty of Forest Resources and Environmental Science, Nanjing Forestry University, Nanjing, Jiangsu, China, **2** Department of Natural Resources and the Environment, University of New Hampshire, Durham, New Hampshire, United States of America, **3** Institute for Tropical Ecosystem Studies, University of Puerto Rico, San Juan, Puerto Rico, United States of America, **4** Administrative Bureau of Wuyishan National Nature Reserve, Wuyishan, Fujian, China, **5** Soil and Water Science Department, University of Florida, Gainesville, Florida, United States of America

Abstract

Stream water dissolved organic carbon (DOC) correlates positively with soil organic carbon (SOC) in many biomes. Does this relationship hold in a small geographic region when variations of temperature, precipitation and vegetation are driven by a significant altitudinal gradient? We examined the spatial connectivity between concentrations of DOC in headwater stream and contents of riparian SOC and water-soluble soil organic carbon (WSOC), riparian soil C:N ratio, and temperature in four vegetation types along an altitudinal gradient in the Wuyi Mountains, China. Our analyses showed that annual mean concentrations of headwater stream DOC were lower in alpine meadow (AM) than in subtropical evergreen broadleaf forest (EBF), coniferous forest (CF), and subalpine dwarf forest (SDF). Headwater stream DOC concentrations were negatively correlated with riparian SOC as well as WSOC contents, and were unrelated to riparian soil C:N ratio. Our findings suggest that DOC concentrations in headwater streams are affected by different factors at regional and local scales. The dilution effect of higher precipitation and adsorption of soil DOC to higher soil clay plus silt content at higher elevation may play an important role in causing lower DOC concentrations in AM stream of the Wuyi Mountains. Our results suggest that upscaling and downscaling of the drivers of DOC export from forested watersheds when exploring the response of carbon flux to climatic change or other drivers must done with caution.

Editor: Xiujun Wang, University of Maryland, United States of America

Funding: This study was supported by the National Science Foundation of China (No. 31170417), Chinese National Public-Welfare Program (No. 200804006), Doctorate Fellowship Foundation of Nanjing Forestry University (No. 163010106) and Scientific and Technological Innovation Foundation for Graduate Student of Jiangsu Province (No. CXZZ_110510). The funders had role in study design, data collection and analysis, decision to publish, or preparation of the manuscript.

Competing Interests: The authors have declared that no competing interests exist.

* E-mail: honghuaruan@gmail.com

Introduction

Stream dissolved organic carbon (DOC) is suggested to be a major constituent of the global carbon cycle [1], [2]. At region to global scales, stream DOC concentrations and flux are largely governed by the quantity and quality of soil organic carbon (SOC). Soil organic carbon storage is an important driver of spatial variation in stream DOC fluxes [3]. At the global scale, DOC flux from watersheds is strongly related to soil C/N ratio [4], and similar relationships hold in some regions [5], [6]. Aitkenhead et al. [7] found that stream DOC concentrations could be predicted by SOC pools at three different scale-catchments with a range of soil types, land use and elevation. At a much smaller scale of investigation, Billett et al. [8] also showed that the spatial pattern of stream DOC concentrations was linked to changes of the SOC content in a small catchment (1.3 km^2). At the stream-soil interface, the importance of the riparian zone in controlling carbon flow to the stream has been highlighted by many works [9], [10], [11]. For example, Dosskey and Bertsch [11] found that riparian soil contributed over 90% of DOC flux in a stream, even though they covered only 6% of the catchment area. Therefore, riparian SOC contents might have a first order control on stream DOC concentrations.

The delivery of SOC to streams can be understood by considering three fundamental processes: (1) partial decomposition of terrestrially fixed carbon producing DOC, (2) hydrologic transport of DOC from the terrestrial environment to streams, and (3) modification of both the quantity and composition of DOC during hydrologic transport [12], [13]. Thus, biotic and abiotic factors controlling DOC production, transport, and modification of quantity during transport, such as temperature [14], [15], precipitation and discharge [16], [17], soil moisture [15], [18], atmospheric nitrogen and SO_4 deposition [19], [20], vegetative cover [21], [22], soil C:N ratio [4] and adsorption of DOC by soil [23], [24], are all likely to influence the relationship between stream DOC concentrations and SOC contents. For example, stream DOC concentration declined after the replacement of native deciduous forest by conifer plantations, suggesting that the change in vegetation resulted in a decline of organic matter decomposition and soil DOC production [21], [22]. Other analyses of data collected in streams have reported positive correlations between DOC and temperature, soil C:N ratio, and antecedent soil moisture [4], [8], [18].

Altitudinal gradients that span considerable variation of temperature, precipitation, vegetation and soil types over short geographic distances are well suited to study the relationship between stream DOC concentrations and SOC contents. However, few studies have shown a connection between stream DOC concentrations and SOC contents such along an altitudinal gradient. Aitkenhead et al. [7] reported that, for a given soil carbon pool, upland catchments with maximum altitude greater than 700 m have lower stream DOC concentrations than low land sites. Because precipitation was relatively constant among the studied watersheds (670.9 to 919.6 mm) [6], it was not thought to be a major driver of DOC concentrations.

The purpose of our study was to determine the relationship between stream DOC concentration and SOC content along an altitudinal gradient with considerable variation of temperature, precipitation, vegetation and soil types. The Wuyi Mountains have a clear vertical zonation of climate, vegetation and soil types in the subtropical monsoon region of China [25]. It provided an ideal site in which to investigate the connection between stream DOC concentrations and SOC contents. We hypothesized that: (1) DOC concentrations in headwater stream are driven by variation in riparian SOC content along the altitudinal gradient where variations of temperature, precipitation, and vegetation and soil types are large; (2) alternatively, stream DOC concentrations are more influenced by the variations of temperature, precipitation, and vegetation and soil types than by riparian SOC contents.

Methods

Site Descriptions and Experimental Design

The experimental sites were located in the Wuyi Mountain National Nature Reserve (27°33′–27°54′N, 117°27′–117°51′E), that occupies 565 km^2 of forested land in the subtropical monsoon region of China [26]. The annual mean temperature is 15.0°C, with annual mean relative humidity of 83.5%, and 100 fog days per years. Annual precipitation reaches 2000 mm, most of which (60%) falls between later spring (April) and summer (July). There are four vegetation types along the altitudinal gradient from low to high elevation: subtropical evergreen broadleaf forest (EBF), coniferous forest (CF), subalpine dwarf forest (SDF) and alpine meadow (AM) (Table 1).

We selected four catchments in our study, one for each of the four vegetation types (Fig. 1) (Jiashe Wang issued the permission for each location of our study. He is the authority responsible for this national nature reserve). These catchments contain relatively pristine ecosystems that are currently not managed for agriculture or forestry. In each catchment, we selected three 1st order streams (Stream orders were determined with the Strahler stream ordering system [29]). Within each 1st order stream watershed, we identified three locations (the stream source, the stream middle – i.e. half way between the source and 2nd older stream, and the junction of the 1st order with the 2nd order stream). At each of these locations, we took stream water and their paired riparian soil samples (within 5 m from the stream bank). This experimental design resulted in 36 stream-water sampling points, 4 vegetation types × 3 streams × 3 locations; and its corresponding 36 riparian soil sampling points (Fig. 1).

Field Sampling and Laboratory Analysis

We sampled stream water and soil every two months from June 2010 to April 2011. Over each 6-day of sampling period, we sampled water on days 1, 3 and 5, and soil samples on day 6. Water samples were collected in 100 ml HDPE bottles. Water samples from three locations of each stream were mixed in field. All water samples were field-filtered through pre-combusted glass fiber filters (0.7 μm Whatman GF/F), kept in a cooler during transportation to the laboratory, and stored at 4°C until analysis within 72 h. Three soil layers were sampled (i.e. 0–10 cm, 10–25 cm and 25–40 cm) from all plots using a 2 cm diameter soil corer.

Soil samples from three locations of each stream riparian zone were composited, sieved (<2 mm) to remove soil macroorganisms, rocks and fine roots, placed in plastic bags and transported in a cooler, thoroughly mixed and then divided into two parts. One part was kept in the refrigerator at 4°C before analysis for water-soluble soil organic carbon (WSOC) content, and the other part was air-dried and sieved (0.25 mm) for SOC and total nitrogen (TN) analysis.

Temperature of stream water and soil at 5 cm depth was measured using an electronic thermometer. Soil moisture was gravimetrically determined using the difference in weight before and after drying a soil sample (24 hours, 105°C). The particle-size was determined by wet sieving and sedimentation using the pipette sampling technique.

Stream DOC (measured as non-purgeable organic carbon) was analyzed via high temperature combustion on a Shimadzu TOC/TN–V_{CPN} analyzer (Shimadzu Corp. Japan).

Water-soluble soil organic carbon was extracted from 20 g of field moist soil samples (<2 mm sieved) with distilled water (1:2 soil-water ratio) [30], [31]. The suspension was shaken for 30 min, centrifuged for 20 min at 4000×g, and filtered through pre-combusted glass fiber filters (0.7 μm Whatman GF/F). Organic carbon concentration of the extracts was determined with a Shimadzu TOC-V_{CPN} analyzer (measured as the difference between total carbon and inorganic carbon). The WSOC content of soil sample was expressed on the basis of equivalent oven dry weight.

Soil organic carbon and TN content were determined by a C/N/S-Analyzer (Vario EL III, Elementar, Germany). The method detection limits for SOC and TN are 0.03 mg C and 0.03 mg N, respectively.

Statistical Analysis

All statistical analyses were performed using SPSS 17.0. Due to the small replications (n = 3), we employed Kruskal-Wallis test to determine if there were statistically significant differences in stream DOC concentrations, riparian SOC contents, riparian WSOC contents, riparian soil C:N ratio and temperature among vegetation types. Linear regression models were used to evaluate relationships between stream DOC concentrations and temperature, riparian SOC contents, riparian WSOC contents, riparian soil C:N ratio, as well as the relationship between riparian WSOC contents and SOC contents. When we performed liner regression, data from all four sites were analyzed together. Before performing linear regression, we used Durbin-Watson, histogram of errors and scatter diagram of errors to test the validity of these linear regressions. Significance was defined as $p<0.05$.

Results

Annual mean concentrations of stream DOC were lower in alpine meadow (AM) than in subtropical evergreen broadleaf forest (EBF), coniferous forest (CF), and subalpine dwarf forest (SDF), exhibiting a decreasing pattern with increasing altitude (Fig. 2a, Table 2). In contrast, annual mean contents of riparian SOC were higher in AM than in other vegetation types in all three soil layers, which showed an increasing trend with altitude (Fig. 2b, Table 2). Annual mean contents of riparian WSOC were also

Table 1. Description of the four study sites located in the Wuyi Mountains, China.

Vegetation type	Altitude (m)	Soil type	Dominant species	AMT (°C)	AMP (mm)	Litter mass (t hm^{-2} y^{-1})	Soil depth (cm)	Soil bulk density (g cm^{-3})	Soil pH	Soil silt plus clay (%)	Soil microbial biomass (g kg^{-1})	Fine root biomass (kg m^{-3})
EBF	690	Humic Acrisols	*Castanopsis carlesii, Castanopsis eyrei*	18.5	1700	5.36a	0–10	0.87	4.54	31.17a	1.14a	3.06a
							10–25	0.89	4.82	36.79a	0.68a	0.75a
							25–40	0.96	4.94	40.12a	N/A	N/A
CF	1140	Humic Alisols	*Pinus taiwanensis, Oligostachyum oedogonatum*	14.5	2000	8.08b	0–10	0.64	4.64	38.07b	1.05a	7.01b
							10–25	0.77	4.79	37.79a	0.65a	1.99b
							25–40	0.87	4.88	45.24b	N/A	N/A
SDF	1750	Dystric Cambisols	*Symplocos paniculata, Stewartia sinensis*	11.2	2200	2.92c	0–10	0.61	4.59	41.93bc	1.46b	3.48a
							10–25	0.79	4.84	45.22b	0.71b	0.62a
							25–40	0.82	4.91	45.06b	N/A	N/A
AM	2060	Cambic Umbrisols	*Calamagrostis brachytrich, Miscanthus sinensis, Lycopodium clavatum*	9.7	3100	1.94d	0–10	0.54	4.86	47.85c	3.34c	9.73c
							10–25	0.65	5.13	51.33c	1.97c	1.21c
							25–40	0.80	5.22	60.77c	N/A	N/A

Note: AMT: annual mean temperature; AMP: annual mean precipitation. Datasets of annual mean temperature and annual mean precipitation are obtained from [25] and [26]. Datasets of soil bulk density and soil pH are obtained from [27]. Datasets of litter biomass, microbial biomass and fine root biomass are obtained from [28]. Different lowercase letters indicate significant differences among four study sites in the same soil layer.

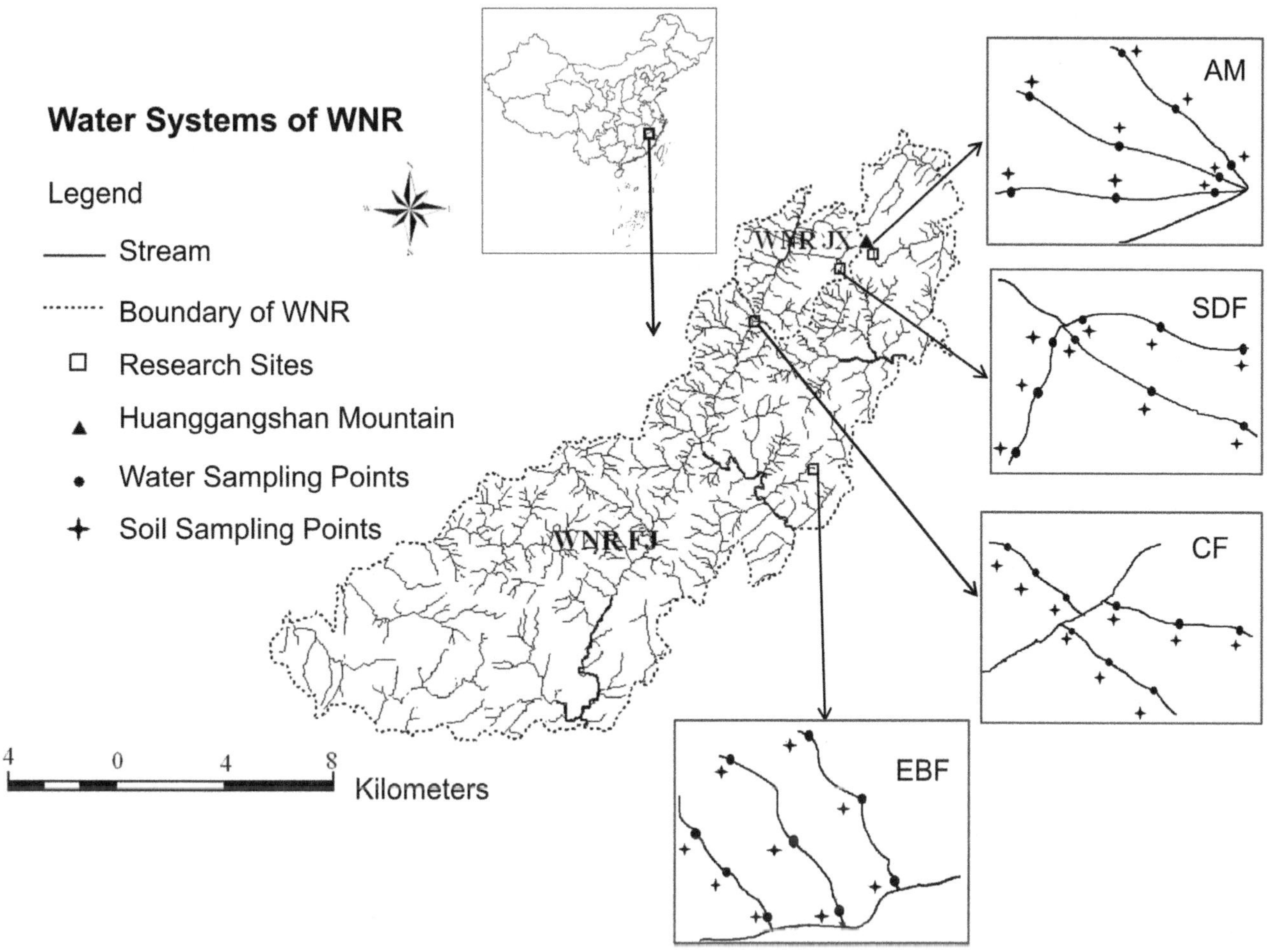

Figure 1. Map of headwater streams in the evergreen broadleaf forest (EBF), coniferous forest (CF), subalpine dwarf forest (SDF) and alpine meadow (AM) in the Wuyi Mountain National Nature Reserve of southeastern China (WNR, Wuyi Mountain National Nature Reserve; FJ, Fujian Province; JX, Jiangxi Province).

higher in AM than in other vegetation in 0–10 cm and 10–25 cm soil layers (Fig. 2c, Table 2). Riparian soil C:N ratio was significantly higher from AM and EBF than for samples from SDF and CF in all three soil layers (Fig. 2d, Table 2), but did not show an altitudinal trend. Annual mean stream water temperature and riparian soil temperature at 5 cm depth were higher in the EBF and CF catchments than in SDF and AM catchments (Fig. 2e).

Headwater stream DOC concentrations were correlated negatively with both riparian SOC and WSOC contents across all four vegetation types along the altitudinal gradient (Fig. 3a, b). There was no significant correlation between riparian soil C:N ratio and headwater stream DOC concentrations (Fig. 3c). Positive correlations were found between headwater stream DOC concentrations and stream temperature or riparian soil temperature at 5 cm depth (Fig. 3d). Contents of riparian WSOC were positively correlated with contents of riparian SOC (Fig. 3e).

Discussion

Contrary to patterns observed in past studies, the DOC concentration of twelve headwater streams in the Wuyi Mountains of China decreased as a function of SOC content, rather than showing the increase in DOC that would be expected from earlier work such as Aitkenhead et al. [7]. Although we studied only riparian SOC and WSOC content in our attempt to link soil and stream DOC concentration, previous work in the Wuyi Mountains also shows very similar overall trends for non-riparian soils. Xu et al. [32] found SOC and its labile fractions, i.e. readily oxidizable carbon, WSOC and microbial biomass carbon increased with altitude. Thus stream DOC concentrations were negatively related to many measures of soil carbon and carbon fractions, and are also unrelated to soil C/N ratio, which was suggested to drive spatial variability in stream DOC flux at the global scale [4].

There are several possible explanations for the negative correlation between stream DOC and soil organic carbon that are opposite to those observed in other studies. We can rule out spatial variability in soils, as we sampled soils extensively in the riparian zones that are thought to preferentially supply streams with DOC [33], [34], and past work at the site shows that watershed soils show strong patterns in carbon with elevation identical to those we have quantified in riparian soils [32]. It also seems unlikely that our stream sampling, which occurred multiple times throughout the year (a total of 6 individual samples), would be insufficient to capture the basic pattern in concentrations of stream DOC as a function of watershed elevation. Our sampling was not sufficient to capture all the hydrologic extremes at a

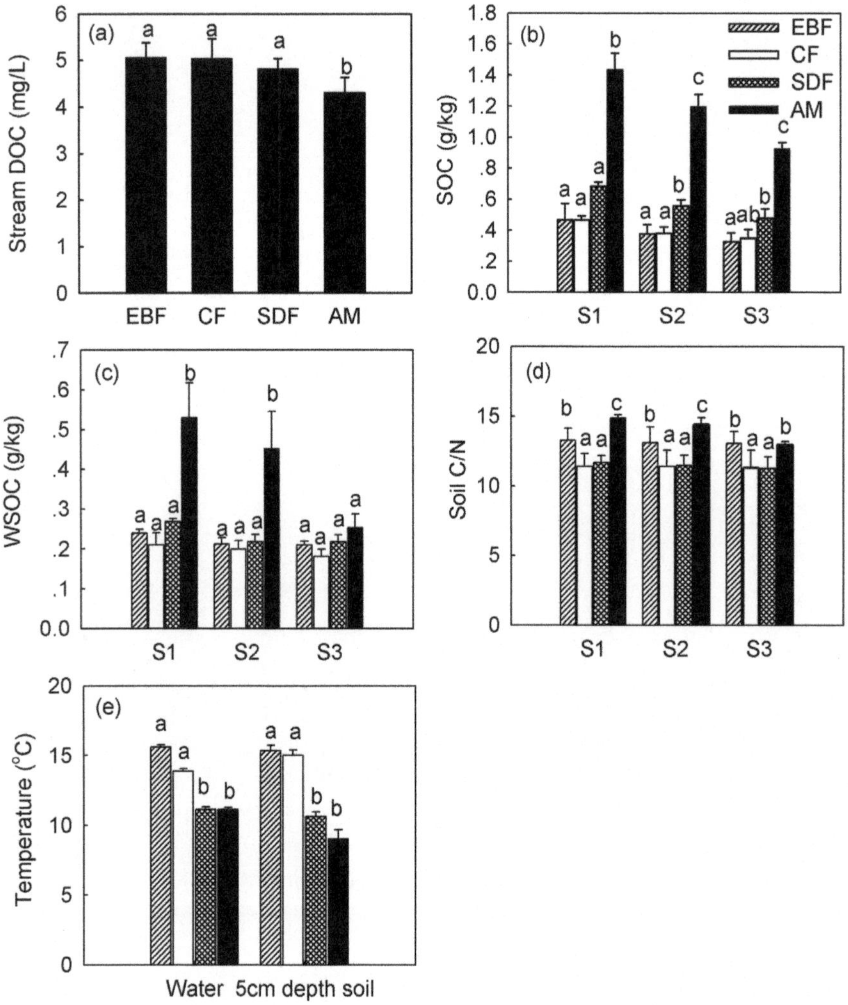

Figure 2. Annual mean values in (a) headwater stream dissolved organic carbon (DOC) concentrations, (b) riparian soil organic carbon (SOC) contents, (c) riparian water-soluble organic carbon (WSOC) contents, (d) riparian soil C:N ratio, and (e) temperature (mean±SD, n = 3) along an altitudinal gradient in the Wuyi Mountains of China (S1:0–10 cm soil layer; S2:10–25 cm soil layer; S3:25–40 cm soil layer; EBF: evergreen broadleaf forest; CF: coniferous forest; SDF: subalpine dwarf forest; AM: alpine meadow; DOC: dissolved organic carbon; SOC: soil organic carbon; WSOC: water-soluble soil organic carbon). Significant differences between the means are marked with different letters.

sampling site, but because it was consistent across elevation, with all elevations sampled on each sampling date, the patterns in stream chemistry should be relatively insensitive to hydrologic variation.

There are three fundamental processes that control the transport of carbon from soils to streams: DOC production in soil, hydrologic transport of DOC from soil to stream, and modification of DOC during transport [12]. In soils, DOC

Table 2. Annual mean data of stream DOC concentration and SOC character (mean±SD, n = 3).

Vegetation type	Stream DOC concentration (mg L^{-1})	SOC (g kg^{-1})			WSOC (g kg^{-1})			Soil C/N		
		0–10 cm	10–25 cm	25–40 cm	0–10 cm	0–25 cm	25–40 cm	0–10 cm	0–25 cm	25–40 cm
EBF	5.06±0.32a	0.47±0.11a	0.38±0.06a	0.33±0.06a	0.24±0.01a	0.21±0.02a	0.21±0.01a	13.30±0.85b	13.11±1.14b	13.09±0.83b
CF	5.03±0.44a	0.47±0.03a	0.38±0.04a	0.35±0.06a	0.21±0.03a	0.20±0.02a	0.18±0.02a	11.40±0.92a	11.40±1.16a	11.28±1.28a
SDF	4.80±0.23a	0.68±0.11a	0.56±0.08b	0.48±0.04b	0.27±0.01a	0.22±0.02a	0.22±0.02a	11.70±0.50a	11.45±0.76a	11.25±0.83a
AM	4.30±0.32b	1.43±0.19b	1.20±0.25c	0.93±0.13c	0.53±0.08b	0.45±0.09b	0.25±0.03a	14.88±0.23c	14.43±0.50c	12.99±0.21b

Different lowercase letters indicate significant differences among four study sites.

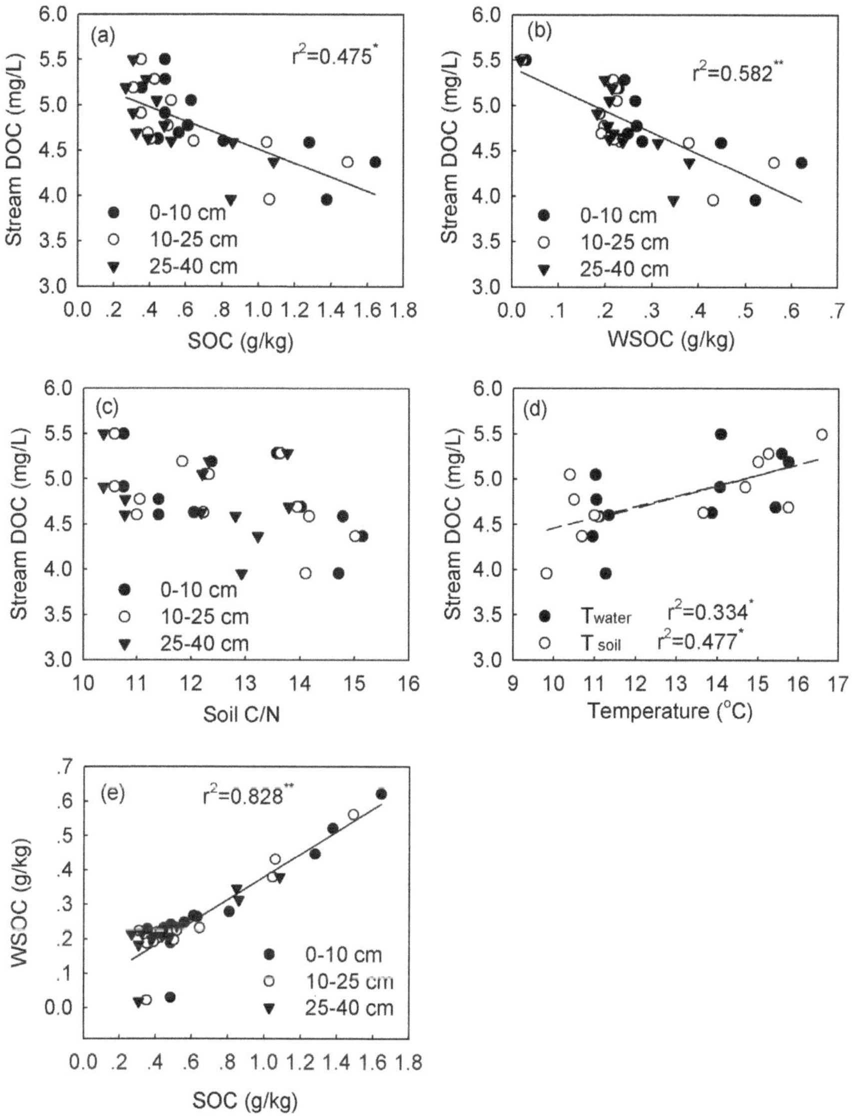

Figure 3. Correlations between annual mean headwater stream dissolved organic carbon (DOC) concentrations and (a) riparian soil organic carbon (SOC) contents, (b) riparian water-soluble organic carbon (WSOC) contents, (c) soil C:N ratio, (d) temperature, and (e) the link between riparian WSOC contents and riparian SOC contents along an altitudinal gradient in the Wuyi Mountains of China (T_{water}: water temperature; T_{soil}: soil temperature; n = 12. *and **indicated significant levels at 0.05 and 0.01 level, respectively).

production depends on environmental factors such as climate (temperature, precipitation), soil properties, and vegetation types [35], [36]. The hydrologic transportation of DOC is influenced by precipitation [37]. And the modification of quantity of DOC during transport appears to be controlled by soil properties [23], [24]. In our study, the negative relationship between stream DOC concentrations and riparian SOC contents may be due to the large changes in precipitation across our study sites (from 1700 mm in EBF to 3100 mm in AM) and differences in soil clay plus silt contents among soils from the four different vegetation types.

Increasing precipitation could lead to a dilution effect on DOC in soil and stream water, and consequently to decrease DOC concentrations in streams [38]. The increased precipitation with altitude [26] suggests that DOC concentrations in riparian soil and stream would nearly be diluted by a factor of 2 at the highest elevation catchment when compared with the lowest.

The capacity of catchment soils to adsorb DOC is an important driver of stream DOC concentrations [23], [24]. Adsorption capacities of soil typically vary as a function of their clay contents [24]. AM soil had higher clay plus silt contents than other vegetation types in all three soil layers. This higher clay plus silt content in AM soil might lead to higher adsorption capacity than in other vegetation types. This might result in lower export of soil DOC in AM than in other vegetation types for a given soil DOC concentration at the same temperature.

Armstrong et al. [39] suggested that vegetation controlled the DOC concentrations of drain water through affecting microbial assemblages, root exudates and litter quantity and quality associated with different plant functional types. Our data show that DOC production in soils is highest in the AM (with its high WSOC), but that loss of DOC along hydrologic flow paths is sufficiently large to lead to the lowest stream DOC concentrations. Fine root and microbial biomass in AM were also higher than those in other three vegetation types. Although the highest litter biomass occurred in CF, litter in CF was slow to decompose due to the large amount of waxes, resins and lignin contents [40]. These fine root and microbial biomass, and the quality and quantity of litter in the Wuyi Mountains reflect, in part, that soil DOC production varies with vegetation types. But survey of our study showed that stream DOC concentrations and WSOC contents presented oppositely spatial pattern along the elevation gradient. Thus we conclude that soil DOC production, which is associated with vegetation, isn't the most important driver of stream DOC concentrations in the Wuyi Mountains.

Vegetation may also affect the transport of soil DOC by affecting the chemical composition of DOC, soil solution pH, and soil physical structure that influence soil water infiltration [41], [42], [43]. The hot-water extractable organic matter from CF soils of Wuyi Mountains contains more highly condensed aromatic and hetero cyclic compounds than other three vegetation types, leading to being preferentially retained in the upper soil layer [27]. But DOC concentration in CF stream wasn't significantly lower than the other three vegetation types. Thus, the spatial pattern of stream DOC concentrations in our study is not likely to be controlled by the influence of vegetation on soil DOC transportation by affecting the chemical composition of DOC. Hardine et al [43] demonstrated that changes in soil solution pH could affect the adsorption efficiency of DOC in minerals. The differences in soil pH across four different vegetation types were rather small. It is unlikely that these differences in soil pH could explain the observed altitudinal pattern in stream DOC concentrations. Vegetation can also affect soil physical properties thus soil water infiltration rate, through producing root tunnels and influencing soil microbial and faunal community structure and their activities [42]. Unfortunately, we do not have data on either water infiltration rates or the biomass of plant large roots and soil fauna to evaluate vegetation influences on soil DOC adsorption and transport.

Temperature is known to stimulate biological activity and decomposition rates, affecting the production of DOC from soil organic matter [44], [45]. However, our data showed both SOC and WSOC were highest where temperature and stream DOC concentrations were lowest, in the high elevation streams. Thus, we suggest that temperature isn't the primary driver of the negative association between stream DOC concentration and SOC along an altitudinal gradient in the Wuyi Mountains.

Conclusion

We found a negative relationship between stream DOC concentrations and SOC contents under four different vegetation types along an altitudinal gradient in the Wuyi Mountains of China. The highest stream DOC concentration was in EBF and lowest in AM. Variations in precipitation and soil clay plus silt content in AM appear to be the dominant control on spatial variation of stream DOC concentrations along the altitudinal gradient in Wuyi Mountains. Our data from the Wuyi Mountains show that the drivers of spatial variability in stream DOC concentrations along a mountain transect can be directly opposite to those obtained from regional or global analyses. This strong relationship between standing condition and DOC export reflects some of the inherent difficulties in extrapolating the results from regional or global models of stream DOC to the steep environmental gradients occurring in mountainous terrain. Under such a situation, extrapolation of the drivers inferred from global comparisons to individual sites must be done with great caution. Furthermore, our results suggest that predicting changes in DOC concentrations in response to changing climate at a given site must be also done with considerable caution, as it is unclear whether changes in soil carbon would be expected to increase or decrease the delivery of DOC to streams from watershed soils. Further research effort is required to establish multi-factors regression model, which include physical (climate and soil) and biotic (chemistry and quantity of organic inputs) variables, to predict stream DOC concentration.

Acknowledgments

We gratefully acknowledge Althea N. Marks, Yiling Luan, Zikun Xu, Shanyan Chi, Jun Gen, Tao Xie, Zilong Ma and Chenpeng Du for help with field and lab work.

Author Contributions

Conceived and designed the experiments: HR WHM WH. Performed the experiments: WH JW LL. Analyzed the data: WH XZ. Wrote the paper: WH XZ. Revised the manuscript critically: WHM.

References

1. Cole JJ, Prairie YT, Caraco NF, McDowell WH, Tranvik LJ, et al. (2007) Plumbing the global carbon cycle: integrating inland waters into the terrestrial carbon budget. Ecosystems 10: 171–184.
2. Balcarczyk KL, Jones Jr JB, Jaffé R, Maie N (2009) Stream dissolved organic matter bioavailability and composition in watersheds underlain with discontinuous permafrost. Biogeochemistry 94: 255–270.
3. Hope D, Billett MF, Milne R, Brown TAW (1997) Exports of organic carbon in British rivers. Hydrol Process 11: 325–344.
4. Aitkenhead JA, McDowell WH (2000) Soil C:N ratio as a predictor of annual riverine DOC flux at local and global scales. Global Biogeochem Cycles 14: 127–138.
5. Aitkenhead-Peterson JA, Aitkenhead MJ, Clair TA (2005) Dissolved organic carbon and d issolved organic n itrogen export from forested watersheds in Nova Scotia: Identifying controlling factors. Global Biogeochem Cycles 19: doi:10.1029/2004GB002438.
6. Aitkenhead-Peterson JA, Smart RP, Aitkenhead MJ, Cresser MS, McDowell WH (2007) Spatial and temporal variation of dissolved organic carbon export from gauged and ungauged watersheds of Dee Valley, Scotland: Effect of land cover and C:N. Water Resour Res 43: doi:10.1029/2006WR004999.
7. Aitkenhead JA, Hope D, Billett MF (1999) The relationship between dissolved organic carbon in stream water and soil organic carbon pools at different spatial scales. Hydrological Processes 13: 1289–1302.
8. Billett MF, Deacon CM, Palmer SM, Dawson JJC, Hope D (2006) Connecting organic carbon in stream water and soils in a peatland catchment. J Geophys Res 111: doi: 10.1029/2005JG000065.
9. Fiebig DM, Lock MA, Neal C (1990) Soil water in the riparian zone as a source of carbon for a headwater stream. J Hydrol 116: 217–237.
10. Fiebig DM, Lock MA (1991) Immobilization of dissolved organic matter from groundwater recharging through the streambed. Freshwater Biol 26: 45–55.
11. Dosskey MG, Bertsch PM (1994) Forest sources and pathways of organic-matter transport to a Blackwater Stream – a hydrologic approach. Biogeochemistry 24: 1–19.
12. Brooks PD, McKnight DM (1999) The relationship between soil heterotrophic activity, soil dissolved organic carbon (DOC) leachate, and catchment-scale DOC export in headwater catchments. Water Resour Res 35: 1895–1902.
13. Harrison AF, Taylor K, Scott A, Poskitt J, Benham D, et al. (2008) Potential effects of climate change on DOC release from three different soil types on the Northern Pennines UK: examination using field manipulation experiments. Global Change Biol 14: 687–702.
14. Weyhenmeyer GA, Karlsson J (2009) Nonlinear response of dissolved organic carbon concentrations in boreal lakes to increasing temperatures. Limnol Oceanogr 54: 2513–2519.
15. Preston MD, Eimers MC, Watmough SA (2011) Effect of moisture and temperature variation on DOC release from a peatland: conflicting results from laboratory, field and historical data analysis. Sci Total Environ 409: 1235–1242.
16. Eimers MC, Watmough SA, Buttle JM, Dillon PJ (2008) Examination of the potential relationship between droughts, sulphate and dissolved organic carbon at a wetland-draining stream. Global Change Biol 14: 938–948.
17. Ågren AM, Haei M, Blomkvist P, Nilsson MB, Laudon H (2012) Soil frost enhances stream DOC concentrations during episodic spring snowmelt from boreal mires. Global Change Biol 18: 1895–1903.

18. Turgeon JML, Courchesne F (2008) Hydrochemical behaviour of dissolved nitrogen and carbon in a headwater stream of the Canadian Shield: relevance of antecedent soil moisture conditions. Hydrol Process 22: 327−339.
19. Findlay SEG (2005) Increased carbon transport in the Hudson River: unexpected consequence of nitrogen deposition? Front Ecol Environ 3: 133−137.
20. Monteith DT, Stoddard JL, Evans CD, de Wit HA, Forsius M, et al. (2007) Dissolved organic carbon trends resulting from changes in atmospheric deposition chemistry. Nature 450: 537−540.
21. Amiotte-Suche P, Linglois N, Leveque J, Andreux F (2007) ^{13}C composition of dissolved organic carbon in upland forested catchments of the Morvan Mountains (France): influence of coniferous and deciduous vegetation. J of Hydrol 335: 354−363.
22. Yamashita Y, Kloeppel BD, Knoepp J, Zausen GL, Jaffé R (2011) Effects of watershed history on dissolved organic matter characteristics in headwater streams. Ecosystems 14: 1110−1122.
23. McDowell WH, Wood T (1984) Podzolization: soil processes controlled dissolved organic carbon concentrations in stream water. Soil Science 137: 23−32.
24. Nelson PN, Baldock JA, Oades JM (1993) Concentration and composition of dissolved organic carbon in streams in relation to catchment soil properties. Biogeochemistry 19: 27−50.
25. He J, Lan S, Liu C, Li L (1994) Wuyishan Research Series. Xiamen: Xiamen University Press. 263p (in Chinese).
26. Wang S, Ruan H, Han Y (2010) Effects of microclimate, litter type, and mesh size on leaf litter decomposition along an elevation gradient in the Wuyi Mountains, China. Ecol Res 25: 1113−1120.
27. Bu X, Wang L, Ma W, Yu X, McDowell WH, et al. (2010) Spectroscopic characterization of hot-water extractable organic matter from soils under four different forests along an elevation gradient in the Wuyi Mountains. Geoderma 159: 139−146.
28. He R, Wang G, Wang J, Xu B, Wang K, et al. (2009) Seasonal variation and its main affecting factors of soil microbial biomass under different vegetations along an elevation gradient in Wuyi Mountains of China. Chinese Journal of Ecology 28: 394–399 (in Chinese with English abstract).
29. Strahler NA (1957) Quantitative analysis of watershed geomorphology. American Geophysical Union Transactions 38: 912−920.
30. Jones DL, Willett VB (2006) Experimental evaluation of methods to quantify dissolved organic nitrogen (DON) and dissolved organic carbon (DOC) in soil. Soil Biol Biochem 38: 991−999.
31. Jiang P, Xu Q, Xu Z, Cao Z (2006) Seasonal changes in soil labile organic carbon pools within a Phyllostachys praecox stand under high rate fertilization and winter mulch in subtropical China. Forest Ecol Manag 236: 30−36.
32. Xu X, Cheng X, Zhou Y, Luo Y, Ruan H, et al. (2010) Variation of soil labile organic carbon pools along an elevational Gradient in the Wuyi Mountains, China. J Resour Ecol 1: 368−374.
33. Meyer JL (1990) Production and utilization of dissolved organic carbon in riverine ecosystems. In: Perdue EM, Gjessing ET, editors. Organic Acids in Aquatic Ecosystems. 281−299.
34. Nakagawa Y, Shibata H, Satoh F, Sasa K (2008) Riparian control on NO_3^-, DOC, and dissolved Fe concentrations in mountainous streams, northern Japan. Limnology 9: 195−206.
35. Moore TR, Paré D, Boutin R (2008) Production of dissolved organic carbon in Canadian forest soils. Ecosystems 11: 740−751.
36. Fröberg M, Hansson K, Kleja DB, Alavi G (2011) Dissolved organic carbon and nitrogen leaching from Scots pine, Norway spruce and silver birch stands in southern Sweden. Forest Ecol Manag 262: 1742−1747.
37. Dillon PJ, Molot LA (2005) Long-term trends in catchment export and lake retention of dissolved organic carbon, dissolved organic nitrogen, total iron and total phosphorus: the Dorset, Ontario study, 1978–1998. J Geoph Res 110: doi:1029/2004JG000003.
38. Clark JM, Lane SN, Chapman PJ, Adamson JK (2007) Export of dissolved organic carbon from an upland peatland during storm events: implications for flux estimates. J Hydrol 347: 438−447.
39. Armstrong A, Holden J, Luxton K, Quinton JN (2012) Multi-scale relationship between peatland vegetation type and dissolved organic carbon concentration. Ecol Eng 47: 182–188.
40. Swift M, Heal OW, Anderson JM (1979) Decomposition in terrestrial ecosystems. Blackwell: Oxford. 372p.
41. Cronan CS, Aiken GR (1985) Chemistry and transport of soluble humic substances in forested watersheds of the Adirondack Park, New York. Geochim Cosm Ac 49: 1697–1705.
42. Lee KE, Foster RC (1991) Soil fauna and soil structure. Aust J Soil Res 29: 745–775.
43. Jardine PM, McCarthy JF, Weber NL (1989) Mechanisms of dissolved organic carbon adsorption on soil. Soil Sci Soc Am J 53: 1378–1385.
44. Moore TR, Dalva M (2001) Some controls on the release of dissolved organic carbon by plant tissues and soil. Soil Science 166: 38−47.
45. Neff JC, Hooper DU (2002) Vegetation and climate controls on potential CO_2, DOC and DON production in northern latitude soils. Global Change Biol 8: 872−884.

3

Changes in the Bacterial Community of Soil from a Neutral Mine Drainage Channel

Letícia Bianca Pereira, Renato Vicentini, Laura M. M. Ottoboni*

Center for Molecular Biology and Genetic Engineering (CBMEG), State University of Campinas - UNICAMP, Campinas, SP, Brazil

Abstract

Mine drainage is an important environmental disturbance that affects the chemical and biological components in natural resources. However, little is known about the effects of neutral mine drainage on the soil bacteria community. Here, a high-throughput 16S rDNA pyrosequencing approach was used to evaluate differences in composition, structure, and diversity of bacteria communities in samples from a neutral drainage channel, and soil next to the channel, at the Sossego copper mine in Brazil. Advanced statistical analyses were used to explore the relationships between the biological and chemical data. The results showed that the neutral mine drainage caused changes in the composition and structure of the microbial community, but not in its diversity. The *Deinococcus/Thermus* phylum, especially the *Meiothermus* genus, was in large part responsible for the differences between the communities, and was positively associated with the presence of copper and other heavy metals in the environmental samples. Other important parameters that influenced the bacterial diversity and composition were the elements potassium, sodium, nickel, and zinc, as well as pH. The findings contribute to the understanding of bacterial diversity in soils impacted by neutral mine drainage, and demonstrate that heavy metals play an important role in shaping the microbial population in mine environments.

Editor: Ahmed Moustafa, American University in Cairo, Egypt

Funding: This work was supported by CNPq/Vale grant 550240/2010-1. LBP received a fellowship from FAPESP (FAPESP 2012/05429-2). The funders had no role in study design, data collection and analysis, decision to publish, or preparation of the manuscript.

Competing Interests: The authors have declared that no competing interests exist.

* E-mail: ottoboni@unicamp.br

Introduction

The mining of metal ores and coal can lead to a variety of environmental problems, including deforestation, soil erosion, and the flooding of low-lying areas. One of the most critical issues in mine environments is the natural oxidation (chemical and biological) of sulfide mineral tailings that are exposed to water, oxygen, and microorganisms. This oxidation is responsible for the generation of mine drainage that compromises the quality of soil, surface water, and sub-surface water bodies, hence affecting overall biodiversity [1,2].

Mine drainage can be highly acidic or alkaline, depending on the complex interactions of hydrological, chemical, and biological processes. In addition, mine drainages are rich in heavy metals and sulfur, which are present in the minerals, but poor in nutrients. As a result, mine drainage is a multi-factor pollutant (considering aspects such as acidity or alkalinity, salinization, metal toxicity, and sedimentation processes), with the importance of each factor depending on the characteristics of the environment affected [2,3,4].

Investigation of the microbial communities of these environments can reveal undiscovered species able to provide a pool of genes and proteins whose potential is still unknown. Microorganisms in mine drainages have been studied for purposes including the bioremediation of mine sites [5] and the development of bioleaching consortia [6,7]. In addition, the metabolic diversity of the microbial community present in mine drainage makes these environments ideal for studies of genomes, ecology, evolution, tolerance mechanisms, and the interactions between bacteria and environmental factors [8,9,10].

Due to the extreme pH conditions (<3.0) and the high content of heavy metals, the microbial community of acid mine drainage (AMD) has been extensively studied. Evaluations have been made of the microbial communities in acid mine drainage [11,12], river sediments contaminated with AMD [13], subsurface mine environments [14], and AMD sediments [11].

In many cases, due to the neutralizing capacity of the waste minerals, or human intervention such as the spreading of limestone to precipitate metals, the drainage can have higher pH values (4.5 to 8.5), and is then called neutral mine drainage (NMD). This can cause severe environmental problems in mine environments because heavy metals, which are often present at high concentrations, can remain soluble at alkaline pH under suitable redox conditions [1,15].

Previous studies have reported that NMD contains great bacterial diversity, while that of AMD is typically low. Using metagenomic and metaproteomic approaches, Halter *et al.* [16] found a wide range of different bacteria in slightly alkaline French mine sediments, including iron-oxidizing and heterotrophic organisms. Reis *et al.* [17] reported a high diversity index for alkaline river sediments contaminated with heavy metals released from a Brazilian arsenic mine. However, NMD is an environment that has been poorly explored, and little is known about the structure and diversity of the bacterial community in soil contaminated with this kind of drainage [18,19].

The interaction between the bacterial community and a heavy metal-contaminated soil is also incompletely understood. Guo *et al.* [19] showed that the heavy metal-contaminated soils of two abandoned copper mines in Australia did not have any impact on microbial diversity. However, Marcin *et al.* [20] reported that a high content of heavy metals in forest soils had a negative influence on bacterial activity and diversity. It is noteworthy that the structure and diversity of bacteria communities in soil is not only influenced by heavy metal concentrations, but also by season [21], pH [22], organic matter [23], and interactions between these factors.

The behavior of the soil bacterial community exposed to different contaminants in extreme environments can be characterized using a combination of high-throughput sequencing, knowledge of the soil chemical parameters, and advanced statistical analysis. In this work, intensive parallel pyrosequencing of 16S rDNA was used to evaluate the microbial community in soil impacted by neutral mine drainage. The influence of soil chemical parameters on the composition, structure, and diversity of the bacterial community was also evaluated.

Materials and Methods

Sample collection and chemical analysis

Samples were collected at the Sossego copper mine (6°25'45"S, 50°3'58"W), in Canaã dos Carajás, Brazil. The mine has been operated by Vale since 1997 and the collection was authorized by this Company. In the mine, a drainage flows continuously from an ore deposit along a channel in the soil that starts at an ore heap and terminates at a runoff area. Along the course of the channel, the drainage is treated with limestone to increase the pH and precipitate heavy metals. Six drainage samples (D1-6, see Table S1 for coordinates) were collected aseptically from the upper soil layer (0–10 cm). Six uncontaminated soil samples (S1-6, see Table S1 for coordinates) were also collected (at 0–10 cm), adjacent to each drainage sample. The samples were stored at −20°C.

The chemical parameters Cd, Ca, Pb, Cu, Cr, S, Fe, P, Mg, Mn, Ni, K, Na, and Zn were measured in each sample prepared according to US EPA SW-846 method 3051A, with detection using inductively coupled plasma atomic emission spectrometry (ICP-AES) [24]. The organic matter (OM) content and pH (soil/water = 1:2.5, w/v) were determined as described by Camargo *et al.* [25].

DNA isolation, 16S rDNA library construction, and pyrosequencing

Prior to the isolation of DNA, the heavy metals were removed from the environmental samples using the method of Sánchez-Andrea *et al.* [13], with modifications. For this, the samples (0.5 g) were resuspended in 1 mL of phosphate-buffered saline (PBS), agitated, and centrifuged at 12879 x g for 10 min. The supernatant was discarded and the samples were resuspended in 1 mL of 0.5 M EDTA, at pH 8, and incubated at 4 °C for 2 h. After the incubation, the samples were centrifuged at 12879 x g for 10 min, and the supernatant was discarded. The wash with EDTA was repeated until the metals had been completely removed. The material was then washed with PBS, and the DNA was isolated from a 0.25 g aliquot using the Power Soil DNA Isolation Kit (MoBio Laboratories, Carlsbad, USA), according to the manufacturer's instructions.

The amplification of the V3–V4 region of the 16S rDNA was performed using the primers 338F and 806R [26,27]. The reaction mixture (final volume 25 μL) consisted of 3–10 ng of DNA, 0.6 μM of each primer, 0.5 U of AccuPrime Pfx DNA Polymerase (Invitrogen), and AccuPrime Pfx 1X reaction mix (Invitrogen). Amplification was performed under the following conditions: initial denaturation at 94°C for 1 min, 20 cycles at 94°C for 15 s, 52°C for 30 s, and 72°C for 30 s, and a final extension at 72°C for 2 min. After the first amplification, a further five cycles were performed, with a final volume of 50 μL and using the same conditions, in order to add specific barcodes to each sample and the adaptors A and B. The purification of the amplicons was accomplished using the GFX PCR DNA and Gel Band Purification Kit (GE Healthcare). An equimolar mixture of the 12 samples was made in order to obtain 2400 ng of PCR product with a concentration of 16.48 ng/μL. The amplicons were sequenced on a 454 GS Junior platform (Roche Company, Branford, CT, USA). The pyrosequencing data were submitted to MG-RAST [28] (ID 4521082.3 to 4521090.3 and 4521341.3 to 4521343.3) and to NIH Sequence Read Archive (BioProject PRJNA239576, accession numbers SRR1178540 and SRR1179196).

Sequence processing and statistical analysis

The QIIME package [29] was used to analyze the quality of the sequences and to group them into operational taxonomic units (OTUs, similarity of 97% or greater). Sequences with a quality score ≥ 25 and size between 400 and 470 bp were used in the analysis. Ambiguous bases and mismatches in primer sequences were not admitted. Chimeras were checked and removed with the ChimeraSlayer algorithm in QIIME. The taxonomic classification was performed using the RDP (Ribosomal Database Project, http://rdp.cme.msu.edu/classifier/classifier.jsp) at the 80% threshold.

The Student's t-test was employed to compare the drainage and soil samples in terms of the chemical parameters, using Statistica v.10 software [30]. Analysis of similarities (ANOSIM) was used to test the significance of the differences observed between the drainage and soil samples, based on the Bray-Curtis distance, considering the OTU composition of the samples. The t-test (with 95% confidence intervals) was used to determine whether the means of the Simpson, Shannon, and Berger-Parker diversity indices were statistically different, in terms of species richness and dominance, between the drainage and soil samples. Non-metric multidimensional scaling (n-MDS) was carried out using the Bray-Curtis distance matrix to plot the distances between the samples, considering the OTU distribution and the chemical parameters. Similarity percentage (SIMPER) analysis was used to identify the taxa that were mainly responsible for the differences observed between the drainage and soil samples. All the analyses were performed with normalized data, by sample length, using the PAST software [31].

Multivariate regression tree (MRT), aggregated boosted tree (ABT), and Spearman rank correlation analyses were carried out to correlate the chemical parameters with the biological data. MRT [32] was used to explore the relationship of the chemical parameters with the relative abundance of the dominant phyla and standardized diversity indices. This analysis was performed using the mvpart package within the R statistical environment, with default parameters. ABT [33] was performed (with 5000 trees, 10-fold cross validation, and three-way interaction) to evaluate the relative importance of the chemical parameters to the relative abundance of the dominant phyla, using the gbmplus package within the R environment. Non-parametric Spearman rank correlation was used to test the significance ($p<0.5$) of the relationships between the chemical parameters and phyla abundance, using the Statistica v.10 software.

Results

Chemical characterization of samples

A total of 16 chemical parameters were examined for each sample. Significant differences between the drainage and soil samples were observed for copper, nickel, potassium, sodium, and zinc (Table 1). The concentrations of the heavy metals were higher in the drainage samples, while those of sodium and potassium were higher in the soils. The pH of the soil samples ranged from 4.86 to 8.66, while pH values between 6.08 and 8.05 indicated that the drainage was a neutral mine drainage. Also, high concentrations of calcium were observed in samples S3, S5, and S6. This was due to the addition of limestone along the drainage channel of the Sossego mine, which also increased the pH and the calcium concentration of the soil next to the drainage (see Table S1 for details). High sulfur concentrations were observed in samples D3 and S2, which can be explained by the fact that the drainage channel flows over uneven ground, resulting in the accumulation of certain elements in some of the drainage and nearby soil samples.

Taxonomic composition

The pyrosequencing resulted in 102655 quality sequences from the 12 samples analyzed. The number of sequences per sample ranged from 3248 to 21198 (8554.58±5390.96). A total of 5901 OTUs were found, and approximately 45% were singletons. Of these OTUs, 5636 (95.5%) could be assigned to at least one particular phylum, and 24 OTUs were assigned to *Archea.* The Good's coverage ratio of the samples ranged from 83.8 to 99% (93.22±4.67%) (see Table S2 for details).

The phyla *Proteobacteria* (25.3%) and *Deinococcus/Thermus* (25.4%) were dominant in the drainage samples (Figure 1), while in soil the dominant phyla were *Proteobacteria* (26.6%), *Chloroflexi* (17.7%), *Acidobacteria* (15%), *Gemmatimonadetes* (14.2%), and *Actinobacteria* (13.1%) (Figure 2). Among the *Proteobacteria,* the most dominant taxa in both environments were *Alphaproteobacteria* (44.7% in drainage and 51.7% in soil), followed by *Betaproteobacteria* (26.5% in drainage and 19% in soil). In general, variation of the relative abundance of different phyla was more notable in the soil samples, while the drainage environment was dominated by a few phyla.

The drainage samples presented a total of 3023 OTUs, while the soil samples presented 4031 OTUs. Only 1153 OTUs (19.5%) were shared between both environments; most OTUs were found exclusively in one of the environments analyzed. All these results suggest that the composition of the bacterial community differed between the mine drainage and soil environments.

Diversity analysis

The analysis of similarities (ANOSIM), which considered the OTU compositions, revealed significant differences between the drainage and soil bacterial communities ($p = 0.0087$). No significant difference was found when the drainage samples were compared to each other ($p = 1$), and the same was observed for the soil samples.

Non-metric multidimensional scaling (n-MDS) (Figure 3) was performed to group the samples by associating the chemical parameters and the OTU frequencies. The results indicated a clear separation between the drainage and soil samples. The soil samples showed more scatter, suggesting greater heterogeneity, and the concentrations of sodium and potassium were the chemical parameters that contributed most to the separation. The drainage samples showed a more homogenous distribution, closely related to the amounts of copper, cadmium, nickel, and zinc. These results were in accordance with the statistical analysis (Student's t-test) showing that the amounts of the chemical parameters described above were different for the two environments. The ANOSIM and n-MDS analyses indicated that the structure of the bacterial community changed when the environment was directly impacted by the mine drainage.

Table 1. Chemical parameters of the drainage and soil samples, and P-values derived from the Student's t-test.

	Drainage Mean values	Soil Mean values	P-value
Cadmium	0.60±0.15	0.47±0.14	0.15
Calcium	40266.67±74148.52	28633.33±50360.92	0.76
Lead	5.35±2.86	5.85±4.49	0.82
Copper	31436.33±22707.81	6287.17±8012.76	0.03*
Chromium	22.75±10.29	26.45±9.86	0.54
Sulfur	24391.67±53721.94	28865.00±69631.71	0.90
Iron	27603.50±15210.08	21695.83±9405.91	0.43
Phosphorus	11286.67±23115.07	538.33±253.57	0.28
Magnesium	7608.33±4871.39	4883.33±4937.37	0.36
Manganese	324.83±139.01	446.50±413.48	0.51
Nickel	112.25±42.2	64.95±26.24	0.04*
Potassium	9.32±8.44	268.50±157.6	0.002**
Sodium	1.82±0.91	102.35±45.13	0.00**
Zinc	31.17±18.34	13.22±3.13	0.04*
pH	7.50±0.71	7.25±1.59	0.73
OM	22.50±9.12	22.33±17.22	0.98

Metal concentrations are expressed in mg/kg. The organic matter (OM) content is expressed in g/dm^3.
* Statistically significant at 5%.
** Statistically significant at 1%.

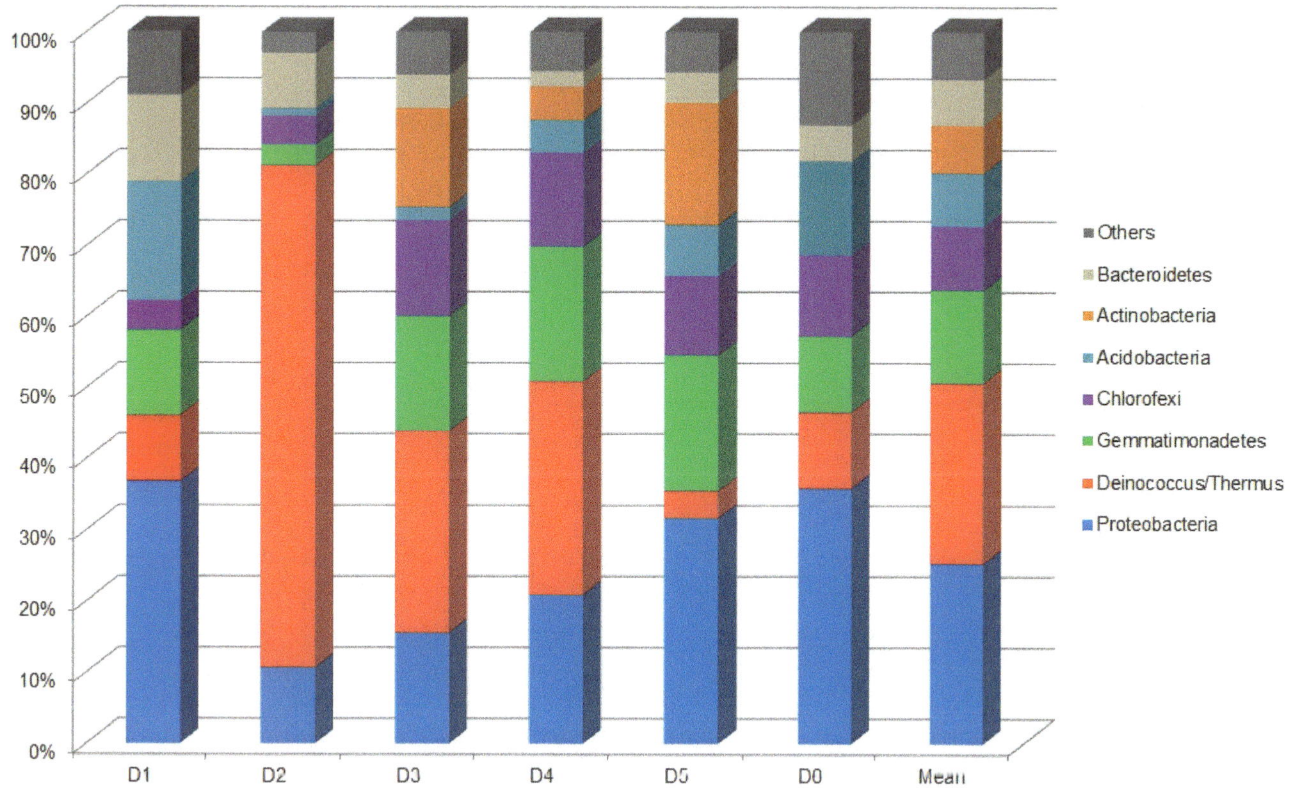

Figure 1. Relative abundance of bacterial phyla in drainage samples. Others: Armatimonadetes, BRC1, Chlamydiae, Chlorobi, Cyanobacteria, Elusimicrobia, Firmicutes, Nitrospirae, OD1, Planctomycetes, Spirochaetes, Synergistetes, Verrucomicrobia, TM6, TM7, WPS-2, WS3, and unclassified Bacteria.

The Student's t-test applied to the diversity indices revealed that, considering all the samples, there were no significant differences between the drainage and soil in terms of diversity ($p = 0.210$, Shannon index) and dominance ($p = 0.26$, Berger-Parker index; $p = 0.25$, Simpson index).

Since a difference in the bacterial community was observed between the environments, SIMPER analysis was used to show which OTUs contributed to this result. The five main OTUs are shown in Table 2. The OTU that contributed the most to the differences between the soil and drainage communities was 824 (12.7% of the total dissimilarity), composed of the genus *Meiothermus*, which was highly abundant in the drainage samples. The OTUs 3607 and 4974 were classified as *Gemmatimonas*, and were more abundant in the soil samples. Together, these OTUs accounted for 4.43% of the difference. The OTU 1366 was identified as *Acidobacteria*, a phylum containing genera commonly found in soil and drainage [14]. The OTU 303 was classified as *Bacteria*, and presented 95% identity with soil bacteria. The high contribution of OTU 824 is particularly interesting, because it was the dominant OTU in all the drainage samples and was most important in the species dominance index (Berger-Parker index, data not shown). These results revealed that the presence of the mine drainage had a positive influence on the abundance of this OTU in the environment.

Influence of the chemical parameters on microbial composition and diversity

The MRT analysis was conducted by associating the chemical parameters with both the relative phylum abundance and the standardized diversity indices. For the distribution of the dominant phylum, the environmental data provided a tree with three splits, based on copper, potassium, and iron (Figure 4). The tree explained 70% of the variance of the relative phylum abundance. In this tree, copper was responsible for the first split, appearing to be a strong predictor in this environment since it explained 58.4% of the variance. *Deinococcus/Thermus* were the most abundant phyla in samples exhibiting a high content of copper, while other phyla were less abundant.

The second and third splits were determined by potassium (8.2% of the variance) and iron (3.5% of the variance), respectively. *Proteobacteria* was abundant in samples that contained a smaller amount of potassium and were richer in iron. In contrast, *Acidobacteria* and *Actinobacteria* were prevalent in samples containing higher amounts of potassium, and the *Actinobacteria* population decreased when the amount of iron increased. *Gemmatimonadetes* showed a homogeneous distribution along the tree.

The tree for the standardized diversity indices (Figure 5) showed one split determined by pH. This tree explained 58.1% of the variance. As shown by the split, the Berger-Parker index decreased at pH above 6.16, while the Shannon and Simpson indices showed higher values, suggesting less dominance and greater diversity in the alkaline environments. Additionally, the single regression tree with the diversity indices (Figure S1) showed that the Shannon index was explained more by potassium and pH, while the Berger-Parker and Simpson indices were explained by pH.

The ABT models showed how the relative importance of the chemical parameters influenced the relative abundance of the dominant phyla (Figure 6). In general, the analysis indicated that

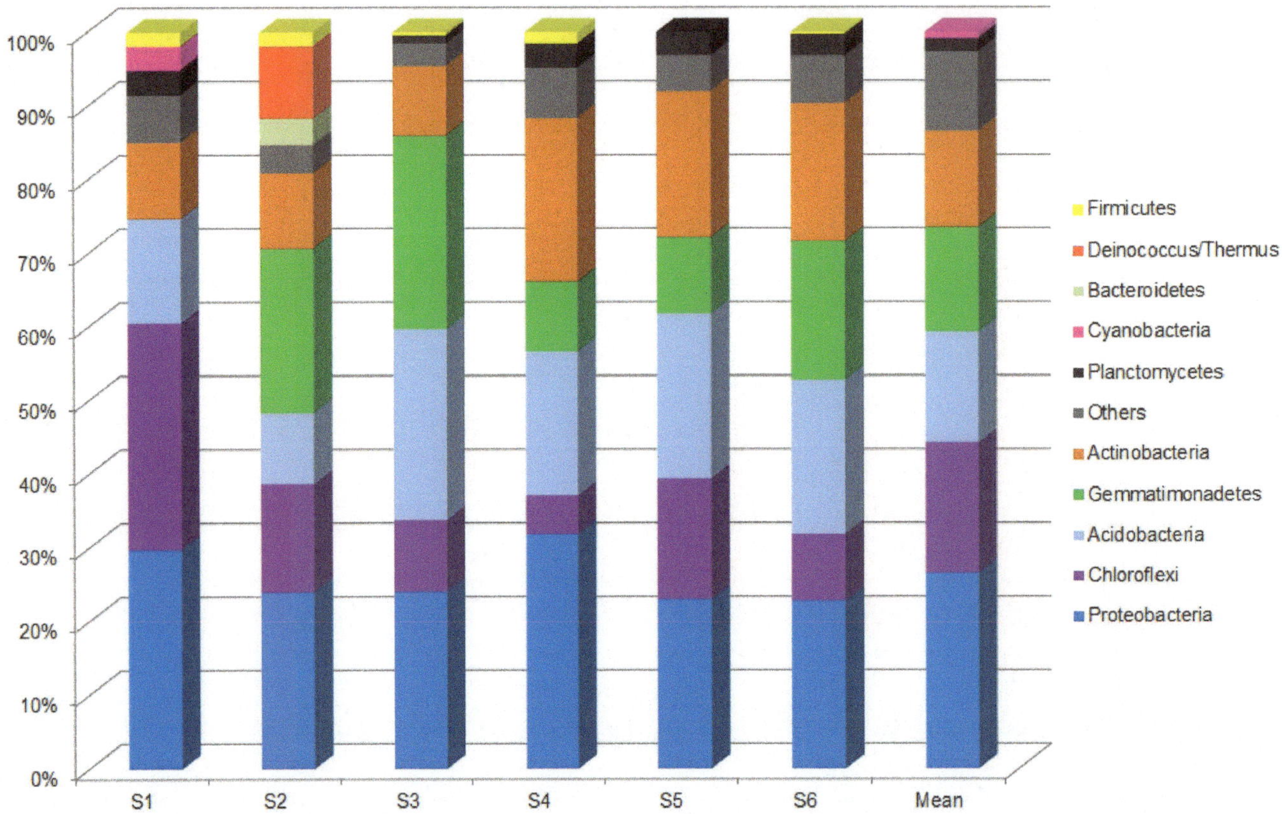

Figure 2. Relative abundance of bacterial phyla in soil samples. Others: Acidobacteria, AD3, Armatimonadetes, Chlamydiae, Chlorobi, Elusimicrobia, Fusobacteria, GAL15 Nitrospirae, OD1, Spirochaetes, Tenericutes, Verrucomicrobia, TM6, TM7, WPS-2, WS3, WYO, and unclassified Bacteria.

copper and potassium were the principal parameters that exerted a strong influence on these phyla. Copper was the principal influence on *Proteobacteria* (59.74%) and *Deinococcus/Thermus* (62.42%), and an intermediate influence on *Acidobacteria* (34.16%) and *Gemmatimonadetes* (3.12%). This metal showed no influence on *Actinobacteria.* Potassium was the principal influence on the distribution of the *Acidobacteria* (44.14%) and *Actinobacteria* (62.24%) communities, and showed an intermediate influence on *Proteobacteria* (13.58%) and *Deinococcus/Thermus* (18.65%). These findings supported the MRT results, showing the importance of these chemical parameters in shaping the microbial communities in the mine environments. Chromium, calcium, iron, and nickel were also important parameters that influenced the distribution of the principal phyla (see Table S3 for complete relative influence data).

Spearman rank correlation analysis (Table S4) revealed a slightly positive correlation between *Acidobacteria* and *Actinobacteria/Proteobacteria. Gemmatimonadetes* presented significant correlations with cadmium, chromium, and sulfur, while *Proteobacteria* was negatively correlated with calcium and pH. The *Thermus* phyla was highly positively correlated with copper, potassium, and zinc, and slightly correlated with sulfur, iron, and nickel, reinforcing the previous findings for the relationships between the chemical and biological data.

Discussion

This work evaluated the impact of a neutral mine drainage flow on the soil microbial community. Statistical analysis of the measured chemical parameters revealed significantly higher levels of copper, nickel, and zinc in the drainage samples, compared to the soil samples. The increase of these metals in the drainage was expected because they were part of the composition of the minerals in the Sossego mine ore deposit, which after oxidation became soluble and flowed along the soil. In contrast, the soil samples contained higher amounts of sodium and potassium. These chemical components could have played an important role in shaping the microbial community in these environments. However, it is important to note that both the drainage and soil samples presented high contents of heavy metals, iron, and sulfur, in addition to the calcium and magnesium that increase the normal pH of the soil. Therefore, both environments (soil and drainage) presented conditions that were adverse to microbial communities, and deserved to be investigated.

Analysis of microbial composition revealed that the bacterial community in the drainage samples was dominated by only a few phyla, with only two phyla (*Proteobacteria* and *Deinococcus/Thermus*) accounting for 50% of the total diversity. The microbial community of the soil samples presented more phyla with similar relative abundances. In areas contaminated with heavy metals, it has been observed that the abundance of tolerant bacteria increases, while that of the more sensitive organisms decreases [34]. Therefore, the higher content of metals such as copper, nickel, and zinc in the drainage samples could have acted to select the resistant bacteria that dominated this environment. *Proteobacteria* was the phylum with the highest relative abundance in both drainage and soil samples. *Proteobacteria* has been found to be the

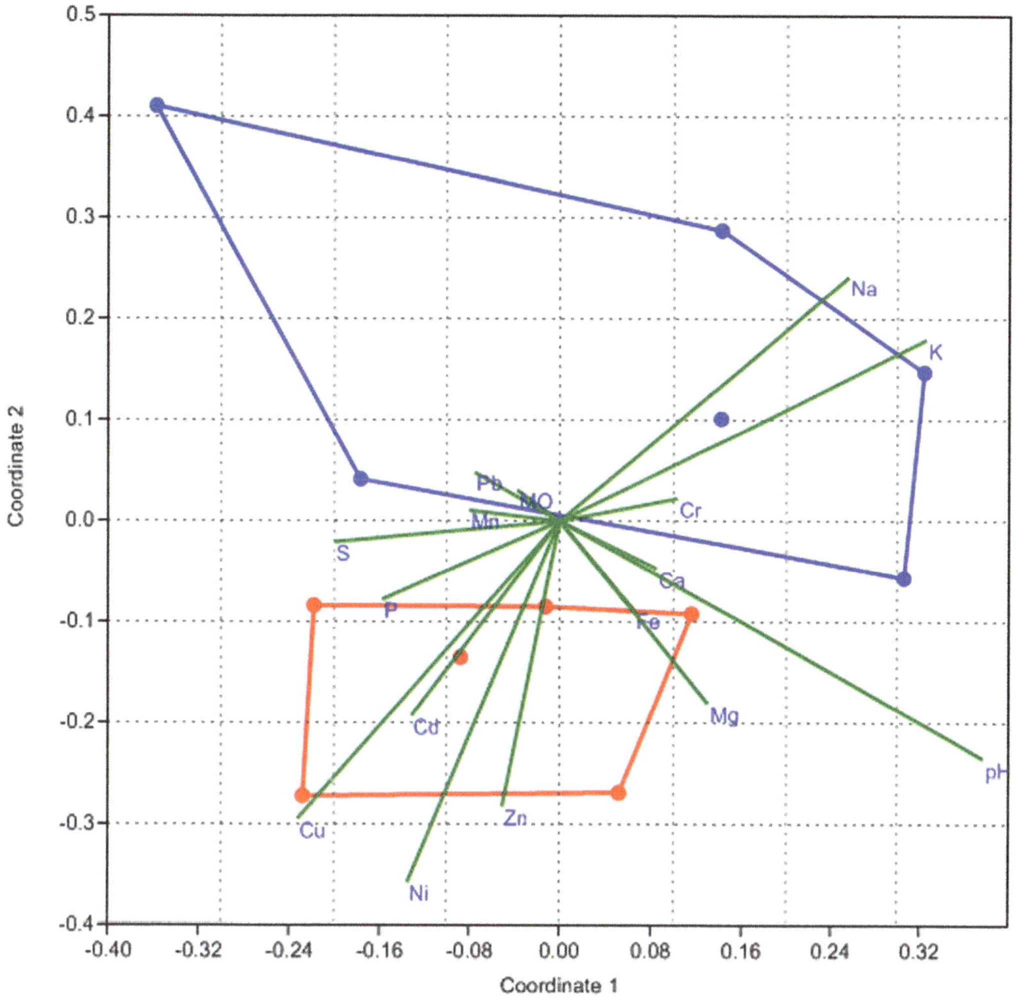

Figure 3. Non-metric multidimensional scaling. Non-metric multidimensional scaling using the Bray-Curtis dissimilarity plots of the first two components for drainage (red) and soil (blue) samples. The stress value is 0.095.

predominant phylum in many mine environments, including arsenic mine sediment [16], gold mine stream sediments [17], and a river contaminated with metals [35]. This phylum exhibits a complex lifestyle and can degrade a variety of complex organic molecules, enabling it to adapt to many different environments [23].

The small number of OTUs shared between samples supports the hypothesis of the existence of large differences between the bacterial community compositions of the drainage and soil samples. The n-MDS and ANOSIM analyses showed that there were differences between the two environments in terms of the structure of the communities. The high heterogeneity observed among the soil samples reflected the wide variety of phyla in this environment, as observed in the community composition results. In contrast, the drainage samples exhibited less diversity, suggesting that in this environment, the bacterial community was more homogeneous because fewer organisms could adapt to the extreme conditions.

Table 2. Principal OTUs responsible for differences between the drainage and soil samples.

OTU	Dissimilarity contribution (%)	Abundance Drainage (%)	Abundance Soil (%)	RDP classifier [bootstrap value]
824	12.7	23.3	1.09	*Meiothermus* [99%]
3607	2.69	2.5	3.31	*Gemmatimonas* [99%]
303	1.8	0.19	3.11	*Bacteria* [100%]
1366	1.78	3.45	0.37	*Acidobacteria* [100%]
4974	1.74	0.012	3.06	*Gemmatimonas* [89%]

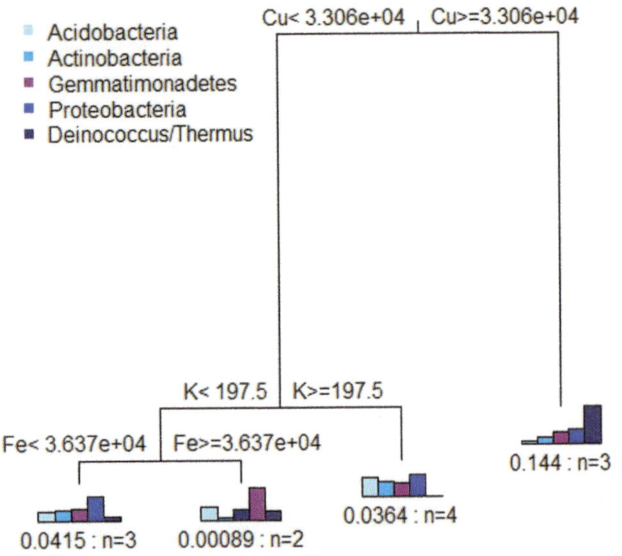

Figure 4. Multivariate regression tree of the abundance of phylum. Multivariate regression tree of the relation between relative abundance of dominant phylum and chemical parameters. The bar plots show the mean relative abundance of each phylum at the terminal nodes. The numbers under the bar plots indicate the error and the number of samples (n) within each group. Error: 0.299.

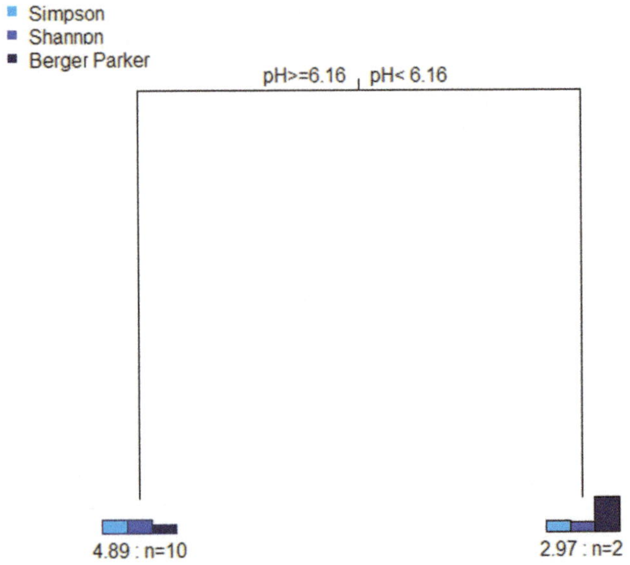

Figure 5. Multivariate regression tree of the diversity indices. Multivariate regression tree of the relation between standardized diversity indices and chemical parameters. The bar plots show the mean of the diversity indices at the terminal nodes. The numbers under the bar plots indicate the error and the number of samples (n) within each group. Error: 0.744.

Comparison of the Simpson, Shannon, and Berger-Parker indices revealed that there was no significant difference between the drainage and soil samples in terms of diversity. According to Kandeler *et al.* [36], enrichment of the soil with heavy metals reduces the biomass and activity of bacteria. However, recent studies have shown that after a century of exposure to acid drainage, the bacterial community tends to stabilize and increase its diversity, resulting in an environment with a microbial diversity similar to that of non-impacted areas [17,23]. The Sossego mine has been in operation for 16 years, and soil contamination is considered to be relatively recent. Therefore, it was expected that diversity would be lower at the drainage sites because only a few groups of bacteria would be resistant to the extreme conditions of this environment. However, it has been suggested that while acute exposure has a negative influence on bacterial diversity, in chronically contaminated environments, the proliferation of resistant bacteria can increase diversity [37,38]. This situation is common in mines such as the Sossego mine, where the drainage flows constantly from the ore deposit. Furthermore, high concentrations of nutrients can improve diversity in sediments contaminated with heavy metals [17,39]. At the Sossego mine, the drainage flows over the soil to a run-off area located about 700 m from the ore deposit, and the vegetation in this area provides high levels of organic matter. The natural nutrient content of this soil might therefore play a key role in the promotion of bacterial diversity.

The SIMPER results revealed that OTU 824, associated with *Meiothermus*, accounted for much of the difference between the drainage and soil communities, because it was much more abundant in the drainage samples. The optimum growth temperatures of these bacteria lie between 50 and 65°C, optimum pH is around 8.0, and they are usually found in warm and nutrient-poor environments, such as geothermal and anthropogenically-influenced areas [40,41,42]. OTU 824 was the taxa with the highest relative importance in all the drainage samples and was responsible for the dominance index in this environment. Other researchers have reported the *Meiothermus* phylum, *Deinococcus/Thermus*, in river sediments affected by an arsenic mine [16] as well as in acid mine drainage [12], albeit in smaller proportions compared to the present work. This is the first time that the genus has been found in great abundance in mine environments, suggesting a broad adaptive capacity of the *Meiothermus* species.

OTUs 3607 and 4974 were classified as *Gemmatimonas* and were more frequently found in the soil samples. Together, these OTUs accounted for 4.43% of the difference observed between the soil and drainage samples. The genus *Gemmatimonas* has been reported in studies of diversity in mine environments that were either uncontaminated or contaminated with acid mine drainage. Mendez *et al.* [43] used 16S rDNA libraries to study the bacterial composition of the Klondyke mine in the USA. *Gemmatimonadetes* was only found in uncontaminated samples, where the pH was around 8.0. Reis *et al.* [17] studied sediments affected by heavy metals from mining areas in Brazil and found that the phylum *Gemmatimonadetes* was only present in contaminated environments. It is therefore possible that these bacteria were widely present in the soil of the Sossego mine region and resisted the impact of the drainage, although their abundance decreased slightly due to the prevalence of other bacteria more resistant to heavy metals.

The OTU 1366 was identified as *Acidobacteria*, a phylum that includes genera commonly found in soil and drainage environments [14], and OTU 303 was identified only as bacteria present in soils.

The MRT and ABT analyses revealed a major influence of copper on the bacterial community of the Sossego mine. Although neutral pH decreases the available copper, its presence can still interfere with the microbial population. Kunito *et al.* [34] evaluated the influence of different forms of copper on the soil microorganism community, and found no significant correlation between total copper and the tolerance of the bacteria. Nonetheless, in alkaline soils a small (but important) proportion (about 4%

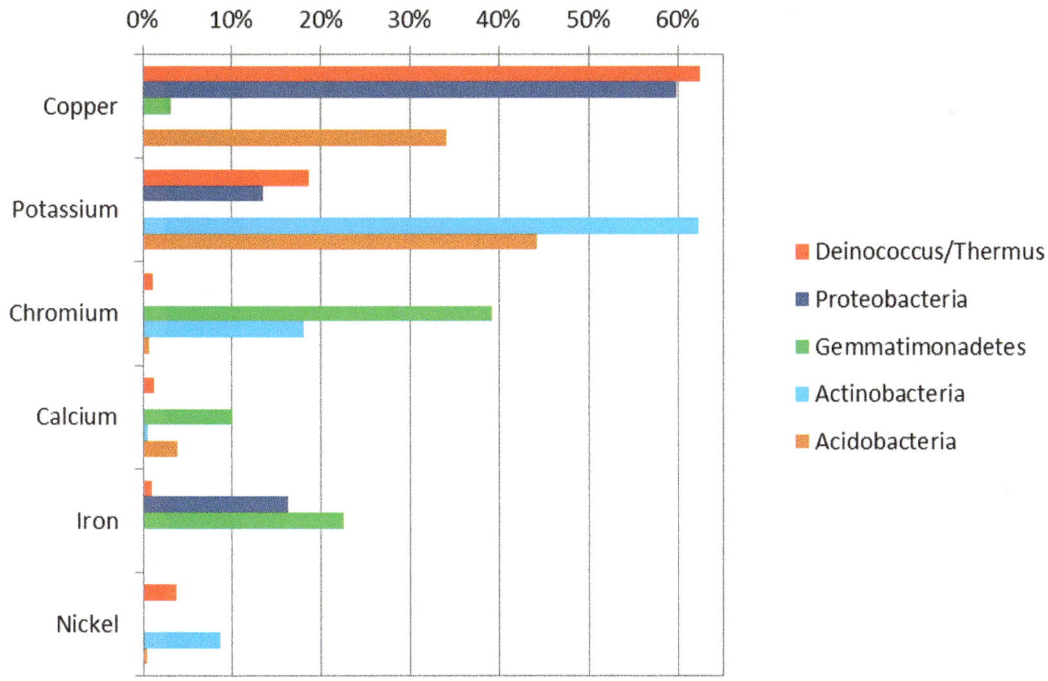

Figure 6. Aggregated boosted tree (ABT). The ABT analysis show the relative influence of the principal chemical parameters for the dominant phyla in the Sossego mine.

of the total copper) can remain adsorbed onto soil particles and/or associated with organic matter. The copper fraction associated with organic matter is slightly soluble, and could have a negative impact on the bacterial community. Furthermore, the soluble and adsorbed fractions of copper are in equilibrium; although the soluble copper reflects the toxicity, the adsorbed copper reflects the toxicity capacity, because it is the source of the soluble copper. In soil with a high content of heavy metals that compete for adsorption sites, copper is the cation most likely to be retained by organic matter, clay, and oxides. This competition process mainly occurs in soil where there are large inputs of limestone, due to the increase in soil sorption capacity [44]. It is therefore possible that copper influenced the structure and composition of the bacterial communities of the Sossego mine, because the soil was exposed to different conditions of rainfall, redox condition, and limestone application.

The most interesting effect of copper was observed for the *Deinococcus/Thermus* phylum, which showed a strong positive correlation with this metal, as well as with other metals, such as zinc and nickel (see the Spearman rank correlation coefficients). This phylum includes the genera *Deinococcus*, *Meiothermus*, *Thermus*, and *Trupera*, which have shown resistance to extreme conditions (high irradiation, oxidation, high temperature, and desiccation) [45,46]. Although they have a broad adaptation capacity, little is known about their survival in the presence of high levels of heavy metals, because (until the present study) this phylum has only been detected at low abundance in mine environments. The ecological importance of *Deinococcus/Thermus* was only proposed recently [47], and their presence in mine environments could provide new opportunities for the study of these microorganisms.

Another important factor that affected bacterial composition in the Sossego mine was the potassium concentration. Both potassium and sodium were present at high levels in the soil samples, and were highly inter-correlated ($r = 0.94$). Both elements could have been influenced by the plant rhizosphere, which was present in most of the soil samples. Potassium is a plant macronutrient that is easily leached because of its mobility in soils. Bacteria can affect the solubility and availability of this and other nutrients, which affects the growth of plants as well as the selection of specific bacteria associated with potassium [48].

The MRT analysis revealed high bacterial diversity in samples with elevated pH. Since there was high exposure to limestone, the bacterial community of the drainage and soil could have become adapted to neutral/alkaline conditions, which would explain the high abundance of genera generally associated with high pH, such as *Gemmatimonas* and *Meiothermus*.

In conclusion, it was found that a chronic impact of neutral mine drainage in the soil environment shifted the composition and structure of the bacterial community, but did not affect bacterial diversity. The high diversity found in both environments could have been due to constant exposure to the drainage, as well as the high availability of organic matter that promoted the growth of an adapted microbial community. The high levels of heavy metals, as well as the concentrations of nutrients such as potassium and sodium, could account for the different microbial compositions of the drainage and soil samples. The *Deinococcus/Thermus* phylum was strongly affected by heavy metals (especially copper), and the *Meiothermus* genus was dominant in the drainage communities. This suggests that the full diversity of these taxa remains unexplored.

Supporting Information

Figure S1 Single regression tree. Showing the relation between standardized diversity indices and chemical parameters. A: Berger-Parker; B: Simpson; C: Shannon.

Table S1 Locations and chemical parameters of the drainage and soil samples. Metal concentrations are expressed in mg/kg. The organic matter (OM) content is expressed in g/dm^3.

Table S2 Sample sequencing and diversity index information.

Table S3 Relative influence (%) of chemical parameters in phylum abundance determined by aggregated boosted tree analysis.

Table S4 Spearman rank correlation results.

Acknowledgments

The authors are grateful to André Luis Martinez Ferreira and Vale for collection of the environmental samples.

Author Contributions

Conceived and designed the experiments: LBP RV LMMO. Performed the experiments: LBP. Analyzed the data: LBP RV LMMO. Contributed reagents/materials/analysis tools: LBP RV LMMO. Wrote the paper: LBP RV LMMO.

References

1. Banks D, Younger PL, Arnesen RT, Iversen ER, Banks SB (1997) Mine-water chemistry: the good, the bad and the ugly. Environ Geol 32: 157–174.
2. Jain RK, Karera AC (2009) Mining operations and environmental monitoring. Minenvis 63: 1–5.
3. Gray NF (1997) Environmental impact and remediation of acid mine drainage: a management problem. Environ Geol 30: 62–71.
4. Pancholi DA, Sharma N, Singh G (2009) Problems of acid mine drainage and the remedial measures at Panandhro lignite mines, Kutch, W. India. Minenvis 63: 6–12.
5. Govarthanan M, Lee KJ, Cho M, Kim JS, Kamala-Kannan S, et al. (2012) Significance of autochthonous Bacillus sp. KK1 on biomineralization of lead in mine tailings. Chemosphere 8: 2267–2272.
6. Cárdenas JP, Valdés J, Quatrini R, Duarte F, Holmes DS, et al. (2010) Lessons from the genomes of extremely acidophilic bacteria and archaea with special emphasis on bioleaching microorganisms. Appl Environ Microb 88: 605–620.
7. Liu Y, Yin H, Liang Y, Shen L, Liu Y, et al. (2011) Changes in the composition of an acid mine drainage microbial community upon successive transfers in medium containing low-grade copper sulfide. Bioresource Technol 102: 9388–9394.
8. Baker BJ, Banfield JF (2003) Microbial communities in acid mine drainage. FEMS Microbiol Ecol 43: 139–152.
9. Lauber CL, Hamady M, Knight R, Fierer N (2009) Pyrosequencing-based assessment of soil pH as a predictor of soil bacterial community structure at the continental scale. Appl Environ Microb 75: 5111–5120.
10. Denef VJ, Mueller RS, Banfield JF (2010) AMD biofilms: using model communities to study microbial evolution and ecological complexity in nature. ISME J 4: 599–610.
11. Bruneel O, Volant A, Gallien S, Chaumande B, Casiot C, et al. (2006) Diversity of microorganisms in Fe-As-rich acid mine drainage waters of Carnoule's, France. Appl Environ Microb 72: 551–556.
12. Kuang JL, Huang LN, Chen LX, Hua ZS, Li SJ, et al. (2012) Contemporary environmental variation determines microbial diversity patterns in acid mine drainage. ISME J 7: 1038–1050.
13. Sánchez-Andrea I, Rodríguez N, Amils R, Sanz JL (2011) Microbial diversity in anaerobic sediments at Río Tinto, a naturally acidic environment with a high heavy metal content. Appl Environ Microb 77: 6085–6093.
14. Johnson DB (2011) Geomicrobiology of extremely acidic subsurface environments. FEMS Microbiol *Ecol* 81: 2–12.
15. Lindsay MBJ, Condon PD, Jambor JL, Lear KG, Blowes DW, et al. (2009) Mineralogical, geochemical, and microbial investigation of a sulfide-rich tailings deposit characterized by neutral drainage. Appl Geochem 24: 2212–2221.
16. Halter D, Cordi A, Gribaldo S, Gallien S, Goulhen-Chollet F, et al. (2011) Taxonomic and functional prokaryote diversity in mildly arsenic-contaminated sediments. Res Microbiol 162: 877–887.
17. Reis MP, Barbosa FAR, Chartone-Souza E, Nascimento AMA (2013) The prokaryotic community of a historically mining-impacted tropical stream sediment is as diverse as that from a pristine stream sediment. Extremophiles 17: 301–309.
18. Liao M, Xie XM (2007) Effect of heavy metals on substrate utilization pattern, biomass, and activity of microbial communities in a reclaimed mining wasteland of red soil area. Ecotox Environ Safe 66: 217–223.
19. Guo Z, Megharaj M, Beer M, Ming H, Rahman MM, et al. (2009) Heavy metal impact on bacterial biomass based on DNA analyses and uptake by wild plants in the abandoned copper mine soils. Bioresource Technol 100: 3831–3836.
20. Marcin C, Marcin G, Justyna MP, Katarzyna K, Maria N (2013) Diversity of microorganisms from forest soils differently polluted with heavy metals. Appl Soil Ecol 64: 7–14.
21. Streten-Joyce C, Manning J, Gibb KS, Neilan BA, Parry DL (2013) The chemical composition and bacteria communities in acid and metalliferous drainage from the wet–dry tropics are dependent on season. Sci Total Environ 443: 65–79.
22. Kock D, Schippers A (2008) Quantitative microbial community analysis of three different sulfidic mine tailing dumps generating acid mine drainage. Appl Environ Microb 74: 5211–5219.
23. Bouskill N, Barker-Finkel J, Galloway TS, Handy RD, Ford TE (2010) Temporal bacterial diversity associated with metal-contaminated river sediments. Ecotoxicology 19: 317–328.
24. U.S. Environmental Protection Agency (EPA) (1986) Test Methods for Evaluating Solid Waste, Physical/Chemical Methods. SW-846. Washington D.C.: US EPA.
25. Camargo AO, Moniz AC, Jorge JA, Valadares JMAS (2009) Methods of Chemical Analysis, Physical and Mineralogical Soil Agronomic Institute of Campinas. Campinas: Agronomic Institute. 77p.
26. Huse SM, Huber JA, Morrison HG, Sogin ML, Welch DM, et al. (2008) Accuracy and quality of massively parallel DNA pyrosequencing. Genome Biol 8: R143.
27. Masoud W, Takamiya M, Vogensen FK, Lillevang S, Al-Soud WA, et al. (2011) Characterization of bacterial populations in Danish raw milk cheeses made with different starter cultures by denaturating gradient gel electrophoresis and pyrosequencing. Int Dairy J 21: 142–148.
28. Meyer F, Paarmann D, D'Souza M, Olson R, Glass EM, et al. (2008) The metagenomics RAST server - a public resource for the automatic phylogenetic and functional analysis of metagenomes. BMC Bioinformatics 9: 386.
29. Caporaso JG, Kuczynski J, Stombaugh J, Bittinger K, Bushman FD, et al. (2010) QIIME allows analysis of high-throughput community sequencing data. Nat Methods 7: 335–336.
30. StatSoft Inc (2003) STATISTICA (data analysis software system). Version 6. Available: www.statsoft.com. Accessed 2013 Sept 20.
31. Hammer O, Harper DAT, Ryan PD (2001) PAST: Paleontological statistics software package for education and data analysis. Palaeontol Electron 4: 1–9.
32. De'ath G (2002) Multivariate regression trees: a new technique for modeling species–environment relationships. Ecology 83: 1105–1117.
33. De'ath G (2007) Boosted trees for ecological modeling and prediction. Ecology 88: 243–251.
34. Kunito T, Saeki K, Oyaizu H, Matsumoto S (1999) Influences of copper forms on the toxicity to microorganisms in soils. Ecotox Environ Safe 44: 174–181.
35. Serkebaeva YM, Kim Y, Liesack W, Dedysh SN (2013) Pyrosequencing-based assessment of the bacteria diversity in surface and subsurface peat layers of a northern wetland, with focus on poorly studied phyla and candidate divisions. PLoS ONE 8: e63994.
36. Kandeler E, Tscherko D, Bruce KD, Stemmer M, Hobbs PJ, et al. (2000) Structure and function of the soil microbial community in microhabitats of a heavy metal polluted soil. Biol Fert Soils 32: 390–400.
37. Roesch LFW, Fulthorpe RR, Riva A, Casella G, Hadwin AKM, et al. (2007) Pyrosequencing enumerates and contrasts soil microbial diversity. ISME J 1: 283–290.
38. Sun MY, Dafforn KA, Johnston EL, Brown MV (2013) Core sediment bacteria drive community response to anthropogenic contamination over multiple environmental gradients. Environ Microbiol doi:10.1111/1462-2920.12133.
39. Zhou J, Kia B, Treves DS, Wu LY, Marsh TL, et al. (2002) Spatial and resource factors influencing high microbial diversity in soil. Appl Environ Microb 68: 326–334.
40. Mori K, Iino T, Ishibashi J, Kimura H, Hamada M, et al. (2012) *Meiothermus hypogaeus* sp. nov., a moderately thermophilic bacterium isolated from a hot spring. Int J Syst Evol Micr 62: 112–117.
41. Mackenzie R, Pedros-Alio C, Dıez B (2013) Bacterial composition of microbial mats in hot springs in Northern Patagonia: variations with seasons and temperature. Extremophiles 17: 123–136.
42. Masurat P, Fru EC, Pedersen K (2005) Identification of *Meiothermus* as the dominant genus in a storage system for spent nuclear fuel. J Appl Microbiol 98: 727–740.
43. Mendez MO, Neilson JW, Maier RM (2008) Characterization of a bacterial community in an abandoned semiarid lead-zinc mine tailing site. Appl Environ Microbiol 74: 3899–3907.
44. Jalali M, Moradi F (2013) Competitive sorption of Cd, Cu, Mn, Ni, Pb and Zn in polluted and unpolluted calcareous soils. Environ Monit Assess DOI 10.1007/s10661-013-3216-1.

45. Costa MS, Rainey FE, Nobre MF (2006) The genus *Thermus* and relatives. In: Dworkin M, Falkow S, Rosenberg E, Schleifer KH and Stackebrant E, editors. The prokaryotes: a handbook on the biology of bacteria. New York: Springer. pp. 797–812.
46. Tian B, Hua Y (2010) Carotenoid biosynthesis in extremophilic *Deinococcus–Thermus* bacteria. Trends in Microbiology 18: 512–520.
47. Woese CR (2006) How we do, don't and should look at bacteria and bacteriology. In: Dworkin M., Falkow S., Rosenberg E., Schleifer K.H. and Stackebrant E, editors. The prokaryotes: a handbook on the biology of bacteria. New York: Springer. pp. 3–23.
48. Miransari M (2013) Soil microbes and the availability of soil nutrients. Acta Physiol Plant. DOI 10.1007/s11738-013-1338-2.

4

Evaluating Status Change of Soil Potassium from Path Model

Wenming He[1,2], Fang Chen[1,3]*

1 Key Laboratory of Aquatic Botany and Watershed Ecology, Wuhan Botanical Garden, Chinese Academy of Sciences, Moshan, Wuchang, Wuhan, Hubei Province, China, 2 Graduate University of Chinese Academy of Sciences, Beijing, China, 3 International Plant Nutrition Institute, Wuhan, China

Abstract

The purpose of this study is to determine critical environmental parameters of soil K availability and to quantify those contributors by using a proposed path model. In this study, plot experiments were designed into different treatments, and soil samples were collected and further analyzed in laboratory to investigate soil properties influence on soil potassium forms (water soluble K, exchangeable K, non-exchangeable K). Furthermore, path analysis based on proposed path model was carried out to evaluate the relationship between potassium forms and soil properties. Research findings were achieved as followings. Firstly, key direct factors were soil S, ratio of sodium-potassium (Na/K), the chemical index of alteration (CIA), Soil Organic Matter in soil solution (SOM), Na and total nitrogen in soil solution (TN), and key indirect factors were Carbonate (CO_3), Mg, pH, Na, S, and SOM. Secondly, path model can effectively determine direction and quantities of potassium status changes between Exchangeable potassium (eK), Non-exchangeable potassium (neK) and water-soluble potassium (wsK) under influences of specific environmental parameters. In reversible equilibrium state of $wsK \underset{\beta}{\overset{\alpha}{\rightleftharpoons}} neK \underset{\chi}{\overset{\gamma}{\rightleftharpoons}} eK$, K balance state was inclined to be moved into β and χ directions in treatments of potassium shortage. However in reversible equilibrium of $wsK \underset{\theta}{\overset{\varepsilon}{\rightleftharpoons}} eK \underset{\omega}{\overset{\lambda}{\rightleftharpoons}} neK$, K balance state was inclined to be moved into θ and λ directions in treatments of water shortage. Results showed that the proposed path model was able to quantitatively disclose moving direction of K status and quantify its equilibrium threshold. It provided a theoretical and practical basis for scientific and effective fertilization in agricultural plants growth.

Editor: Raffaella Balestrini, Institute for Plant Protection (IPP), CNR, Italy

Funding: The study was supported by the National Natural Science Foundation of China (41171243), the National Key Technology R&D Program (2012BAD15B01) and the Cooperated Program with the International Plant Nutrition Institute (IPNI-HB-33). The funders had no role in study design, data collection and analysis, decision to publish, or preparation of the manuscript.

Competing Interests: The authors have declared that no competing interests exist.

* E-mail: fchen@ipni.ac.cn

Introduction

The status and transformation of potassium in the soil is of great significance to crop growth [1,2]. Investigations on potassium status gained a numerous achievements[3,4]. The effectiveness of potassium in soil is controlled by four forms, e.g., mineral potassium, non-exchangeable potassium (neK), exchangeable potassium (eK) and water-soluble potassium (wsK) which can be transformed into each other [5–8]. Absorption capacity of crops and application of farming fertilizer have influence on the potassium forms, thus impact on the release and fixation of potassium in the soil [9]. Within these different forms of soil potassium, there exists a complex and dynamic chemical balance, and it greatly depends on the situation of each form and its environment conditions [10–13]. However, there are few experimental studies on ternary systems involving K^+ in spite of the plant nutritional importance of K^+ in soils [3,5,14]. Research indicated that the amount of K ions retained on the solid phase of soil was much larger than that dissolved in the soil solution. It suggested that a root segment become easily permeable to Ca^{2+} followed by Na^+ and K^+ with aging [15,16]. Thus, it is indispensable to take into account the cation exchange processes in modeling the response of the soil solution to fertilizer application and transport of cationic solutes in soils.

The transform of soil non-exchangeable potassium to soil exchangeable potassium and soil soluble potassium is a slowly process. It is hard to be determined by using the routine method of ion exchange, although it shows that sodium tetraphenylborate method is more accurate than conventional methods to reflect changes in soil potassium through long-term field fixed experiments of potash fertilization [17,18]. Path analysis is a statistical technique that distinguishes coefficient and causation by partitioning coefficients into direct and indirect effects. Researchers used path analysis to analysis phosphorus retention capacity in allophanic and non-allophanic andisols and soil organic matter effects on phosphorus sorption is successful [19,20]. The method of path coefficients, which was proposed by Wright (1921), was effective in disclosing relationships between variables by diagrams [21–25]. However, path coefficients approach has not yet applied to the study on soil potassium. The potassium absorb efficiency of plant root is affected by multiple factors [26], so it is not suitable to make estimation by use a single parameter [27–29]. The above mentioned studies focused on qualitatively descriptive analysis of single factor or multifactor, and failed to quantitatively disclose the dynamic process of potassium status [30].

In this study, we conducted soil K ions exchange experiments for soil samples and further calculated their selectivity coefficients, and explored potassium status on basis of path model to determine dynamic changes of potassium status and specific environmental parameters. The model assumptions were related to the adsorption mechanism at molecular level. The first step, we attempted to examine different indicators of potassium in the soils, and then to analyze direct and indirect correlations between water-soluble potassium (wsK), non-exchangeable potassium (neK) and exchangeable potassium (eK), and further to calculate path coefficients of different potassium status, finally to construct path model for describing the changes of potassium status. The model was calibrated and validated by using 48 cotton rhizosphere/nonrhizosphere soils samples to test its accuracy. We quantitatively investigated on interaction process of plant-soil-microorganisms and relationship between environmental parameters and potassium release through path coefficient model, and the model was used to predict the multi-component ion exchange equilibrium in soil.

Materials and Methods

Hereby, I, along with coauthor, confirm that no specific permissions were required for our experiment locations/activities since this experiment field belongs to our institute and for scientific research only. And the field studies did not involve endangered or protected species.

In this study, meadow soil was selected as experimental agents in order to investigate potassium nutrient supply and physiological mechanism (Table 1). Rhizosphere soil samples were collected from 90 days cotton plant. When collecting soil samples, we firstly loosed root zone soil to collect rhizosphere soil, and uprooted whole plant cotton, and then gently shook off root zone soil, and finally get root surface adhesion soil. The non-rhizosphere soil samples were collected 10–15 cm depth from surface.

For purpose of this study, we proposed a detail research scheme to evaluate status changes of soil potassium from path model (Fig. 1). We conducted soil K ions exchange experiments for soil samples, and further calculated their selectivity coefficients, and explored potassium status on basis of path model to determine dynamic changes of potassium status and specific environmental parameters. Soil samples were **N**on-**R**hizosphere soil in **OPT**imum of **K** of **H**igh efficiency genotype cotton (NROPTH); **N**on-**R**hizosphere soil in **OPT**imum of **K** of **L**ow efficiency genotype cotton (NROPTL); **N**on-**R**hizosphere soil in **S**hortage of **K** of **H**igh efficiency genotype cotton (NRSKH); **N**on-**R**hizosphere soil in **S**hortage of **K** of **L**ow efficiency genotype cotton (NRSKL); **N**on-**R**hizosphere soil in **S**hortage of **W**ater of **H**igh efficiency genotype cotton (NRSWH); **N**on-**R**hizosphere soil in **S**hortage of **W**ater of **L**ow efficiency genotype cotton (NRSWL); **N**on-**R**hizosphere soil in **S**hortage of **W**ater and **K** of **H**igh efficiency genotype cotton (NRSWKH); **Non-R**hizosphere soil in **S**hortage of **W**ater and **K** of **L**ow efficiency genotype cotton (NRSWKL); **R**hizosphere soil in **OPT**imum of **K** of **H**igh efficiency genotype cotton (ROPTH); **R**hizosphere soil in **OPT**imum of **K** of **L**ow efficiency genotype cotton (ROPTL); **R**hizosphere soil in **S**hortage of **K** of **H**igh efficiency genotype cotton (RSKH); **R**hizosphere soil in **S**hortage of **K** of **L**ow efficiency genotype cotton (RSKL); **R**hizosphere soil in **S**hortage of **W**ater of **H**igh efficiency genotype cotton (RSWH); **R**hizosphere soil in **S**hortage of **W**ater of **L**ow efficiency genotype cotton (RSWL); **R**hizosphere soil in **S**hortage of **W**ater and **K** of **H**igh efficiency genotype cotton (RSWKH); **R**hizosphere soil in **S**hortage of **W**ater and **K** of **L**ow efficiency genotype cotton (RSWKL). Firstly, we divided experiment soil into 8 treatments, and then implemented 256 soil samples test (Fig. 2). We measured 33 parameters of the different attributes of 256 soil samples. Those parameters were selected based on standards of Hashimoto and Kang research methods of phosphorus[19,31]. Humic acid (NHA) in soil which can be extracted by NaOH solution (0.1 mol·L^{-1}, pH = 3), humic acid (PHA) in soil which can be extracted by $Na_2P_2O_4$ solution (0.05 mol·L^{-1}, pH = 9.2), humus linked to iron (HMi) and humic linked to clay (HMc). Key experimental sampling test datasets were listed in Table 2–Table 6. Secondly, statistical analysis was further implemented to investigate relationship between environmental parameters and potassium status. Datasets were normalized, followed by correlation analysis (Table 7), 8 out of 16 parameters were selected thus to determine the most important parameters. Path coefficients of absolute value of represented the size effect on potassium morphology change. The size of the "+" meant the same as the arrow direction with arrow, "−" represents in contrast to the arrow direction with arrow. Finally, path model of potassium status was constructed on basis of direct and indirect correlated parameters. In this model, "e1" represents unknown variable and its impact factor, the straight line with arrows stands for direct impact factor, and double arc arrow is direction of interaction between parameters.

Table 1. Properties of mead soil for experiments.

Soil properties		Average value
Mean compact density (g·cm^{-3})		1.54
Particle composition	**Clay (<0.002 mm, %)**	16.8
	Sand (2-0.05 mm, %)	61.7
	Silt (0.05-0.002 mm, %)	21.4
Mineral composition	**Smectite (%)**	4.0
	Vermienlite (%)	25.0
	Intergrade mineral hydromica (1.4 nm, %)	49.0
	Kaolinite (%)	19.0
Cations composition	**exchangeable Ca (mmol·kg^{-1})**	35.4
	exchangeable Mg (mmol·kg^{-1})	18.1
	exchangeable Na (mmol·kg^{-1})	2.0
	FeO (noncrystalline iron extracted with Tamm's solution, g·kg^{-1})	49.0

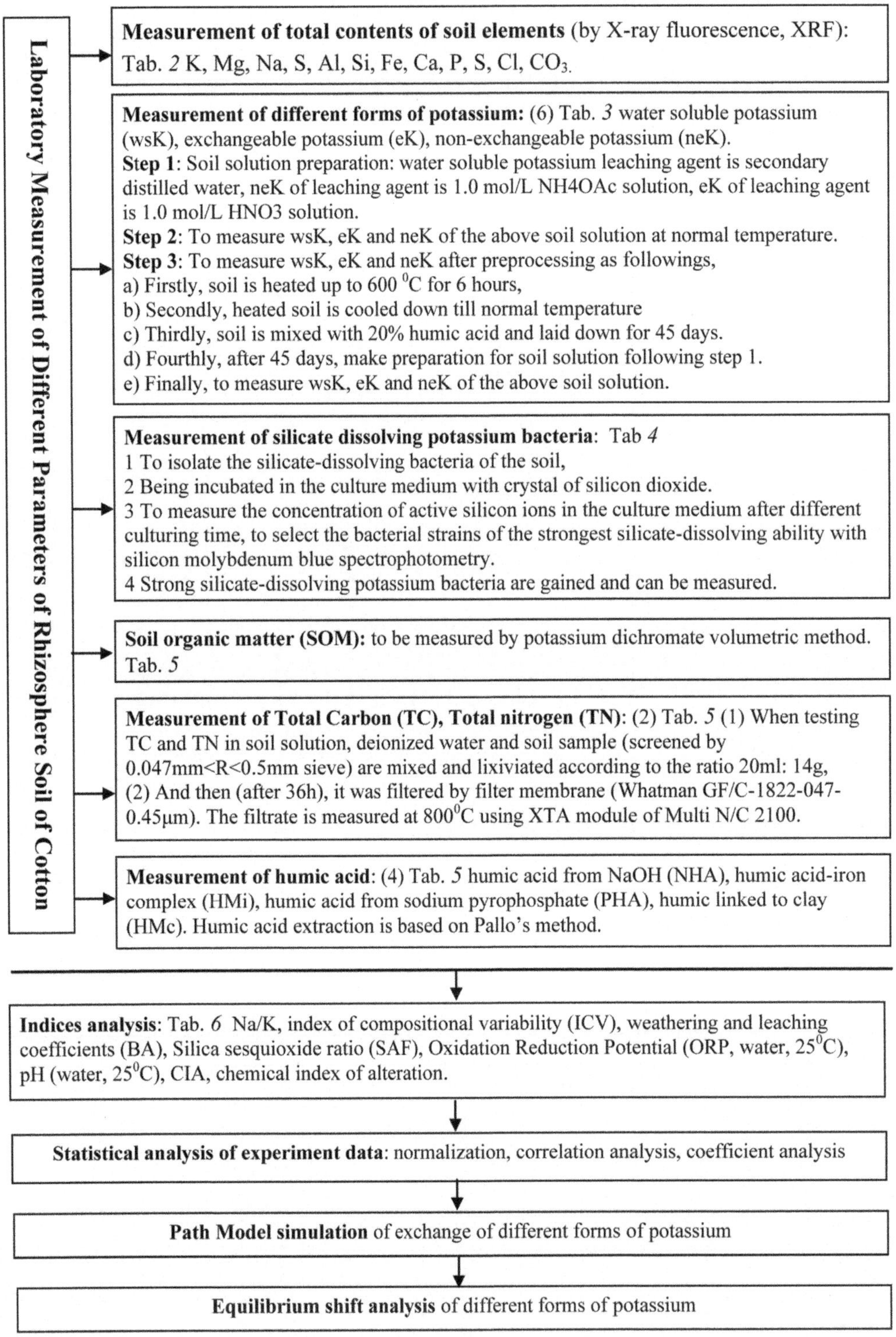

Figure 1. Workflow of evaluating status changes of soil potassium from Path Model.

Results

Path analysis of water-soluble potassium (wsK)

For purpose of path analysis of water-soluble potassium, direct and indirect impact factors of the path analysis were derived from multiple linear regressions coefficients of wsK, and correlation coefficients between soil properties. Direct coefficient variables of wsK include neK, eK, PHA, SOM, S, Na and Na/K; while indirect coefficient variables of wsK include CIA, Na, Mg, CO_3 and pH. Impact factor of path size of water-soluble potassium was determined by pH, CIA, Oxidation Reduction Potential (ORP), S, Na (Fig. 3 and Table 8). In the path model of wsK, direct effects of soil properties on the normalized value of wsK (ZwsK) were represented by single-headed arrows, while coefficients between soil properties were represented by double-headed arrows. Direct and indirect effects were indicated by value and marked with "±".

Figure 2. Diagram of experimental design to evaluate status changes of soil potassium.

Table 2. Total contents of soil elements of Laboratory measurement for different parameters of 256 soil samples (R, rhizosphere soil; NR, non-rhizosphere).

Sample	Na%	Mg%	Al%	Si%	Fe%	Mg%	Ca%	Pppm	CO_3%	Sppm	Clppm
NR1-16	0.44	0.39	6.49	36.86	3.35	1.49	0.37	491.01	49.74	219.34	119.8
R17-32	0.45	0.37	6.53	37.16	3.34	1.49	0.39	507.57	49.38	243.65	228.2
NR33-48	0.44	0.39	6.5	36.92	3.33	1.49	0.38	494.36	49.72	218.56	122.07
R49-64	0.44	0.39	6.51	36.94	3.34	1.49	0.39	483.04	49.64	251.93	217.78
N65-80	0.44	0.39	6.46	36.93	3.32	1.47	0.36	488.97	49.77	220.97	68.15
R81-96	0.47	0.42	6.3	36.38	3.29	1.46	0.4	515.89	50.45	248.77	80.32
NR97-112	0.44	0.38	6.49	37.04	3.35	1.48	0.37	506.34	49.62	218.38	82.48
R113-128	0.44	0.38	6.53	37.09	3.35	1.48	0.4	518.35	49.48	243.94	89.5
NR129-144	0.44	0.39	6.38	36.84	3.29	1.48	0.36	481.3	49.95	224.13	152.4
R145-160	0.45	0.39	6.39	36.94	3.27	1.47	0.38	512.73	49.84	237.7	201.2
NR161-176	0.43	0.38	6.49	37.09	3.32	1.49	0.37	484.59	49.58	225.39	142.88
R177-192	0.45	0.38	6.51	37.01	3.33	1.48	0.39	496.2	49.57	236.79	189.7
NR193-208	0.44	0.39	6.43	36.92	3.31	1.47	0.36	502.75	49.83	223.08	71.6
R209-224	0.45	0.38	6.43	37.08	3.32	1.46	0.39	491.22	49.64	245.74	81.48
NR225-240	0.44	0.38	6.46	37	3.31	1.47	0.38	509.11	49.7	228.9	90.5
R241-256	0.44	0.37	6.46	37.28	3.34	1.47	0.39	498.72	49.4	239.9	87.4

Table 3. Different forms of potassium of Laboratory measurement for different parameters of 256 soil samples (R, rhizosphere soil; NR, non-rhizosphere).

Sample	K(g/Kg)	wsK(ug/g)	neK(ug/g)	eK(ug/g)	wsK(ug/g, (weathered)	neK(ug/g) (weathered)	eK(ug/g) weathered
NR1-16	14.93	27.1	134.9	58.64	67.57	174.48	62.9
R17-32	14.9	12.03	54.21	14.32	59.36	155.15	66.59
NR33-48	14.9	20.01	134.13	50.84	76.97	177.27	55.87
R49-64	14.87	9.22	58.27	12.21	123.72	170.67	71.24
N65-80	14.7	24.8	121.55	40.27	53.02	190.22	59.93
R81-96	14.57	5.65	39.17	6.28	61.31	153.27	57.35
NR97-112	14.75	41.28	101.36	27.46	56.24	167.15	69.44
R113-128	14.8	6.05	42.83	5.23	58.41	149.91	52.24
NR129-144	14.77	38.55	158.29	56.41	68.09	190.55	65.03
R145-160	14.73	15.92	78.62	23.18	60.89	167.13	52.79
NR161-176	14.87	88.78	162.24	61.11	70.46	181.02	55.25
R177-192	14.83	18.61	70.07	19.27	64.82	185.07	59.89
NR193-208	14.73	13.26	99.74	30.21	63.65	171.35	62.95
R209-224	14.63	6.41	38.74	7.63	61.96	162.21	52.09
NR225-240	14.73	25.27	105.8	29.54	64.59	182.24	59.77
R241-256	14.73	6.56	44.8	5.31	72.96	154.08	60.7

The direct effects of soil properties on the ZwsK were termed path coefficients and were standardized partial regression coefficients for each of the soil properties in the multiple linear regression against the ZwsK. Indirect effects of soil properties on the ZwsK were determined from the product of the simple coefficient between soil properties and the path coefficient (i.e., one two-headed arrow and one single-headed arrow). The coefficient between the ZwsK and soil property was the sum of the entire path connecting two variables, as described by

$$\begin{aligned} r_{\text{neK,wsK}} = {} & P_{\text{wsK,neK}} + r_{\text{neK,PHA}} \cdot P_{\text{wsK,PHA}} + \\ & r_{\text{neK,S}} \cdot P_{\text{wsK,S}} + r_{\text{neK,Na/K}} \cdot P_{\text{wsK,Na/K}} + \\ & r_{\text{neK,CIA}} \cdot P_{\text{wsK,CIA}} + r_{\text{neK,Na}} \cdot P_{\text{wsK,Na}} + \\ & r_{\text{neK,SOM}} \cdot P_{\text{wsK,SOM}} + r_{\text{neK,eK}} \cdot P_{\text{wsK,eK}} \end{aligned} \tag{1}$$

Table 4. Silicate dissolving potassium bacteria of Laboratory measurement for different parameters of 256 soil samples (R, rhizosphere soil; NR, non-rhizosphere).

Sample	Bacteria of seedling stage ($\times 10^4$ CFUg^{-1})	Bacteria of budding stage ($\times 10^4$ CFUg^{-1})	Bacteria of wadding stage ($\times 10^4$ CFUg^{-1})
NR1-16	0.8	2.1	1.6
R17-32	1.5	5.2	1.4
NR33-48	0.7	4.3	1.7
R49-64	1.1	3.4	1.9
N65-80	0.9	17	14
R81-96	1.0	87	30
NR97-112	1.2	26	19
R113-128	1.9	39	24
NR129-144	0.1	1.2	0.5
R145-160	0.4	1.3	0.7
NR161-176	0.9	4.3	2.2
R177-192	0.3	1.7	0.6
NR193-208	0.8	1.9	1.6
R209-224	0.7	7.1	4.0
NR225-240	0.1	2.3	0.9
R241-256	1.2	6.3	2.8

Table 5. Soil organic matter, humic acid, Total Carbon (TC), Total nitrogen (TN) of Laboratory measurement for different parameters of 256 soil samples (R, rhizosphere soil; NR, non-rhizosphere).

Sample	TC%	TN%	SOM(%)	PHA%	NHA%	Hmi%	HMc%
NR1-16	4.51E-03	1.71E-03	7.88	4.53	2.59	1.47	0.92
R17-32	4.41E-03	5.74E-03	5.67	4.84	3.52	0.33	0.33
NR33-48	5.34E-03	2.02E-03	7.32	3.48	3.67	0.42	0.37
R49-64	3.74E-03	1.04E-02	5.54	4.11	2.55	0.45	0.32
N65-80	4.50E-03	1.64E-03	8.01	5.33	3.89	1.35	1.24
R81-96	3.64E-03	4.24E-03	5.79	5.26	4.58	0.42	0.23
NR97-112	3.01E-03	3.93E-03	8.53	3.62	3.28	0.66	0.41
R113-128	2.80E-03	1.31E-02	5.70	4.41	3.68	0.46	0.27
NR129-144	3.52E-03	3.46E-03	8.30	4.46	3.70	0.35	0.28
R145-160	3.22E-03	2.02E-02	5.38	5.00	3.82	0.30	0.30
NR161-176	3.18E-03	3.97E-03	7.84	4.90	1.75	0.88	0.45
R177-192	3.12E-03	2.31E-02	6.79	6.89	3.95	0.34	0.26
NR193-208	3.57E-03	3.24E-03	8.45	4.57	3.82	0.25	0.19
R209-224	3.79E-03	2.67E-02	5.79	5.43	4.22	0.44	0.36
NR225-240	4.30E-03	3.83E-03	8.01	4.94	4.61	0.52	0.32
R241-256	4.37E-03	2.75E-02	6.23	4.70	4.08	0.36	0.18

Table 6. Weathering indices of soil minerals of Laboratory measurement for different parameters of 256 soil samples (R, rhizosphere soil; NR, non-rhizosphere).

Sample	ORP(eV)	pH	ba	Na/K	ICA	saf	CIA
NR1-16	17.30	7.27	0.45	0.29	1.03	8.73	76.08
R17-32	18.65	6.84	0.44	0.30	1.03	8.77	75.82
NR33-48	13.80	7.20	0.45	0.29	1.03	8.75	76.02
R49-64	30.24	6.5	0.45	0.30	1.04	8.74	75.95
N65-80	22.66	6.79	0.45	0.30	1.03	8.8	76.18
R81-96	22.68	6.73	0.48	0.32	1.08	8.86	75.08
NR97-112	10.22	6.94	0.44	0.30	1.03	8.78	76.21
R113-128	65.80	6.09	0.45	0.30	1.03	8.74	76
NR129-144	18.50	6.93	0.45	0.30	1.04	8.88	75.88
R145-160	80.84	5.79	0.46	0.31	1.04	8.90	75.67
NR161-176	31.46	6.48	0.44	0.29	1.02	8.80	76.16
R177-192	62.82	5.85	0.45	0.30	1.03	8.75	75.92
NR193-208	19.33	6.87	0.45	0.30	1.03	8.83	76.06
R209-224	52.48	6.19	0.45	0.31	1.04	8.86	75.76
NR225-240	13.72	6.98	0.45	0.30	1.03	8.83	75.99
R241-256	20.54	6.86	0.45	0.30	1.04	8.87	75.84

where, r_{ij} is the simple correlation between the ZwsK and a parameter of soil property, P_{ij} is the path coefficient between the ZwsK and a parameter of soil property, and $r_{ij}P_{ij}$ is the indirect effect of a parameter of on the ZwsK. An uncorrelated residue (e_{wsK}) that represented the unexplained part of an observed variable in the path model was

$$e_{wsk} = (1 - R_{wsk}{}^2)^{0.5} = 0.2 \tag{2}$$

where, $R_{wsk}{}^2$ is the coefficient of determination of the multiple regression equation between the ZwsK and the eight soil properties. Those parameters of soil properties were neK, PHA, S, Na/K, CIA, Na, SOM, eK. Backward regression analysis was performed with the ZwsK as the dependent variable and the eight soil properties were regarded as independent variables. Backward regression was a multiple regression procedure in which all the independent variables were entered into the regression equation at the beginning, and variables that did not contribute significantly to the fit of regression model were successively eliminated until only statistically significant variables remain. Fig. 4 and Fig. 5 were similar expressions.

In the path of water-soluble potassium, path contribution coefficients were quite different. Coefficient values were 2.10, −1.52 and 0.20 for wsK to neK, neK to wsK, and other variables to wsK respectively. It disclosed that most part of wsK was changed into neK (with values of 2.10). In the path of neK to wsK, coefficient was negative (with value of −1.52); it indicated that there was an inverse change from wsK to neK, a lot of wsK was changed into neK. Those dynamic changes indicated that wsK and eK reached equilibrium in soil solution, and this equilibrium represented its quick reaction process.

These changes impacted greatly on nutrients dynamic equilibrium of rhizosphere soil. This changed nutrients recycling process of the soil environment since continuous-release of humic acid had direct effects on absorption and uptakes of those variables including neK, eK and PHA, SOM, S, Na and Na/K. It was clear that rhizosphere effects were likely to result in changes in root exudates, pH and ORP (eV). Consequently, it led to changes in pH values and population, quality and activity of microorganisms while pH change was caused by root exuded organic acids. Under conditions of different pH values, humic acid significantly correlated to adsorption and desorption of potassium. Humic acid of potassium adsorption and desorption presented an upward trend when it increased in initial concentration (pH was 4.0–8.0), however, desorption rate declined. The migration distance also showed very significant linear relationship within ranges of migrations between water-soluble potassium and exchangeable potassium. Organic matter nutrients and chemical efficiency of humic acid altered different variables such as physical, chemical, and biotic in root-soil interface.

Path analysis of non-exchangeable potassium (neK)

For path analysis of non-exchangeable potassium, we further investigated its direct and indirect impact factors. Results showed that direct coefficient variables of neK include eK, wsK, SOM, S, Na, Na/K, TN, CIA; and indirect coefficient variables of neK include Mg, CO_3. Impact factor of path size of neK was determined by eK, CO3, pH (Fig. 4 Table 9). Table 9 showed there was significant correlation between eK and neK (with coefficient of 0.81). In the Path of non-exchangeable potassium, contribution coefficient was 0.01 for wsK to neK, although wsK was determined by general effects of N, Na, S and CIA. The SOM was also an important contributor to eK and wsK of rhizosphere soil. Coefficients were 0.69 and 0.55 between SOM and eK, SOM and wsK, respectively.

The correlation between the ZneK and a soil property was the sum of the entire path connecting two variables, as described by

Table 7. Correlation analysis of environmental parameters of soil potassium.

	Z(SOM)	Z(CIA)	Z(Na/K)	Z(TC)	Z(TN)	Z(wsK)	Z(neK)	Z(eK)	Z(bacteria)	Z(ORP)	Z(pH)	Z(PHA)	Z(HMi)
Z(CIA)	0.61a												
Z(Na/K)	−0.52a	0.829b											
Z(TN)	−0.61a												
Z(wsK)	0.58a		−0.52a										
Z(neK)	0.80b	0.56a	−0.66b		−0.64b	0.77b							
Z(ek)	0.74b	0.51a	−0.67b		−0.62a	0.73b	0.98b						
Z(bacteria)	0.55a						0.53a						
Z(ORP)	−0.59a			−0.56a	0.66b								
Z(pH)	0.58a			0.66b	−0.68b					−0.95b			
Z(NHA%)											−0.57a		
Z(PHA%)		−0.51a	0.64			−0.65b		−0.51a					
Z(HMi%)								0.54a					
Z(HMc%)													0.93b
Z(CO$_3$)		−0.62b	0.56						−0.50a				
Z(K)		0.53a	−0.79b									−0.77b	

(Note: statistically, probabilities of a and b: a, p<0.01; b, p<0.05)

$$r_{\mathrm{neK,eK}} = P_{\mathrm{neK,eK}} + r_{\mathrm{neK,SOM}} \cdot P_{\mathrm{neK,SOM}} + r_{\mathrm{neK,S}} \cdot P_{\mathrm{neK,S}} + r_{\mathrm{neK,Na/K}} \cdot P_{\mathrm{neK,Na/K}} + r_{\mathrm{neK,TN}} \cdot P_{\mathrm{neK,TN}} + r_{\mathrm{neK,Na}} \cdot P_{\mathrm{neK,Na}} + r_{\mathrm{neK,SOM}} \cdot P_{\mathrm{neK,SOM}} + r_{\mathrm{neK,wsK}} \cdot P_{\mathrm{neK,wsK}} \quad (3)$$

where, r_{ij} is the simple correlation coefficient between the ZneK and a soil property, P_{ij} is the path coefficient between the ZneK and a soil property, and $r_{ij}P_{ij}$ is the indirect effect of a soil property on the ZneK. An uncorrelated residue (e) that represents the unexplained part of an observed variable in the path model was

$$e_{nek} = (1 - R_{nek}{}^{2})^{0.5} = 0.01 \quad (4)$$

where $R_{nek}{}^{2}$ is the coefficient of determination for the multiple regression equation between the ZneK and those soil properties. Those soil properties were wsK, SOM, S, Na/K, TN, Na, SOM, eK.

It disclosed that plant nutrition was indirectly influenced by nutrients release of microbial decomposition and organic matter. It was particularly obvious in nutrition release of low molecular weight organics in rhizosphere soil. High contents of associated ions (Na^{+}) were limit factors that control crop yields and soil nutrients use. Ammonium nitrogen was likely to reduce fixation of potassium ions of soil, and increase risk of potassium leaching. $NH_4{}^{+}$ and K^{+} were competitive ions for absorption of ORP (eV) of eK and neK in rhizophere soil, and different status of nitrogen effects on efficient use of potassium. Therefore, suitable proportion of N and K is important for plant. Soil nitrogen availability determines potassium absorption of root surface and cell membrane. Root membrane could directly affected by $NH_4{}^{+}$, $NO_3{}^{-}$, K^{+} when it absorbs potassium ions, and indirect affected by potential balance from assimilation of $NH_4{}^{+}$. Strong affinity of K^{+}, high efficiency and fast speed of absorption of potassium are likely to improve efficient absorption of potassium.

Path analysis of exchangeable potassium (eK)

As to exchangeable potassium, it was quite different situation in comparison with those of water soluble and non-exchangeable potassium. It showed that direct coefficient variables of eK include neK, wsK, SOM, S, Na, Na/K, TN, CIA, HMi, PHA; and indirect coefficient variables of eK include Mg, CO_3. Impact factor of path size of eK was determined by neK, CO_3, pH, ORP, Na (Table 10).

From Fig. 5, it showed that neK was closely related to wsK (with correlation coefficients of 0.72), and there was an inverse proportion between ZneK and ZwsK (the normalized value of neK and wsK) in condition of changes of micro-bioorganic and geochemical environmental factors in rhizosphere soil. Path contribution coefficients were 1.10 and 0.04 for neK to eK, and wsK to eK. It suggests that neK is more likely to be changed into eK while the possibility of wsK to eK is relatively low.

The correlation between the normalized value of eK (ZeK) and a soil property is the sum of the entire path connecting two variables, as described by

$$r_{\mathrm{neK,eK}} = P_{\mathrm{eK,neK}} + r_{\mathrm{neK,SOM}} \cdot P_{\mathrm{eK,SOM}} + r_{\mathrm{neK,Na/K}} \cdot P_{\mathrm{eK,Na/K}} + r_{\mathrm{neK,Na}} \cdot P_{\mathrm{eK,Na}} + r_{\mathrm{neK,HMi}} \cdot P_{\mathrm{eK,HMi}} + r_{\mathrm{neK,TN}} \cdot P_{\mathrm{eK,TN}} + r_{\mathrm{neK,wsK}} \cdot P_{\mathrm{eK,wsK}} + r_{\mathrm{neK,S}} \cdot P_{\mathrm{eK,S}} \quad (5)$$

where r_{ij} is the simple correlation coefficient between the ZeK and a soil property, P_{ij} is the path coefficient between the ZeK and a soil property, and $r_{ij}P_{ij}$ is the indirect effect of a soil property on the ZeK. An uncorrelated residue (e) that represents the unexplained part of an observed variable in the path model was

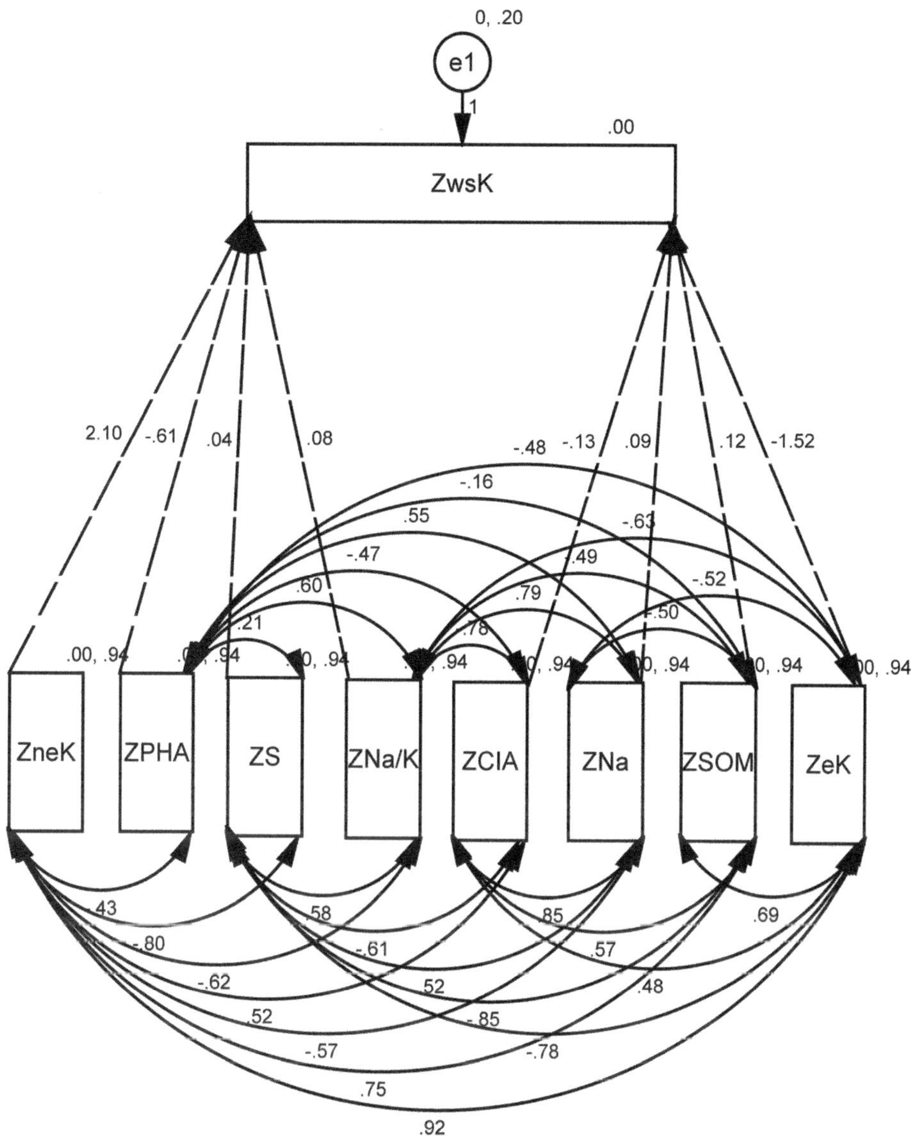

Figure 3. Path model of water-soluble potassium (wsK, solid line for double arrow, dashed-line for single arrow, and single arrow flows into ZwsK).

$$e_{ek} = \left(1 - R_{ek}{}^{2}\right)^{0.5} = 0.01 \qquad (6)$$

Where, $R_{ek}{}^2$ is the coefficient of determination of the multiple regression equation between the ZwsK and the eight soil properties. Those soil properties are neK, SOM, Na/K, Na, HMi, TN, wsK, S. Backward regression analysis was performed with the ZeK as the dependent variable and the eight soil properties as independent variables. Backward regression is a multiple regression procedure in which all the independent variables are entered into the regression equation at the beginning, and variables that do not contribute significantly to the fit of the regression model are successively eliminated until only statistically significant variables remain.

Cation exchange capacity (CEC) or the number of exchangeable cation depends on the type and number of clay and organic matter in the soil. To a certain extent, TN, SOM, HMi and the PHA and the CIA are able to well reflect type and number of clay and organic matte in rhizosphere soil. Proper pH value of solution, hot water pressure, concentration increases of magnesium carbonate and ion are useful to exchangeable potassium (neK) when it is absorbed by crops. A plenty of CO_2 could be produced by root respiration and soil microbe respiration, and is fixed in form of $CO_3{}^{2-}$. For stable pH values, it is beneficial for organic acid to reach in equilibrium of absorption in soil solution. The exchangeable potassium and fixed potassium which comes from layers of secondary clay mineral are released when there occurs hydrolysis and cation substitution resulted by chemical element H^+. Those variables, such soil organic matter (ZSOM), iron-binding Humic (ZHMi), and water-soluble nitrogen (ZTN), not only provide nutrition for cotton root, but also changed soil chemical compositions and absorption characteristics of potassium in rhizosphere soil. Additionally, concomitant phenomenon (ZS) appears between S (the lost chemical element in soil) and potassium. Antagonism occurred between Na and K, and it

Table 8. Direct and indirect correlation coefficients of water-soluble potassium (*wsK*).

Variables	Direct correlation coefficients	Indirect correlation coefficients	Path contrition coefficients
Z(PHA)	−0.65	−0.32	−0.38
Z(S)	−0.56	0.17	−0.8
Z(Na)	−0.55	−1.15	−0.78
Z(Na/K)	−0.52	−0.01	−0.01
Z(TN)	−0.41	0.23	−0.38
Z(ORP)	−0.22	0.02	−1.28
Z(Mg)	−0.14	−0.79	−0.37
Z(CO_3)	−0.07	−0.53	−0.24
Z(pH)	0.14	−0.35	−1.87
Z(HMi)	0.39	0.2	0.29
Z(K)	0.41	0.02	−0.01
Z(CIA)	0.47	2.02	−1.61
Z(SOM)	0.58	0.06	0.25
Z(eK)	0.73	0.05	0.14
Z(neK)	0.77		

implied that sodium salt was able to have significant effect on plant (ZNa/K, ZNa).

From above analysis, we conclude that key direct factors of wsK, neK, and eK were S, Na/K, the CIA, SOM, Na, and TN, and key indirect factors were CO_3, Mg, pH, Na, S, and SOM. Whereas, direct impact factors of water soluble potassium (wsK) were PHA, S, Na, Na/K, K, the CIA, SOM, eK and neK. Each of its impact value was −0.65, −0.56, −0.55, −0.52, 0.41, 0.47, 0.58, 0.73, and 0.77 respectively. Indirect impact factors were Na, Mg, CO_3, pH, PHA, S, HMi, TN and CIA. Each of its impact value was −1.15, −0.79, −0.53, −0.35, −0.32, 0.17, 0.2, 0.23, and 2.02, respectively. Non exchangeable potassium direct impact factors were: S, Na/K, Na, TN, the CIA, wsK, SOM and eK. Each of its impact value was −0.86, −0.66, −0.64, −0.61, 0.56, 0.77, 0.8, and 0.98 respectively. Indirect impact factors were CO_3, eK, S, SOM, K, pH, Na and Mg. Each of its impact value was 1.18, 0.76, 0.19, −0.11, −0.14, −0.16, −0.34, and −0.67 respectively. Exchangeable potassium direct impact factors were S, Na/K, Na, TN, HMi, wsK, SOM and neK. Each of its impact value was −0.83, −0.67, −0.62, −0.55, 0.54, 0.73, 0.74, and 0.98 respectively. Indirect impact factors were CO_3, S, SOM, K, pH, Na, Mg and neK. Each of its impact value was −1.7, −0.28, 0.18, 0.21, 0.23, 0.46, 0.95, and 1.49 respectively. Absolute values of path coefficients represent the change ability of the potassium morphology, and the bigger value means the stronger ability to change potassium morphology. And "+"with and the arrow direction represent change direction of potassium morphology. In this way, we can quantitatively describe the K elements morphological changes.

Path model of soil potassium status changes of cotton

It is of great importance to further investigate on the equilibrium movement in soil. The following part will discuss the shift of dynamic balance among wsK, eK and neK, and their flow path. To test the reliability of the models, we collect another 48 cotton soil samples in the same experiment, using stepwise regression method, observation of potassium mobility and morphological change.

We carried out a plot experiment of potash application with high K-efficiency genotype (HEG) and low K-efficiency genotype (LEG) cotton. The soil in each treatment weights 8.5 kg. Detail fertilizer schemes are listed in Table 11. LEG and HEG cottons' planting effects is shown in Fig. 6.

(1) Equilibrium shift in the path for non-exchangeable potassium

Supposed that $V_{wsK \rightarrow nek}$ is path for wsK to neK, $V_{ek \rightarrow nek}$ is path for eK to neK, therefore, equilibrium shift in path for non-exchangeable potassium can be expressed by contribution coefficient functions as below,

$$\begin{cases} y_{wsk} = 0.60 \cdot x_{neK} - 0.37 \cdot x_{PHA} + 5.87 \times 10^{-7} (R^2 = 0.87) \\ y_{neK} = 1.67 \cdot x_{wsk} + 0.62 \cdot x_{PHA} - 9.87 \times 10^{-7} \\ y_{eK} = 0.98 \cdot x_{neK} + 1.23 \times 10^{-6} (R^2 = 0.92) \\ y_{neK} = 1.02 \cdot x_{eK} - 1.25 \times 10^{-6} \end{cases} \quad (7)$$

Equilibrium occurs when X_{wsK}, X_{PHA}, X_{eK} satisfy with Function 8,

$$\begin{cases} 1.67 \cdot x_{wsk} + 0.62 \cdot x_{PHA} - 9.87 \times 10^{-7} = 1.02 \cdot x_{eK} - 1.25 \times 10^{-6} \\ x_{wsk} + 0.37 \cdot x_{PHA} = 0.61 \cdot x_{eK} - 1.57 \times 10^{-7} \end{cases} \quad (8)$$

Here, y_{nek} is the equilibrium threshold. And the value of x_{wsK}, x_{PHA}, x_{eK} is determined by soil characteristics. Calculated results from Equation 7 were listed in Table 12. In the table, column 1 and 2 were calculated from Equations 7. We used the wsK, eK, neK to construct dynamic equilibrium equations ($wsK \underset{\beta}{\overset{\alpha}{\rightleftharpoons}} neK \underset{\gamma}{\overset{\chi}{\rightleftharpoons}} eK$).The column 3 was normalized values of neK and the column 4 was the direction of balance movement of neK. These three column datasets were presented in Fig. 6.

In the reversible equilibrium of $wsK \underset{\beta}{\overset{\alpha}{\rightleftharpoons}} neK \underset{\chi}{\overset{\gamma}{\rightleftharpoons}} eK$, soil samples RSWKL, RSWH, RSWL, NRSWKL, NRSWL, K balance cycle were moved in α, χ direction movement; soil

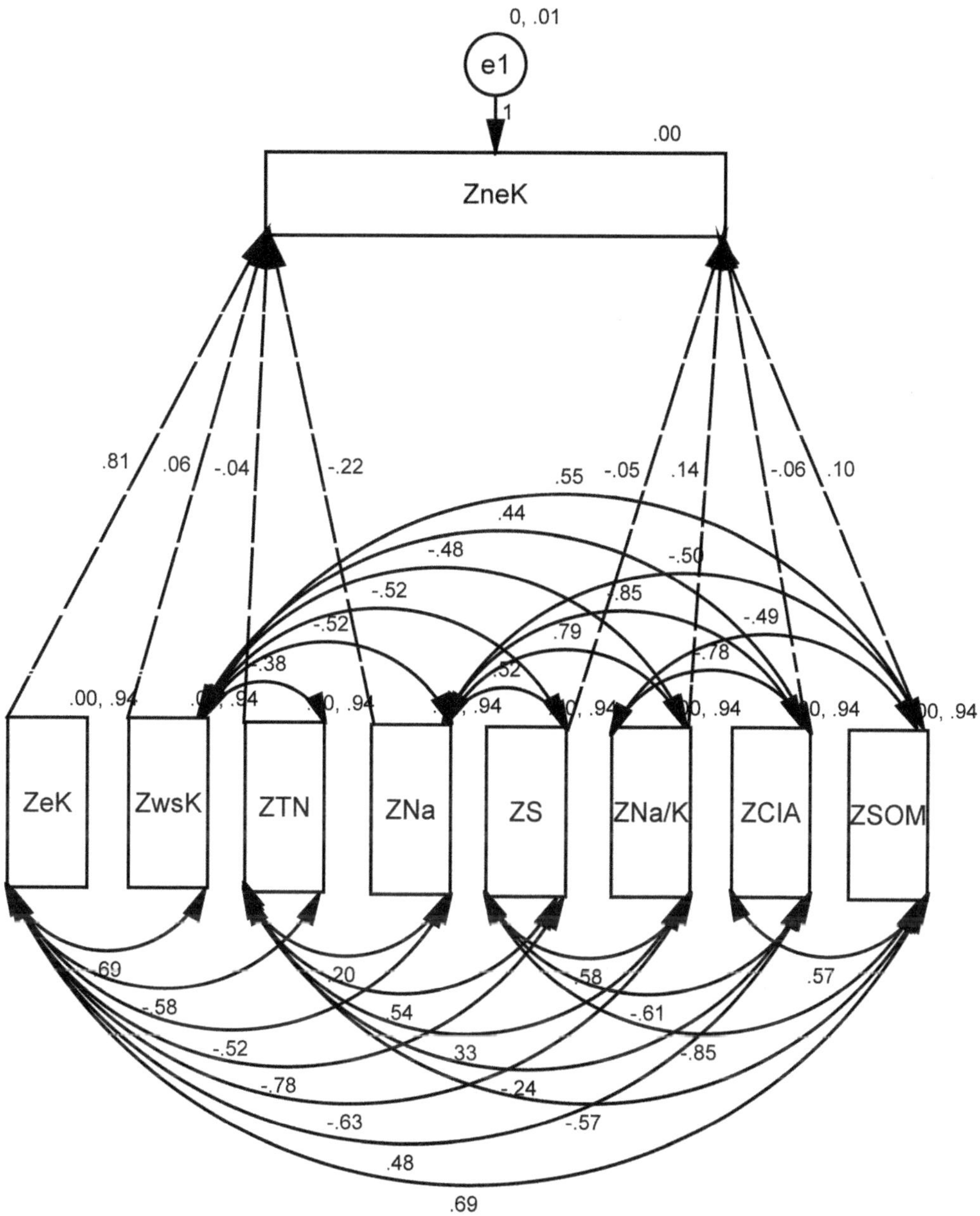

Figure 4. Path model of non-exchangeable potassium (neK, solid line for double arrow, dashed-line for single arrow, and single arrow flows into ZwsK).

samples RSWKH, RSKH, RSKL, ROPTH, ROPTL, NRSWKH, NRSWH, NROPTH, NROPTL, K balance cycle were moved in β and γ direction movement; soil samples NRSKH, NRSKL, K balance cycle were moved in β, χ direction movement.

In Fig. 6, neK reaches its equilibrium in following soil types which includes LEG in water stress treatment (SW), HEG after potassium stress treatment (SK), LEG after optimum fertilization treatment (OPT), and LEG after water stress treatment (SW). However, all the value in those soils with HEG in SW, LEG in SK, HEG in SW and SK, the values were all greater than the equilibrium, wsK reached its saturation. For other treatment soils, when wsK was lower than the equilibrium and PHA and eK was greater than equilibrium, neK changed into wsK. It proves that eK equales to total K in matured crops. When eK is below a certain level in soil, plants can no longer get eK. When eK was at low levels, it was absorbed by other ions with stronger absorption. This reduces chance of potassium to enter into the soil solution. In this condition, neK begins to release through low concentrations of wsK and eK. However, the release capacity of neK depends on situations of crop types

(2) Equilibrium shift in the path for exchangeable potassium

Supposed that $V_{wsK \rightarrow ek}$ is path for wsK to eK, $V_{nek \rightarrow ek}$ is path for neK to eK, therefore, equilibrium shift in path for exchangeable potassium can be expressed by contribution coefficient functions as below,

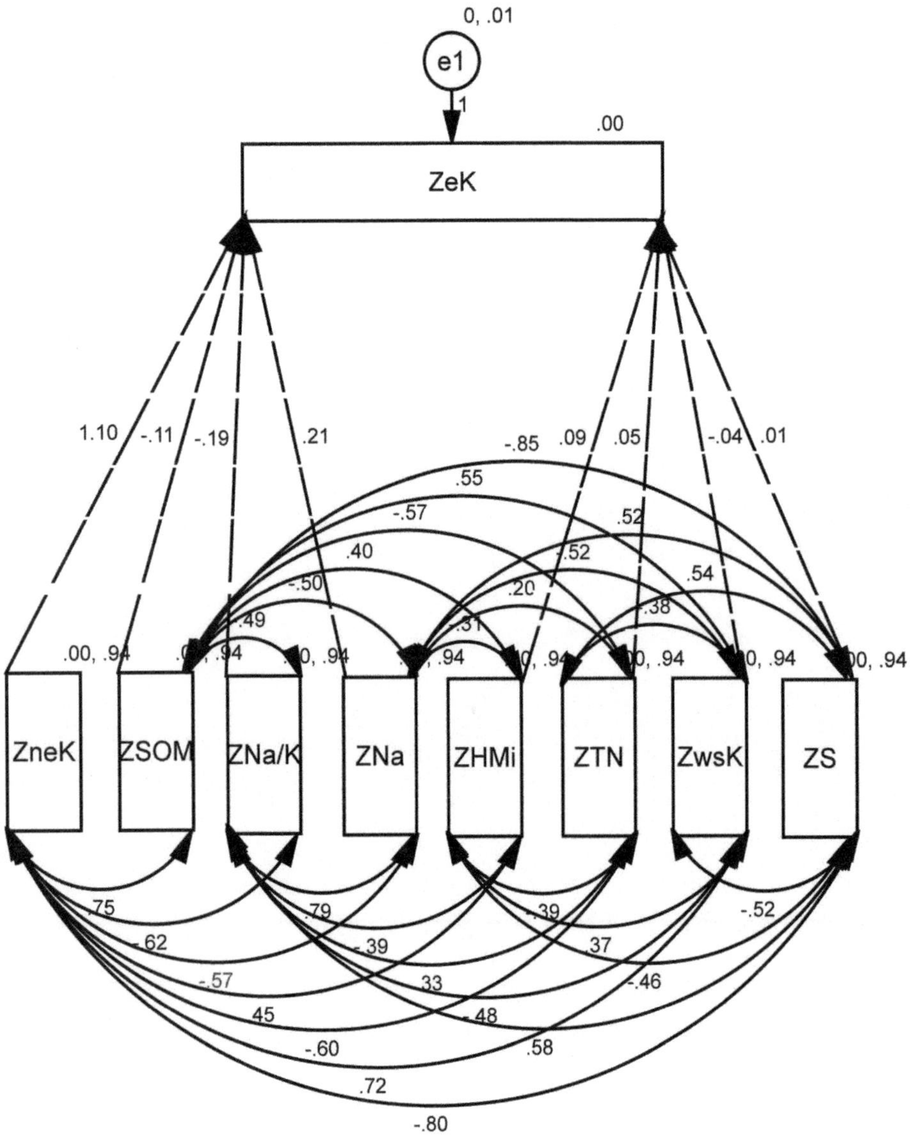

Figure 5. Path model of exchangeable potassium (eK, solid line for double arrow, dashed-line for single arrow, and single arrow flows into ZwsK).

$$\begin{cases} y_{wsk} = 0.98 \cdot x_{eK} + 3.328 \times 10^{-16} \quad (R^2 = 0.96) \\ y_{eK} = 1.02 \cdot x_{wsk} - 3.39 \times 10^{-16} \\ y_{neK} = 0.60 \cdot x_{eK} - 0.37 \cdot x_{PHA} + 7.89 \times 10^{-17} \quad (R^2 = 0.70) \\ y_{eK} = 1.67 \cdot x_{neK} + 0.62 \cdot x_{PHA} - 1.32 \times 10^{-16} \end{cases} \tag{9}$$

Equilibrium occurs when x_{wsK}, x_{PHA}, x_{neK} satisfy with Functions 11,

$$x_{wsk} = 1.64 \cdot x_{neK} + 0.61 \cdot x_{PHA} + 2.07 \times 10^{-16} \tag{10}$$

Here, y_{ek} is the equilibrium threshold. And the value of x_{wsK}, x_{PHA}, x_{neK} is determined by soil characteristics. Calculated results from Equations 9 and 10 are listed in Table 12. In this table, column 5 and 6 are calculated from Equations 9 and 10. We use the wsK, eK, neK constructs the dynamic equilibrium equations ($wsK \underset{\theta}{\overset{\varepsilon}{\rightleftharpoons}} eK \underset{\omega}{\overset{\lambda}{\rightleftharpoons}} neK$).The column 7 is normalized values of neK and the column 8 is the direction of movement of balance of eK. These three column datasets are described in Fig. 7.

In the reversible equilibrium of $wsK \underset{\theta}{\overset{\varepsilon}{\rightleftharpoons}} eK \underset{\omega}{\overset{\lambda}{\rightleftharpoons}} neK$, soil samples SWH, NRSWL, K balance cycle were moved in ε, λ direction movement; soil samples RSWKL, RSWH, RSWL, K balance cycle were moved in θ, λ direction movement; soil samples RSWKH RSKH, RSKL ROPTH, ROPTL, NRSWKH, NRSKH, NRSKL, NROPTH, NROPTL K balance in soil samples, K balance cycle were moved in θ, ω direction movement respectively.

As is shown in Fig. 7, neK reaches its equilibrium in following soil plots, including LEG (SW), HEG (SK), LEG (OPT), and LEG (SW). And the exchangeable K reaches its equilibrium in soil types of LEG (SW). For other treatment soils, such as LEG (SW and

Table 9. Direct and indirect correlation coefficients of non-exchangeable potassium (*neK*).

Variables	Direct correlation coefficients	Indirect correlation coefficients	Path contrition coefficient
Z(S)	−0.86	0.19	−0.25
Z(Na/K)	−0.66	0.02	0.03
Z(TN)	−0.64	0.03	−0.03
Z(Na)	−0.61	−0.34	−0.37
Z(PHA)	−0.46	0.03	0.16
Z(ORP)	−0.41	0.10	−0.42
Z(Mg)	0.04	−0.67	−0.34
Z(CO3)	0.10	1.18	0.56
Z(K)	0.44	−0.14	0.2
Z(pH)	0.46	−0.16	−0.50
Z(HMi)	0.48	0.07	0.06
Z(CIA)	0.56	0.05	−0.07
Z(wsK)	0.77	0.01	0.01
Z(SOM)	0.8	−0.11	−0.14
Z(eK)	0.98	0.76	0.69

SK), LEG (SW), LEG (SK), HEG (SW and SK), HEG (OPT), neK changes to eK more or less. In soil of HEG (SW and SK), eK changes to wsK when wsK was lower than equilibrium. In soil of LEG (SW and SK), neK changes to eK when wsK of soil was greater than that of crops. Soils of both genotypes cotton (HEG and LEG) can efficiently absorb exchangeable potassium.

During a lot of vermiculite in the soil, those negative charges caused by vermiculite isomorphous substitution are near p-site, the electrostatic attraction of potassium ion became bigger, the adsorption capacity much more than other type (2:1) minerals, therefore, wsK and neK are fixed and the soil changed into deficiency. In general, drying accelerated the fixation of wsK adsorption, but neK still moved in the direction of eK in the reversible equilibrium in HEG rhizosphere soil because rhizosphere soil exchangeable potassium content was low to meet the needs of the cotton growth in the SW and SWK plant. When moisture was adequate, redox reaction of soil became strengthened and oxidation reduction potential was reduced. For example, Fe^{2+} and Mn^{2+} ions were increased rapidly, Fe^{2+} and Mn^{2+} replaced p-site potassium of soil colloids and part of i-site potassium, and this improved the content of rapidly-available potassium and biological effective parts. In addition, the dissolved iron and manganese mineral alteration and also increased the release of mineral K.

Table 10. Direct and indirect correlation coefficients in exchangable potassium (*eK*).

Variables	Direct correlation coefficients	Indirect correlation coefficients	Path contrition coefficient
Z(S)	−0.83	−0.28	0.36
Z(Na/K)	−0.67	−0.03	−0.04
Z(TN)	−0.62	−0.04	0.04
Z(Na)	−0.55	0.46	0.54
Z(PHA)	−0.51	−0.06	−0.24
Z(ORP)	−0.38	−0.16	0.6
Z(Mg)	0.07	0.95	0.49
Z(CO3)	0.11	−1.7	−0.81
Z(pH)	0.47	0.23	0.71
Z(K)	0.48	0.21	−0.3
Z(CIA)	0.51	−0.07	0.1
Z(HMi)	0.54	−0.09	−0.09
Z(wsK)	0.73	−0.01	−0.02
Z(SOM)	0.74	0.18	0.2
Z(neK)	0.98	1.49	1.44

Table 11. Fertilizing rates of different treatments (unit: g).

Treatments	OPT	SW	SK	SWK
Urea	6.65	6.65	6.65	6.65
$Na_2PO_4.2H_2O$	11.72	11.72	11.72	11.72
KCl	6.83	6.83	0	0
$CaCO_3$	2.13	2.13	2.13	2.13
H_3BO_3	1.00	1.00	1.00	1.00
$ZnSO_4.7H2O$	1.87	1.87	1.87	1.87
$MgSO_4.7H2O$	2.61	2.61	2.61	2.61

OPT, optimum fertilization treatment; *SW*, water limited; *SK*, potassium limited; *SWK*, water & potassium limited.

Discussion and Conclusions

In this study, changes of potassium status in soil were investigated through our proposed path model on basis of laboratory experimental datasets. Through this investigation, following findings were achieved.

Firstly, changes of potassium status in the rhizosphere soil were controlled by different environmental variables. wsK, neK and eK were controlled by some common factors and have their speculiar characteristics respectively. Those parameters, such as soil organic matter (SOM), Na and Na/K and S, were significant direct coefficient factors which control dynamic process of wsK, neK and eK. CO_3 and Mg are significant indirect coefficient factors which control dynamic of wsK, neK and eK. The pH is the primary factor which controls path of the wsK, neK and eK. Organic acids of HEG root system secretion increased soil acidity; the amount of fixed potassium was reduced. PHA surface is porous structure, PHA increased the adsorption of wsK in the experiment, and HMi could promote the moving to wsK, eK in the K form exchange. Both of them were negative correlation significantly negative correlation. First, under acid condition, hydroxyl aluminum and aluminum ions can occupy the potassium fixation point stronger. At the same time, the larger diameter of hydroxy aluminium ions entered into the interlayer of mineral, formed as the "island" and played a supporting role, provided the potassium ion diffusion of mobile channel in the interlayer. These reduced the soil potassium fixation. Second, water protonated in acidic condition, the radius of H_3O^+ 12–13.3 nm were similar to K^+, H_3O^+ can produced competitive adsorption in soil where fixed on K^+ parts, so potassium reduced fixation. Those dynamic processes, such as eK to wsK, low concentration wsK and eK to neK, are likely to occur, in condition that wsK is lower than equilibrium value and PHA and eK are greater than equilibrium values. eK is changed neK when wsK is greater than equilibrium value. In soil of HEG (SW and SK), eK is changed into wsK when plant wsK is lower than soil wsK. In soil of LEG (SW and SK), neK is changed into eK when plant wsK is higher than soil wsK. Both genotype cottons can effectively absorb eK in treatment of stress of potassium (SK).

Secondly, it discloses that SOM, S, TN, HMi and PHA in rhizosphere soil determined dynamic balance of potassium status which includes adsorption and desorption, precipitation and dissolution, complexion and chelating. Special characteristics of humus, e.g., acidic and hydrophilic, cationic exchange, complexing capacity and high absorption capacity, could improves the sustained-release effect of potassium fertilizer. Many variables, pH, ORP, etc., are closely related to transformation and utilization of potassium fertilizer. The pH change in soil is mainly caused by coupling effects of nutrient uptake by plant roots and the secretion of organic acid and absorption imbalance of positive ion and negative ion leads to rhizosphere pH change, however, different factors affect imbalance of absorption of positive ion and negative ion. Dynamic equilibrium of potassium is affected by equilibrium constants, temperature, and products of this equilibrium system (wsK). Therefore, it is useful for cotton planting to improve to appropriate temperature, to alter SOM, TN and pH in soil, and to reduce the equilibrium constant to positive reaction. Meanwhile, it is good to choose HEG cotton for better use soil wsK. It is important to understand ways to slow down the fast response and accelerate speed of slow response in potassium balance system.

Finally, path model can effectively determine direction and quantities of potassium status changes between eK, neK and wsK under influences of specific environmental parameters. The proposed path model was able to quantitatively disclose moving

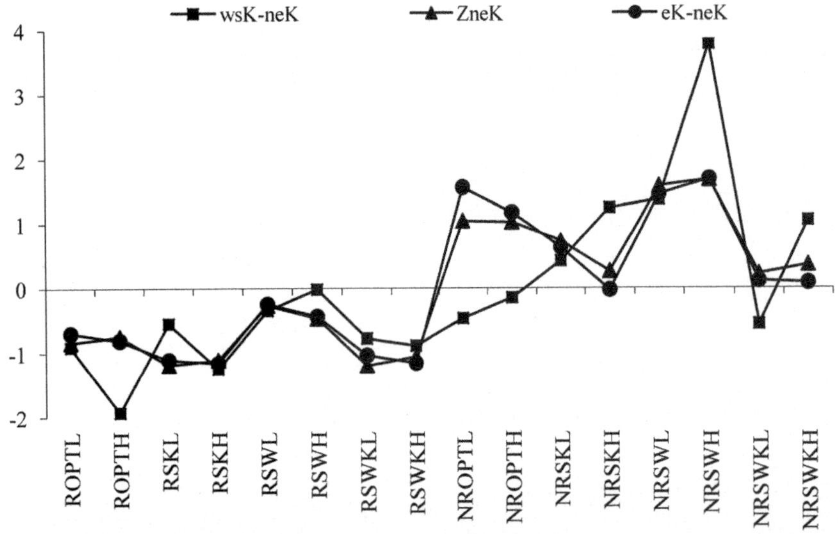

Figure 6. Dynamic equilibrium of non-exchangeable potassium in rhizosphere/non-rhizosphere soils.

Table 12. Dynamic equilibrium of exchangeable/non-exchangeable potassium (*neK/eK*) in rhizosphere/non-rhizosphere soils.

Variables	$wsK \underset{\beta}{\overset{\alpha}{\rightleftarrows}} neK \underset{\chi}{\overset{\gamma}{\rightleftarrows}} eK$			Movement direction at balance state	$wsK \underset{\theta}{\overset{\varepsilon}{\rightleftarrows}} eK \underset{\omega}{\overset{\lambda}{\rightleftarrows}} neK$			Movement direction at balance state
	$V_{wsk\to neK}$	$V_{ek\to neK}$	ZneK		$V_{wsk\to eK}$	$V_{nek\to eK}$	ZeK	
RSWKH	−0.89	−1.16	−1.06	β,χ	−0.78	−1.39	−1.14	θ, ω
RSWKL	−0.78	−1.04	−1.2	α, χ	−0.78	−1.51	−1.02	θ, λ
RSWH	−0.02	−0.44	−0.47	α, χ	−0.19	−0.5	−0.44	θ, λ
RSWL	−0.35	−0.25	−0.27	α, χ	−0.32	−0.28	−0.24	θ, λ
RSKH	−1.25	−1.16	−1.11	β, χ	−0.8	−1.79	−1.14	θ, ω
RSKL	−0.55	−1.11	−1.19	β, γ	−0.82	−1.19	−1.09	θ, λ
ROPTH	−1.93	−0.81	−0.75	β, χ	−0.65	−2.11	−0.79	θ, ω
ROPTL	−0.91	−0.70	−0.84	β, γ	−0.51	−1.48	−0.69	θ, λ
NRSWKH	1.05	0.08	0.36	β, χ	0.14	1.43	0.08	ε, λ
NRSWKL	−0.56	0.11	0.22	α, χ	−0.45	0.54	0.11	θ, ω
NRSWH	3.78	1.69	1.68	β, χ	3.24	1.28	1.66	ε, λ
NRSWL	1.37	1.45	1.58	α, χ	0.79	2.73	1.42	θ, λ
NRSKH	1.24	−0.03	0.26	β, χ	0.92	0.17	−0.03	ε, λ
NRSKL	0.42	0.63	0.73	β, χ	0.11	1.45	0.61	θ, ω
NROPTH	−0.15	1.17	1.02	β, γ	−0.12	1.76	1.14	θ, ω
NROPTL	−0.47	1.56	1.04	β, γ	0.23	0.90	1.53	θ, ω

direction of K status and quantify its equilibrium threshold. It provided a theoretical and practical basis for scientific and effective fertilization in agricultural plants growth. The significance of this study is that we are able to implement investigation on the equilibrium movement of potassium status through path model analysis. It discloses the gradual dynamic process of potassium status, and to decouple the sophisticated interactions between different variables. It is useful to guide the use of potassium fertilizer for cotton crops in practical. The model provides a biogeochemical theory basis of controlling moisture in cotton plant, applying potash fertilizer and organic fertilizer, and appropriately reducing nitrogen fertilizer.

However, environmental parameters such as soil moisture and temperature are some of the most important factors that influenced crop growth. Observation of dynamic change of soil moisture and temperature in every other time-step is quite sophisticated for us to design the control experiments while it involves in more than 4 times treatments, which means that we have more than 500 pots. By limitation, we had to set the soil moisture into two states, water sufficient (>35%) and deficient (<25%), and temperature were set to indoor temperature.

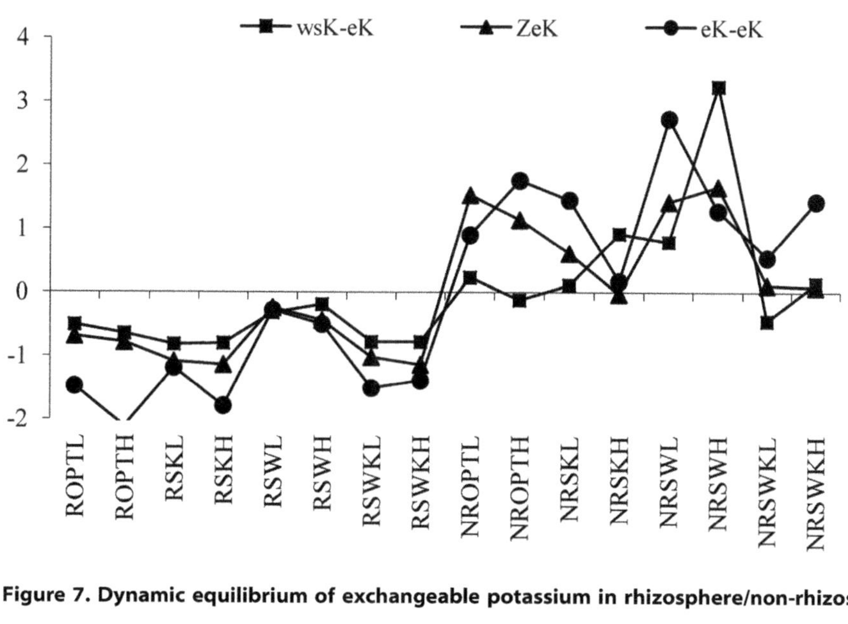

Figure 7. Dynamic equilibrium of exchangeable potassium in rhizosphere/non-rhizosphere soils.

Anyway, we were very aware of the importance of soil moisture and temperature impacts on control experiment potassium status. The next step of our work is to further investigate efficient utilization of potassium and the influence of soil moisture and temperature on exchange of potassium forms. We hope this study is able to gain knowledge of promoting high efficient use of potassium for crops.

Acknowledgments

We are grateful to invaluable review comments of the editor and two anonymous reviewers to make the paper perfect. We would like to express our special thanks to Dr. Feng WU of the Institute of Earth Environment, Chinese Academy of Sciences for his kind technical support.

Author Contributions

Conceived and designed the experiments: WH FC. Performed the experiments: WH. Analyzed the data: WH. Contributed reagents/materials/analysis tools: FC WH. Wrote the paper: WH FC.

References

1. Mahmood-Ul-Hassan M, Rashid M, Rafique E (2011) Nutrients transport through variably structured soils. Soil Science and Plant Nutrition 57: 331–340.
2. Osaki M, Matsumoto M, Shinano T, Tadano T (1996) A root-shoot interaction hypothesis for high productivity of root crops. Soil science and plant nutrition 42: 289–301.
3. Wada S-I, Seki H (1994) Ca-K-Na exchange equilibria on a smectitic soil: Modeling the variation of selectivity coefficient. Soil Science and Plant Nutrition 40: 629–636.
4. Puente M, Li C, Bashan Y (2004) Microbial Populations and Activities in the Rhizoplane of Rock-Weathering Desert Plants. II. Growth Promotion of Cactus Seedlings. Plant Biology 6: 643–650.
5. Wada S-I, Masuda K (1995) Control of salt concentration of soil solution by the addition of synthetic hydrotalcite. Soil Science and Plant Nutrition 41: 377–381.
6. Kobayashi H, Masaoka Y, Takahashi Y, Ide Y, Sato S (2007) Ability of salt glands in Rhodes grass (Chloris gayanaKunth) to secrete Na+and K+. Soil Science and Plant Nutrition 53: 764–771.
7. Han MY, Zhang LX, Fan CH, Liu LH, Zhang LS, et al. (2011) Release of nitrogen, phosphorus, and potassium during the decomposition of apple (Malus domestica) leaf litter under different fertilization regimes in Loess Plateau, China. Soil Science and Plant Nutrition 57: 549–557.
8. He H, Zhou J, Wu Y, Zhang W, Xie X (2008) Modeling the interaction of urbanization and surface water quality environment. Environmental Forensics 9: 215–225.
9. Eick MJ, Sparks DL, Bar-Tal A, Feigenbaum S (1990) Analyses of adsorption kinetics using a stirred-flow chamber: II. Potassium-calcium exchange on clay minerals. Soil Science Society of America Journal 54: 1278–1282.
10. McCune B, Caldwell BA (2009) A single phosphorus treatment doubles growth of cyanobacterial lichen transplants. Ecology 90: 567–570.
11. Schneider A-K, Schröder B (2012) Perspectives in modelling earthworm dynamics and their feedbacks with abiotic soil properties. Applied Soil Ecology 58: 29–36.
12. He H, Jim C (2012) Coupling model of energy consumption with changes in environmental utility. Energy Policy 43: 235–243.
13. He H, Jim C (2010) Simulation of thermodynamic transmission in green roof ecosystem. Ecological Modelling 221: 2949–2958.
14. Brouder S, Cassman K (1994) Evaluation of a mechanistic model of potassium uptake by cotton in vermiculitic soil. Soil Science Society of America Journal 58: 1174–1183.
15. Nakahara O, Wada S-I (1995) Surface Complexation Model of Cation Adsorption on Humic Soils. Soil Science and Plant Nutrition 41: 671–679.
16. Uroz S, Calvaruso C, Turpault M-P, Frey-Klett P (2009) Mineral weathering by bacteria: ecology, actors and mechanisms. Trends in microbiology 17: 378–387.
17. Puente M, Bashan Y, Li C, Lebsky V (2004) Microbial populations and activities in the rhizoplane of rock-weathering desert plants. I. Root colonization and weathering of igneous rocks. Plant Biology 6: 629–642.
18. Hinsinger P, Fernandes Barros ON, Benedetti MF, Noack Y, Callot G (2001) Plant-induced weathering of a basaltic rock: experimental evidence. Geochimica et Cosmochimica Acta 65: 137–152.
19. Hashimoto Y, Kang J, Matsuyama N, Saigusa M (2012) Path Analysis of Phosphorus Retention Capacity in Allophanic and Non-allophanic Andisols. Soil Science Society of America Journal 76: 441–448.
20. Kang J, Hesterberg D, Osmond DL (2009) Soil Organic Matter Effects on Phosphorus Sorption: A Path Analysis. Soil Science Society of America Journal 73: 360.
21. Lambers H, Mougel C, Jaillard B, Hinsinger P (2009) Plant-microbe-soil interactions in the rhizosphere: an evolutionary perspective. Plant and Soil 321: 83–115.
22. Wright S (1921) Correlation and causation. Journal of agricultural research 20: 557–585.
23. Wright S (1920) The relative importance of heredity and environment in determining the piebald pattern of guinea-pigs. Proceedings of the National Academy of Sciences of the United States of America 6: 320.
24. Wright S (1921) Systems of mating. I. The biometric relations between parent and offspring. Genetics 6: 111.
25. Wright S (1921) Systems of mating. II. The effects of inbreeding on the genetic composition of a population. Genetics 6: 124.
26. Kulahci F (2011) A risk analysis model for radioactive wastes. J Hazard Mater 191: 349–355.
27. Berthrong ST, Jobbágy EG, Jackson RB (2009) A global meta-analysis of soil exchangeable cations, pH, carbon, and nitrogen with afforestation. Ecological Applications 19: 2228–2241.
28. Kocev D, Naumoski A, Mitreski K, Krstić S, Džeroski S (2010) Learning habitat models for the diatom community in Lake Prespa. Ecological Modelling 221: 330–337.
29. He C, Cui K, Duan A, Zeng Y, Zhang J (2012) Genome-wide and molecular evolution analysis of the Poplar KT/HAK/KUP potassium transporter gene family. Ecol Evol 2: 1996–2004.
30. Kamewada K (1996) Application of "Four-Plane Model" to the Adsorption of K+, NO3-, and SO42-from a Mixed Solution of KNO3and K2SO4on Andisols. Soil Science and Plant Nutrition 42: 801–808.
31. Kang J, Hesterberg D, Osmond DL (2009) Soil organic matter effects on phosphorus sorption: A path analysis. Soil Science Society of America Journal 73: 360–366.

5

Biochar-Induced Changes in Soil Hydraulic Conductivity and Dissolved Nutrient Fluxes Constrained by Laboratory Experiments

Rebecca T. Barnes*¤a, Morgan E. Gallagher¤b, Caroline A. Masiello, Zuolin Liu, Brandon Dugan

Department of Earth Science, Rice University, Houston, Texas, United States of America

Abstract

The addition of charcoal (or biochar) to soil has significant carbon sequestration and agronomic potential, making it important to determine how this potentially large anthropogenic carbon influx will alter ecosystem functions. We used column experiments to quantify how hydrologic and nutrient-retention characteristics of three soil materials differed with biochar amendment. We compared three homogeneous soil materials (sand, organic-rich topsoil, and clay-rich Hapludert) to provide a basic understanding of biochar-soil-water interactions. On average, biochar amendment decreased saturated hydraulic conductivity (*K*) by 92% in sand and 67% in organic soil, but increased *K* by 328% in clay-rich soil. The change in *K* for sand was not predicted by the accompanying physical changes to the soil mixture; the sand-biochar mixture was less dense and more porous than sand without biochar. We propose two hydrologic pathways that are potential drivers for this behavior: one through the interstitial biochar-sand space and a second through pores within the biochar grains themselves. This second pathway adds to the porosity of the soil mixture; however, it likely does not add to the effective soil *K* due to its tortuosity and smaller pore size. Therefore, the addition of biochar can increase or decrease soil drainage, and suggests that any potential improvement of water delivery to plants is dependent on soil type, biochar amendment rate, and biochar properties. Changes in dissolved carbon (C) and nitrogen (N) fluxes also differed; with biochar increasing the C flux from organic-poor sand, decreasing it from organic-rich soils, and retaining small amounts of soil-derived N. The aromaticity of C lost from sand and clay increased, suggesting lost C was biochar-derived; though the loss accounts for only 0.05% of added biochar-C. Thus, the direction and magnitude of hydraulic, C, and N changes associated with biochar amendments are soil type (composition and particle size) dependent.

Editor: Zhi Zhou, Purdue University, United States of America

Funding: This work was supported by National Science Foundation (NSF-EAR-0911685) to CAM & BD (www.nsf.gov), US Department of Agriculture (2011-67009-20074) to CAM (www.usda.gov), Rice University's Shell Center for Sustainability to CAM (shellcenter.rice.edu). The funders had no role in study design, data collection and analysis, decision to publish, or preparation of the manuscript.

Competing Interests: The authors have declared that no competing interests exist.

* Email: becca.barnes@coloradocollege.edu

¤a Current address: Environmental Program, Colorado College, Colorado Springs, Colorado, United States of America
¤b Current address: Center on Global Change, Duke University, Durham, North Carolina, United States of America

Introduction

Woolf et al. [1] estimate that 1.8 Pg CO_2-carbon equivalents can be sequestered each year through the sustainable production and application of 0.9 Pg of biochar to agricultural land which sequesters carbon (C), reduces CH_4 and N_2O emissions, and results in avoided CO_2 emissions. Biochar, charcoal intentionally produced by humans through pyrolysis for soil amendment, is a type of black carbon, like soot or charcoal [2,3]. Adding 0.9 Pg of biochar to the agricultural landscape would correspond to a 4–20 fold increase in global black carbon production (0.04 to 0.194 Pg yr−1; [4]). Recent work illustrates the likely mobility of biochar added to soil: charcoal in soils can be released into rivers [5,6], where, given its aromatic structure, it can be photo-oxidized [7] and/or transported downstream where it has the potential to alter ecosystem processes [6]. Given the potential magnitude of ecosystem perturbations from full-scale implementation of biochar-C sequestration, it is critical to determine the effects of biochar soil amendment on water and biogeochemical cycling.

Biochar soil amendment can increase crop productivity [3,8], potentially by improving the hydrologic properties of the soils [9]. Biochar can also increase soil water-holding capacity, and available water content [8,10–13], plant available water [14–16], alter soil hydrophobicity [17], and change soil hydraulic conductivity [18–22]. Biochar is predicted to cause sandy soils to drain more slowly [23] and clay-rich soils to drain more rapidly [24]. However, past results have not been consistent, likely due to confounding factors such as biochar characteristics (i.e. feedstock and pyrolysis temperatures), application rates, and soil characteristics. Given the importance of hydraulic conductivity in determining the partitioning of precipitation between infiltration and overland flow (i.e. infiltration rates; [25]), which impacts water storage in the subsurface and thus plant available water, it is necessary to understand the effects of biochar on the hydraulic properties of different soil types.

The C and nitrogen (N) content of biochar varies with feedstock and production conditions [26]. These conditions and the C:N ratio of biochar influence its stability [27,28] as well as possible soil C and N losses [29]. While dependent on production conditions, biochar tends to have a high cation exchange capacity [30] and anion sorption ability [31], allowing for adsorption of dissolved organic matter (DOM) [32] and N [33], and can alter greenhouse gas emissions [34–36]. While biochar amendment adds C and N to soils (which may be available for leaching), it is also able to sequester additional C and nutrients in the soil due to its sorptive properties. Therefore the addition of biochar to soil could result in a net increase or decrease in dissolved C and N losses [37]. Worldwide analysis of dissolved black carbon (DBC), which includes derivatives of charcoal and biochar, exported from rivers indicates that, on average, DBC contributes 10% of the global total dissolved organic carbon (DOC) flux [6]. In addition, analysis of the bioavailability of biochar extracts in natural stream water suggests that some biochar-C molecules have turnover rates on the order of days to a month [38], indicating that at least a portion of biochar-C is not recalcitrant. In addition to the dissolution of biochar within soils, particles are also transported. Recent studies illustrate that the movement of biochar particles is related to particle size and surface chemistry, as well as pore water salt content and pH [39,40]. Collectively this research points to the uncertainty in the fate of biochar and biochar-C and their down-gradient effects on aquatic ecosystems.

Using column experiments, we quantify the effects of a 10% (by mass) biochar amendment on the saturated hydraulic conductivity (K) of sand, clay-rich, and organic-rich soil materials as well as report the effects of this soil amendment on C and N leaching. Using simple, homogeneous soil materials with different grain sizes and surface chemistry allowed us to examine biochar-soil interactions and to compare our results to established soil hydrology models. These data begin to address an important knowledge gap by providing new quantitative constraints on how biochar amendments change K and the chemistry of soil leachate; this work points to the need for more mechanistic studies to examine biochar-soil-water interactions.

Materials and Methods

Soil and Biochar

Three soil materials were used to gauge soil property responses to biochar amendment. Sand (Pavestone Natural Play Sand) and organic-rich soil (Micro-Gro Organic Rich Garden Soil, with no N fertilizer) were purchased from Home Depot. The organic-rich soil was texturally similar to a sandy loam. A clay-loam Hapludert, characterized by its poor drainage [41], was collected from Rice University's campus. All materials were oven dried at 60°C to remove any moisture prior to dry sieving. Dry materials were mixed, and then oven dried at 100°C for 24 hours to create homogenous mixtures with initial water content of zero; 100°C facilitates water loss but minimizes chemical impacts as it is significantly lower than our pyrolysis temperature [42]. We determined the grain size of the three soils and the biochar using seven sieves (38 μm to 500 μm).

We produced biochar from mesquite wood (*Prosopis* sp.), ground to smaller than 20 mesh (850 μm). Batches of mesquite (70–80 g) were pyrolyzed using the reactor design described in Kinney et al. [17], by heating in a muffle furnace at 6°C min^{-1} and holding at 400°C for 4 hours. On average these pyrolysis conditions provided a biochar mass yield of 40.4%. The produced biochar had a pH of 6.5±0.1, ash content of 3.33%±0.04%, and liming equivalent of 4%, determined using protocols outlined by the International Biochar Initiative [43–45].

Column Experiments

To test the response of K and dissolved nutrient fluxes to biochar amendment we conducted falling head experiments [46] across six materials: sand, sand+biochar, organic soil, organic soil+biochar, clay, and clay+biochar. We packed 150 g of each mixture into three replicate columns, 50 g per column, with 54 μm polyester mesh screen (Small Parts Inc.) at the bottom. Materials were packed with a consistent force into columns in four equal increments to achieve uniform bulk density [47] and the initial soil length was recorded. Bulk density for soil materials and soil+biochar mixtures was determined using the dry mass and column dimensions (height of soil materials, diameter of column) at the start of the experiment. Grain size distributions of soil+biochar mixtures were estimated using the proportional masses of each material (i.e. 10% biochar, 90% soil material) and appropriate grain size data (Table 1).

Biochar constituted 10% of the total mass in the columns that contained biochar. This represents a 133 tons ha^{-1} (95 Mg C ha^{-1}) application rate with a 10 cm tillage depth. We chose a high amendment rate to ensure that we altered the soil-water system in a way that would allow us to detect any effects across the three contrasting soil materials. Specifically, we were concerned about the potential of soil biochar amendments increasing C exports to surface waters, as well as any unforeseen consequences on soil hydrologic properties. This amendment rate is above what is likely to be added to an agricultural field, though within the range reported for positive or neutral productivity effects (up to 140 Mg C ha−1; [3]), similar to the application made by Chan et al. [48], and well below what was shown to be an upper limit to biochar-induced benefits to plants (200 tons C ha−1; [15]).

To saturate the soils, we capped the column bottoms, added 150 mL of 18 MΩ-cm MilliQ water, and allowed the columns to sit for 48 h before drainage. Six consecutive falling head experiments [i.e. flushing events; [46] were conducted on saturated soils using 150 mL of MilliQ water, with leachate collected at the end of each experiment for all columns. The same experimental set up was used for all six materials to allow for inter-comparison. The leachate was weighed, filtered through a pre-combusted glass fiber filter (Whatman GF/F) and kept at 4°C until analysis. This process was repeated without allowing columns to dry between flushing events. Evaporation was monitored daily (via the net change in the mass of water in a beaker) and water throughput (leachate volume) was corrected for evaporation by adding the product of the daily evaporative loss (mg hr^{-1}) and duration of flushing event (hrs).

Saturated hydraulic conductivity (K) from falling head data was calculated using equation 1 [49]:

$$K = (L/\Delta t) * ln(h_2/h_1) \tag{1}$$

where L is soil sample length (m), Δt is time elapsed (s), and h_1 and h_2 are the initial and final water heights (m), respectively (data available in Table S1). Separate experiments on sand-only systems confirm consistent measures of K from top-saturated and bottom-saturated falling head and constant head experiments, suggesting full saturation and steady-state conditions were achieved using this falling-head technique.

Table 1. Physical and elemental properties of soil materials and biochar.

Material	Grain Size			Bulk Density	%C	%N
	d_{10} (μm)	d_{50} (μm)	d_{90} (μm)	ρ_d (g cm^{-3})		
sand	70	160	380	1.68±0.18	0.4	0.01
organic-rich	95	400	480	0.43±0.002	37.9	0.54
clay-rich	45	115	460	1.72±0.04	0.9	0.03
biochar (mesquite)	75	320	470	0.36±0.03	71.6	0.84

Soil and sample analysis

We used a Shimadzu TOC-VCN to determine the DOC and total dissolved nitrogen (TDN) concentrations of the filtered leachate. Sample replicates indicate a 0.08 mg L^{-1} and 0.04 mg L^{-1} precision for DOC and TDN, respectively. Mass loss of C and N (dissolved flux) were determined by multiplying the DOC and TDN concentrations by the water volume of the sample, respectively. In contrast, water throughput was calculated using the evaporation-corrected water volumes. The aromaticity of the dissolved C was determined by calculating the specific UV absorbance at 254 nm (SUVA254; [50]). The UV absorption (m^{-1}) of filtered leachate was measured on a Cary UV-Vis spectrophotometer and divided by the DOC concentration (mg C L^{-1}) following the protocol developed by Weishaar et al. [50] to calculate $SUVA_{254}$ (L mg C^{-1} m^{-1}).

After six flushing events, soil mixtures were removed from the columns, weighed, dried at 100°C for at least 24 hours, and reweighed to determine water content at field capacity by mass. We measured C and N content on original soils, biochar, and dried, post-experiment soils using a Costech 4010 CHNS/O Elemental Analyzer. Replicate analysis shows a precision of 0.6% and 0.02% for C and N measurements, respectively.

Statistical analyses

We used two-sample t-tests to determine statistical differences in the soil and leachate characteristics between treatments and controls. Paired t-tests were used to determine the statistical differences of *K* over the course of the experiment. Finally, a general linear model with flush number as a covariate, was used to determine if *K* changed significantly over the course of the experiment as a result of the biochar amendment. All statistical tests were done in the RStudio environment (v0.98.507, 2014 RStudio, Inc.) and results were considered significant when $p < 0.05$.

Results and Discussion

Soil physical characteristics

The addition of biochar to the soil materials changed a number of physical properties, e.g. grain size distribution and bulk density, which likely affected water movement. The grain size distribution of biochar differed from that of the three soils and was most similar to the organic soil (Table 1). When biochar was added to sand and clay soils, the d_{50} of the mixtures increased; however, when biochar was added to the organic soil the d_{50} decreased. Given the similarity in d_{10} of sand and biochar, the addition of biochar did not appreciably change the proportion of fines in the sand versus the sand+biochar mixture. The addition of 10% biochar changed soil bulk density (ρ_b) (Table 2), though these changes were not always significant; the addition of biochar decreased the ρ_b of sand and clay by 17% (p = 0.056) and 20% (p = 0.052), respectively. Biochar addition to clay lowered ρ_b enough to bring it within the range recommended by the National Soil Conservation Service to allow for adequate root growth (<1.47 g cm^{-3}, USDA, 2008). In contrast, when biochar was added to organic soil the mixture ρ_b increased 10% (p = 0.018), despite the biochar having a lower ρ_b (organic: 0.43±0.002 g cm^{-3}; biochar: 0.36±0.03 g cm^{-3}; Table 1); this is likely related to the smaller relative grain size of biochar and grain arrangement during packing of the columns (Table 1).

Soil hydraulic characteristics

Saturated hydraulic conductivity (*K*) describes the ease of fluid flow through saturated porous media and it can be directly measured with flow through experiments [46,51] or estimated using theoretical or empirical models. Theoretical models, such as the Kozeny equation [52,53] require significant knowledge of porosity, tortuosity, pore shape, grain density, and specific surface area of solid grains. Because of the difficulty constraining all these parameters, others have developed empirical models that relate porosity and *K* [54] or grain size and *K* [55]. Here we compare our *K* results with these empirical relationships using our bulk density, grain size, and porosity data. Changes in *K* accompanying biochar amendment for both clay- and organic-rich soils follow the empirical relationships discussed above; specifically the change in *K* is inversely related to the change in bulk density (or positively related to the change in porosity) caused by the biochar amendment. The addition of biochar to clay-rich soil resulted in a ρ_b decrease of 20% (a porosity increase), an estimated d_{50} increase of 18%, and an increase in *K* of over 300% (3.26×10^{-8} m s^{-1} to 1.16×10^{-7} m s^{-1}; Figure 1c, Table 2). In contrast, when biochar was added to organic-rich soil, the mixture was 10% denser (lower porosity) with a 2% smaller d_{50}, and *K* decreased by 67% (2.23×10^{-6} m s^{-1} to 7.79×10^{-7} m s^{-1}; Figure 1b, Table 2). Similarly, experiments using silt loams reported an increase in *K* with biochar amendment, though there were not always corresponding decreases in ρ_b (Table 3, [20]). Previous work in organic-rich soils has not documented changes in *K* in response to biochar amendment (0.5 to 2% by mass) (Table 3; [56]). Differences in results are attributable to different amendment rates, biochar grain size, soil properties, and/or threshold effects of amendment rate or grain size.

Several studies have documented threshold effects on porosity and permeability when fine-grained particles are added to a soil. Boadu [57] found that porosity decreased with increasing fines up to a threshold of ~5% (by mass) at which point fine grain additions increased the porosity. Similarly, Crawford et al. [58] noted that porosity is lowest when the fine particle volume equals the pore space of the coarse grains; however, if fine grains are removed or added, the porosity increases.

Table 2. The mean (and standard deviation) of physical, hydraulic, and nutrient properties of the three replicates of each soil and soil+biochar treatment.

Soil	Hydraulic conductivity (K) (m s^{-1})		Bulk Density (ρ_d) (g cm^{-3})		Water Content at Field Capacity (fraction water)		Cumulative DOC loss (mg)		Cumulative TDN loss (mg)		$SUVA_{254}$ (L mg C^{-1} m^{-1})	
	Soil	+biochar	Soil	+biochar	Soil	+biochar	Soil	+biochar	Soil	+biochar	Soil	+biochar
sand	2.9×10^{-6} (6.3×10^{-7})	2.3×10^{-7} (5.9×10^{-8})	1.69 (0.18)	1.39 (0.06)	0.15 (0.02)	0.30 (0.03)	1.71 (0.08)	3.29 (0.35)	0.25 (0.04)	0.19 (0.01)	1.86 (0.83)	2.75 (0.41)
p-value	*<0.001*		*0.056*		*0.002*		*0.017*		*0.081*		*0.043*	
organic-Rich	2.1×10^{-6} (1.9×10^{-6})	7.8×10^{-7} (7.0×10^{-7})	0.43 (0.002)	0.47 (0.008)	0.65 (0.01)	0.66 (0.02)	105.45 (3.74)	95.6 (3.07)	5.52 (0.26)	5.99 (1.25)	3.48 (0.96)	3.38 (0.73)
p-value	*0.038*		*0.018*		*0.757*		*0.039*		*0.565*		*0.442*	
clay-rich	3.2×10^{-8} (1.9×10^{-8})	1.2×10^{-7} (1.2×10^{-8})	1.72 (0.04)	1.38 (0.14)	0.27 (0.01)	0.33 (0.004)	4.86 (2.22)	4.03 (0.13)	0.66 (0.30)	0.25 (0.01)	2.54 (0.70)	3.75 (0.74)
p-value	*<0.001*		*0.052*		*0.003*		*0.587*		*0.078*		*<0.001*	

Two-tailed t-tests were conducted to determine statistical differences between control and +biochar treatments; p-values are shown in italics below mean and standard deviation values for each treatment.

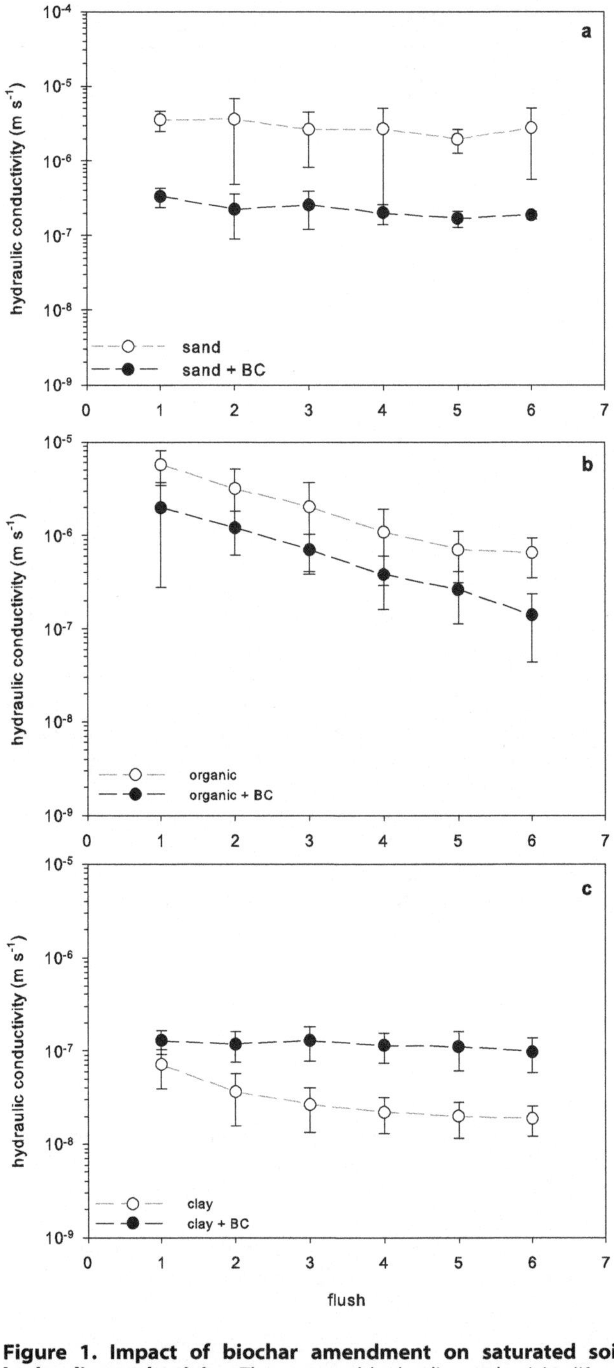

Figure 1. Impact of biochar amendment on saturated soil hydraulic conductivity. The saturated hydraulic conductivity (K), as measured using falling head experiments, for six soil treatments over subsequent flushing events: (a) sand and sand+biochar, (b) organic and organic+biochar, and (c) clay and clay+biochar. Note: the different soil treatment flushing events varied in duration with the clay (c) taking up to 10× longer to drain than the sand (a) or organic soil (b). Saturated hydraulic conductivity data and flushing duration for each flushing experiment available in Table S1.

Models relating K to porosity and grain size suggest that amending sand with biochar should increase K given the greater porosity, overall grain size, and decreased ρ_b. A review of research reveals that this inverse relationship between ρ_b and K is preserved in most cases (Table 3). However, in our experiments biochar amendment to sand decreased the ρ_b and decreased K by 92%

Table 3. A comparison of studies that examined the impact of biochar amendments on soil saturated hydraulic conductivity (K) and bulk density (ρ_d).

Feedstock	Temperature (°C)	Application rate[a]	Application rate (tons biochar ha^{-1})[b]	Experiment duration (d)	Soil type and/or % sand/silt/clay	Response of soil bulk density to biochar addition	Response of K to biochar addition	Reference
mixed hardwood lump charcoal (e.g. oak & hickory)	NR	5, 10, 20 g kg^{-1}	5.5, 11, 22 t ha^{-1}	500	Mesic Typic Hapludolls	decreased	no effect detected	Laird et al. 2010
mesquite	*400*	*10% by dry weight mass*	*133 t ha^{-1}*	*1–20*	*organic rich*	*increased*	*decreased*	*this study*
commercial biochar, wood residue (e.g. teak and rosewood)	NR	4, 8, 16 t ha^{-1}	4, 8, 16 t ha^{-1}	60	18/34/48	NR	increased	Asai et al. 2009
					27/45/28	NR	no significant change	
corn stover	350 & 550	7.18 t C ha^{-1}	350°C = 11.3 t ha^{-1}	295	Alfisol (silt loam)	decreased	increased with both biochars	Herath et al. 2013
			550°C = 10.0 t ha^{-1}		Andisol (silt loam)	no change	increased with 350°C, no change for 550°C	
mesquite	*400*	*10% by dry weight mass*	*133 t ha^{-1}*	*3–23*	*clay-loam Hapludert*	*decreased*	*increased*	*this study*
dairy manure	300, 500, 700	5% by dry weight mass	61 t ha^{-1}	180	loamy, 40/35/25	decreased	increased, greater increases for biochars at higher temperatures	Lei & Zhang 2013
woodchip						decreased	increased, greater increases for biochars at higher temperatures	
charcoal produced in kilns	~300–500	ambient levels - beneath charcoal kilns	NA	NA	Haplic Acrisols	no significant change	increased	Oguntude et al. 2008
black locust	300, 400, 500	10, 20 Mg ha^{-1}	10, 20 t ha^{-1}	27	sandy	no change	decreased, greatest decrease seen for 500°C biochar & higher application rate	Uzoma et al. 2011
powdered wood charcoal	NR	0.5%, 1.5%, 2.5%, 5% by dry weight mass	6.3, 18.8, 31.3, 62.5 t ha^{-1}	60	sandy loam	decreased	decreased with increasing biochar concentration	Devereaux et al. 2012
acacia green waste	NR	47 Mg ha^{-1}	47 t ha^{-1}	900	Planosol, 72.8/16.8/10.4	decreased	increased near saturated K, no effect on unsaturated K	Hardie et al. 2014
mesquite	*400*	*10% by dry weight mass*	*133 t ha^{-1}*	*1–4*	*sand*	*decreased*	*decreased*	*this study*

[a] Biochar application rate reported in paper.
[b] Biochar application rate converted using bulk density of soils or column materials provided in the paper and assuming tillage depth of 10 cm.
NR: not reported.
NA: not applicable.
Biochar amendment rates are provided two ways: the units provided by the study and in tons biochar ha^{-1}. The conversion assumed a tillage depth of 10 cm and the bulk density of the soil or column materials provided in the paper. Studies are organized by soil type, top to bottom: organic-rich soils, clay- and silt-rich soils, and sandy soils. Results from this study are in italics.

(2.88×10^{-6} m s^{-1} to 2.28×10^{-7} m s^{-1}; Figure 1a, Table 2). Similar changes in K and ρ_b were reported by Deveraux et al. ([22]; Table 3), though this may be due to a decrease in porosity attributed to the relatively smaller d_{50} of the biochar. In our experiments, the median grain size of biochar was larger than the sand and therefore the decrease in ρ_b was accompanied by an increase in porosity.

In addition to grain size, how the biochar is mixed into the soils is also important. The incorporation of charcoal into sandy, Haplic Acrisols underlying kilns, resulted in a non-significant change (9%) in ρ_b and an 88% increase in K ([18]; Table 3). It is likely that the location of the charcoal within the soil column (only in the surface layer), the soil-charcoal layering, and the charcoal properties, significantly affected infiltration rates and other hydrologic properties. Furthermore, in cases where biochar amendment results in bulk density decreases and/or porosity increases, concomitant with a decrease in K (e.g. our results and [21]; (Table 3), changes in physical soil properties are not sufficient to explain the observed hydrologic changes.

The observed decreases in K despite the increased porosity and decreased ρ_b are likely due to the internal structure of biochar. The biochar had an average pore volume of 1.18 cm^3 g^{-1}, porosity of 0.62, and surface area of 6×10^5 cm^2 g^{-1} [59]. Thus the biochar has a greater porosity (typical sand has a porosity between 0.17 and 0.33 [60]) and surface area than our sand (based on d_{50}, sand = 140 cm^2 g^{-1}). The highly porous structure of biochar [59,61] creates two theoretical flow pathways, one in the interstitial space within the biochar-sand matrix and a second connecting the pores within the biochar. According to measurements made by Brewer et al. [59] the biochar we used is dominated (99%) by macropores (0.05 to 1000 µm), and therefore includes many pores larger than the diameter of a water molecule (0.28 nm). However, this second pathway likely has greater tortuosity and smaller median pore throat size due to the size of the smallest pores as well as their lack of complete connectivity [59]. While these pores contribute to the bulk density and total porosity of the mixture, they may not contribute to the effective porosity. In addition, the biochar grains likely create torturous interstitial space between the sand and biochar grains, further decreasing K. This mechanism assumes that the internal pores and surface of biochar are not hydrophobic. While we did not make hydrophobicity measurements on the mesquite biochar, multiple studies have reported that biochar amendments do not result in greater soil hydrophobicity [17,20]. Furthermore, Briggs et al. [12] found that laboratory-produced charcoals leached with distilled water and naturally-produced charcoal collected beneath leaf litter were less water repellent than non-leached and surficial charcoal, respectively; suggesting that the hydrophobic surface compounds may be easily removed.

A second mechanism driving decreases in K could be related to the high field capacity of biochar [8]; i.e. water may have continued to sorb to biochar particles, contributing to the apparent decrease in K for some of the soil mixtures (Figure 1a). While partial saturation of column materials is possible, the measured K of sand-only systems from constant-head and falling-head experiments were similar, suggesting full saturation and equilibrium conditions for both experiments. Saturated hydraulic conductivity changed with repeated flushing events in four of the six soil treatments ($p<0.01$) (Figure 1). The change in K with flushing events was only dependent on the presence of biochar in the case of the organic-rich soil ($F = 7.366$, $p = 0.01$), i.e. while K changed over time in the unamended organic soil, the change was greater in organic+biochar columns (Figure 1b). These shifts in K could be related to physical mechanisms, such as swelling and grain segregation, leading to the clogging of pores, decrease in pore radii, and possibly a variation in bulk density and sample heterogeneity over the course of the experiment. Comparing soil column heights from before and after flushing events revealed that all materials swelled between 6% (sand) and 21% (organic+ biochar) suggesting that bulk density changed over the course of the experiments; but biochar addition did not result in statistically different amounts of swelling. Visual observations indicate that biochar particles moved upwards within the columns during the experiment, consistent with observations made by Wang et al. [62]; however, these changes were not quantified and require further study to determine how this movement is related to flow in soil systems and how it impacts hydraulic properties.

The addition of biochar significantly ($p<0.01$) increased the water content at field capacity of the sand and clay. On average biochar amendment doubled the water content at field capacity in the sand and increased it by 20% in clay (Table 2). Several studies have reported increased water content in biochar-amended soils: e.g. organic-rich Mollisols [56], Amazonian *terra preta* [8], sandy loams [13,16,21,22], and silt loams [20]. The increased water holding capacity is likely the result of the internal porosity of the biochar and grain-to-grain interactions; interactions that are highly dependent on soil type and biochar production conditions and are likely the reason that other studies point to mixed results [63]. The organic soil's water content at field capacity did not change with biochar addition, likely due to the already high water holding capacity of the organic soil (Table 2).

Soil carbon and nitrogen dynamics

The addition of biochar significantly ($p<0.05$) increased the %C, by mass in all soil materials: sand increased from 0.40 ± 0.35 to 6.98 ± 0.5%C, clay-rich from 0.90 ± 0.06 to 8.33 ± 0.21%C, and organic-rich from 37.85 ± 3.25 to 42.47 ± 0.30%C. Similar patterns were seen for the N content of the soil materials, though not all changes were significant: %N, by mass, increased from 0.01 ± 0.0 to 0.06 ± 0.03%N for sand, 0.03 ± 0.0 to 0.11 ± 0.02%N for clay, and 0.51 ± 0.01 to 0.54 ± 0.06%N for organic-rich soil.

While biochar addition increased the amount of C in all soils, it did not have a universal effect on the C loss as soil leachate (i.e. DOC flux), suggesting that the release of DOC from biochar-amended soils varies with soil type. When added to the C-poor sand, biochar amendments significantly increased the aromaticity and cumulative loss of DOC (Table 2). The increase in aromaticity of the leached DOC from the biochar treatments suggests that biochar-derived C, and not native C, was lost. DOC fluxes decreased with each subsequent flushing event, suggesting that the easily leachable biochar-C was quickly depleted.

In contrast to sand, the addition of biochar to the organic- and clay-rich soils did not increase the cumulative DOC loss (Table 2). When added to C-rich organic soil, biochar significantly decreased the cumulative DOC loss and did not change the aromaticity (Table 2). This suggests that while the biochar added leachable C to the soil (as observed in the sand+biochar treatment); it is also capable of sorbing soil C. While there was no significant change in the magnitude of DOC lost when biochar was added to clay-rich soil, the aromaticity of the leachate was significantly greater when biochar was present, suggesting that while biochar-derived leachable DOC was lost, additional soil-derived C was retained within the soil-biochar matrix.

The partitioning of biochar-C between soil sequestration and leachate is partially dependent on soil type. Approximately 0.05% of the added biochar-C was lost from the sand+biochar mixture, an amount on the low end of the cumulative losses (0.1 to 1.7% of added C) reported by Mukherjee and Zimmerman [37] for similar short-term laboratory experiments. Furthermore, the $SUVA_{254}$ measurements indicate that while biochar-C is preferentially lost with water flow-through in sandy soils, there was also a net sorption of soil-DOC in both the clay and organic-rich soils. Thus the risk of a net increase of DOC export associated with biochar soil amendments may be minimal in soils with moderate amounts of clay, silt or organic material, a conclusion that mirrors field trials and modeling results [24,64,65].

Given the potential use of biochar to improve soil fertility in arid, drought-prone environments with C-poor sandy soils (e.g. [21] and references therein) it is important to quantify the partitioning of C between soil and leachate in sandy soils. In particular, while our results suggest that the majority of biochar-C remains in the sandy soil, other studies have shown that this partitioning is dependent on pore water and biochar characteristics [37,66]. For example, Bruun and others [66] found that there is a tradeoff between increased water holding capacity and increased biochar-C leachate in sandy soils; the amount of C lost was dependent on production conditions, with greater losses seen when fast- versus slow-pyrolysis biochar was added to soils. Given the results from these experiments and recent analyses of DBC in rivers [5,6] biochar amendments could change the chemical composition of DOC exported to downstream aquatic systems. The ecosystem impacts of this change in the chemical composition of the DOC pool remain unclear [6] and require further study.

Many studies have shown that biochar is able to retain inorganic N, reducing the nitrate flux from soils [29,33,67–69], and biochar-amended sand and clay retained more N in our experiments, resulting in a 24% ($p = 0.081$) and 62% ($p = 0.078$) reduction in N losses (Table 2); though these differences were not statistically significant. The well documented sorptive properties of biochar are attributed to surficial carboxylic and other acid functional groups that provide the cation and anion exchange capacity of biochar [32,70,71] and are dependent on production temperatures and feedstock [30]. The lack of statistical difference with biochar amendment in our study is likely due to the relatively low N content of the soil materials and biochar (<0.8%, Table 1), the only sources of N within our columns. Thus while the results of our experiments are in line with past research, it is important to note that they do not represent the full potential of biochar to mitigate N leaching from agricultural fields.

Conclusions

Biochar-associated changes in K and field capacity have implications for infiltration rates and plant water availability. As shown by our experiments, the addition of biochar to coarser soils decreases K, indicating the potential to decrease crop water stress and reduce nutrient loss below the rooting zone [29]. Conversely, biochar is able to increase porosity and permeability in fine-grained clay soils, making them more suitable for crop growth by increasing infiltration rates. Our results, combined with those of other studies, strongly support the argument that biochar addition increases the water holding capacity in coarse-grained soils, likely improving plant water availability. The saturated hydraulic conductivity (K) results of these laboratory and field experiments should be considered short-term, as reported changes in soil structure and hydrology are likely to evolve with time, impacting K. Brodowksi et al. [72] showed that biochar degrades into silt-sized particles over time likely changing the porosity and K of the amended soil. Other mechanisms may additional act to alter K on longer timescales, including increased microbial activity and increased bioturbation. Increased microbial activity and the addition of OM associated with biochar amendments has been shown to increase soil aggregation and macropore volume, thereby increasing K [73]. Increased bioturbation associated with biochar amendment can also increase K [74]. These examples provide further impetus for examining the effects of biochar particle size on soil hydraulic properties in the near- and long-term and in both laboratory/greenhouse and field settings.

Our experiments also illustrate that biochar addition to soils can add and sorb leachable DOC, and potentially add aromatic DOC to rivers [5]. Furthermore, biochar has the potential to reduce TDN in leachates, mitigating environmental impacts of agricultural N pollution. These changes in biogeochemical cycling, as well as alterations in greenhouse gas fluxes, plant productivity, and microbial activity are inherently linked to soil hydraulic properties. It is therefore crucial that future research addresses the complex interactions between biochar amendment and soil hydrology, C and N cycling, as well as how different soil types, biochars (varying feedstock, particle size, and production conditions) and amendment rates control these processes.

Supporting Information

Table S1 Average saturated hydraulic conductivity results for soil materials with and without biochar. The average and standard deviation saturated hydraulic conductivity (K) for each flushing experiment (n = 3 for each soil material) in m/s. The duration of the experiment (in hours) is also provided.

Acknowledgments

We thank F. Liljestrand, A. Woda, A. Hilton, and D. Leahy for help with experiments, soil processing and data collection; and V. Chuang for biochar production.

Author Contributions

Conceived and designed the experiments: RTB MEG CAM BD. Performed the experiments: RTB MEG ZL. Analyzed the data: RTB MEG. Contributed reagents/materials/analysis tools: CAM BD. Wrote the paper: RTB. Significant help revising the manuscript: MEG CAM BD.

References

1. Woolf D, Amonette JE, Street-Perrott FA, Lehmann J, Joseph S (2010) Sustainable biochar to mitigate global climate change. Nature Communications 1: 1–9.
2. Masiello CA (2004) New directions in black carbon organic geochemistry. Marine Chemistry 92: 201–213.
3. Lehmann J, Gaunt J, Rondon M (2006) Bio-char sequestration in terrestrial ecosystems - A review. Mitigation and Adaptation Strategies for Global Change 11: 403–427.
4. Kuhlbusch TAJ, Crutzen PJ (1996) Black Carbon, the Global Carbon Cycle, and Atmospheric Carbon Dioxide. In: Levine JS, editor. Biomass Burning and Global Change: Remote Sensing, Modeling and Inventory Development, and Biomass Burning in Africa. Cambridge, MA: Massachusetts Institute of Technology. pp. 160–169.
5. Dittmar T, de Rezende CE, Manecki M, Niggemann J, Ovalle ARC, et al. (2012) Continuous flux of dissolved black carbon from a vanishing tropical forest biome. Nature Geoscience 5: 618–622.
6. Jaffe R, Ding Y, Niggemann J, Vahatalo AV, Stubbins A, et al. (2013) Global Charcoal Mobilization from Soils via Dissolution and Riverine Transport to the Oceans. Science 340: 345–347.
7. Stubbins A, Spencer RGM, Chen H, Hatcher PG, Mopper K, et al. (2010) Illuminated darkness: Molecular signatures of Congo River dissolved organic matter and its photochemical alteration as revealed by ultrahigh precision mass spectrometry. Limnology and Oceanography 55: 1467–1477.
8. Glaser B, Lehmann J, Zech W (2002) Ameliorating physical and chemical properties of highly weathered soils in the tropics with charcoal - a review. Biology and Fertility of Soils 35: 219–230.

9. Kammann C, Ratering S, Eckhard C, Müller C (2012) Biochar And Hydrochar Effects On Greenhouse Gas (carbon Dioxide, Nitrous Oxide, And Methane) Fluxes From Soils. J Environ Qual 41: 1052–1066.
10. Novak JM, Busscher WJ, Watts DW, Amonette JE, Ippolito JA, et al. (2012) Biochars Impact on Soil-Moisture Storage in an Ultisol and Two-Aridisols. Soil Science 177: 310–320.
11. Gaskin JW, Speir A, Morris LM, Ogden L, Harris K, et al. (2007) Potential for Pyrolysis Char to Affect Soil Moisture and Nutrient Status of a Loamy Sand Soil. March 27–29; University of Georgia.
12. Briggs C, Breiner JM, Graham RC (2012) Physical and chemical properties of *Pinus ponderosa* charcoal: implications for soil modification. Soil Science 177: 263–268.
13. Tryon EH (1948) Effects of Charcoal on Certain Physical, Chemical, and Biological Properties of Forest Soils. Ecological Monographs 18: 81–115.
14. Baronti S, Vaccari FP, Miglietta F, Calzolari C, Lugato E, et al. (2014) Impact of biochar application on plant water relations in *Vitis vinifera* (L.). European Journal of Agronomy 53: 38–44.
15. Kammann CI, Linsel S, Gößling JW, Koyro H-W (2011) Influence of biochar on drought tolerance of *Chenopodium quinoa* Willd and on soil-plant relations. Plant and Soil.
16. Bruun EW, Petersen CT, Hansen E, Holm JK, Hauggaard-Nielsen H (2014) Biochar amendment to coarse sandy subsoil improves root growth and increases water retention. Soil Use and Management 30: 109–118.
17. Kinney TJ, Masiello CA, Dugan B, Hockaday WC, Dean MR, et al. (2012) Hydrologic properties of biochars produced at different temperatures. Biomass and Bioenergy 41: 34–43.
18. Oguntunde PG, Abiodun BJ, Ajayi AE, van de Giesen N (2008) Effects of charcoal production on soil physical properties in Ghana. Journal of Plant Nutrition and Soil Science 171: 591–596.
19. Asai H, Samson BK, Stephan HM, Songyikhangsuthor K, Homma K, et al. (2009) Biochar amendment techniques for upland rice production in Northern Laos 1. Soil physical properties, leaf SPAD and grain yield. Field Crops Research 111: 81–84.
20. Herath HMSK, Camps-Arbestain M, Hedley M (2013) Effect of biochar on soil physical properties in two contrasting soils: An Alfisol and an Andisol. Geoderma 209–210: 188–197.
21. Uzoma KC, Inoue M, Andry H, Zahoor A, Nishihara E (2011) Influence of biochar application on sandy soil hydraulic properties and nutrient retention. Journal of Food, Agriculture & Environment 9: 1137–1143.
22. Deveraux RC, Sturrock CJ, Mooney SJ (2012) The effects of biochar on soil physical properties and winter wheat growth. Earth and Environmental Science Transactions of the Royal Society of Edinburgh 103: 13–18.
23. Atkinson CJ, Fitzgerald JD, Hipps NA (2010) Potential mechanisms for achieving agricultural benefits from biochar application to temperate soils: a review. Plant and Soil 337: 1–18.
24. Major J, Lehmann J, Rondon M, Goodale CL (2010) Fate of soil-applied black carbon: downward migration, leaching and soil respiration. Global Change Biology 16: 1366–1379.
25. Mishera SK, Tyagi JV, Singh VP (2003) Comparison of infiltration models. Hydrological Processes 17: 2629–2652.
26. Krull ES, Baldock JA, Skjemstad JO, Smernik RJ (2009) Characteristics of biochar: organo-chemical properties. In: Lehmann J, editor. Biochar for environmental management science and technology. London: Earthscan. pp. 53–65.
27. Baldock JA, Smernik RJ (2002) Chemical composition and bioavailability of thermally altered *Pinus resinosa* (Red pine) wood. Organic Geochemistry 33: 1093–1109.
28. Schneider MPW, Hilf M, Vogt UF, Schmidt MWI (2010) The benzene polycarboxylic acid (BPCA) pattern of wood pyrolyzed between 200°C and 1000°C. Organic Geochemistry 41: 1082–1088.
29. Major J, Steiner C, Downie A, Lehmann J (2009) Biochar effects on nutrient leaching. In: Lehmann J, Joseph S, editors. Biochar for environmental management, science, and technology. London: Earthscan. pp. 271–287.
30. Lehmann J (2007) Bio-energy in the black. Frontiers in Ecology and the Environment 5: 381–387.
31. Cheng C-H, Lehmann J, Engelhard MH (2008) Natural oxidation of black carbon in soils: Changes in molecular form and surface charge along a climosequence. Geochimica et Cosmochimica Acta 72: 1598–1610.
32. Liang B, Lehmann J, Solomon D, Kinyangi J, Grossman J, et al. (2006) Black carbon increases cation exchange capacity in soils. Soil Science Society of America Journal 70: 1719–1730.
33. Steiner C, Glaser B, Teixeira WG, Lehmann J, Zech W (2008) Nitrogen retention and plant uptake on a highly weathered Amazonian Ferralsol amended with compost and charcoal. Journal of Plant Nutrition and Soil Science 171: 893–899.
34. Yu L, Tang J, Zhang R, Wu Q, Gong M (2013) Effects of biochar application on soil methane emission at different soil moisture levels. Biology and Fertility of Soils 49: 119–128.
35. Mukome FND, Six J, Parikh SJ (2013) The effects of walnut shell and wood feedstock biochar amendments on greenhouse gas emissions from a fertile soil. Geoderma 200–201: 90–98.
36. Angst TE, Patterson CJ, Reay DS, Anderson P, Peshkur TA, et al. (2013) Biochar Diminishes Nitrous Oxide and Nitrate Leaching from Diverse Nutrient Sources. Journal of Environmental Quality 42: 672–682.
37. Mukherjee A, Zimmerman AR (2013) Organic carbon and nutrient release from a range of laboratory-produced biochars and biochar-soil mixtures. Geoderma 193–194: 122–130.
38. Norwood MJ, Louchouarn P, Kuo L-J, Harvey OR (2013) Characterization and biodegradation of water-soluble biomarkers and organic carbon extracted from low temperature chars. Organic Geochemistry 56: 111–119.
39. Zhang W, Niu J, Morales VL, Chen X, Hay AG, et al. (2010) Transport and retention of biochar particles in porous media: effect of pH, ionic strength, and particle size. Ecohydrology 3: 497–508.
40. Wang D, Zhang W, Hao X, Zhou D (2013) Transport of Biochar Particles in Saturate Granular Media: Effects of Pyrolysis Temperature and Particle Size. Environmental Science & Technology 47: 821–828.
41. Wheeler FF (1976) Soil Survey of Harris County, Texas. United States Department of Agriculture, Soil Conservation Service. pp. 152.
42. ASTM International (2010) D2216-10, Standard Test Methods for Laboratory Determination of Water (Moisture) Content of Soil and Rock by Mass.
43. ASTM International (2007) D1762-84, Standard Test Method for Chemical Analysis of Wood Charcoal.
44. Rayment GE, Lyons DJ (2010) Soil Chemical Methods. Melbourne: CSIRO.
45. Rajkovich S, Enders A, Hanley K, Hyland C, Zimmerman A, et al. (2011) Corn growth and nitrogen nutrition after additions of biochars with varying properties to a temperate soil. Biology and Fertility of Soils.
46. ASTM International (2010) D5084-10 Standard Test Methods for Measurement of Hydraulic Conductivity of Saturated Porous Materials Using a Flexible Wall Perimeter.
47. Oliveira IB, Demond AH, Salehzadeh A (1996) Packing of sands for the production of homogeneous porous media. Soil Science Society of America Journal 60: 49–53.
48. Chan KY, Van Zwieten L, Meszaros I, Downie A, Joseph S (2007) Agronomic values of greenwaste biochar as a soil amendment. Australian Journal of Soil Research 45: 629–634.
49. Klute A (1986) Methods of soil analysis. Part 1. Physical and mineralogical methods. Madison, WI: American Society of Agronomy.
50. Weishaar JL, Aiken GR, Bergamaschi BA, Fram MS, Fujii R, et al. (2003) Evaluation of specific ultraviolet absorbance as an indicator of the chemical composition and reactivity of dissolved organic carbon. Environmental Science & Technology 37: 4702–4708.
51. ASTM International (2006) D2434, Standard Test Method for Permeability of Granular Soils (Constant Head) In: Materials ASfTa, editor. West Conshohocken, PA.
52. Kozeny J (1927) Über kapillare Leitung des Wassers im Boden – Aufstieg, Versickerung und Anwendung auf die Bewässerung. Sitzungsberichte Akad Wiss Wien 136: 271–306.
53. Carmen PC (1937) Flow through a granular bed. Transactions of Institute of Chemical Engineers, London 15: 150–156.
54. Neuzil CE (1994) How permeable are clays and shales? Water Resources Research 30: 145–150.
55. Alyamani MS, Sen Z (1993) Determination of hydraulic conductivity from complete grain-size distribution curves. Groundwater 31: 551–555.
56. Laird DA, Fleming P, Davis DD, Horton R, Wang B, et al. (2010) Impact of biochar amendments on the quality of a typical Midwestern agricultural soil. Geoderma 158: 443–449.
57. Boadu FK (2000) Hydraulic conductivity of soils from grain-size distribution: New models. Journal of Geotechnical and Geoenvironmental Engineering 126: 739–746.
58. Crawford BR, Faulkner DR, Rutter EH (2008) Strength, porosity, and permeability development during hydrostatic and shear loading of synthetic quartz-clay fault gouge. Journal of Geophysical Research 113.
59. Brewer CE, Chuang VJ, Masiello CA, Gonnermann H, Gao X, et al. (2014) New approaches to measuring biochar density and porosity. Biomass and Bioenergy 66: 176–185.
60. Holtz RD, Kovacs WD (1981) An Introduction to Geotechnical Engineering. Upper Saddle River, NJ: Prentice Hall.
61. Sun H, Hockaday WC, Masiello CA, Zygourakis K (2012) Multiple Controls on the Chemical and Physical Structure of Biochars. Industrial and Engineering Chemistry Research 51: 3587–3597.
62. Wang C, Walter MT, Parlange J-Y (2013) Modeling simple experiments of biochar erosion from soil. Journal of Hydrology 499: 140–145.
63. Novak JM, Lima IM, Xing B, Gaskin JW, Steiner C, et al. (2009) Characterization of designer biochar produced at different tempeartures and their effects on a loamy sand. Annals of Environmental Science 3: 195–2006.
64. Jones DL, Rousk J, Edwards-Jones G, DeLuca TH, Murphy DV (2012) Biochar-mediated changes in soil quality and plant growh in a three year field trial. Soil Biology and Biochemistry 45: 113–124.
65. Foereid B, Lehmann J, Major J (2011) Modeling black carbon degradation and movement in soil. Plant and Soil 345: 223–236.
66. Bruun EW, Petersen CT, Strobel BW, Hauggaard-Nielsen H (2012) Nitrogen and carbon leaching in repacked sandy soil with added fine particulate biochar. Soil Science Society of America Journal 76: 1142–1148.
67. Novak JM, Busscher WJ, Watts DW, Laird DA, Ahmedna MA, et al. (2010) Short-term CO_2 mineralization after additions of biochar and switchgrass to a Typic Kandiudult. Geoderma 154: 281–288.

68. Bell MJ, Worrall F (2011) Charcoal addition to soils in NE England: A carbon sink with environmental co-benefits? Science of the Total Environment 409: 1704–1714.
69. Clough TJ, Condron LM (2010) Biochar and the Nitrogen Cycle: Introduction. Journal of Environmental Quality 39: 1218–1223.
70. Singh BP, Hatton BJ, Singh B, Cowie AL, Kathuria A (2010) Influence of biochars on nitrous oxide emission and nitrogen leaching from two contrasting soils. Journal of Environmental Quality 39: 1224–1235.
71. Asada T, Ishihara S, Yamane S, Toba T, Yamada A, et al. (2002) Science of bamboo charcoal: Study of carbonizing temperature of bamboo charcoal and removal capability of harmful gases. Journal of Health Sciences 48: 473–479.
72. Brodowski S, Amelung W, Haumaier L, Zech W (2007) Black carbon contribution to stable humus in German arable soils. Geoderma 139: 220–228.
73. Lei O, Zhang R (2013) Effects of biochars derived from different feedstocks and pyrolysis temperatures on soil physical and hydraulic properties. Journal of Soils and Sediments.
74. Hardie M, Clothier B, Bound S, Oliver G, Close D (2014) Does biochar influence soil physical properties and soil water availability? Plant and Soil 376: 347–361.

6

Gypsophile Chemistry Unveiled: Fourier Transform Infrared (FTIR) Spectroscopy Provides New Insight into Plant Adaptations to Gypsum Soils

Sara Palacio[1]*, Matt Aitkenhead[2], Adrián Escudero[3], Gabriel Montserrat-Martí[4], Melchor Maestro[4], A. H. Jean Robertson[2]

1 Instituto Pirenaico de Ecología (IPE-CSIC), Jaca, Huesca, Spain, **2** The James Hutton Institute, Aberdeen, Scotland, United Kingdom, **3** Biodiversity and Conservation Group, E.S.C.E.T., Universidad Rey Juan Carlos, Móstoles, Madrid, Spain, **4** Instituto Pirenaico de Ecología (IPE-CSIC), Zaragoza, Spain

Abstract

Gypsum soils are among the most restrictive and widespread substrates for plant life. Plants living on gypsum are classified as gypsophiles (exclusive to gypsum) and gypsovags (non-exclusive to gypsum). The former have been separated into wide and narrow gypsophiles, each with a putative different ecological strategy. Mechanisms displayed by gypsum plants to compete and survive on gypsum are still not fully understood. The aim of this study was to compare the main chemical groups in the leaves of plants with different specificity to gypsum soils and to explore the ability of Fourier transform infrared (FTIR) spectra analyzed with neural network (NN) modelling to discriminate groups of gypsum plants. Leaf samples of 14 species with different specificity to gypsum soils were analysed with FTIR spectroscopy coupled to neural network (NN) modelling. Spectral data were further related to the N, C, S, P, K, Na, Ca, Mg and ash concentrations of samples. The FTIR spectra of the three groups analyzed showed distinct features that enabled their discrimination through NN models. Wide gypsophiles stood out for the strong presence of inorganic compounds in their leaves, particularly gypsum and, in some species, also calcium oxalate crystals. The spectra of gypsovags had less inorganic chemical species, while those of narrow gypsum endemisms had low inorganics but shared with wide gypsophiles the presence of oxalate. Gypsum and calcium oxalate crystals seem to be widespread amongst gypsum specialist plants, possibly as a way to tolerate excess Ca and sulphate. However, other mechanisms such as the accumulation of sulphates in organic molecules are also compatible with plant specialization to gypsum. While gypsovags seem to be stress tolerant plants that tightly regulate the uptake of S and Ca, the ability of narrow gypsum endemisms to accumulate excess Ca as oxalate may indicate their incipient specialization to gypsum.

Editor: Heidar-Ali Tajmir-Riahi, University of Quebect at Trois-Rivieres, Canada

Funding: SP was funded by a Juan de la Cierva contract (Ministerio de Ciencia e Innovación, Spain) and projects [CGL2011-26654] (Ministerio de Economía y Competitividad, Spain) and [ARBALMONT/786-2012] (Organismo Autónomo de Parques Nacionales, Ministerio de Agricultura, Alimentación y Medio Ambiente, Spain). The funders had no role in study design, data collection and analysis, decision to publish, or preparation of the manuscript.

Competing Interests: The authors have declared that no competing interests exist.

* Email: s.palacio@ipe.csic.es

Introduction

The ability of plants to survive in substrates with limiting conditions for plant life has intrigued biologists since early times [1,2]. Gypsum soils are amongst the most widespread special substrates, extending over 100 million hectares [3,4], but yet they have received comparatively less attention than other substrates such as serpentines, saline or calcicolous soils [5]. Gypsum soils develop from gypsic rocks in arid and semi-arid areas where low precipitation prevents gypsum from being leached [6]. Together with the arid conditions, gypsum soils have particularly stressful physical and chemical properties for plant life including the presence of hard soil crusts, high mechanical instability, low soil porosity, extreme nutritional deficits, high concentration of sulphates and moderate salinity [7,8,9,10]. As a consequence, they are among the most restrictive soils for plants [6]. Nevertheless, the adverse conditions of gypsum soils contrast with the rich and specialized flora they shelter, comprising diverse arrays of narrow endemic and rare plants in arid and semiarid regions, many of which are threatened or endangered and constitute a global conservation biodiversity concern [6,11].

Depending on their specificity to gypsum soils, plants can be classified as gypsophiles, i.e. plants growing exclusively on gypsum substrates, or gypsovags, i.e. plants growing both in and out of gypsum [11]. Despite the recent efforts devoted to understand plant life on gypsum, the mechanisms displayed by plants to become competitive on gypsum soils are still not fully understood [5]. Early investigations showed that the chemical composition of gypsophiles and gypsovags differs, with the former showing higher concentration of certain nutrients (N, P, Ca, S) and total ashes than the latter [12,13,14]. Previous studies also showed that gypsophile seedlings (like *H. squamatum* and *L. subulatum*) show a higher ability than gypsovags to surpass the physical soil crust characteristic of gypsum soils, and hence are better adapted to cope with the physical limitations of gypsum soils [15]. More recently, however, it was found that gypsophiles could be segregated in two different groups of plants, with distinct chemical composition and different putative ecological strategies [16]. These authors found differences in the composition of narrowly and widely distributed gypsophiles, the former being more similar to the chemical composition of gypsovags. According to Gankin and Major's interpretation of edaphism origin [17], narrowly distributed gypsophiles (similarly to gypsovags) would fit the *refuge* model, being stress tolerant species not specifically adapted to gypsum soils that avoid competition in marginal soils, while widely distributed gypsophiles would fit the *specialist* model, being specifically adapted to gypsum and dramatically lowering their performance in other soils [16]. The increased concentration of Ca and S of widely distributed gypsophiles suggests the existence of certain (still unexplored) physiological adaptations to cope with the excess of calcium sulphate in gypsum soils [13,16]. The high concentrations of certain nutrients (such as N or P) of wide gypsophiles are also intriguing, as gypsum soils are inherently nutrient poor [18,19].

The results of the study by Palacio *et al.* [16] are crucial since they show that various chemical strategies are feasible for plants living exclusively on gypsum soils. However, they are solely based on the elemental composition of plant species. To know if the generality of their conclusions stands with more comprehensive analyses of plant biochemistry, biochemical fingerprinting techniques allowing for the identification of the functional chemical groups of plants are needed. Such approaches could also shed light on the biochemical and physiological adaptations of different groups of plants to survive on gypsum soils. For example, they could inform on the biochemical mechanisms of wide gypsophiles to accumulate S, Ca, N or P. Such information is critical to understand plant life on gypsum substrates.

Fourier transform infrared (FTIR) spectroscopy is a powerful tool for the chemical analysis of biological samples [20]. This technique offers a fast, cost-effective, non-destructive way of obtaining a biochemical fingerprint of samples, where the main functional groups and bonds can be identified, thus giving structural information on the chemical compounds present rather than just elemental information. Crucially, both organic and inorganic compounds have features in the FTIR spectrum, allowing not only identification of the main organic constituents of plant material but also characterisation of the forms of inorganic compounds present in the plants [21]. In addition, by comparing the replication of FTIR spectra for different plants of the same species grown in the same conditions and localities, the technique provides a means of assessing the variability of the chemical profile among individuals.

FTIR spectroscopy is a widely used tool in plant biological studies [20]. It has been used to detect changes in plant chemistry in relation to fertilization [22,23,24] or heavy metal addition [25,26,27]. FTIR spectra analysed through artificial neural networks (NN) show a high potential to discriminate biological samples. For instance, [28], [29] and [30] applied NN to FTIR spectral data in order to distinguish between plant species, using a range of different kinds of source material. Despite the huge potential and increasing use of FTIR in plant studies, it has never been applied to the study of edaphic endemisms before.

The aim of this study was to compare the identity of the main chemical groups present in the leaves of plants with different specificity to gypsum soils and to explore the ability of FTIR spectra analyzed with NN to discriminate groups of gypsum plants. We hypothesize that: 1) data obtained by FTIR spectroscopy will show a high correlation with the elemental composition of plants. Consequently, owing to the observed differences in the mineral composition of gypsum plants in previous studies, we further hypothesize that 2) narrow gypsum endemisms will show a similar FTIR spectral composition to gypsovags, while widely distributed gypsophiles will show distinct spectral features.

Materials and Methods

Species and study sites

Species and study sites generally followed [16], except for the addition of the gypsovag *Thymelaea tinctoria*, and included fourteen species for analysis: five gypsovags and nine gypsophiles. The latter comprised most woody gypsophile species of study areas, including five widely distributed gypsophiles and four narrow gypsum endemics (Table 1). The distinction between both types of gypsophiles was made according to the extent of their distribution area: widely distributed gypsophiles were species that showed a wide distribution range in the Iberian Peninsula, comprising almost every gypsum outcrop in this large territory, whereas narrow gypsum endemics were species that showed a limited number of populations, growing only in one of the gypsum areas of the Iberian Peninsula. In order to minimize bias due to variations in plant functional strategies, all study species were shrubs or sub-shrubs, which are prevalent growth forms in gypsum outcrops [6,31], and had a similar branch morphology and architecture (for further details see [16]). None of the species included in this study is endangered or protected [32].

Plant species were collected from the two more massive and distinctive gypsum outcrops of the Iberian Peninsula: Central Spain (Middle Tajo Basin, near Madrid) and NE Spain (Middle Ebro Basin, near Zaragoza). Samples were not replicated across both areas owing to the absence of qualitative differences in the spectra of samples of *Ononis tridentata* from different sites (Fig. S1) and the previously reported similar elemental composition of samples collected from both areas [16]. Study species and sampling sites are shown in Table 1, while climatic and edaphic features of each sampling site are in Table 2. All permits required for plant collection at Madrid Community were requested by AE. Permit for sampling of plant material in Aragón was issued by the Government of Aragón (Diputación General de Aragón, DGA) to GMM and SP.

Plant and soil sampling and processing

To avoid variability in the chemical composition of plants due to phenology, all plant material was collected during winter, when chemical concentrations of nutrients in the leaves of study plants are steady [16]. Five adult individuals were harvested in each studied population (Table 1). Once in the laboratory, a 15 g sample of the leaves of the short branches of each individual was collected and oven-dried to a constant weight at 60°C. Dry and damaged leaves were excluded from the analyses.

Chemical analyses

Samples were ground in a ball mill (Retsch Mixer MM301, Leeds, UK) to a fine powder. N and C concentrations were analyzed with an elemental analyzer (Elementar VarioMax N/CN, Hanau, Germany). Subsamples were burnt at 550°C for 4 hours and ash was dissolved in HNO3-HCl-H20 (1:3:9) and filtered. Concentrations of Na and K were measured in the soluble (silica-free) ash by flame photometry, Ca and Mg concentrations were determined by complexometry [33] and P concentration was assessed by vanado-molybdate colorimetry [34]. Total sulphur was analyzed by a turbidimetric method with barium chloride [33].

Table 1. Location and characteristics of study species and sites.

Species	Species code	Species type[a]	N[b]	Sampling sites[c]	UTM	Altitude (m a.s.l.)	Date of sampling	Subst.[d]
Centaurea hyssopifolia Vahl.	Ch	NE	4	Chinchón (M)	VK6247	640	15-Mar-10	G
Gypsophila struthium L. subsp. *hispanica* (Wilk.) G. López	Gh	G	5	Villamayor (Z)	XM8820	300	19-Feb-10	G
Helianthemum marifolium (L.) Mill. subsp. *conquense* Borja and Rivas Goday	Hco	NE	7	Tendilla (Cu)	WK4214	780	3-Mar-03	G
Helianthemum squamatum (L.) Pers.	Hsq	G	5	Villamayor (Z)	XM8718	290	21-Jan-03	G
Herniaria fruticosa L.	Hf	G	C	Villamayor (Z)	XM8820	300	20-Jan-03	G
							19-Feb-10	
					XM8718	290	21-Jan-03	G
Lepidium subulatum L.	Lsb	G	5	Villamayor (Z)	XM8820	320	13-Jan-03	G
							19-Feb-10	
Linum suffruticosum L.	Lsf	GV	5	Villamayor (Z)	XM8820	320	7-Jan-03	G
Ononis tridentata L.	Ot	G	5	Villamayor (Z)	XM8820	300	20-Jan-03	G
							19-Feb-10	
Rosmarinus officinalis L.	Ro	GV	5	Peñaflor (Z)	XM8626	300	24-Jan-03	G
Salvia lavandulifolia Vahl.	Sl	GV	5	Villamayor (Z)	XM8820	320	21-Jan-03	G
Teucrium polium L. subsp. *capitatum* (L.) Arcangelli	Tc	GV	5	Chinchón (M)	VK6247	640	6-Feb-03	G
Teucrium pumilum L.	Tp	NE	6	Chinchón (M)	VK6247	640	6-Feb-03	G
Thymelaea tinctoria (Pourr.) Endl. subsp. *tinctoria*				Villamayor (Z)			19-Feb-10	
Thymus lacaitae Pau	Tl	NE	5	Alcalá de Henares (M)	VK8282	710	28-Jan-03	G

[a]NE = Narrow endemism, G = gypsophile, GV = gypsovag.
[b]N = number of sampled individuals, C = composite sample.
[c]M = Madrid (Central Spain, Middle Tajo Basin), Cu = Cuenca (Central Spain, Middle Tajo Basin), Z = Zaragoza (NE Spain, Middle Ebro Basin).

Table 2. Climatic and edaphic characteristics of study sites.

Study sites	Area[a]	Climate		Soil				
		T (°C)	P (mm)	pH	SOM (%)	N (%)	$CaCO_3$ (%)	Gypsum (%)
Villamayor (Z)	Z	14.2	370	7.7	1.9	0.14	20.0	55.5
Peñaflor (Z)	Z	14.2	370	8.00	0.88	0.051	13.04	66.30
Alcalá de Henares (M)	M	13.2	350	7.8	1.8	0.15	20.6	32.0
Chinchón (M)	M	13.8	422	7.7	2.1	0.12	9.7	59.3
Tendilla (Cu)	M	13.0	410	7.8	2.4	0.17	22.3	29.3

Abbreviations: T = Mean annual temperature; P = total annual rainfall; SOM = soil organic matter; N = nitrogen.
[a]M = Madrid (Central Spain, Middle Tajo Basin), Z = Zaragoza (NE Spain, Middle Ebro Basin).

FTIR Spectroscopic Analysis

As for the other chemical analyses, dried samples for FTIR analysis were prepared by grinding in a ball mill (Retsch Mixer MM301, Leeds, UK) to a fine powder. FTIR spectra were recorded on a Bruker Vertex 70 FTIR spectrometer (Bruker, Ettlingen, Germany) fitted with a potassium bromide beam splitter and a deutroglycine sulphate detector from three replicates of each species and, where the replicates did not match closely, spectra were recorded for two further replicates. A Diamond Attenuated Total Reflectance (DATR) sampling accessory, with a single reflectance system, was used to produce "transmission-like" spectra. Samples were placed directly on a DATR/KRS-5 crystal, and a flat tip powder press was used to achieve even distribution and contact. Spectra were acquired by averaging 200 scans at 4 cm^{-1} resolution over the range 4000 – 370 cm^{-1}. A correction was made to the ATR spectra to allow for differences in depth of beam penetration at different wavelengths, using OPUS software (Bruker, Ettlingen, Germany, version 6.0). The spectra were also baseline corrected. No correction was required for water vapour and CO_2 as the spectrometer is continuously flushed with dry air.

Statistical analyses

Predictions of chemical properties from the FTIR spectra. The relationship between FTIR spectra and the results of the chemical analyses was evaluated through Partial Least Squares (PLS) regression, using the OPUS QUANT version 6.0 software (Bruker, Ettlingen, Germany). Data were available for C (%), N (mg/g), C/N, Na (mg/g), K (mg/g), Ca (%), Mg (mg/g), P (mg/g), S (%) and Ash (%) (Table S1). The validation was done by cross validation, which allows the maximum representation of the samples, essential for such a small dataset. Correlation analyses between "predicted" and "true" values were used to assess the relationships between spectral information and elemental composition of plant samples (R^2, RMSECV (Root Mean Square Error of Cross Validation), bias and RPD (SD/SECV), Regression Point Displacement, which equals the standard deviation in the field data values divided by the standard error in the Cross Validation error values). There was insufficient representation of all types of species to enable robust calibrations to be developed for prediction of unknown samples, but the data should give a good indication of the extent to which these parameters are related to the IR spectral features.

Neural network analysis. The ability of FTIR spectra to discriminate groups of plants with different specificity to gypsum was evaluated by neural network (NN) analyses. Neural networks are particularly useful when the relationships between the input data (in this case the FTIR spectra) and the output parameters (plant species classification) are complex and/or unknown (e.g. [35]). They are also useful when applied to datasets with a large number of input parameters, such as infrared spectra [36]. In order to train the neural network model, the FTIR spectral data was pre-processed to make it more acceptable for NN training. This was carried out in two stages. Firstly, a moving window averaging process was applied to each spectrum, with the moving average at each point subtracted from the actual value. This had the effect of 'flattening' the spectra and setting the value of flatter sections of the curve to zero (Fig. S2). The moving window width was set at 101 (averaged over a window that extended 50 values to the left and 50 to the right from each value), following trial and error to optimise the performance of the neural network model. Secondly, the maximum absolute value within each derived spectrum was determined, and the value at each point was divided by this. This had the effect of normalising the spectra within the range [−1, 1] (Fig. S2).

The neural network architecture had four layers, with one input layer of 1882 nodes, two hidden layers each of 100 nodes and one output layer of 3 nodes. The neural network was fully connected (i.e. each node in the input layer was connected to each node in the first hidden layer, and so on). The training rate of the network was set at 0.05, and a total of 10,000 training epochs were used to train the network. These parameter values were selected using trial and error to optimise the model performance. The training algorithm used was the feed-forward back-propagation, a standard method that has been applied in many cases. For further details on this algorithm as applied to environmental data, see [37].

The training data consisted of 45 spectra, which is a relatively small number of data points for training a neural network. To ensure that overfitting of the data was not an issue, a separate neural network was trained for each of the data points, using all of the data except that point and using the one left out for testing purposes. This 'leave-one-out' approach ensures that training and testing data do not overlap, and provides more accurate information about how well the model is performing under conditions of low training data numbers. For each data point, the network was trained to predict a value of 1 for output node corresponding to the plant type being presented, and a value of 0 for the other output nodes.

Results

Description of FTIR spectra of the different groups

Although the spectra of each species were different and were interpreted individually, this provided a large amount of potentially useful information. Therefore, the results have been summarised for each of the three groups of species analyzed, i.e. "wide gypsophiles", "narrow gypsum endemisms" and "gypsovags", and in Table 3 the frequencies and assignments of the main peaks identified are presented.

Overall, the major difference between the spectra of wide gypsophiles and the other two groups is the presence of appreciable gypsum in their spectra (with one exception - see below) and often considerable oxalate too (Figs. 1, 2, and 3, Table 3). In general, the spectra of gypsovags had a lowest proportion of inorganic compounds present in them, while the spectra of narrow gypsum endemisms had a higher proportion, and mostly show the presence of oxalate (Figs. 1, 2, and 3, Table 3). However, for a number of species there was very little difference between the gypsovag and narrow gypsum endemism categories (see below for details).

Gypsovags. Replication of the FTIR spectra obtained from gypsovags was generally good (data not shown) with subtle variations in intensities of bands, indicating similar overall chemical profiles within each species. They all exhibited sharp CH_2 stretching bands, (indicating the presence of long carbon chains) and the same basic spectral patterns with features relating to protein (amide), polysaccharide (cellulose, pectin), lignin, lipids and other long chain hydrocarbons, as might be expected in plant spectra (Fig. 1, Table 3). A feature of the spectra of *R. officinalis* and *S. lavandulifolia* was a distinct band at 1686 cm^{-1}, which is possibly an aromatic ester (Fig. 1, Table 3). Notably, in comparison with the spectra of the species in the other two groups, the spectra of gypsovags had the least bands attributable to inorganic components (although all species may have traces). Among the few exceptions were the evidence for calcite, and possibly ammonium sulphate, in the spectrum of *L. suffruticosum*, and oxalate in the spectrum of *T. polium* subsp. *capitatum* (Fig. 1, Table 3).

Narrow gypsum endemisms. Replication of the FTIR spectra of narrow gypsophiles was also good, again implying consistent chemical composition within each species. The spectra of this group of species generally appeared to have greater evidence for the presence of inorganic compounds than those of the gypsovag group, although the spectra still showed predominantly features expected from plant material (Fig. 2, Table 3). Apart from *C. hyssopifolia*, where it was not readily detected, the spectra of narrowly distributed gypsophile species all appeared to show the presence of some oxalate. Despite this common feature, there was variation among narrow gypsum endemism in the intensity of the CH_2 stretching bands in their spectra (indicative of long carbon chains and sometimes referred to as a "waxy" character, [38]). Two of the species (*C. hyssopifolia* and *H. marifolium* subsp. *conquense*) had much weaker CH_2 absorption in their spectra compared with the spectra of the other two (*T. pumilum* and *T. lacaitae*) which appeared to have stronger absorption, much more similar to the spectra of species in the gypsovag group (Figs. 1–2, Table 3). In fact, *T. pumilum* and *T. lacaitae* had spectra generally quite closely related to those of gypsovags; in particular the spectrum of *T. pumilum* quite closely resembled that of its congener *T. polium* subsp. *capitatum* (including the presence of oxalate). The spectrum of *T. lacaitae* also resembled that of *T. polium* subsp. *capitatum*, and it had also evidence for a band at 1686 cm^{-1}(sh), compatible with an aromatic ester, as indicated above for the spectra of other Lamiaceae (*R. officinalis* and *S. lavandulifolia*) of the gypsovag group (Figs. 1–2, Table 3). *C. hyssopifolia* had a spectrum more reminiscent of that of *L. suffruticosum* than the spectra of the other narrow gypsophile species (with a high proportion of protein/amide bands and other bands present which could be assigned to ammonium sulphate or carboxylate). Finally, *H. marifolium* subsp. *conquense* had a spectrum that generally differed from the other narrow gypsophiles, and spectral searches suggested that the presence of tannic acid is a possibility. However it had the oxalate present and relatively strong protein/amide bands, as seemed to be common in the rest of plants from this group (Figs. 1–2, Table 3).

Wide gypsophiles. The replication of spectra within this group was noticeably less tight, with variations in intensity of bands, particularly those arising from inorganic components, rather than different bands being present (i.e. same chemistry but different proportions). A distinct feature of the spectra of this group of plants was the relatively high dominance of bands related to inorganic compounds (Fig. 3, Table 3), with clear bands for gypsum in all but one of them (*L. subulatum*). The presence of crystalline gypsum could be confirmed, even in the presence of other sulphates, by shoulders on the broad OH stretch at 3518 and 3398 cm^{-1} (Fig. 3, Table 3). These bands were similar to those detected in the spectra of soil samples collected at the sampling site (data not shown) and had not been identifiable in any of the previous spectra of the gypsovag and narrow gypsophile groups. Oxalate was also strongly present in the spectra of two of the wide gypsophiles analyzed (*H. fruticosa* and *G. struthium* subsp. *hispanica*) and some small traces were also visible in *H. squamatum* and *O. tridentata*, with the most diagnostic band being at 1315 cm^{-1} (Fig. 3, Table 3). Although *L. subulatum*, had a spectrum where inorganic compounds were not so immediately obvious, a broad absorption band in its spectrum at ~1600 cm^{-1} could possibly be ammonium sulphate or carboxylate, and sulphates other than gypsum may also be present (Fig. 3, Table 3). This spectrum also had the strongest bands relating to amide/protein.

Bands expected from plant material (such as cellulose, lignin, pectin, protein etc) were also present in spectra of wide gypsophiles, but were often obscured by the inorganic compounds

Table 3. Frequencies and assignments of the main peaks identified in the analysis of FTIR spectra.

Wave number, cm^{-1}	Assignment	Characterisation
3522,3400	O-H stretching	Gypsum[1, 2]
3340	O-H stretching	Cellulose, in samples with defined 3340 peak[3]
2920	antisymmetric CH_2 stretching	Fats, wax, lipids[3]
2850	symmetric CH_2 stretching	Fats, wax, lipids[3]
1740-1720	C=O stretch of COOR	Esters[3, 4]
1707-1710	C=O stretch of COOH	Carboxylic acids[3, 4]
1653	C=O of amide I	Proteinaceous origin[3]
1615	C-O stretching	Calcium oxalate[5]
1600-1650 (1610)	Aromatic C=C stretching and/or asymmetric C-O stretch in COO-	Lignin and other aromatics, or aromatic or aliphatic carboxylates[3]
1550	N-H in plane (amide-II)	Proteinaceous origin[3]
1505-1515	Aromatic C=C stretching	Lignin/Phenolic backbone[3, 4]
1450-1410	C-O stretching	Calcium carbonate[4, 6]
1426	Symmetric C-O stretch from COO- or stretch and OH deformation (COOH)	Carboxylate/Carboxylic structures (humic acids)[3]
1371, 1450	C-H deformations	Phenolic (lignin) and aliphatic structures[3]
1312	C-O stretching	Calcium oxalate[5]
1265 (approx.)	C-O stretching of phenolic OH and/or arylmethylethers	Indicative of lignin backbone[3]
1265 -1240	C-O-C stretching	Esters
	C-N stretching	Amide III[4]
1140-1080	S-O stretching	Sulphates[4]
1100-1000	P-O stretching	Phosphates[6]
1100-950	Si-O stretching	Silicates[4, 6]
1050 (1030-1080)	Combination of C-O stretching and O-H deformation	Polysaccharides[3]
874	C-O in plane bending	Calcium carbonate[4, 6]
835	Aromatic CH out of plane	Lignin[3]
780	COO bending	Calcium oxalate[5]
720	CH_2 wag	Long chain (> C4) alkanes[3]
715	C-O in plane bending	Calcium carbonate[4, 6]
680-610	S-O bending	Sulphates[4]
669,597	S-O bending	Gypsum[1, 2]

[1][39]; [2] [56]; [3] [38]; [4] [21]; [5] [57]; [6] [58].

(Fig. 3, Table 3). Absorption in the region between 1200 - 1000 cm^{-1} could arise from a range of inorganic compounds rather than polysaccharide but this region is very hard to interpret as sulphate, phosphate, polysaccharide and silicate groups all absorb in this frequency range and often have overlapping bands [21,38,39].

Correlation between FTIR spectral data and chemical analyses

In general, FTIR data adjusted reasonably well (very well for some cases) to the concentration of most nutrients in plants (Table 4). Correlation coefficients were above 90.0 for ashes, carbon and sulphur, and above 79.0 for Mg, P, Ca, Na and N. The different forms of N in different species, particularly those in the wide gypsophile group, may result in poorer correlation overall than might be the case if they are separated out, but there were not enough cases to represent each type of N form separately and not enough replicates to separate out each group (Table S1). The main exception was K, which implies that there was not such a clear cut relationship between spectral features and the variance in values of K for the samples analysed.

Suitability of FTIR spectra to discriminate ecological strategies of gypsum plants

The neural network (NN) approach was successful in correctly assigning the majority (39 out of 44) of the cases analyzed (Table 5). Predictive accuracies were assessed in terms of the error between target and actual values at each output node (RMSE), and are given in Table 5. RMSE values for gypsovag and narrow endemisms were high. However, they were not as high as would be obtained through purely random output values, for which RMSE values greater than 0.4 would be expected. The RMSE value for gypsophiles was much lower. In general, the NN models built were able to discriminate between the three groups of plants (Table 5). The wide gypsophile category was most accurately predicted with 15 data points correctly predicted out of 15 points analysed. It was followed by gypsovags, with 15 data points correctly predicted out of 17, and narrow gypsum endemisms or narrow gypsophiles, with

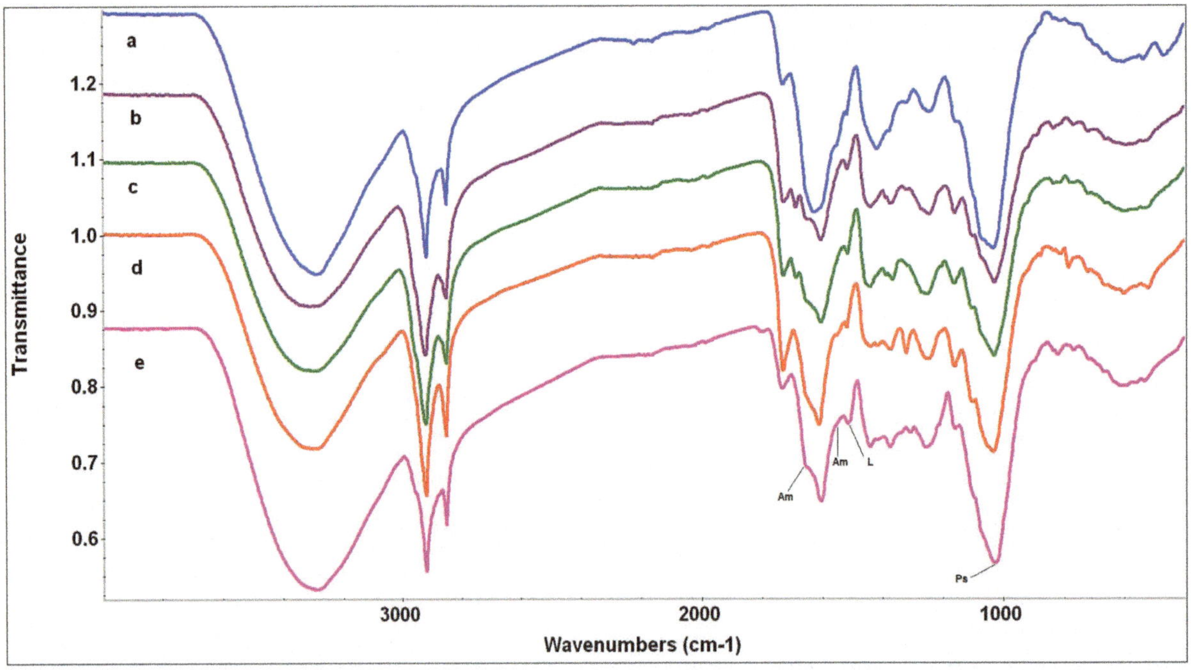

Figure 1. FTIR spectra of gypsovag species. The scale of the Y axis is offset for illustration purposes. a) *Linum suffruticosum* (blue), b) *Rosmarinus officinalis* (purple), c) *Salvia lavandulifolia* (green), d) *Teucrium polium* subsp. *capitatum* (red), e) *Thymelaea tinctoria* (pink). Abbreviations of compound peaks: Am: Amide, L: Lignin/Phenolic backbone, Ps: Polysaccharides.

9 data points correctly predicted out of 12, and therefore the least accurately predicted group.

Of the five data points that were incorrectly identified, none were from the wide gypsophile category and no other species from either of the other two categories was incorrectly classified as a

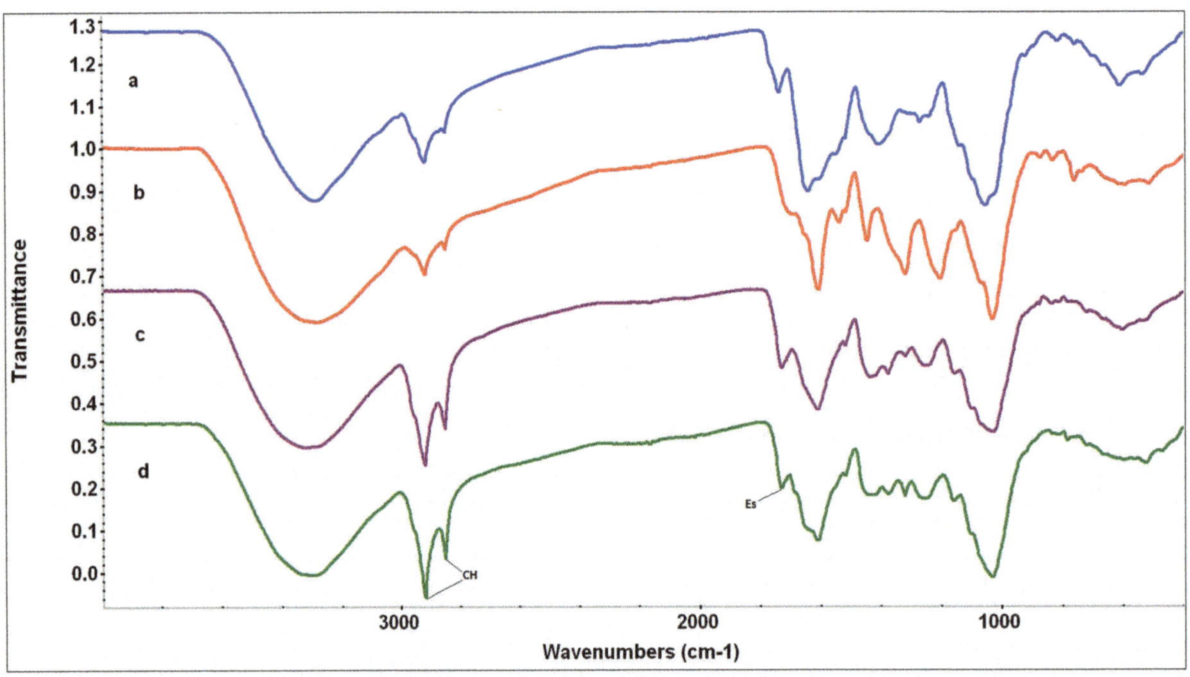

Figure 2. FTIR spectra of narrow gypsum endemisms. The scale of the Y axis is offset for illustration purposes. a) *Centarurea hyssopifolia* (blue), b) *Helianthemum marifolium* subsp. *conquense* (red), c) *Teucrium pumilum* (purple), d) *Thymus lacaitae* (green). Abbreviations of compound peaks: CH: CH-stretching, Es: Esters.

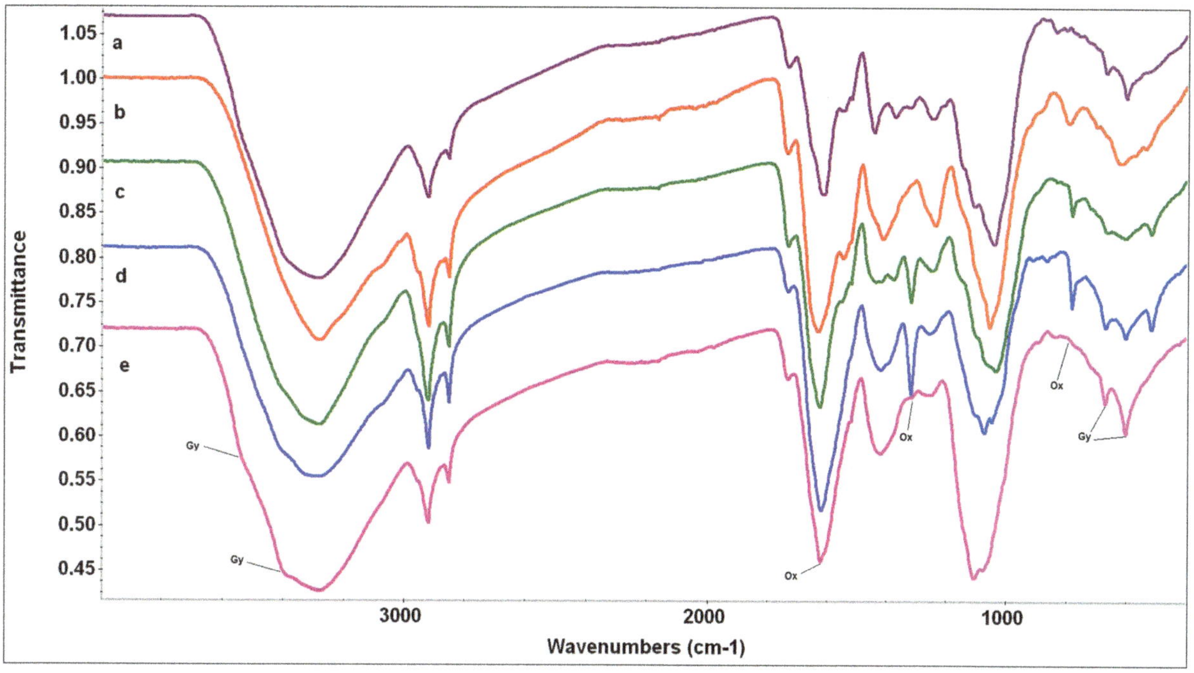

Figure 3. FTIR spectra of wide gypsophiles. The scale of the Y axis is offset for illustration purposes. a) *Helianthemum squamatum* (purple), b) *Lepidium subulatum* (red), c) *Herniaria fructicosa* (green), d) *Gypsophyla struthium* subsp. *hispanica* (blue), e) *Ononis tridentata* (pink). Abbreviations of compound peaks: Gy: Gypsum, Ox: Oxalate.

wide gypsophile. There was, however, some crossover between the spectra of gypsovags and narrow gypsophiles, with five cases being assigned to the wrong group, although generally they could be accurately distinguished. The cases of mis-classification mostly appear to relate directly to the spectral features noted above, i.e. the strong similarity, including the presence of oxalate, between the spectra of *Teucrium polium* subsp. *capitatum* (a gypsovag misclassified as narrow gypsophile) and those of *Teucrium pumilum* and *Thymus lacaitae* (misclassified as gypsovags).

Visual examination of the input datasets averaged over each group of species (Fig. 4) does not allow much distinction to be drawn between the three types of spectra, although it can be seen that between the frequencies of 3200 and 3600 cm^{-1}, and between 1500 and 1900 cm^{-1} (Fig. 5), wide gypsophiles do demonstrate differences to narrow endemisms and gypsovags, which have more subtle differences to each other in these ranges. These ranges, as described above for the FTIR results (Table 3), correspond to bands relating to gypsum present in wide gypsophiles, but not detectable in the other two groups. Also there is variation in oxalate and ester concentrations, with generally lower oxalate in the gypsovags, but higher ester (Fig. 4, Table 3).

Examination of the averaged input datasets over smaller ranges however, shows that, while at some ranges narrow gypsum endemisms show similar trends to gypsovags (Figs. 5 and 6), with higher ester (1740 cm^{-1}) than wide gypsophiles and no gypsum (1620 cm^{-1}), at other ranges narrow gypsum endemisms input value trends are more similar to wide gypsophiles with noticeable (although lower) presence of oxalate (1315 cm^{-1} and 1606 cm^{-1}), absent in most gypsovags (Figs. 5, 6). Therefore, it appears that

Table 4. Results of PLS regression analyses between FTIR spectra and the C (%), N (mg/g), C/N, Na (mg/g), K (mg/g), Ca (%), Mg (mg/g), P (mg/g), S (%) and Ash (%) concentrations of the samples analysed.

Variable	R^2	RPD values (Std Dev/Std error)
C (%)	95.7	4.81
N (mg/g)	79.8	2.24
Na (mg/g)	86.2	2.23
K (mg/g)	65.1	1.69
Ca (%)	88.1	2.69
Mg (mg/g)	79.9	2.9
P (mg/g)	85.5	2.62
S (%)	91.9	3.51
Ash (%)	94.5	4.28

Table 5. Confusion matrix of plant types, based on neural network model prediction.

Predicted groups	Observed groups		
	Gypsovag	Narrow endemism	Gypsophile
Gypsovag	15	2	0
Narrow endemism	3	9	0
Gypsophile	0	0	15
Model fit			
MAE	0.211	0.196	0.068
RMSE	0.305	0.287	0.098

Values give the number of each type given in the left-hand column that are categorised as each type given in the top row. Bottom rows indicate mean absolute error (MAE) and RMSE (Root Mean Squared Error) values for each output node in the neural network models, in relation to the output errors (range [0, 1]).

narrow gypsum endemisms, in terms of input values as developed using this FTIR-based approach, show some features both of gypsovags and gypsophiles and can thus be successfully discriminated from them.

Discussion

The FTIR spectra of the three groups of species analyzed (wide gypsophiles, narrow gypsum endemisms and gypsovags) showed distinct features that enabled their discrimination through neural network (NN) models with relatively high accuracy. This indicates that, although as predicted, elemental and spectral data correlated reasonably well, the analysis of FTIR spectra with NN provided an increased discriminating ability.

As expected, wide gypsophiles stood out for the strong presence of inorganic compounds in their leaves, particularly gypsum. This is in agreement with previous studies on the elemental composition of gypsum plants [12,13,14,16], finding high total ash, Ca and S (mostly in the form of sulphate) concentrations. Some of these studies suggested most of such S could be accumulated in the form of calcium sulphate [13,14], however, the widespread presence of mineral gypsum (i.e. $CaSO_4{\cdot}2H_2O$) in gypsophiles has never been reported before. With the sole exception of *L. subulatum*, all the wide gypsophiles analyzed accumulated mineral gypsum in their

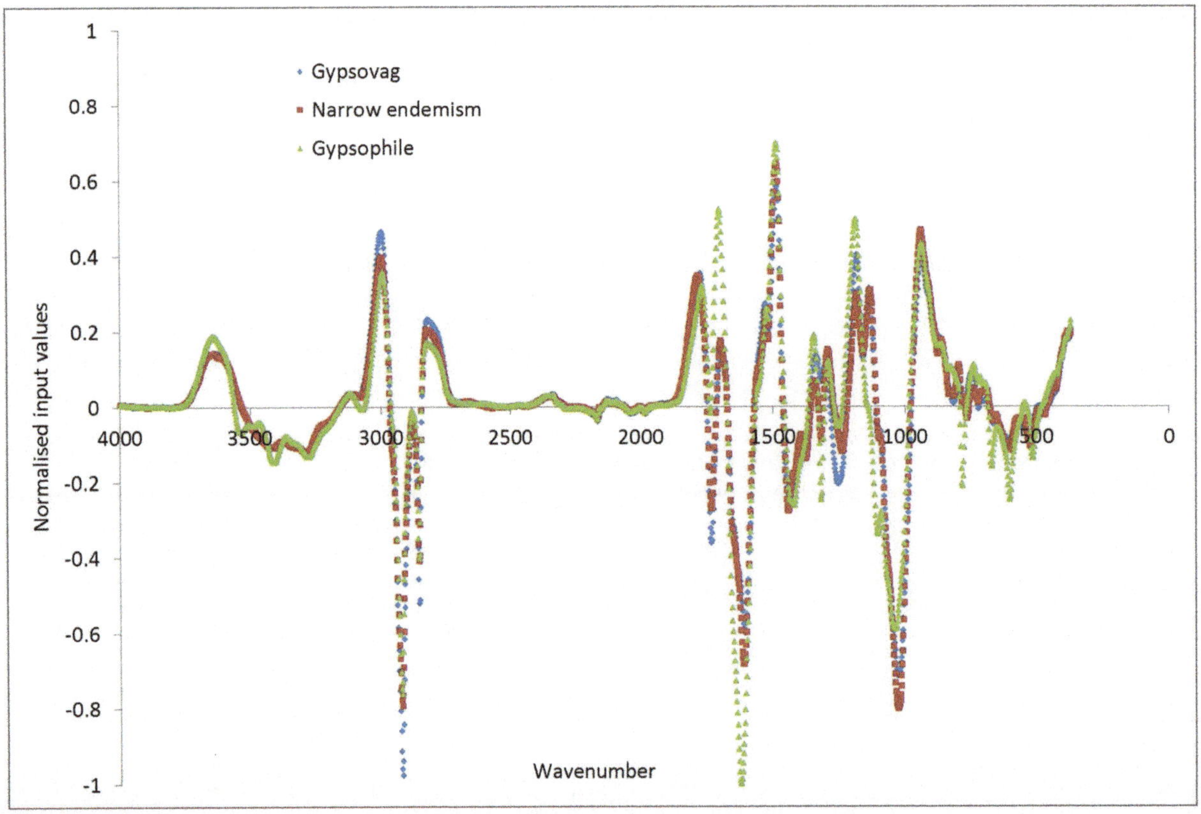

Figure 4. Input datasets averaged over each of the gypsum plant types (gypsovags in blue, narrow gypsum endemisms in red, wide gypsophiles in green), for the full wavenumber range.

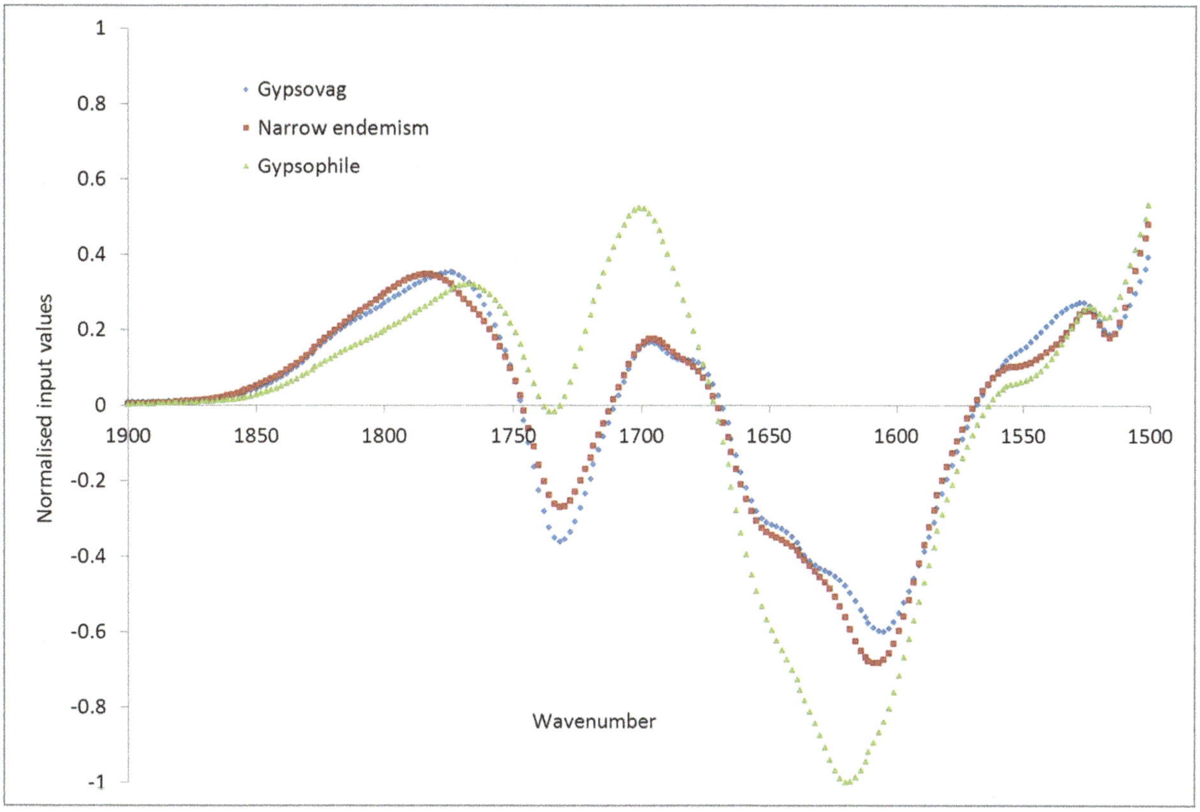

Figure 5. Input datasets averaged over each of the plant types (gypsovags in blue, narrow gypsum endemisms in red, wide gypsophiles in green), for the input wavenumber range 1500 to 1900 cm^{-1}.

leaves in the form of crystals. Although previous studies had reported the presence of crystals in some wide gypsophiles like *H. fruticosa* or *G. struthium* subsp. *hispanica* [40,41], this is the first evidence that most wide gypsophiles studied show such a distinct biochemical trait. Furthermore, the composition of crystals had only been analysed in gypsophiles growing on saline gypsum soils, like *Limonium* spp. or *Frankenia thymifolia* [6,42]. This is, therefore, the first empirical confirmation that some of the crystals observed in gypsophiles from non-saline soils, such as those considered in this study, are made of gypsum. Wide gypsophiles are commonly considered as "accumulators" of compounds such as S, Ca or Mg, being highly permeable to them and hence able to tolerate high concentrations (toxic for most plant species) in their leaves [12,13,14,43]. Soil gypsum contents above 20–30% are generally considered to be toxic for most plant species [44], but gypsum contents often surpass 50% in the soils where these plants were collected (Table 2). Similarly, normal average concentrations of S in plant leaves are ≈ 0.2% [45], but concentrations between 3 and 6% are common in wide gypsophiles [13,14,16,46]. The ability to "sequester" such excess S in the form of gypsum crystals that accumulate harmlessly in plant cells could be an adaptive mechanism of gypsum-specialist plants. Although gypsum crystals are rare among vascular plants [47], a recent study on the Australian endemic *Acacia robeorum* growing on high S soils, showed that this species was able to accumulate calcium sulphate crystals, presumably as a way to remove excess Ca and S [48]. The ability to accumulate excess gypsum in crystals may be widespread among different taxonomic groups, since the four species that showed gypsum crystals in their leaves in this study belong to three different taxonomic families (i.e. Cariophyllaceae, Fabaceae and Cistaceae). This ability is similar to the common mechanism of halophytes to excrete salts. Although it has been suggested that both mechanisms could be related [41], the question remains if lineages of halophytes and gypsophiles excreting salts are interconnected (i.e. share a common origin), or if both strategies have evolved independently. The general finding of gypsum in the leaves of wide gypsophiles opens up interesting questions regarding the regulatory mechanisms and the processes involved in the metabolism of gypsum in these plants, which is mostly unexplored.

Nevertheless, the formation of gypsum crystals is not the only strategy of wide gypsophiles to cope with high gypsum contents in the soil. *L. subulatum*, a gypsophile common to disturbed gypsum soils of the Iberian Peninsula [31,49], shows high S concentrations [13,16,46], but notably no gypsum. According to our results, this species seems to accumulate S in the form of other sulphates, possibly in combination with N. *L. subulatum* is known for its remarkably high N, amino acid and protein content [16,46,50], and high amide and protein bands were found for this species. It is hence possible that part of the S of this species is accumulated as ammonium sulphate (which presence could not be ruled out) or else in the form of S-rich proteins, peptides (glutathione), amino acids (methionine, cysteine) or other organic compounds rich in both N and S such as secondary metabolites (e.g. glucosinolates), which are common among the Brassicaceae. Indeed, the ability of *L. subulatum* to incorporate S into organic compounds has been suggested as a metabolic process explaining *L. subulatum* adaptation to gypsum soils [46].

The spectra of wide gypsophiles were also differentiated by their high calcium oxalate bands, a trait they shared with narrow gypsum endemisms. Calcium oxalate deposition is a common

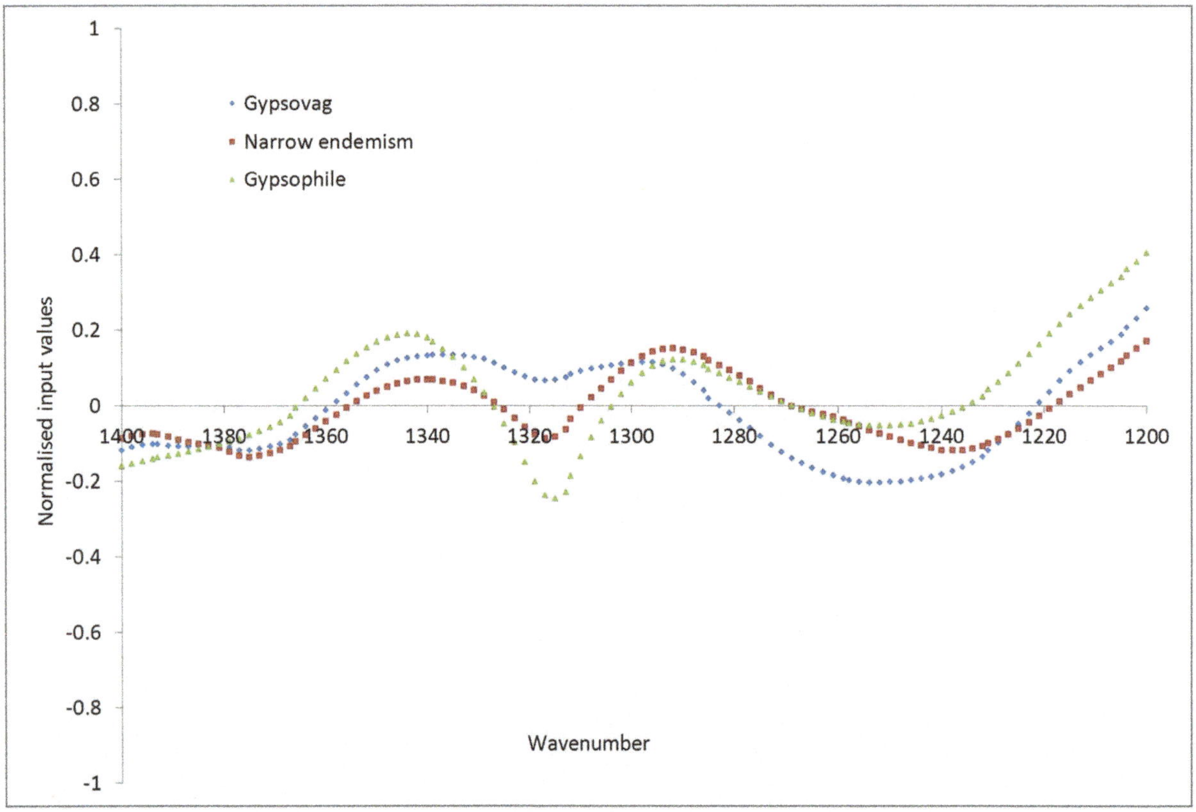

Figure 6. Input datasets averaged over each of the plant types (gypsovags in blue, narrow gypsum endemisms in red, wide gypsophiles in green), for the input wavenumber range 1200 to 1400 cm^{-1}.

phenomenon in plants, being present in more than 215 plant families [51]. The formation of calcium oxalate crystals in plant tissues is related to various important plant functions, among which the regulation of calcium concentrations seems primary [52]. Accordingly, several studies using a variety of plants have shown that the size and number of Ca oxalate crystals are responsive to changes in the soil concentrations of Ca ([52] and references therein). Previous studies indicated that gypsum specialists are Ca-accumulators [12,16,43], a trait that was interpreted as suggestive of the accumulation of mineral gypsum in this type of plants [14]. However, our results indicate that the accumulation of Ca in wide gypsophiles cannot solely be explained by the presence of mineral gypsum. Wide gypsophiles seem to show a further excess of Ca that is accumulated in the form of calcium oxalate. Interestingly, this is a common trait to most narrow gypsum endemisms, which have been suggested to be less specialized to live on gypsum than wide gypsophiles [16]. We therefore suggest that, given the widespread ability to form calcium oxalate crystals in different plants, the liberation of excess Ca in calcium oxalate could be a relatively easy to implement mechanism, as part of the process of adaptation to living on gypsum, of narrow gypsum endemisms.

Iberian gypsovags seem to show physiological mechanisms to block calcium and sulphate uptake by roots. This does not seem to be the case in all gypsovags from other regions of the world. For example, Boukhris and Lossaint [14,53] found that some species of Tunisian gypsovags were able to accumulate mineral S in their tissues when growing on gypsum, sometimes reaching similar (and even higher) values than gypsophiles. Similarly, Borer et al. [54] identified several strategies to accumulate and excrete excess Ca among gypsovags growing on gypsum soils in the Chihuahuan Desert, possibly involving the formation of calcium oxalate. Although the presence of gypsum was not evaluated by these authors, this discrepancy indicates results from Iberian gypsovags should be extrapolated with care. An increase in glutathione synthesis in leaves has been reported as a signalling factor to regulate sulphate uptake by roots in plants not adapted to gypsum soils [55]. Such a mechanism could also be operating in Iberian gypsovags and gypsum endemisms, but the sulphate metabolism of gypsum plants remains unexplored. The ability of Iberian gypsovags to grow in and out of gypsum soils could be related to their capacity to regulate the uptake of excess nutrients in the soil (particularly S and Ca). Nevertheless, such "blocking" of soil nutrients may also have negative consequences on their ability to uptake other nutrients, such as N and P, which are inherently poor in gypsum soils [18,19]. This suggestion agrees with the stress tolerant nature of Iberian gypsovags and their reported low nutrient concentrations [16].

Different to previous analyses based only on the elemental composition of gypsum plants [16] where narrow gypsum endemisms were found to be more similar to gypsovags than to wide gypsophiles, our results indicate that local gypsum endemisms share spectral features with the other two groups of gypsum plants that identify them as a separate group. These results have important implications for the understanding of plant adaptation to gypsum substrates as they suggest that narrow gypsum endemisms are not just stress-tolerant plants that find refuge from competition on gypsum without particular specialization to this special substrate. Although, like gypsovags, they do not seem to be able to cope with sulphate accumulation, they share the ability to eliminate excess Ca in calcium oxalate crystals with wide gypsophyles. The narrow gypsum endemisms studied might have recently evolved from stress-tolerant gypsovag species or ecotypes, and could be in the process of specializing to gypsum soils, developing adaptive mechanisms, such as the accumulation of oxalate, to survive on gypsum.

The analysis of FTIR spectra with NN provided an accurate tool to separate the three groups of gypsum plants analyzed. The discrimination ability was decreased for closely related species, such as *T. polium* subsp. *capitatum*, *T. pumilum* or *T. lacaitae*. However, the analysis worked generally well and only 5 out of 44 cases were misclassified. Therefore, this methodology arises as a promising tool for the analysis of plant adaptation to special substrates. Additional future work in this area could focus on determining the chemical composition of plant samples, rather than attempting to discriminate between different functional groups of species. This would provide useful information for nutrition studies and would allow us to determine relationships between soil and plant chemical composition. A NN model that could extract composition information (such as N or S) from FTIR spectra would be more useful than one that provides a categorisation, as it would provide more fundamental information about the samples. To achieve this, it would be necessary to develop calibrations for a wider range of samples and plants from different gypsum areas.

To conclude, FTIR spectroscopy linked to NN analysis is an efficient tool to discriminate plants with different specificity to gypsum substrates. Our results present evidence of the widespread presence of gypsum and calcium oxalate crystals in most gypsum specialist plants studied, although other mechanisms such as the accumulation of sulphates in organic molecules are also compatible with plant specialization to gypsum. While gypsovags seem to be stress tolerant plants that tightly regulate the uptake of S and Ca by their roots, narrow gypsum endemisms share the ability to accumulate excess Ca as oxalate with gypsophiles, possibly indicating their incipient specialization to live on gypsum. Further studies should focus on evaluating the generality of these conclusions and their extrapolation to gypsum plants from other regions of the world.

Supporting Information

Figure S1 Comparison of FTIR spectra of *Ononis tridentata* from different sites: a) red line, Chinchón (Madrid), b) blue line, Villamayor (Zaragoza).

Figure S2 Normalisation of FTIR spectra. Top – raw spectrum, bottom – after moving average preprocessing and normalisation to within range [-1, 1].

Table S1 Average chemical composition of study species.

Acknowledgments

The authors thank M.J. Albert, R. Milla, P. Fustero and A. Lamana for help with plant sampling and processing, to R. Hill for running the FTIR spectrometer, to E. Pérez-Fernández for assistance with FTIR figures and to J. F. Mota and an anonymous Referee for helpful comments on earlier versions of the manuscript.

Author Contributions

Conceived and designed the experiments: SP AE GM-M. Performed the experiments: SP AE GM-M MM. Analyzed the data: MA AHJR. Contributed reagents/materials/analysis tools: SP MA AE GM-M MM AHJR. Contributed to the writing of the manuscript: SP MA AE GM-M AHJR.

References

1. Mason H (1946) The edaphic factor in narrow endemism. The nature of environmental influences. Madroño 8: 209–226.
2. Raven PH (1964) Catastrophic selection and edaphic endemism. Evolution 18: 336–338.
3. Eswaran H, Gong ZT (1991) Properties, genesis, classification, and distribution of soils with gypsum. In: Nettleton WD, editor. Occurrence, characteristics, and genesis of carbonate, gypsum, and silica accumulations in soils.Madison: Soil Science Society of America. pp. 89–119.
4. Verheye WH, Boyadgiev TG (1997) Evaluating the land use potential of gypsiferous soils from field pedogenic characteristics. Soil Use and Management 13: 97–103.
5. Escudero A, Palacio S, Maestre FT, Luzuriaga AL (2014) Plant life on gypsum: an overview of its multiple facets. Biological Reviews (in press) DOI: 10.1111/brv.12092
6. Parsons RF (1976) Gypsophily in plants. A review. American Midland Naturalist 96: 1–20.
7. Guerrero Campo J, Alberto F, Hodgson J, García Ruiz JM, Montserrat Martí G (1999) Plant community patterns in a gypsum area of NE Spain. I- Interactions with topographic factors and soil erosion. Journal of Arid Environments 41: 401–410.
8. Guerrero Campo J, Alberto F, Maestro Martínez M, Hodgson J, Montserrat Martí G (1999) Plant community patterns in a gypsum area of NE Spain. II.- Effects of ion washing on topographic distribution of vegetation. Journal of Arid Environments 41: 411–419.
9. Meyer SE, García-Moya E (1989) Plant community patterns and soil moisture regime in gypsum grasslands of north central Mexico. Journal of Arid Environments 16: 147–155.
10. Meyer SE, García-Moya E, Lagunes-Espinoza LC (1992) Topographic and soil surface effects on gypsophile plant community patterns in central Mexico. Journal of Vegetation Science 3: 429–438.
11. Meyer SE (1986) The ecology of gypsophile endemism in the Eastern Mojave desert. Ecology 67: 1303–1313.
12. Duvigneaud P, Denaeyer-De Smet S (1968) Essai de classification chimique (éléments minéraux) des plantes gypsicoles du bassin de l'Ebre. Bulletin de la Société Royale de Botanique de Belgique 101: 279–291.
13. Duvigneaud P, Denaeyer-De Smet S (1966) Accumulation du soufre dans quelques espèces gypsophiles d'Espagne. Bulletin de la Société Royale de Botanique de Belgique 99: 263–269.
14. Boukhris M, Lossaint P (1970) Sur la teneur en soufre de quelques plantes gypsophiles de Tunisie. (Note préliminaire). Acta Oecologica 5: 345–354.
15. Romao RL, Escudero A (2005) Gypsum physical soil crusts and the existence of gypsophytes in semi-arid central Spain. Plant Ecology 181: 127–137.
16. Palacio S, Escudero A, Montserrat-Marti G, Maestro M, Milla R, et al. (2007) Plants living on gypsum: beyond the specialist model. Annals of Botany 99: 333–343.
17. Gankin R, Major J (1964) *Arctostaphylos myrtifolia*, its biology and relationship to the problem of endemism. Ecology 45: 792–808.
18. Delgado-Baquerizo M, Covelo F, Gallardo A (2011) Dissolved organic nitrogen in Mediterranean ecosystems. Pedosphere 21: 309–318.
19. Salmerón-Sánchez E, Martínez-Nieto MI, Martínez-Hernández F, Garrido-Becerra JA, Mendoza-Fernández AJ, et al. (2014) Ecology, genetic diversity and phylogeography of the Iberian endemic plant *Jurinea pinnata* (Lag.) DC. (Compositae) on two special edaphic substrates: Dolomite and gypsum. Plant and Soil 374: 233–250.
20. Berthomieu C, Hienerwadel R (2009) Fourier transform infrared (FTIR) spectroscopy. Photosynthesis Research 101: 157–170.
21. Smidt E, Böhm K, Schwanninger M (2011) The Application of FT-IR Spectroscopy in Waste Management. In: Nikolic G, editor. Fourier Transforms - New Analytical Approaches and FTIR Strategies. Rijeka, Croatia: InTech.pp. 405–430.
22. Euring D, Löfke C, Teichmann T, Polle A (2012) Nitrogen fertilization has differential effects on N allocation and lignin in two *Populus* species with contrasting ecology. Trees - Structure and Function 26: 1933–1942.
23. Jones R, Causton D, Emmett B, Mur L, Gwynn-Jones D (2012) Latent soil effects of grazing and ammonium deposition on *Deschampsia flexuosa* tillers inserted and grown in heather moorland soil. Environmental and Experimental Botany 81: 72–78.
24. Pitre FE, Lafarguette F, Boyle B, Pavy N, Caron S, et al. (2010) High nitrogen fertilization and stem leaning have overlapping effects on wood formation in poplar but invoke largely distinct molecular pathways. Tree Physiology 30: 1273–1289.

25. McNear Jr DH, Chaney RL, Sparks DL (2010) The hyperaccumulator *Alyssum murale* uses complexation with nitrogen and oxygen donor ligands for Ni transport and storage. Phytochemistry 71: 188–200.
26. Wei ZL, Dong L, Tian ZH (2009) Fourier transform infrared spectometry study on early stage of cadmium stress in clover leaves. Pakistan Journal of Botany 41: 1743–1750.
27. Xue S, Zhu F, Ye S, Wang J, Wu X (2011) Physiological response of *Mirabilis jalapa* Linn. to lead stress by FTIR spectroscopy. Shengtai Xuebao/Acta Ecologica Sinica 31: 6143–6148.
28. Dziuba B (2011) Identification of selected *Leuconostoc* species with the use of FTIR spectroscopy and artificial neural networks. Acta Scientiarum Polonorum, Technologia Alimentaria 10: 275–285.
29. Hong Q, Yu R, Zheng R, Wang H, Cheng C (2011) Identification between *Fimbristylis miliacea* (L.) Vahl and *Fimbristhlis stauntonii* Debeaux et Franch. by CWT-FTIR-RBFNN. Chinese Journal of Chemistry 29: 1024–1030.
30. Dharmaraj S, Gam LH, Sulaiman SF, Mansor SM, Ismail Z (2011) The application of pattern recognition techniques in metabolite fingerprinting of six different *Phyllanthus* spp. Spectroscopy 26: 69–78.
31. Braun-Blanquet J, Bolòs O (1957) Les groupements végétaux du Bassin Moyen de l'Ebre et leur dynamisme. Anales de la Estación Experimental de Aula Dei 5: 1–266.
32. Devesa JA, Ortega A (2004) Especies vegetales protegidas en España: plantas vasculares. Madrid: Ministerio de Medio ambiente.
33. Allen SE, editor (1989) Chemical analysis of ecological materials. Oxford: Blackwell Scientific Publ.
34. Becker (1961) Análisis y valoración de piensos y forrajes. Zaragoza, Spain: Ed. Acribia.
35. Aitkenhead MJ, Coull M, Towers W, Hudson G, Black HIJ (2013) Prediction of soil characteristics and colour using data from the National Soils Inventory of Scotland. Geoderma 200–201: 99–107.
36. Mouazen AM, Kuang B, De Baerdemaeker J, Ramon H (2010) Comparison among principal component, partial least squares and back propagation neural network analyses for accuracy of measurement of selected soil properties with visible and near infrared spectroscopy. Geoderma 158: 23–31.
37. Aitkenhead MJ, McDonald AJS, Dawson JJ, Couper G, Smart RP, et al. (2003) A novel method for training neural networks for time-series prediction in environmental systems. Ecological Modelling 162: 87–95.
38. Artz RRE, Chapman SJ, Jean Robertson AH, Potts JM, Laggoun-Défarge F, et al. (2008) FTIR spectroscopy can be used as a screening tool for organic matter quality in regenerating cutover peatlands. Soil Biology and Biochemistry 40: 515–527.
39. Farmer V (1974) The infrared spectra of minerals. London, UK: Mineralogical Society.
40. Grigore MN, Toma C, Zamfirache MM, Boscaiu M (2011) Anatomical considerations on Spanish gypsophytes. Where is their place within plant ecology? Scientific Annals of Alexandru Ioan Cuza University of Iasi New Series, Section 2 Plant Biology 57: 31–38.
41. Mota JF, Sánchez-Gómez P, Guirado JS, editors (2011) Diversidad vegetal de las yeseras ibéricas. El reto de los archipiélagos edáficos para la biología de la conservación. Almería: ADIF- Mediterráneo Asesores Consultores. 636 pp.
42. Denayer-De Smet S (1970) Note sur la composition chimique des sels recrétés par diverses espèces gypsohalophytes d'Espagne. Bulletin de la Société Royale de Botanique Belge 103: 273–278.
43. Merlo ME, Mota JF, Cabello J, Alemán MM (1998) La gipsofilia en plantas: un apasionante edafismo. Investigación y Gestión 3: 103–112.
44. FAO, editor (1990) Management of Gypsiferous Soils. Rome, Italy: FAO.
45. Tcherkez G, Tea I (2013) $^{32}S/^{34}S$ isotope fractionation in plant sulphur metabolism. New Phytologist 200: 44–53.
46. Ruiz JM, López-Cantarero I, Rivero RM, Romero L (2003) Sulphur phytoaccumulation in plant species characteristic of gypsipherous soils. International Journal of Phytoremediation 5: 203–210.
47. He H, Veneklaas EJ, Kuo J, Lambers H (2014) Physiological and ecological significance of biomineralization in plants. Trends in Plant Science 19: 166–174.
48. He HH, Bleby TM, Veneklaas EJ, Lambers H, Kuo J (2012) Morphologies and elemental compositions of calcium crystals in phyllodes and branchlets of *Acacia robeorum* (Leguminosae: Mimosoideae). Annals of Botany 109: 887–896.
49. Mota JF, Sola AJ, Dana ED (2003) Plant succession in abandoned gypsum quarries in SE Spain. Phytocoenologia 33: 13–28.
50. Alvarado JJ, Ruiz JM, López-Cantarero I, Molero J, Romero L (2000) Nitrogen metabolism in five plant species characteristic of gypsiferous soils. Journal of Plant Physiology 156: 612–616.
51. McNair JB (1932) The intersection between substances in plants: essential oils and resins, cyanogen and oxalate. American Journal of Botany 19: 255–271.
52. Nakata PA (2003) Advances in our understanding of calcium oxalate crystal formation and function in plants. Plant Science 164: 901–909.
53. Boukhris M, Lossaint P (1975) Aspects écologiques de la nutrition minérale des plantes gypsicoles de Tunisie. Revue d'Ecologie et de Biologie du Sol 12: 329–348.
54. Borer CH, Hamby MN, Hutchinson LH (2012) Plant tolerance of a high calcium environment via foliar partitioning and sequestration. Journal of Arid Environments 85: 128–131.
55. Herschbach C, Rennenberg H (1994) Influence of glutathione (GSH) on net uptake of sulphate and sulphate transport in tobacco plants. Journal of Experimental Botany 45: 1069–1076.
56. Jones F (2012) Infrared investigation of barite and gypsum crystallization: Evidence for an amorphous to crystalline transition. CrystEngComm 14: 8374–8381.
57. Ito K, Bernstein HJ (1956) The vibrational spectra of the formate, acetate and oxalate ions. Canadian Journal of Chemistry 34: 170–178.
58. Wilson MJ (1994) Clay mineralogy: spectroscopic and chemical determinative methods. London: Chapman &Hall.

7

Early Response of Soil Properties and Function to Riparian Rainforest Restoration

Rose Gageler[1], Mark Bonner[1], Gunnar Kirchhof[1], Mark Amos[2], Nicole Robinson[1], Susanne Schmidt[1]*, Luke P. Shoo[3]

1 School of Agriculture and Food Sciences, The University of Queensland, St Lucia, Australia, 2 Lake Baroon Catchment Care Group, Maleny, Australia, 3 School of Biological Sciences, The University of Queensland, St Lucia, Australia

Abstract

Reforestation of riparian zones is increasingly practiced in many regions for purposes of biodiversity conservation, bank stabilisation, and improvement in water quality. This is in spite of the actual benefits of reforestation for recovering underlying soil properties and function remaining poorly understood. Here we compare remnant riparian rainforest, pasture and reforestation plantings aged 2–20 years in an Australian subtropical catchment on ferrosols to determine the extent to which reforestation restores key soil properties. Of the nine soil attributes measured (total nitrogen, nitrate and ammonium concentrations, net nitrification and ammonification rates, organic carbon, bulk density, fine root biomass and water infiltration rates), only infiltration rates were significantly lower in pasture than remnant riparian rainforest. Within reforestation plantings, bulk density decreased up to 1.4-fold and infiltration rates increased up to 60-fold with time post-reforestation. Our results suggest that the main outcome of belowground processes of early reforestation is the recovery of the soils' physical structure, with potential beneficial ecosystem services including reduced runoff, erosion and associated sediment and nutrient loads in waterways. We also demonstrate differential impacts of two commonly planted tree species on a subset of soil properties suggesting that preferential planting of select species could accelerate progress on specific restoration objectives.

Editor: Kurt O. Reinhart, USDA-ARS, United States of America

Funding: LPS was supported by funding from the Australian Government's National Environmental Research Program (Tropical Ecosystems and Environmental Decisions Hubs). The funders had no role in study design, data collection and analysis, decision to publish, or preparation of the manuscript.

Competing Interests: The authors have declared that no competing interests exist.

* Email: Susanne.Schmidt@uq.edu.au

Introduction

Globally, riparian zones are widely recognized for their critical role in water regulation [1–5], and conservation of biodiversity [6–9]. Clearing of vegetation from stream banks can cause increased soil erosion and decreased water quality through the loss of filtration services, in addition to the loss of critical species habitat [10]. In direct response, degraded stream-banks are increasingly being restored to reinstate ecological functions [11]. Riparian zones are also regarded as foci of restoration opportunity as they are often preferentially renounced from agricultural production over agronomically more valuable areas [9].

Soil physical, chemical and functional properties change with succession from pasture to secondary forest including reforestation plantings in the tropics, subtropics and temperate regions [12–16]. Soil properties such as bulk density, soil water content, fine root biomass, porosity, stocks of nitrogen, carbon and phosphorus, nitrogen cycling, availability of macro- and micronutrients, as well as toxins (metals, salts), have been studied in context of land use change [17–19]. However, most studies focus on mid- to long-term effects of reforestation with less consideration of the short-term effects (<10–20 years) of reforestation plantings, and associated reinstatement of soil functions that might motivate landholders to become actively engaged in restoration. Another knowledge gap is the relative contribution of tree species to outcomes of riparian restoration plantings, and such information could inform decisions on species selection and reforestation goals.

In Australia, riparian reforestation is a major focus of habitat restoration in the tropics and subtropics; over 70% of rainforest restoration projects have targeted banks of rivers and streams [20]. However, much evaluation of the success of reforestation plantings has focused on aboveground attributes [21–23], with less regard to belowground processes including soil function (but see Paul [14] for the recovery of soils via different rainforest restoration pathways). Understanding belowground processes is important to quantify the capacity of reforestation to recover ecosystem function.

Soil properties and functions would intuitively be expected to change over time following reforestation towards a state convergent on intact rainforest ecosystems. Studying riparian soil under remnant rainforest, pasture, and within 20 years post reforestation, we addressed three questions (i) how do soil attributes differ between pasture and remnant riparian rainforest, (ii) how do soils change with time following reforestation, and (iii) do tree species affect soils differentially following reforestation?

The study was motivated by the need for restoration of forest cover in water catchments due to the improvements in water quality that can be achieved [24]. The region targeted for our study is located in the Australian subtropics and has experienced large-scale clearing of rainforest since the mid 19th century. The region is an important water catchment that supplies drinking water to more than 100000 people in the region [25], and numerous land owners and stakeholders are committed to ongoing restoration of riparian zones. Our research aimed to generate knowledge of soil properties to direct future reforestation activities.

Materials and Methods

2.1. Study site

This study focused on riparian areas in the Maleny region in southeastern Queensland, Australia (26°45′S, 152°48′E). The climate is classified as subtropical with an annual rainfall of 1900 mm and annual daily mean temperature maximum and minimum of 23.2°C and 14.2°C, respectively. The soils in the region are classified as permeable red clay-loam ferrosols [26,27] or Nitisol in the US soil taxonomy. Prior to European settlement in the mid-1800s and subsequent land clearing for timber production and agriculture (especially dairy production), the region was dominated by cool-subtropical rainforest classified as complex notophyll vine forest [28,29].

2.2. Sampling design

Field measurements and soil sampling were carried out between May and June 2013. The sampling design was based around riparian reforestation projects established in the region by the Lake Baroon Catchment Care Group (LBCCG). Reforestation sites were grouped into five spatial clusters across the Maleny Plateau (range of elevation 196 to 441 m a.s.l.) to capture broad variation in landform and land use history. A subset of one to three reforestation sites ranging in age between 2 and 20 years were then selected from each cluster for a total of 10 sites (Table 1). Sites were also established in remnant riparian rainforest where available (2 of 5 clusters) and samples were collected from pasture (adjacent to plantings or remnant forest) at all sites in each cluster.

At each reforestation or remnant site, five subplots were established. Each subplot was deliberately located under a different tree species. The five tree species targeted here were derived from a pool of species commonly used in local rainforest plantings, which in turn were selected to include early to late successional species, nitrogen fixing and/or waterlogging tolerant species. The tree species chosen to investigate species-specific effects on soil properties included *Acacia melanoxylon* R.Br. (Fabaceae, N_2 fixer, early succession species), *Ficus coronata* Spin (Moraceae, nitrophile, waterlogging tolerant, early succession species), *Flindersia schottiana* F. Muell. (Rutaceae, mid-late succession), *Homalanthus nutans* (G.Forst.) Guill. (Euphorbiaceae, early succession species) and *Podocarpus elatus* R.Br. ex Endl (Podocarpaceae, mid-late succession, gymnosperm). In the event that one of the chosen tree species was absent from a site, a closely related species was substituted or, if not available, subplots were located under a random tree within the site. Subplots were positioned under the canopy of each selected tree (mean distance from stem 89.5 cm, SD 45.2 cm).

At each reforestation or remnant forest site, we performed the following procedures: (1) collected a single 8 cm diameter core to a depth of 30 cm (hereafter "textural core") and divided it into 0–15 cm and 15–30 cm fractions for textural analysis; (2) performed an infiltration test; and, (3) collected two samples of the surface soil using a 12.5 cm diameter, 8.5 cm tall, 1043 cm^3 volume bulk density ring (hereafter "bulk density ring") from each of five subplots for analysis of physical and chemical properties of the soil. In addition, a single texture core and two to four bulk density ring subplots were collected from adjacent pasture and infiltration measurements performed. For the purposes of analysis, samples and infiltration measures from pasture were calculated as means for each cluster.

Overall, we sampled 10 reforestation sites, 2 remnant riparian rainforest sites (Table 1) and 5 pasture clusters comprising 24 cores for soil texture analysis, 150 cores for soil chemical and physical analyses and 24 infiltration tests. All sites were private land holdings and owners approval was obtained by Mark Amos (Lake Baroon Catchment Care Group, Maleny, info@lbccg.org.au). Upon returning from the field, all soil samples were stored at 4 °C in sealed plastic bags prior to respective analyses.

Table 1. GPS coordinates and ages of reforestation and remnant sites.

Site type	Age (years)	Site location
Reforestation	3	26°46′02 S, 152°49′38 E
Reforestation	3	26°46′17 S, 152°49′51 E
Reforestation	4	26°45′06 S, 152°50′46 E
Reforestation	6	26°42′01 S, 152°53′58 E
Reforestation	7	26°46′10 S, 152°49′57 E
Reforestation	8	26°44′05 S, 152°50′54 E
Reforestation	11	26°44′46 S, 152°51′13 E
Reforestation	11	26°46′12 S, 152°49′39 E
Reforestation	12	26°46′08 S, 152°49′43 E
Reforestation	20	26°42′01 S, 152°53′58 E
Remnant	-	26°45′07 S, 152°51′02 E
Remnant	-	26°44′50 S, 152°51′11 E

All sites were located on private land in the Maleny region. Pasture sites were located immediately adjacent to each forested site. Ages are based on the year that they were planted relative to 2013.

2.3. Texture analysis

Changes in soil texture mostly occur due to weathering over very long time periods and as demonstrated by Maloney [13]. Unless there is erosion or deposition of soil material, texture is unlikely to vary over time since reforestation. For this reason, the purpose of conducting a soil texture analysis at each site was not to determine changes due to reforestation, but to characterize and evaluate background variation in soil types among sites that might warrant differential treatments of sites in the analysis. Soil samples were not dispersed prior to analysis and as such, the calculated percentages represent the relative size of secondary particles (aggregates) present rather than primary particle size *sensu stricto*. This was considered a more relevant measure of soil function than particle size determined using standard methods, as it gives a relative measure of soil aggregation and structure. Furthermore, the ferrosols in this study are recognized as being subplastic, with clay-sized particles being primarily composed of sesquioxides that do not readily break down into primary particles [30]. These analyses of soil aggregation were conducted on air-dried soil from 0–15 and 15–30 cm depth cores according to the hydrometer method for soil texture analysis [31,32]. For each sample, 40 g of air-dried soil was ground and mixed with 100 ml 50 g l^{-1} Sodium hexametaphosphate solution and filled to 500 ml with deionized water. Samples were shaken overnight and then emptied into a 1 l cylinder for analysis. Cylinders were filled to 1 l with deionized water, mixed with a plunger and left to settle. Temperature and specific gravity readings were taken 40 seconds and 7 hours after plunging [32]. Using temperature and specific gravity readings, the fraction of sand, silt and clay sized particles in each soil sample was then calculated as a percentage.

2.4. Soil physical properties: bulk density and infiltration rates

Soil bulk density was determined from the second bulk density ring in each subplot. Fresh soil was oven dried to a constant weight at 105 °C. Soil bulk density is expressed as grams of dry soil per volume. To quantify fine root biomass, soil from one bulk density ring per subplot was passed through a sieve (8 mm mesh) and roots were manually removed. Roots were then oven dried at 65 °C to a constant weight, and results expressed as grams of fine roots per kilogram of dry soil.

Infiltration rates were measured using a ponded disk permeameter [33]. Soils were assumed to be at field capacity as the region had received in excess of 30 mm in the week preceding the first set of measurements and over 100 mm in the week prior to the second sampling visit. Infiltration tests were performed at a random, relatively flat area representative of the site. Tests in the adjacent pasture were performed at a random point beyond the farthest extent of the neighbouring forest canopy. Upon setting up of the permeameter, grass and other ground covers were clipped to just above the soil surface and the soil pre-moistened with rainwater to reduce the time needed to reach steady state infiltration. The permeameter was filled with rainwater collected from water tanks on local landholder properties to ensure that water quality was similar to that occurring under natural condition. Once the disk permeameter was in place, measurements of water level changes were recorded until the time interval between measurements was stable. As soils were close to field capacity at the start of measurements, steady state was reached quickly. Infiltration rate was calculated as the average volume change per second and then converted into millimeters per hour.

2.5. Soil chemical properties: nitrogen and organic carbon

Total concentrations of soil nitrogen (TN) and organic carbon (SOC) were determined using a TruSpec CHN analyzer (LECO Australia Pty. Ltd., Castle Hill, NSW 2154, Australia), using dried, ground and sieved soil (2 mm sieve). Nitrate (NO_3^-) and ammonium (NH_4^+) were extracted with potassium chloride (KCl) and quantified via colourimetric assays. To remove NO_3^- and NH_4^+ from soil exchange sites, 15 g of fresh sieved soil was mixed with 1 M KCl in a 15 ml falcon tube and shaken on an orbital shaker for 1 hour. Samples were centrifuged at 4000 rpm for 4 minutes and 1 ml of the supernatant was frozen until analysis via colourimetric assays according to standard procedures [34,35]. Net nitrification and ammonification rates were calculated from the difference in concentration between fresh soil samples and those incubated for 20 days at 28°C. Briefly, 15 g of soil were weighed into 50 ml falcon tubes and incubated for 20 days. Soils were kept at field moisture by watering daily with deionized water. Seeds that germinated during the incubation period were removed. Concentrations of NO_3^- and NH_4^+ in the incubated soils following KCl extraction were quantified using colourimetric assays as described above. Net nitrification and ammonification rates are expressed as mg nitrate or ammonium generated per kg dry soil per day.

2.6. Statistical analysis

Statistical analyses were performed in R version 3.0.1 [36]. To avoid non-independence among subsamples, we derived the aggregated mean of soil attributes for each site (reforestation or remnant rainforest) or cluster (pasture) and used these values as inputs in all analyses, excepting those determining species-specific effects which used values from a single subplot within each site. Pairwise correlation tests (Spearman's) were performed to examine associations between tested soil attributes. Soil attributes under both pasture and remnant riparian rainforest were directly compared using a one-way analysis of variance (ANOVA). Developmental trajectories of soil attributes within reforested sites (site level and separately for specific tree species within sites) were examined using linear regressions between either age or log-transformed age and individual soil attributes. The age term was log-transformed to account for potential non-linear developmental trajectory of reforestation plantings; often showing a rapid change immediately post planting followed by diminishing returns as the system reaches natural equilibrium. For the purposes of quantifying response trajectories with age, we assume that pasture represents 0 years post reforestation.

Results

Of the 56 subplots within sites, 47 (84%) were associated with one of the five target trees. *Homalanthus nutans* was the species most commonly absent, being present in 50% of reforested sites. Soil aggregation showed only minor variation among sites and land uses so was not considered further in subsequent analyses (sand sized aggregates: mean = 60%, range = 46–76%, clay sized aggregates: mean = 30%, range = 14–46%).

3.1. Correlations between variables

Infiltration rates were inversely correlated with bulk density ($P = 0.005$) but were not correlated with soil particle size (percentage sand, silt and clay) at either 0–15 cm or 15–30 cm depth ($P > 0.5$, not shown). Bulk density was correlated with both SOC and TN ($P = 0.02$ and 0.01 respectively) (Table 2). Pairwise

Table 2. Pairwise correlations (Spearman's) between soil attributes across pasture, remnant rainforest and reforestation sites (n = 5, 2, 10 respectively).

Variable	Variable	Spearman's correlation coefficient	P value
Bulk density	SOC	−0.554	0.023
Bulk density	TN	−0.615	0.010
Infiltration	Bulk density	−0.654	0.005
TN	SOC	0.784	<0.001
Ammonification rate	Nitrification rate	−0.664	0.005
Ammonification rate	Fine root biomass	0.662	0.005
Nitrate	Ammonification rate	−0.529	0.031
Nitrate	TN	0.495	0.045
Nitrate	SOC	0.630	0.008

Only significant pairwise correlations (P<0.05) are shown.

correlations revealed strong associations between soil nitrogen-related variables.

3.2. Differences between pasture and remnant rainforest

There was no apparent difference between pasture and remnant riparian rainforest in the tested soil attributes except infiltration rates (Table 3). Infiltration rates were significantly greater in the remnant rainforest (1421±995 mm h^{-1}) than in pasture sites (220±151 mm h^{-1}) ($F = 9.526$, $DF = 5$, $P = 0.027$, Table 3).

3.3. Recovery of soil properties under reforestation plantings

Only bulk density and infiltration rates showed a relationship with age within 20 years post reforestation (Table 4). In each case, models with a log-transformed age term explained more of the variation than models with an un-transformed age term. Bulk density of the soils ranged from 0.312 to 0.991 g cm^{-3}, and decreased significantly with time post reforestation ($F = 6.312$, $DF = 13$, $P = 0.026$, Figure 1). Conversely, infiltration rates significantly increased ($F = 10.56$, $DF = 13$, $P = 0.006$) within 20 years post reforestation (Figure 1). Infiltration rates in reforestation sites ranged from 149 mm h^{-1} at a 7-year-old reforestation site to over 1800 mm h^{-1} at an 11-year-old reforestation site. Infiltration rates of pasture sites varied from 3 mm h^{-1} to over 650 mm h^{-1}, although this variation was slightly less when data was aggregated by cluster.

3.4. Species specific effects on the recovery of soils under reforestation plantings

Species-specific effects were apparent for two of five tree species when tested soil attributes were modeled against log-transformed age (Figure 2). Ammonium concentrations in soil associated with *Flindersia schottiana* decreased with time post reforestation ($F = 4.97$, $DF = 8$, $P = 0.056$). Fine root biomass in soil associated with *Ficus coronata* decreased through time within 20 years post reforestation ($F = 4.921$, $DF = 8$, $P = 0.057$).

Discussion

Replanting riparian areas can be driven by a variety of objectives including increasing landscape connectivity for wildlife movements and biogeochemical functions such as trapping nutrients, sediment, pesticides, bank stabilisation, improved water quality and flood mitigation [10]. Here, we studied the trajectory of riparian reforestation to identify key changes in soil attributes through the transition from pasture to riparian tree plantings of increasing age and remnant rainforest. We showed that the major

Table 3. ANOVA results from a comparison of soil attributes between remnant riparian rainforest (n = 2) and pasture (n = 5).

Variable	ANOVA P value	Mean pasture ±1 SD	Mean remnant ±1 SD	Contrast remnant (R) vs. pasture (P)
Fine root biomass (g kg^{-1})	0.09	6.64±4.05	13.00±0.38	NS
Bulk density (g cm^{-3})	0.36	0.69±0.08	0.61±0.12	NS
SOC (%)	0.60	6.15±1.83	7.31±4.15	NS
TN (%)	0.85	0.71±0.14	0.74±0.26	NS
Nitrate (mg kg^{-1})	0.26	11.8±7.24	22.7±17.9	NS
Nitrification rate (mg kg^{-1} soil d^{-1})	0.65	0.67±0.75	0.39±0.03	NS
Ammonium (mg kg^{-1})	0.46	4.36±4.27	7.85±8.10	NS
Ammonification rate (mg kg^{-1} soil d^{-1})	0.73	0.65±0.98	0.65±0.98	NS
Infiltration rate (mm h^{-1})	**0.03**	**220±151**	**1421±995**	**R>P**

Bold text indicates significance at 0.05 level, NS not significant if P>0.05, SD standard deviation.

Table 4. Relationship between tested soil attributes and the age of the reforestation sites (n = 15).

Variable	Age				Log Age			
	Slope	R^2 value	F statistic	P value	Slope	R^2 value	F statistic	P value
Fine root biomass (g kg^{-1} soil)	0.02	0.001	0.02	0.90	1.14	0.03	0.34	0.57
Bulk density (g cm^{-3})	−0.009	0.28	5.16	**0.04**	−0.12	0.32	6.31	**0.03**
SOC (%)	−0.03	0.01	0.15	0.70	−0.03	<0.001	0.002	0.97
TN (%)	−0.001	0.004	0.05	0.83	0.005	<0.001	0.004	0.95
Nitrate (mg kg^{-1})	0.09	0.002	0.03	0.86	3.58	0.03	0.38	0.55
Nitrification rate (mg kg^{-1} d^{-1})	0.03	0.13	1.93	0.19	0.30	0.07	0.95	0.35
Ammonium (mg kg^{-1})	−0.09	0.04	0.51	0.49	−1.72	0.11	1.55	0.24
Ammonification rate (mg kg^{-1} d^{-1})	−0.03	0.11	1.54	0.24	−0.47	0.15	2.25	0.16
Infiltration rate (mm h^{-1})	54.9	0.39	9.77	**0.008**	678	0.45	10.6	**0.006**

Pasture is assumed to be 0 years post reforestation. Results from linear models with both age and log-transformed age are shown. P values in bold are significant at the 0.05 level.

belowground outcomes within 3 to 20 years since planting are physical and associated functional attributes of the soil.

4.1. Differences in soil properties between pasture and remnant rainforest

Infiltration rate was the only soil variable studied here that differed significantly between pasture and remnant riparian rainforest. Pasture infiltration rates were ≈15% of those of riparian rainforest, consistent with the finding of an infiltration rate in pasture 20% of that in remnant rainforest by Peñuela and Drew [15]. We detected no statistically significant differences in other soil properties between pasture and remnant rainforest soils. Limited effects of land use on selected chemical properties are consistent with previous findings in Australian on ferrosol soils [14]. However, it was expected that attributes such as bulk density and fine roots would differ between land use types [14]. It is possible that these attributes differ at a scale inscrutable beyond background variability between sites, possibly necessitating longitudinal study of permanent plots following reforestation. Limited availability of remnant forest within sampling clusters, a consequence of extensive clearing of riparian rainforest in the region, was also a contributing factor in the comparatively low statistical power for detecting changes in measured attributes. Nevertheless, the data also indicate that there is little degradation of soil with respect to N and C stock in the studied systems.

4.2. Recovery of soils under reforestation plantings

Bulk density and infiltration rates both changed significantly with time under the reforestation plantings. Bulk density showed a negative trend with age whereas infiltration increased with time in the reforestation plantings. The observed trend for decreasing bulk density is consistent with previous studies globally [9,14,37,38]. The higher bulk density and lower infiltration rates of the pastures may be partially attributed to compaction by livestock and/or machinery [39]. Lower bulk densities are often associated with soil of forests and reforestation plantings because greater presence of roots and soil biota and higher concentrations of organic matter enhance soil porosity. In remnant forest and reforestation plantings, the high turnover of roots results in more open soil pores as roots grow and subsequently die [40]. Additionally, pastures generally have more shallow and non-woody roots while forests have a broader range of fine to large woody roots including with larger, woody roots that leave conduits in the soil post death [41–43].

Higher densities and diversities of soil organisms in forests and reforestation plantings [44,45] contribute to the aeration of soil and reduce bulk density. High soil organic matter is also associated with low bulk density [18], a trend seen in this study as bulk density was negatively correlated with SOC. This was expected as SOC is indicative of litter dynamics, turnover and decomposition rates and is responsible for providing low-density mass to the soil and improving soil structure [15,18,37,46]. The bulk density values in our study (0.65 g cm^{-3}) are low in comparison to average soils (>1.2 g cm^{-3}). Paul [14] also reported low bulk densities (0.6–0.8 g cm^{-3}) for similar ferrosols in reforestation plantings and remnant rainforest, confirming the low bulk density of this soil type. We attribute these low bulk densities to the friable nature of the ferrosols in the study region, high porosity and the high organic matter content in the topsoil. We also propose that the low bulk density contributes to the trend of increasing infiltration rates with time following reforestation as the two variables were inversely correlated.

Similar to prior studies, we found a significant increase in infiltration rates with reforestation age, with infiltration rates of the

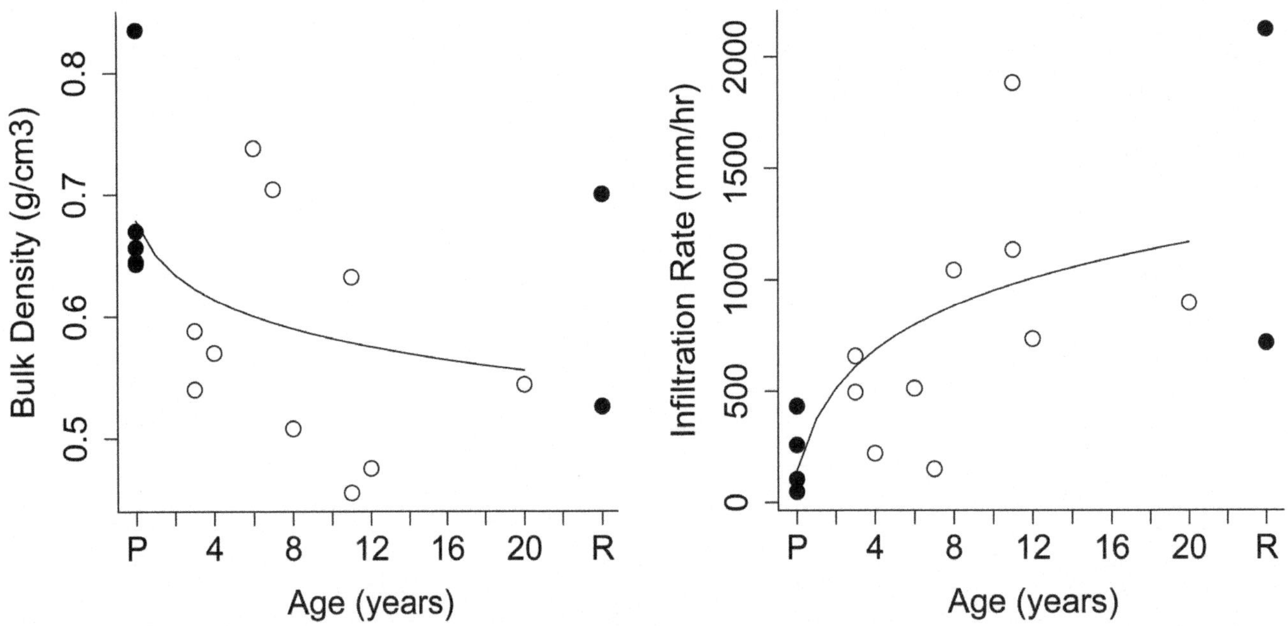

Figure 1. Differences in bulk density (left panel) and infiltration rates (right panel) among pasture (P), reforestation sites aged 2 to 20 years, and remnant rainforest (R) (n = 5, 10 and 2 respectively). Lines show fitted values of the modeled relationship between soil properties and log-transformed age of reforestation. Open circles show reforestation plantings, filled circles indicate pasture clusters and remnant riparian rainforest.

pastures being always lower than those of adjacent reforestation sites. We note a strong positive pattern with age within the first 5–10 years post reforestation, indicating that even young tree plantings are able to promote increased infiltration rates [37,47]. The trajectory in infiltration rates from the plantings (up to 20 years old) indicate that increases of 182 mm h^{-1} can be expected for each year in the first decade since planting, with most gains seen within the first 10 years post-reforestation. This trend is consistent with findings of Bharati [48] of short term changes in infiltration rates under riparian buffers when compared to adjacent cultivated fields and pastures, noting a 5-fold increase in infiltration rates under riparian buffers after 6 growing seasons. Our finding that infiltration rates are strongly enhanced soon after reforestation has meaningful consequences for decision making on land use, water quality and conservation with reduced run-off, slowed surface erosion and reduced non-point source pollution as beneficial outcomes [49].

Here, two of the tested nine soil attributes varied between pasture and reforestation sites. One possible explanation is the recentness of the plantings. Reforestation sites were relatively recent compared to studies that focused on plantings of much greater age (up to 60 years) [13,38,43] and this may have contributed to the uniformity between sites in our study. However, the common lack of differences between the studied pasture and remnant riparian rainforest sites suggests that inherent high soil fertility may be a larger contributor to soil properties than land cover, as has been suggested previously [43,50].

4.3. Species specific effects on the recovery of soils under reforestation plantings

In previous studies, species-specific effects were not used to help elucidate changes in soil properties post-reforestation [14,51], although it is known that species-specific effects can modify soil properties [52]. We detected two species-specific effects in the five tree species and eight variables examined. *Ficus coronata* and *Flindersia schottiana* showed negative relationships with time since reforestation for fine root biomass and soil ammonium levels respectively. We hypothesized that as reforested sites become more established, fine root biomass increases to access nutrients in the surface soil layer of rainforests [41,43,53]. The studied *F. coronata* specimens displayed the opposite trend to that expected, showing a negative fine root relationship with time post-reforestation. This decreasing trend exhibited by *F. coronata* may be explained by a need for greater structural support through large woody roots in which they invest carbon at the expense of fine surface roots, or the ability to access nutrients from the deeper soil, however greater sampling at depth would be needed to confirm these hypotheses. Nevertheless, if verified, an ability to shift root production into in deeper layers could be a useful trait providing physical impediment in the subsoil, particularly where greater soil stability is a desired goal of reforestation.

Another species-specific effect was observed with *F. schottiana* showing a decreasing relationship with ammonium concentrations and time post-reforestation. This result was not entirely unusual as ammonium is readily converted to nitrate in some soils and nitrate may be preferentially used by plants [54]. However, we did not detect higher nitrate concentration in soil under *F. schottiana*, which would be expected if nitrification is simulated. This may be due to high N uptake and demand of *F. schottiana,* which was not investigated here. However, this finding conforms broadly to the results of prior studies [14,38,55], which saw that ammonium levels were highest in pasture and lowest in forest.

It is noteworthy that the species-specific effects detected were not the same as those that were seen at a site level. When means at the site level were considered, only infiltration and bulk density were significant, however at a species level, both *F. schottiana* and *F. coronata* showed significant variations through time for ammonium levels and fine root biomass respectively. This suggests that while individual species may be able to preferentially alter their immediate surroundings, their influence is diminished at a

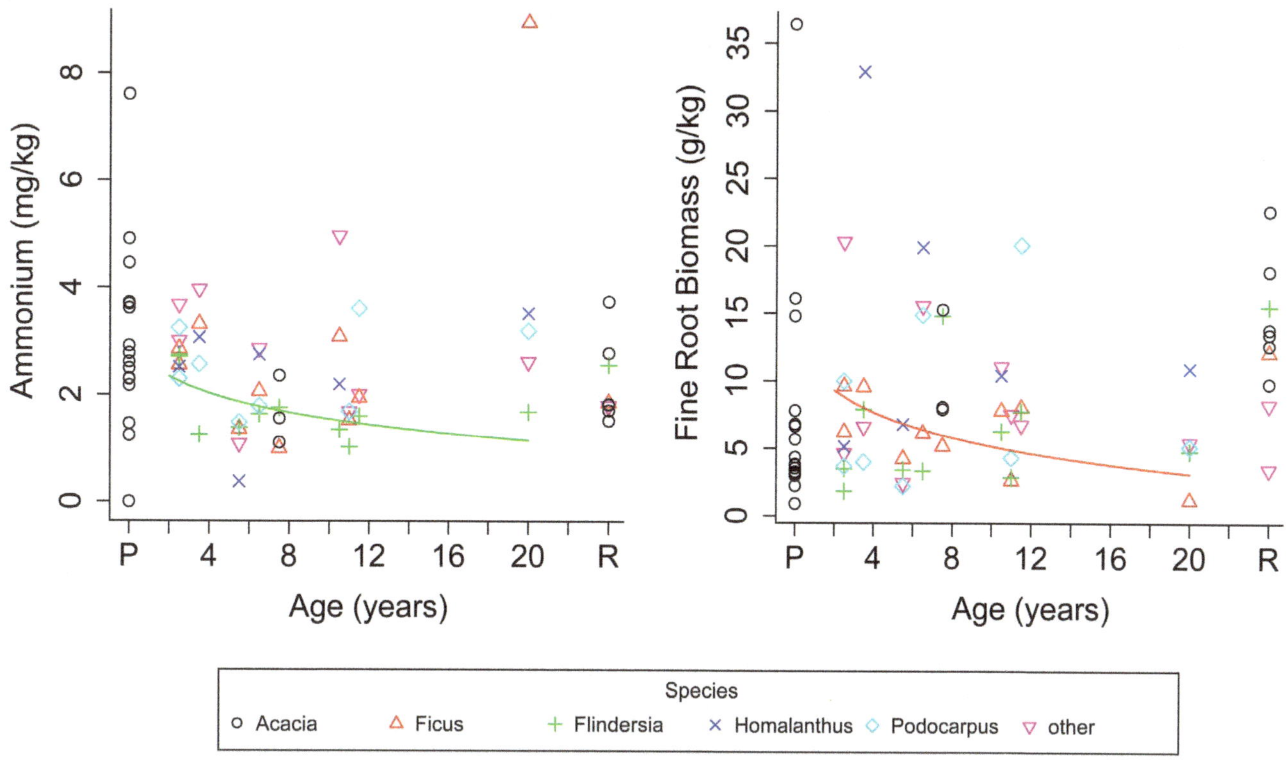

Figure 2. Differences in ammonium (left panel) and fine root biomass (right panel) over time under five tree species in pasture (P), reforestation sites aged 2 to 20 years, and remnant rainforest (R). Genus names refer to the selected tree species under which the soil samples were taken. "Other" refers to soil samples taken from non-specified tree species or pastoral sites. Lines show fitted values of the modeled relationship between individual soil properties and log-transformed age for species with significant models - *Flindersia schottiana* (a) and *Ficus coronata* (b).

site level. A possible explanation is that the diverse mix of native tree species, representing a variety of physiotypes, and the individual species-specific effects on soil are less influential than the effects of the overall plantings on soil properties. Had the reforestation planting been of a simpler species mix or the relative proportion of select species been greater, soil traits may have reflected more strongly particular species-specific effects. This indicates that particular reforestation outcomes may be facilitated by explicit selection of certain tree species. Further studies have to determine the efficacy and benefits of a less species-rich reforestation species mixture.

Conclusion

Reduced bulk density and enhanced infiltration rates were the key immediate outcomes resulting from riparian reforestation plantings. It is noteworthy that the most change occurred within just 5 to 10 years post-planting suggesting that this form of active restoration can reinstate important soil functions within a comparatively short time. Given the linkages between increased infiltration and ecosystem services such as nutrient buffering, filtration of runoff and sediments, we conclude that the reforestation is, by extension, achieving the objective of improving water quality. This finding has consequences for land managers and conservationists interested in lowering runoff, non-point source erosion and limiting loss of topsoil, all of which are major concerns and motivators for reforestation. Further, soil infiltration rates are intricately linked with several soil properties influenced by slight changes in soil structure well before changes are perceivable in other soil properties [56]. This signifies that the reforestation plantings are on a positive trajectory to restoring other soil properties and ecosystem function with infiltration rates as front-runner in these changes.

Acknowledgments

We thank the landholders that participated in the study for generously allowing access to their land. We also thank the Lake Baroon Catchment Care Group (LBCCG) for liaising with landholders, being so knowledgeable and motivated, and accommodating to our needs. We acknowledge the advice and thank Jessica Vogt and Richard Brackin on sample processing.

Author Contributions

Conceived and designed the experiments: RG GK MA SS LPS. Performed the experiments: RG MB MA SS LPS. Analyzed the data: RG GK LPS. Contributed reagents/materials/analysis tools: RG MA GK SS LPS NR. Contributed to the writing of the manuscript: RG MB GK MA SS LPS.

References

1. Connor S, Nelson PN, Armour JD, Hénault C (2012) Hydrology of a forested riparian zone in an agricultural landscape of the humid tropics. Agr Ecosyst Environ. 180: 111–122.
2. Meynendonckx J, Heuvelmans G, Muys B, Feyen J (2006) Effects of watershed and riparian zone characteristics on nutrient concentrations in the River Scheldt Basin. Hydrol Earth Syst Sci Discuss 3: 653–679.
3. Orzetti LL, Jones RC, Murphy RF (2010) Stream condition in Piedmont streams with restored riparian buffers in the Chesapeake Bay watershed. J Am Water Resour As 46: 473–485.
4. Russo TA, Fisher AT, Roche JW (2012) Improving riparian wetland conditions based on infiltration and drainage behavior during and after controlled flooding. J Hydrol 432: 98–111.
5. Vidon P (2012) Towards a better understanding of riparian zone water table response to precipitation: surface water infiltration, hillslope contribution or pressure wave processes? Hydrol Processes 26: 3207–3215.
6. Charron I, Lalonde O, Roy AG, Boyer C, Turgeon S (2008) Changes in riparian habitats along five major tributaries of the saint Lawrence river, Québec, Canada: 1964–1997. River Res Appl 24: 617–631.
7. Griscom HP, Griscom BW, Ashton MS (2009) Forest Regeneration from Pasture in the Dry Tropics of Panama: Effects of Cattle, Exotic Grass, and Forested Riparia. Restor Ecol 17: 117–126.
8. Scott SJ, McLaren G, Jones G, Harris S (2010) The impact of riparian habitat quality on the foraging and activity of pipistrelle bats (Pipistrellus spp.). J Zool 280: 371–378.
9. Rodrigues Nogueira Jr L, de Moraes Goncalves JL, Lex Engel V, Parrotta J (2011) Soil dynamics and carbon stocks 10 years after reforestation of degraded land using Atlantic Forest tree species. Forest Syst 20: 536–545.
10. Naiman RJ, Décamps H, McClain ME (2005) Riparia: ecology, conservation and management of streamside communities. New York: Academic Press.
11. Sweeney BW, Bott TL, Jackson JK, Kaplan LA, Newbold JD, et al. (2004) Riparian deforestation, stream narrowing, and loss of stream ecosystem services. Proc Nat Acad Sci USA 101: 14132–14137.
12. Guo LB, Gifford M (2002) Soil carbon stocks and land use change: a meta analysis. Glob Change Biol 8: 345–360.
13. Maloney KO, Garten CT, Ashwood TL (2008) Changes in soil properties following 55 years of secondary forest succession at Fort Benning, Georgia, USA. Restor Ecol 16: 503–510.
14. Paul M, Catterall CP, Pollard PC, Kanowski J (2010) Recovery of soil properties and functions in different rainforest restoration pathways. Forest Ecol Manage 259: 2083–2092.
15. Peñuela MC, Drew AP (2004) A model to assess restoration of abandoned pasture in Costa Rica based on soil hydrologic features and forest structure. Restor Ecol 12: 516–524.
16. Zhu B, Li Z, Li P, Liu G, Xue S (2010) Soil erodibility, microbial biomass, and physical-chemical property changes during long-term natural vegetation restoration: a case study in the Loess Plateau, China. Ecol Res 25: 531–541.
17. Bronick CJ, Lal R (2005) Soil structure and management: a review. Geoderma 124: 3–22.
18. Murty D, Kirschbaum MUF, McMurtrie RE, McGilvray H (2002) Does conversion of forest to agricultural land change soil carbon and nitrogen? a review of the literature. Glob Change Biol 8: 105–123.
19. Schoenholtz SH, Van Miegroet H, Burger JA (2000) A review of chemical and physical properties as indicators of forest soil quality: challenges and opportunities. Forest Ecol Manag 138: 335–356.
20. Catterall C, Harrison DA (2006) Rainforest Restoration Activities in Australia's Tropics and Subtropics. Cairns, Australia: Cooperative Research Centre for Tropical Rainforest Ecology and Managemen, Rainforest CRC. 0864437692 0864437692.
21. Catterall CP, Kanowski J, Wardell-Johnson GW, Proctor H, Reis T, et al. (2004) Quantifying the biodiversity values of reforestation: perspectives, design issues and outcomes in Australian rainforest landscapes. In: Conservation of Australia's Forest Fauna (ed Lunney, D), Vol. 2. Royal Zoological Society of New South Wales, Sydney 2: 359–393.
22. Kanowski J, Catterall C, Wardell-Johnson G, Proctor H, Reis T (2003) Development of forest structure on cleared rainforest land in eastern Australia under different styles of reforestation. Forest Ecol Manage 183: 265–280.
23. Kanowski J, Catterall CP (2010) Carbon stocks in above-ground biomass of monoculture plantations, mixed species plantations and environmental restoration plantings in north-east Australia. Ecol Manage Restor 11: 119–126.
24. Wickham JD, Wade TG, Riitters KH (2011) An environmental assessment of United States drinking water watersheds. Landscape Ecol 26: 605–616.
25. Australian Bureau of Statistics (2011) Census.
26. Ellison L, Coaldrake JE (1954) Soil mantle movement in relation to forest clearing in Southeastern Queensland. Ecology 35: 380–388.
27. Willmott WF (2007) Rocks and landscapes of the Sunshine Coast. Brisbane: Geological Society of Australia, Queensland Division.
28. Nakamura A, Catterall CP, House APN, Kitching RL, Burwell CJ (2007) The use of ants and other soil and litter arthropods as bio-indicators of the inpacts of rainforest clearing and subsequent land use. J Insect Conserv 11: 177–186.
29. Smith I (2001) Juggling-balancing Tourism and Biodiversity in a Small, Valuable Reserve - Mary Cairncross Scenic Reserve. In: Boyes B, editor. *Biodiversity Conservation "From Vision to Reality"*. Gatton, Queensland: Lockyer Watershed Management Association (LWMA) Inc. - Lockyer Landcare Group. 81–84.
30. McIntyre D (1976) Subplasticity in Australian soils. I. Description, occurrence, and some properties. Aust J Soil Res 14: 227–236.
31. DeForest JL (2011) Soil texture - Hydrometer method. Lab protocols for testing soils Ohio University, Athens, Ohio: The Department of Environmental and Plant Biology.
32. Robertson GP, Coleman DC, Bledsoe CS, Sollins P (1999) Standard soil methods for long-term ecological research. New York: Oxford University Press.
33. Perroux KM, White I (1988) Designs for Disc Permeameters. Soil Sci Soc Am J 52: 1205–1215.
34. Kandeler E, Gerber H (1988) Short-term assay of soil urease activity using colorimetric determination of ammonium. Biol Fert Soils 6: 68–72.
35. Miranda KM, Espey MG, Wink DA (2001) A rapid, simple spectrophotometric method for simultaneous detection of nitrate and nitrite. Nitric Oxide 5: 62–71.
36. R Core Team (2013) R: A language and environment for statistical computing. Vienna, Austria: R Foundation for Statistical Computing.
37. Neris J, Jiménez C, Fuentes J, Morillas G, Tejedor M (2012) Vegetation and land-use effects on soil properties and water infiltration of Andisols in Tenerife (Canary Islands, Spain). Catena 98: 55–62.
38. Reiners WA, Bouwman AF, Parsons WFJ, Keller M (1994) Tropical rain forest conversion to pasture: changes in vegetation and soil properties. Ecol Appl 4: 363–377.
39. Bell LW, Kirkegaard JA, Swan A, Hunt JR, Huth NI, et al. (2011) Impacts of soil damage by grazing livestock on crop productivity. Soil Tillage Res 113: 19–29.
40. Beven K, Germann P (1982) Macropores and water flow in soils. Water Resour Res 18: 1311–1325.
41. Finér L, Ohashi M, Moguchi K, Hirano Y (2011) Factors causing variation in fine root biomass in forest ecosytems. Forest Ecol Manage 261: 265–277.
42. Jiménez EM, Moreno FH, Peñuela MC, Patiño S, Lloyd J (2009) Fine root dynamics for forests on contrasting soils in the Colombian Amazon. Biogeoscience 6: 2809–2827.
43. Powers JS, Perez-Aviles D (2012) Edaphic factors are a more important control on surface fine roots than stand age in secondary tropical dry forests. Biotropica 45: 1–9.
44. Colloff MJ, Pullen KR, Cunningham SA (2010) Restoration of an ecosystem function to revegetation communities: The role of invertebrate mscropores in enhancing soil water infiltration. Restor Ecol 18: 65–72.
45. Nakamura A, Proctor H, Catterall CP (2003) Using soil and litter arthropods to assess the state of rainforest restoration. Ecol Manage Restor 4: S20–S28.
46. Laub BG, McDonough OT, Needelman BA, Palmer MA (2013) Comparison of designed channel restoration and riparian buffer restoration effects on riparian soils. Restor Ecol: 1–9.
47. Wood HB (1977) Hydrologic differences between selected forested and agricultural soils in Hawaii. Soil Sci Soc Am J 41: 132–136.
48. Bharati L, Lee K-H, Isenhart TM, Schultz RC (2002) Soil-water infiltration under crops, pasture, and established riparian buffer in Midwestern USA. Agrofor Syst 56: 249–257.
49. Schultz R, Isenhart T, Simpkins W, Colletti J (2004) Riparian forest buffers in agroecosystems–lessons learned from the Bear Creek Watershed, central Iowa, USA. Agrofor Syst 61: 35–50.
50. Lamb D (1980) Soil nitrogen mineralisation in a secondary rainforest succession. Oecologia 47: 257–263.
51. Paul M, Catterall CP, Pollard PC, Kanowski J (2010) Does soil variation between rainforest, pasture and different reforestation pathways affect the early growth of rainforest pioneer species? Forest Ecol Manage 260: 370–377.
52. Bezemer TM, Lawson CS, Hedlund K, Edwards AR, Brook AJ, et al. (2006) Plant species and functional group effects on abiotic and microbial soil properties and plant–soil feedback responses in two grasslands. J Ecol 94: 893–904.
53. Claus A, George E (2005) Effect of stand age on fine-root biomass and biomass distribution in three European forest chronosequences. Can J Forest Res 35: 1617–1625.
54. Piccolo MC, Neill C, Cerri CC (1994) Net nitrogen mineralization and net nitrification along a topical forest-to-pasture chronosequence. Plant Soil 162: 61–70.
55. Rasiah V, Florentine SK, Williams BL, Westbrooke ME (2004) The impact of deforestation and pasture abandonment on soil properties in the wet tropics of Australia. Geoderma 120: 35–45.
56. Radke JK, Berry EC (1993) Infiltration as a tool for detecting soil changes due to cropping, tillage, and grazing livestock. Am J Alternative Agr 8: 164–174.

Effect of Drying on Heavy Metal Fraction Distribution in Rice Paddy Soil

Yanbing Qi[1,2]*, Biao Huang[3], Jeremy Landon Darilek[3]

1 College of Resources and Environment, Northwest Agriculture and Forestry University, Yangling, Shaanxi, People's Republic of China, **2** Key Laboratory of Plant Nutrition and the Agri-environment in Northwest China, Ministry of Agriculture, Yangling, Shaanxi, People's Republic of China, **3** State Key Laboratory of Soil and Sustainable Agriculture, Institute of Soil Science, Chinese Academy of Sciences, Nanjing, Jiangsu, People's Republic of China

Abstract

An understanding of how redox conditions affect soil heavy metal fractions in rice paddies is important due to its implications for heavy metal mobility and plant uptake. Rice paddy soil samples routinely undergo oxidation prior to heavy metal analysis. Fraction distribution of Cu, Pb, Ni, and Cd from paddy soil with a wide pH range was investigated. Samples were both dried according to standard protocols and also preserved under anaerobic conditions through the sampling and analysis process and heavy metals were then sequentially extracted for the exchangeable and carbonate bound fraction (acid soluble fraction), iron and manganese oxide bound fraction (reducible fraction), organic bound fraction (oxidizable fraction), and residual fraction. Fractions were affected by redox conditions across all pH ranges. Drying decreased reducible fraction of all heavy metals. $Cu_{residual\ fraction}$, $Pb_{oxidizable\ fraction}$, $Cd_{residual\ fraction}$, and $Ni_{residual\ fraction}$ increased by 25%, 33%, 35%, and >60%, respectively. $Pb_{residual\ fraction}$, $Ni_{acid\ soluble\ fraction}$, and $Cd_{oxidizable\ fraction}$ decreased 33%, 25%, and 15%, respectively. Drying paddy soil prior to heavy metal analysis overestimated Pb and underestimated Cu, Ni, and Cd. In future studies, samples should be stored after injecting N_2 gas to maintain the redox potential of soil prior to heavy metal analysis, and investigate the correlation between heavy metal fraction distribution under field conditions and air-dried samples.

Editor: Wen-Xiong Lin, Agroecological Institute, China

Funding: This work was financially supported by the Natural Science Foundation of China (31100516), and the National "863" Plan Project of China (No. 2013AA102401). The funders had no role in study design, data collection and analysis, decision to publish, or preparation of the manuscript.

Competing Interests: The authors have declared that no competing interests exist.

* E-mail: yanbing_qi@sina.com

Introduction

Soil heavy metals Cu, Pb, Ni, and Cd, are regarded as "chemical time bombs" because of their propensity for accumulation in the soil and uptake by crops. This ultimately causes human toxicity in both the short- and long-term [1–2], making farmland ecosystems dangerous to health [3].

Morphological characteristics and processes of heavy metals have been studied to better understand heavy metal occurrence in various environments, transport pathways, and crop uptake. There are a wide range of soil inorganic and organic substances affected by redox conditions. Redox changes the valence of ions, subsequently affecting the forms of various elements and compounds and their transport processes. Furthermore, heavy metal behavior is strongly correlated with redox potential (Eh) [4].

Many studies of river and lake sediments have reported the effect of redox conditions on the distribution of heavy metals. For example, such changes in redox conditions affect heavy metal association with organic matter (OM) [5], and iron (Fe) and manganese (Mn) oxidized fractions are unstable in reduced environments [3,6]. Kelderman and Osman [2] reported that in river sediments, exchangeable and carbonate bound fractions of Cu, Zn, and Pb increased as Eh increased, and organic fractions, oxidizable Cu, and oxidizable Pb decreased. Lu et al. [7] found that in reduced Iraqi River sediment in Changchun City, Mn oxide fractions increased significantly, organic bound Cu decreased by 40%, but decrease in organic bound Pb, Zn, and Ni was less than Cu.

Soil heavy metal studies which include plant available indices, total concentrations, and fraction distributions have also been conducted on many agricultural systems, including rice (*Oryza sativa*) production. Rice has a high water requirement compared to other crops and during the growing season, Eh can reach roughly 300–200 mV [4]. Soil samples, including from rice paddies under submerged conditions, are routinely air-dried prior to heavy metal analysis [8–12]. Zheng and Zhang [13] studied the effect of moisture regimes on paddy soil heavy metals and found that soil moisture did not affect the direction or pathways of fractionation distribution (from active to stable fractions), but did affect the transformation rate. Zheng and Zhang [13] dried rice paddy soil samples after collection and then reconstituted three moisture regimes under controlled conditions in the lab. It does not appear that sample preparation was conducted in an anaerobic environment, therefore this result does not represent in situ soil moisture regime that controls the distribution of heavy metals.

Redox conditions are known to have a significant effect on heavy metal speciation in sediments [14]. As indicated by Calmano et al. [5], if anoxic sediments are exposed to atmosphere, redox condition change and redistribution and transformation of heavy metal fractions in the sediments takes place. A few studies show the effect of redox conditions on heavy metal availability [15–17]. Paddy soil has the anoxic condition during rice growing season similar to the river sediment. Thus, we hypothesize that drying prior to analysis for paddy soil changes the anoxic condition to redox condition as a consequence mobility and uptake of heavy metal by crop are changes. To date, the effect of drying on soil heavy metal fraction distribution in rice paddies is largely unknown.

In order to fill this knowledge gap and work towards a better methodology for sampling and handling paddy soil prior to heavy metal analysis, we conducted a study in Zhangjiagang County, a typical rice production region of China, to 1) understand heavy metal distribution in paddy soil under various redox and pH conditions, and 2) determine the effect of soil air-drying on heavy metal fractions. This study intends to provide useful information for heavy metal assessment in situ. It is hoped that through a better understanding of heavy metal fractions in situ, a better understanding of correlation of heavy metal between soils and crops can be drawn, and a suitable set of corrective measures to prevent heavy metal pollution can be put in place.

Materials and Methods

Ethics Statement

All the sample sites were distributed in the private land and permission were approved by the land owner in each site (Zhigang Wang can be contacted for the future permissions). The field studies did not involve endangered or protected species because all the sample sites were in the farmland with rice planted. The coordinates of sample sites ranged from 120°35′ to 120°42′E and from 31°44′ to 31°51′N.

Sample Collection

Zhangjiagang County was selected as the research area because of two reasons, firstly, the soil is typical paddy soil with rice planting history of more than 100 years, secondly, heavy metals have been accumulated slightly in this area because of sparkly distributed chemical factories. Zhangjiagang County has a humid monsoon climate in the north subtropical zone, with four distinct seasons, and average annual temperature and precipitation is 15.0°C and 1045.9 mm, respectively.

An investigation of soil physical and chemical characteristics of Zhangjiagang County, Jiangsu, was conducted in 2004 [18–19]. In the current study, ten locations from the 2004 investigation, each with a long history of rice production, were selected to represent a wide pH range (4.81–8.16).

Topsoil (0–20 cm) was collected in October, 2007 from rice paddies before the harvest and after a full season of submersion. At the time of collection, there was still a 1–2 cm water layer covering the soil. Each soil sample was a composite of sub-samples taken from 6–8 points within 50 m^2 with a stainless steel soil probe. Half of the sample was quickly placed in polyethylene plastic tubes and the headspace was immediately purged with nitrogen (N) gas [20], pre-treated with acid, and weighed. Once transported to the laboratory, samples were stored under refrigeration at 4°C for the determination of heavy metals fractions under reduction conditions. The remaining sample was placed in polyethylene bags.

Chemical Analysis

Soil moisture was measured using soil from the polyethylene bags by drying at 105°C for 24 h. The remaining soil from the polyethylene bags were air-dried and sieved through a 2 mm sieve. Soil pH was measured with a glass electrode in a 1:2.5 soil: water suspension [21]. Soil OM was determined using the dichromate-wet combustion method [22]. The samples were digested using HNO_3-$HClO_4$-HF for measurement of the total heavy metals [19].

The Community Bureau of Reference (BCR) sequential extraction procedure described by Ure et al. [23] was used to extract heavy metal fractions. For the exchangeable and carbonate bound fraction (acid soluble fraction), acetic acid (20 ml, 0.11 mol l^{-1}) was added to a 50 ml polypropylene centrifuge tube containing 0.5 g sample and shaken for 16 h at ambient temperature (20±1°C) on an end-over mechanical shaker operating at 40 rpm. The extract was centrifuged at 4000 rpm, decanted into a polyethylene container, and stored at −4°C until analysis. For the Fe and Mn oxide bound fraction (reducible fraction), the residue from the step above was shaken with hydroxylamine hydrochloride (20 ml, 0.1 mol l^{-1}, acidified to pH 2 with nitric acid). This was also centrifuged, decanted, and stored as described above. For the organic bound fraction (oxidizable fraction), the residue from reducible fraction was shaken with 30% hydrogen peroxide (20 ml, acidified to pH 2 with nitric acid) and then ammonium acetate (25 ml, 1 mol l^{-1}, acidified to pH 2 with nitric acid). This was also centrifuged, decanted, and stored as above. For the residual fraction, the residue from oxidizable fraction was digested using the method described above for the total digestion of heavy metals.

The entire extraction procedure for soil samples in polyethylene tubes was the same as described for the air-dried samples except that samples were handled in an anaerobic chamber, using N as the purge gas. All analysis of heavy metal concentrations was determined using inductively coupled plasma-atomic emission spectrometry (ICP-AES).

Statistical Analysis

Analysis of variance was performed using the GLM procedure. Mean separations were performed using Duncan's Multiple Range to test soil pH and redox on soil heavy metal fractions. Linear relationship (correlation) within fractions and regression within the percentage distribution of each metal fraction in oxidized and reduced condition were conducted. All statistical analysis was conducted using SPSS (version 13.0).

Results

Soil Physicochemical Properties

Soil physicochemical properties of rice paddies in Zhangjiagang are shown in Table 1. Average soil moisture of submerged soil was 30.2% and ranged from 24.6% to 33.9%. Soil OM ranged from 19.40 g kg^{-1} to 31.30 g kg^{-1} and had a mean of 25.47 g kg^{-1}. The most abundant metal was Pb, which had a mean concentration of 43.27 mg kg^{-1}. The least abundant metal was Cd, which had a mean concentration of 0.20 mg kg^{-1}. The coefficient of variation (CV) for Cd (41.3%) was higher than Cu, Pb, and Ni (25.6%, 8.8% and 9.3%, respectively).

Soil Heavy Metal Fraction Distribution

Heavy metal fraction distribution varied among elements (Fig. 1 and 2). Redox conditions had a significant effect on all Cu fractions (Fig. 1, Table 2), soil pH did not affect Cu fractions, and there was only a significant interaction effect of redox conditions and pH with reducible fraction. The predominant fractions were reducible fraction and oxidizable fraction (35–40%), followed by residual fraction (15–18%) and acid soluble fraction (<1%). Air-dried samples had significantly different Cu fraction distribution. Compared with reduced soil, acid soluble fraction was still the smallest fraction accounting for 5% of the total Cu. Reducible fraction and oxidizable fraction had significantly lower proportions

Table 1. Soil chemical properties of rice paddies in Zhangjiagang County, Jiangsu.

Samples	pH	Moisture (%)	Soil Organic Matter (g kg^{-1})	Cu	Pb	Ni	Cd
				(mg kg^{-1})			
1	4.81	33.93	31.30	34.87	48.96	28.82	0.236
2	5.54	30.75	26.50	28.35	38.10	26.41	0.087
3	5.87	26.74	26.80	28.57	41.64	24.66	0.160
4	6.12	31.21	28.70	26.64	40.10	25.70	0.190
5	6.45	24.62	23.70	46.71	42.92	30.15	0.107
6	6.94	28.10	29.30	25.73	39.13	24.61	0.135
7	7.17	29.77	19.40	32.56	43.46	31.74	0.180
8	7.61	31.79	21.10	30.99	44.81	28.80	0.327
9	8.02	32.66	26.70	53.48	44.26	28.54	0.300
10	8.16	32.53	21.20	30.26	49.34	24.62	0.281
Average	6.67	30.2	25.47	33.82	43.27	27.41	0.200
St dev	1.10	2.9	3.95	8.66	3.79	2.55	0.083
Cv (%)	16.49	9.7	15.53	25.62	8.76	9.31	41.28

while residual fraction had significantly higher proportion and was the predominant fraction.

Redox conditions affected all Pb fraction distributions except reducible fraction, and pH did not affect fraction distribution ($p> 0.05$) (Table 2). There was also no interaction effect of redox conditions and pH with Pb fractions. In reduced soil, the predominant fraction was residual fraction (60%), followed by reducible fraction (20%), oxidizable fraction and acid soluble fraction. In oxidized soil, acid soluble fraction was 3% higher than in the reduced soil, oxidizable fraction was 20% higher and residual fraction was 33% lower, respectively, than in reduced soil, and reducible fraction was not significantly different.

Redox conditions affected all Ni fraction distributions and pH affected all fraction distributions except oxidizable fraction (Table 2). In reduced soil with pH 6 to 7, $Ni_{oxidizable\ fraction}$ had the lowest proportion, and $Ni_{residual\ fraction}$ had the highest (Fig. 1). There was significant interaction of redox conditions and pH with acid soluble fraction. In oxidized soil, acid soluble fraction decreased from 30% to 5% and reducible fraction decreased from 35% to 15%. The residual fraction increased by more than 60% and oxidizable fraction increased less than 5%.

Redox conditions and pH affected $Cd_{oxidizable\ fraction}$ and $Cd_{residual\ fraction}$ (Table 2). In reduced soil the predominant fraction was reducible fraction (55%), followed by oxidizable fraction (25%), acid soluble fraction and residual fraction. In oxidized soil, residual fraction increased by 15% and oxidizable fraction decreased by 15%. Treatments had no significant effect on acid soluble fraction or reducible fraction.

Regression of Reduced and Oxidized Samples

Regression analysis did not reveal significant relationships between soil heavy metal distribution of oxidized and reduced soil. None of the Cu fractions had significant regression results. For Pb, only residual fraction was significant, and for Ni, only reducible fraction was significant (Fig. 3). For Cd fractions, regression of oxidized and reduced samples had R^2 values of 0.26, 0.69, 0.48, and 0.66 for acid soluble fraction, reducible fraction, oxidizable fraction, and residual fraction, respectively (Fig. 4). Regression analysis between relative concentration in oxidized soil and relative concentration in reduced heavy soil showed significant correlation for reducible fraction, oxidizable fraction and residual fraction (Fig. 4). Regression analysis separately by pH and OM did not improve correlation between the dependent and independent variable in Fig. 4.

Discussion

Total Heavy Metal Concentrations in Rice Paddies

Soil heavy metals originate from both the environment and anthropogenic sources. Although all heavy metal concentrations in this study, except three Cd samples, were higher than typical background concentrations [24], they are lower than the critical limits established by the Soil Environmental Quality Standard of China (GB 15618–1995) [25]. The results of our study are consistent with previous research in the area [25]. In rice paddies of China, there is some evidence that this accumulation is a result of heavy metal rich irrigation water [9]. However, low concentrations of heavy metals suggest that although heavy metal concentrations in Zhangjiagang have experienced accumulation from anthropogenic sources, it is not yet a pressing environmental concern.

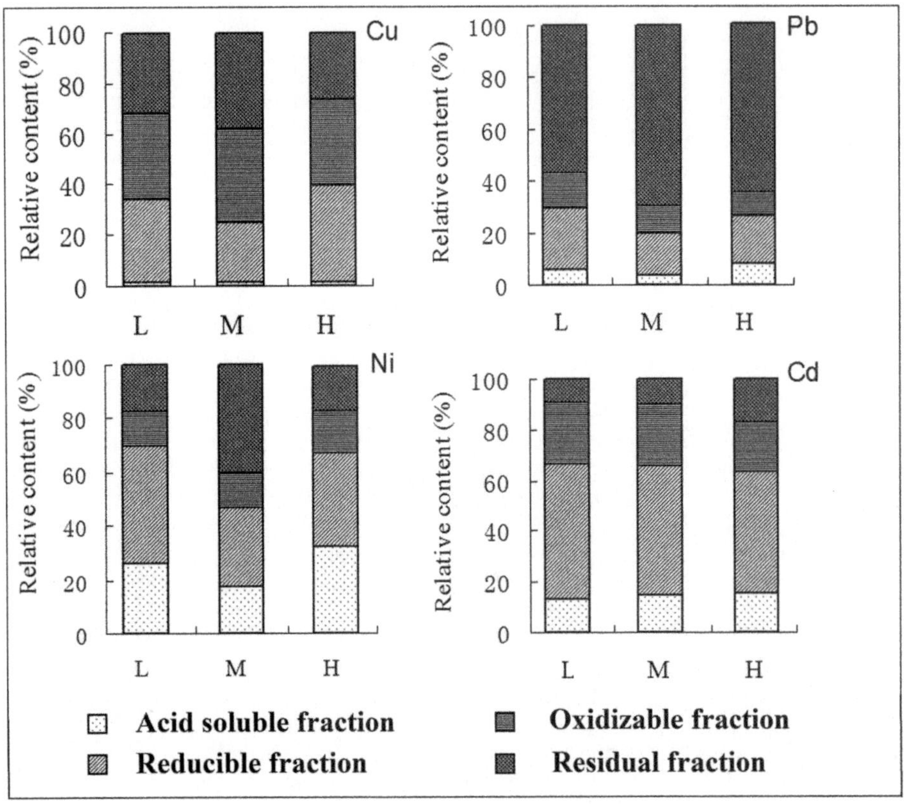

Figure 1. The percentage distribution of each metal fraction in reduced condition in paddy soil samples in Zhangjiagang County. L means lower pH (<6), M means moderate pH (6–7), H means higher pH (>7). Note: significant interactions within and between factors are presented in Table 2.

Heavy Metal Fractionation in Reduced Rice Paddies

Generally, rice paddies are submerged during rice cultivation and have lower Eh values during this part of the year. Reduced soil had lower $Cu_{acid\ soluble\ fraction}$ and $Pb_{acid\ soluble\ fraction}$ and higher $Ni_{acid\ soluble\ fraction}$, $Ni_{reducible\ fraction}$, and $Cd_{reducible\ fraction}$ than air-dried soil. This is similar to heavy metal fractionation changes that occur in various redox states of sewage sludge [26–28]. Förstner and Wittmann [3] explain that for Cu and Pb the most unstable fraction, acid soluble fraction, is replaced by reducible fraction and oxidizable fraction under reduced conditions.

For Ni and Cd, acid soluble fraction accounted for a relatively high proportion (15%–50%), which is not consistent with heavy metal fractionation ranges in sewage sludge. This may be due to Eh differences in rice paddies and sewage sludge, interactions in the rhizosphere of rice paddies, or both. Several studies address changes in soil heavy metal fraction distribution in the rhizosphere [29–31]. High $Ni_{acid\ soluble\ fraction}$ and $Ni_{reducible\ fraction}$ indicated that Ni is adsorbed to solid materials or combined with weak acid and Fe and Mn oxides in reduced soil [27]. High $Cd_{reducible\ fraction}$ was also found in dredged sediments [32].

High $Cu_{reducible\ fraction}$ and $Cu_{oxidizable\ fraction}$ in reduced soil indicates that Fe and Mn oxides play a crucial role in Cu processes, and that Fe and Mn oxides cause this fraction to be relatively stable [33–34]. This result is consistent with many studies which show dominant oxidizable fraction in reduced sediments [26,34–37].

We found the predominant fraction for Pb in rice paddies to be residual fraction. This fraction is generally considered to be the most stable heavy metal fraction in the soil and unavailable to plants because they are found in the lattice of clay mineral [35,38]. In addition to residual fraction, sediment Pb in many studies has been found to have large percentages of reducible fraction [2]. Lower reducible fraction in our study might be because of the higher Eh values, interactions in the rhizosphere of rice paddies, or both.

Effect of Drying on Heavy Metal Fractionation

When soils are routinely dried in the lab prior to chemical analysis, Eh rapidly increases and metals are oxidized. While this may not have a significant effect on heavy metal fractions in agriculture soil which is not submerged, this procedure can drastically change the heavy metal chemistry of reduced soils.

Generally, heavy metal acid soluble fraction increases as Eh increases, however, in our study, Ni and Cd did not respond accordingly. The reason for this is not clear and further study is needed. As Eh increases, reducible fraction generally decreases, most likely because of heavy metal bonding with Fe and Mn oxides [2,4–5,39]. Our study bears this out as well.

Previous research showed that the relationship of Eh with heavy metal oxidizable fraction and residual fraction is not clear [2–5,7,39]. In our study, increasing Eh led to increased $Pb_{oxidizable\ fraction}$ and decreased $Cu_{oxidizable\ fraction}$. Collavini et al. [40] suggested that decreased $Cu_{oxidizable\ fraction}$ as soil oxidizes occurs due to the interaction of copper sulfide oxidation and Fe and Mn oxides. Released Cu^{2+} is then redistributed to other fractions such as acid soluble fraction and residual fraction. Part of the obscurity may be because of the complex nature of bonding in copper sulfides, which

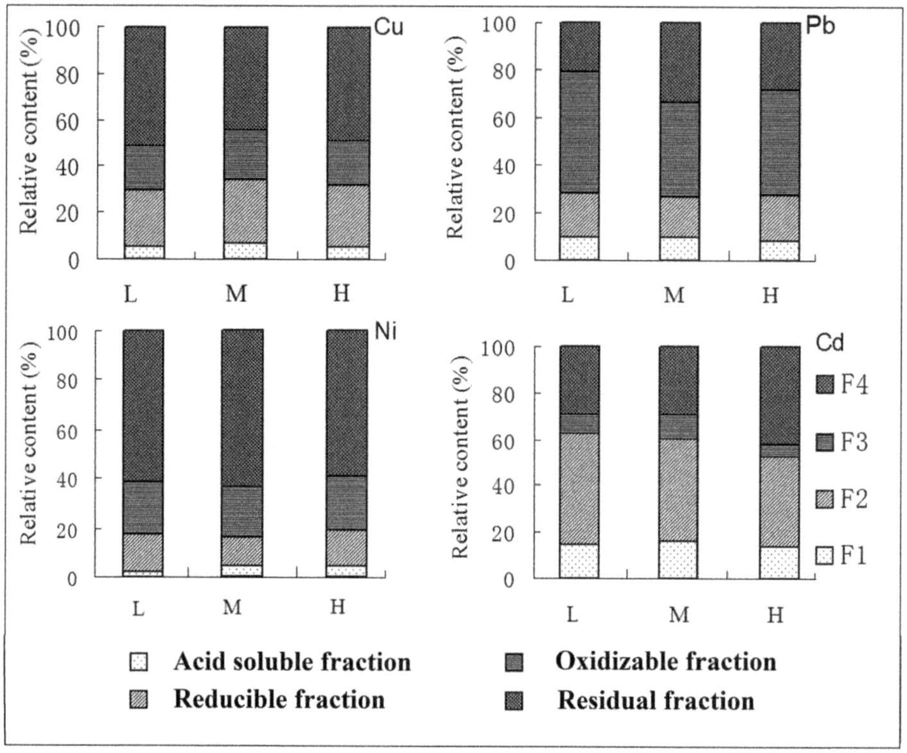

Figure 2. The percentage distribution of each metal fraction in oxidized condition in paddy soil samples in Zhangjiagang County. L means lower pH (<6), M means moderate pH (6–7), H means higher pH (>7). Note: significant interactions within and between factors are presented in Table 2.

have both covalent and ionic bonding characteristics and a high occurrence of delocalized electrons [41–42].

In this study, increasing Eh led to increased $Cu_{reducible\ fraction}$, which is in agreement with Kelderman and Osman [2], who reported a decrease in reducible fraction for river sediments under submerged conditions, and with Saeki et al. [14] who reported an increase in $Cu_{reducible\ fraction}$ after drying of lake sediments.

Drying caused an increase in $Pb_{oxidizable\ fraction}$, which contradicts previous literature [2,7]. As Eh increases, $Pb_{reducible\ fraction}$ slightly decreases, mainly due to bonding of Pb with Fe and Mn oxides [7]. Dried soil had lower $Ni_{reducible\ fraction}$ and $Cd_{oxidizable\ fraction}$ and higher $Ni_{residual\ fraction}$ and $Cd_{residual\ fraction}$ than reduced soil. This was consistent with the results of other studies [2,7,40].

Table 2. P-values from the analysis of variance of redox conditions and pH for heavy metal fraction distributions.

Heavy Metal	Factors	P value			
		Acid soluble fraction	Reducible fraction	Oxidizable fraction	Residual fraction
Cu	Redox	<0.001	0.050	<0.001	<0.001
	pH	0.685	0.109	0.325	0.555
	Interaction	0.308	0.034	0.873	0.111
Pb	Redox	0.001	0.901	<0.001	0.000
	pH	0.230	0.544	0.124	0.068
	Interaction	0.123	0.092	0.345	0.970
Ni	Redox	0.000	0.000	0.000	0.000
	pH	0.023	0.009	0.314	0.009
	Interaction	0.040	0.304	0.356	0.164
Cd	Redox	0.08	0.316	0.000	0.024
	pH	0.810	0.261	0.015	0.142
	Interaction	0.089	0.359	0.277	0.413

P-values <0.05 show significant difference within factors and significant interaction between factors for each heavy metal fraction, separately.

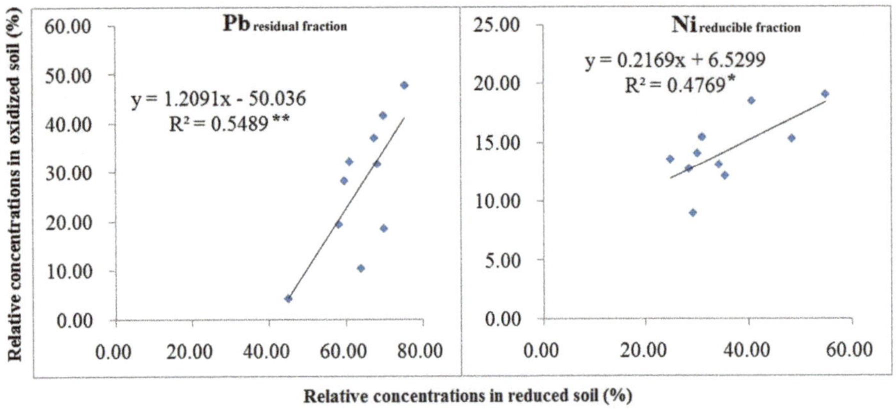

Figure 3. Correlation between the percentage distribution of iron and manganese bound nickel ($Ni_{reducible\ fraction}$) and residual lead ($Pb_{residual\ fraction}$) in oxidized and reduced paddy soil samples in Zhangjiagang County. *, and ** denote the significant correlation at $p \leq 0.05$ and 0.01, respectively.

There is strong evidence that drying significantly affects soil heavy metal fractionation of submerged rice paddy soil samples but effects are different than lake sediment, river sediment, and sewage sludge. One of the main reasons most likely is the differences in Eh of the sample sources. Generally, sediment Eh is <-100mv, while rice paddy soil Eh ranged from 200–300 mv [4]. Abundant roots in rice paddies may be another important factor. Root exudates can affect many soil properties such as pH and microbial activity. Changes in these properties can then indirectly affect soil heavy metal fraction distribution. Chen et al. [43] reported that abundant rhizosphere can significantly increase $Cu_{acid\ soluble\ fraction}$ and significantly decrease $Cd_{acid\ soluble\ fraction}$. Shuman and Wang [44] reported that rice rhizosphere significantly increased $Zn_{reducible\ fraction}$ and $Cu_{oxidizable\ fraction}$. Similarly, Lin et al. [24] reported that rice rhizosphere significantly increased $Cd_{reducible\ fraction}$ and $Cd_{oxidizable\ fraction}$.

Effects of Drying on Heavy Metal Availability

The distribution, migration, and plant availability of soil heavy metals are reflected not in the total concentrations, but in specific

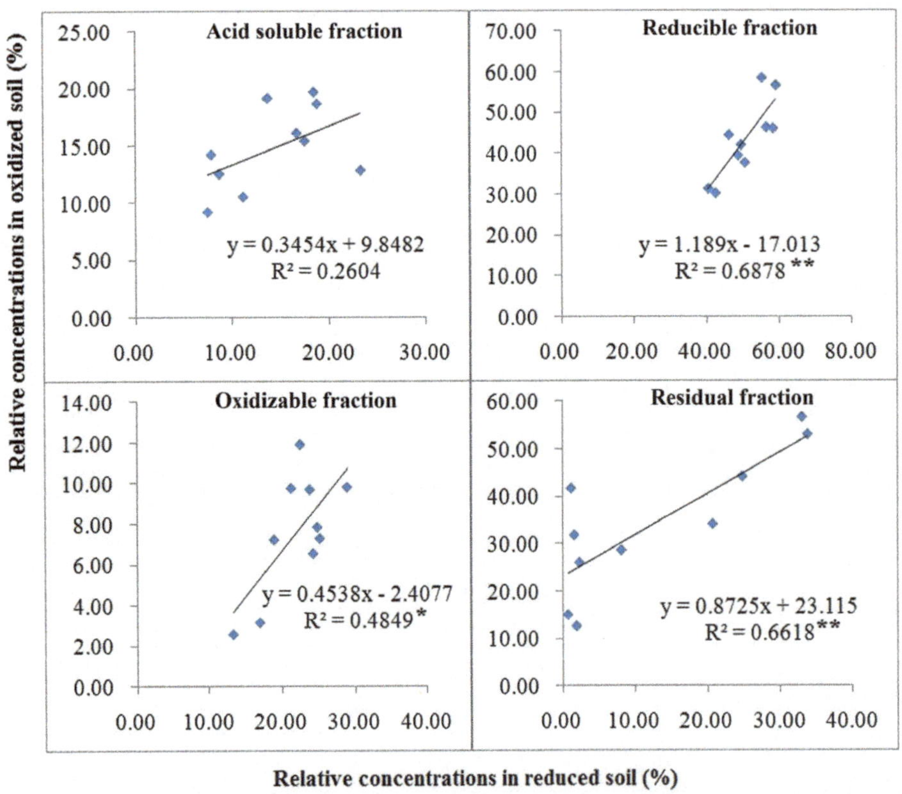

Figure 4. Correlation between the percentage distribution of cadmium in oxidized and reduced paddy soil samples in Zhangjiagang County. *, and ** denote the significant correlation at $p \leq 0.05$ and 0.01, respectively.

fractions. With the BCR sequential extraction method, heavy metal bioavailability decreases with each progressively stronger extraction [45]. acid soluble fraction has the highest bioavailability and residual fraction has the lowest. Acid soluble fraction, reducible fraction, and oxidizable fraction are considered active fractions and each contain some bioavailable heavy metals [46]. In this study, Ni and Cd have large active fractions (80%), followed by Cu (>60%), and Pb (<40%). Drying submerged paddy soil samples caused decreases in the active fractions of Cu, Ni, and Cd and increased active fractions of Pb. Zheng and Zhang [13] found an increased bioavailability with increased Eh across all heavy metals measured (Cu, Pb, Cd, and Hg). Our results suggest that when samples are dried prior to heavy metal bioavailability analysis, Pb is overestimated and Cu, Ni, and Cd are underestimated.

Regression analysis results showed that samples which have been dried prior to analysis are not suitable indicators of reduced soil conditions and cannot be used to predict field conditions. However, adding pH and OM did not improve the predictive power, particularly for Cd (data not shown).

Conclusions

Almost all heavy metal fractions were significantly affected by redox conditions, and few heavy metal fractions were affected by pH. Drying the soil prior to analysis caused metal ions to change valence and a redistribution of heavy metal fractions. Some redistribution occurred as metals moving from active fractions to stable fractions. Therefore, drying decreases the representativeness of *in situ* conditions and the availability of some heavy metals are misestimated. We suggest that soil heavy metal fractionation procedures for rice paddies ensure anaerobic conditions from the time of sampling until analysis. Methodology studies to improve sampling and lab handling techniques would be beneficial, as would further investigation of possible correlation of non-dried sampling and analysis using bioavailable extractions with air-dried samples. For some heavy metals, the processes driving redistribution trends are still unclear. Because the relationship between soil Eh and heavy metal fractionation distribution directly affects heavy metal mobility and bioavailability, a better understanding of the soil heavy metal chemistry under various redox conditions is imperative.

Author Contributions

Conceived and designed the experiments: YQ. Performed the experiments: YQ JLD. Analyzed the data: YQ BH. Contributed reagents/materials/analysis tools: BH. Wrote the paper: YQ.

References

1. Gong ZT, Huang B (1998) Studies on potential "chemical timer bombs" and their igniting factors in soils. Adv. Earth Sci. 23: 184–191.
2. Kelderman P, Osman AA (2007) Effect of redox potential on heavy metal binding forms in polluted canal sediments in Delft (the Netherlands). Water Res. 41: 4251–4261.
3. Förstner U, Wittmann GTW (1983) Metal pollution in the aquatic environment, second ed. Springer, Berlin.
4. Li YG, Xue SG, Wu XY (2004) Transport and transformation of heavy metals in the soil-paddy plants system. Geo. China. 31(supp.): 87–92.
5. Calmano W, Hong J, Förstner U (1993) Binding and mobilization of heavy metals in contaminated sediments affected by pH and redox potential. Water Sci. Tech. 28: 223–235.
6. Stumm W, Morgan JJ (1996) Aquatic chemistry: an introduction emphasizing chemical equilibria in natural water, third ed. Wiley, New York.
7. Lu YZ, Dong DM, Fu Y (2006) Comparison of speciation patterns of heavy metals in Yitong River sediment under reduced and oxidized conditions. Chem. J. Chinese. Uni. 27: 449–453.
8. Andreu V, Gimeno-García E (1999) Evolution of heavy metals in marsh areas under rice farming. Environ. Pollution 104: 271–282.
9. Luo YM, Jiang XJ, Wu LH, Song J, Wu SC, Lu RH, Christie P (2003) Accumulation and chemical fractionation of Cu in a paddy soil irrigated with Cu-rich wastewater. Geoderma, 115: 113–120.
10. Li JX, Yang XE, He ZL, Jilani G, Sun CY, Chen SM (2007) Fractionation of lead in paddy soils and its bioavailability to rice plants. Geoderma, 141: 174–180.
11. Zhao KL, Liu XM, Xu JM, Selim HM (2010) Heavy metal contaminations in a soil-rice system: Identification of spatial dependence in relation to soil properties of paddy fields. J. Hazard. Mater. 181: 778–787.
12. Ngoc MN, Dultz S, Kasbohm J (2009) Simulation of retention and transport of copper, lead and zinc in a paddy soil of the Red River Delta, Vietnam, Agr. Ecosyst. Environ. 129: 8–16.
13. Zheng SA, Zhang MK (2011) Effect of moisture regime on the redistribution of heavy metals in paddy soil. J. Environ. Sci. 23: 434–443.
14. Saeki K, Okazaki M, Matsumoto S (1993) The chemical phase changes in heavy metals with drying and oxidation of the lake sediments. Water Res. 27: 1243–1251.
15. Reddy CN, Patrick WHJ, (1977) Effect of redox potential and pH on the uptake of cadmium and lead by rice plants. J. Environ. Qual. 6: 259–262.
16. Sajwan KS, Lindsay WL (1986) Effects of redox on zinc deficiency in paddy rice. Soil Sci. Soc. Am J. 50: 1264–1269.
17. Kashem MA, Singh BR (2001) Metal availability in contaminated soils: I. Effects of flooding and organic matter on changes in Eh, pH and solubility of Cd, Mi and Zn. Nutr. Cycl. Agroecosys. 61: 247–255.
18. Qi YB, Huang B, Gu ZQ, Sun WX, Zhao YC (2008) Spatial and temporal variation of C/N ratio of agricultural soils in typical area of Yangtze Delta region and its environmental significance. B. Miner. Petro. Geochem. 27: 50–56.
19. Shao XX, Huang B, Gu ZQ, Sun WX, Zhao YC (2006) Spatial-temporal variation of pH values of soil in a rapid economic developing area in the Yangtze Delta Region and their causing factors. B. Miner. Petro. Geochem. 25: 35–38.
20. Moore P, Coale F (2000) Phosphorus fractionation in flooded soils and sediments. In: G. M. Pierzenski, (Eds.), Methods of phosphorus analysis for soils, sediments, residuals, and waters. Available online at http://www.soil.ncsu.edu/sera17/publications/sera17-2/pm_cover.htm.
21. Lu RK (2004) Soil and agro-chemical analytical methods, China Agricultural Science and Technology Press, Beijing.
22. Nelson DW, Sommers LE (1996) Total carbon, organic carbon, and organic matter, In: D. L. Sparks (Eds.), Methods of soil analysis, part 3: Chemical methods, Madison, Wisc.: SSSA and ASA, 961–1010.
23. Ure AM, Quevauviller PH, Muntau H (1993) Speciation of heavy metals in soils and sediments-an account of the improvement and harmonization of extraction techniques undertaken under the auspices of the BCR of the Comission of European Communities. Int. J. Environ. Anal. Chem. 51: 135–151.
24. Lin Q, Zheng CR, Chen HM (1998) Transformation of cadmium species in rhizosphere. Acta Pedologica Sinica 35: 461–467.
25. Shao XX, Huang B, Zhao YC, Sun (2008) WX Pollution assessment of soil heavy metals in a representative area of the Yangtze River Delta region. Environ. Chem. 27: 218–221.
26. Tsai LJ, Yu KC, Chang JS (1998) Fractionation of heavy metals in sediment cores from the Ell-Ren River, Taiwan. Water Sci. Tech. 37: 217–224.
27. Tessier A, Campbell PGC, Bisson M (1979) Sequential extraction procedure for the speciation of particulate trace metals. Anal. Chem. 51: 844–851.
28. Argese E, Bettiol C (2001) Heavy metal partitioning in sediments from the lagoon of Venice (Italy). Toxicol. Environ. Chem. 79: 157–170.
29. Hinsinger P, Plassard C, Tang C, Jaillard B (2003) Origins of root-mediated pH changes in the rhizosphere and their responses to environmental constraints: a review. Plant Soil. 248: 43–59.
30. Tao S, Chen YJ, Xu FL, Cao J, Li BG (2003) Changes in copper speciation in maize rhizosphere soil. Environ Pollution. 122: 447–454.
31. Martínez-Alcalá I, Walker DJ, Bernal MP (2010) Chemical and biological properties in the rhizosphere of *Lupinus albus* alter soil heavy metal fractionation. Ecotox. Environ. Safe. 73: 595–602.
32. Singh SP, Ma LQ, Tack FMG (2000) Trace metal leach ability of land-disposed dredged sediments. J. Environ. Qual. 29: 1124–1132.
33. Wu HS, Zhang AQ, Wang LS (2004) Immobilization study of biosorption of heavy metal ions onto activated sludge. J. Environ. Sci. 16: 640–645.
34. Wang C, Li XC, Ma HT, Qian J, Zhai JB (2006) Distribution of extractable fractions of heavy metals in sludge during the wastewater treatment process. J. Hazard. Mater. 137: 1277–1283.
35. Šurija B, Branica M (1995) Distribution of Cd, Pb, Cu and Zn in carbonate sediments from the Krka River estuary obtained by sequential extraction. Sci. Total Environ. 170: 101–118.
36. Li RY, Yang H, Zhou ZG, Lv JJ, Shao XH, Jin F (2007) Fractionation of heavy metals in sediments from Dianchi lake, China. Pedosphere, 17: 265–272.
37. Shrivastaia SK, Banerjee DK (2004) Speciation of metals in sewage sludge and sludge-amended soils. Water, Air, Soil Pollut. 152: 219–232.
38. Teixeira E, Ortiz L, Alves M, Sanchez J (2001) Distribution of selected heavy metals in fluvial sediments of the coal mining region of Baixo Jacuí, RS, Brazil. Environ Geol. 41: 145–154.

39. Kazi TG, Jamali MK, Kazi GH, Arain MB, Afridi HI, et al. (2005) Evaluating the mobility of toxic metals in untreated industrial wastewater sludge using a BCR sequential extraction procedure and a leaching test. Anal. Bioanal. Chem. 383: 297–304.
40. Collavini F, Zonta R, Novelli AA, Zaggua L (2000) Heavy metals behavior during resuspension of the contaminated reduced sludge of the Venice canals. Toxicol. Environ. Chem. 77: 171–187.
41. Zheng SA, Zheng XQ, Chen C (2012) Leaching Behavior of Heavy Metals and Transformation of Their Speciation in Polluted Soil Receiving Simulated Acid Rain. PLoS ONE 7(11): e49664. doi:10.1371/journal.pone.0049664.
42. Goh SW, Buckley AN, Lamb RN, Rosenberg RA, Moran D (2006) The oxidation states of copper and iron in mineral sulfides, and the oxides formed on initial exposure of chalcopyrite and bornite to air. Geochim. Cosmochim. Ac. 70: 2210–2228.
43. Chen YJ, Tao S, Deng BS, Zhang XQ, Huang Y (2001) Effect of root system on metal fractionation in rhizosphere of contaminated soil. Acta Pedologica Sinica 38: 54–59.
44. Shuman LW, Wang J (1997) Effect of rice variety on zinc, cadmium, iron and manganese contents in rhizosphere and non-rhizosphere soil fractions. Commun. Soil Sci. Plant 28: 23–36.
45. Kashem MA, Singh BR, Kswai S (2007) Mobility and distribution of Cadmium, Nickel, and Zinc in contaminated soil profiles. Earth Environ. Sci. 77: 187–198.
46. Wang CQ, Dai TF, Li B (2007) The speciation and bioavailability of heavy metals in paddy soils under the rice-wheat cultivation rotation. Acta Ecologica Sinica 27: 889–897.

9

Effects of Manure Compost Application on Soil Microbial Community Diversity and Soil Microenvironments in a Temperate Cropland in China

Zhen Zhen[1,2,3], Haitao Liu[1], Na Wang[4], Liyue Guo[1], Jie Meng[1], Na Ding[5], Guanglei Wu[1], Gaoming Jiang[1]*

1 State Key Laboratory of Vegetation and Environmental Change, Institute of Botany, the Chinese Academy of Sciences, Beijing, China, **2** Laboratory of Crop, Guangdong Ocean University, Zhanjiang, China, **3** University of the Chinese Academy of Sciences, Beijing, China, **4** Development and Promotion Center, Shandong Small and Medium-sized Enterprises, Ji'nan, China, **5** State Key Laboratory of Crop Biology, Shandong Agricultural University, Tai'an, China,

Abstract

The long-term application of excessive chemical fertilizers has resulted in the degeneration of soil quality parameters such as soil microbial biomass, communities, and nutrient content, which in turn affects crop health, productivity, and soil sustainable productivity. The objective of this study was to develop a rapid and efficient solution for rehabilitating degraded cropland soils by precisely quantifying soil quality parameters through the application of manure compost and bacteria fertilizers or its combination during maize growth. We investigated dynamic impacts on soil microbial count, biomass, basal respiration, community structure diversity, and enzyme activity using six different treatments [no fertilizer (CK), N fertilizer (N), N fertilizer + bacterial fertilizer (NB), manure compost (M), manure compost + bacterial fertilizer (MB), and bacterial fertilizer (B)] in the plowed layer (0–20 cm) of potted soil during various maize growth stages in a temperate cropland of eastern China. Denaturing gradient electrophoresis (DGGE) fingerprinting analysis showed that the structure and composition of bacterial and fungi communities in the six fertilizer treatments varied at different levels. The Shannon index of bacterial and fungi communities displayed the highest value in the MB treatments and the lowest in the N treatment at the maize mature stage. Changes in soil microorganism community structure and diversity after different fertilizer treatments resulted in different microbial properties. Adding manure compost significantly increased the amount of cultivable microorganisms and microbial biomass, thus enhancing soil respiration and enzyme activities ($p<0.01$), whereas N treatment showed the opposite results ($p<0.01$). However, B and NB treatments minimally increased the amount of cultivable microorganisms and microbial biomass, with no obvious influence on community structure and soil enzymes. Our findings indicate that the application of manure compost plus bacterial fertilizers can immediately improve the microbial community structure and diversity of degraded cropland soils.

Editor: Wen-Xiong Lin, Agroecological Institute, China

Funding: This research was jointly supported by the Key Strategic Project of the Chinese Academy of Sciences (No.KSZD-EW-Z-012-2) and State Key Laboratory of Vegetation and Environmental Change (2014ZDFX04). The funders had no role in study design, data collection and analysis, decision to publish, or preparation of the manuscript.

Competing Interests: The authors have declared that no competing interests exist.

* Email: jianggm@126.com

Introduction

Chemical fertilizers have been extensively applied to sustain global agricultural production since the first Green Revolution [1,2]. Modern high-intensity agricultural ecosystems are characterized by the excessive use of chemical fertilizers, pesticides, and herbicides [1]. Chemical fertilizers directly enhance crop yield because plants directly or indirectly assimilate the nutrients provided by these inorganic fertilizers. However, on one hand, the production and use of these chemicals impart various negative effects on the agricultural ecosystem such as degradation of the soil, loss of crop genetic diversity, reduction in soil microbial diversity, contamination of ground-water resources, and pollution of the atmosphere [3,4]. On the other hand, with the intensive development of animal husbandry, animal dung has become one of the major pollution sources in China. There are a lot of organic matter and nutrients such as nitrogen and phosphorus in animal manure, especially through composting. Application of manure compost may also enhance soil microbial activities that improve the crop growth, and restrain the pests and diseases Compared with chemical fertilizers, manure compost has been comprehensively tested and determined as effective in increasing nutrient availability to crops, thus improving grain yield in a cost-effective and environmentally friendly manner [5,6]. The addition of manure compost can also increase the levels of organic matter and improve soil porosity, structural stability, moisture, and nutrient availability, as well as biological activity [7,8]. Hence, it is becoming a more popular practice to add manure compost to the soils if the degraded cropland is considered to be restored.

Soil microorganisms play important roles in ecological functions such as nutrient cycling and formation of soil aggregates through the decomposition of organic matter [9]. The stability of the microbial community structure has important implications for the rates of soil processes. For example, variations in microbial community structure in the soil influence rates of de-nitrification, nitrification, and nitrogen fixation [10,11]. Organic and inorganic

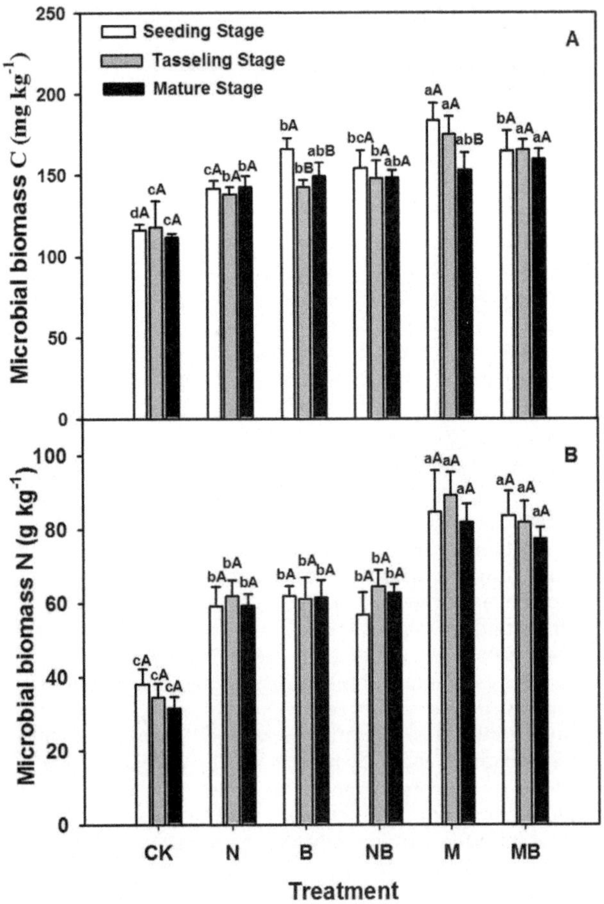

Figure 1. Soil microbial biomass C and soil microbial biomass N in various treatments during maize growth stages. CK: no fertilizer; N: N fertilizer; B: bacteria fertilizer; NB: N fertilizer + bacteria fertilizer; M: manure compost; MB: manure compost + bacteria fertilizer. Data are means ± SD (n = 3). Different lower case letters indicate significant differences (ANOVA, LSD test, $p<0.05$) among fertilizer treatments; different upper case letters indicate significant differences among maize growth stages.

amendments can significantly affect soil microorganisms [12]. Changes in microbial activity and composition, for instance, can influence plant growth by enhancing nutrient turnover and suppressing or mitigating disease incidence [13]. In addition, soil microbial biomass, activity, and community structure are useful indicators of soil quality and health because these parameters are sensitive to changes in cropland management practices [14]. Thus, the adaption of soil microbial structure and functions to the environment are considered essential for sustainable agricultural production [15]. Nevertheless, methods of increasing the soil microbiological diversity and/or restoring the microorganism communities when the soils have been seriously degraded after the long-term application of chemical fertilizers remain largely unknown.

Previous studies have focused on the influence of fertilizers on soil microbial communities, resulting in positive or neutral effects. Some studies have reported that soils in organic farming regimes had higher microbial functional diversity than those in conventional farming systems [16]. According to some investigators, the bacterial diversity was always higher in manure compost-amended farmyard soils regardless of land use patterns or seasons [17].

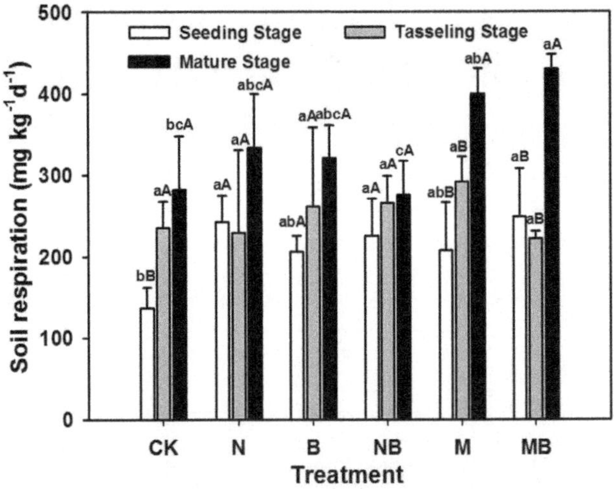

Figure 2. Soil respirations rate in various treatments during maize growth stages. CK: no fertilizer; N: N fertilizer; B: bacteria fertilizer; NB: N fertilizer + bacteria fertilizer; M: manure compost; MB: manure compost + bacteria fertilizer. Data are means ± SD (n = 3). Different lower case letters indicate significant differences (ANOVA, LSD test, $p<0.05$) among fertilizer treatments; different upper case letters indicate significant differences among maize growth stages.

Although fertilization has resulted in increases in crop yield, this application was not sufficient in triggering a significant improvement in the soil microbial properties [8]. Even some argued [18] that there were no significant differences in bacterial communities between improved and unimproved grassland [18]. While numerous studies have investigated the effects of organic fertilization on microbial communities [19,20], the dynamic patterns of bacteria, fungi, and actinomycetes under different types of fertilization have remained elusive.

Though some studies have recorded the long-term effects of different fertilization practices on soil microbial properties, few attempts have been made to improve the degraded farmland by enhancing soil microbial properties, especially under conditions that do not include cropland resting. Understanding how soil microbial properties respond to manure compost may help to strengthen agricultural ecosystem health practices. In this study, we hypothesized that adding manure compost and bacterial fertilizers to the soils that have been experienced with chemical fertilizers for several decades may rapidly improve soil microbial diversity. Our objectives were: (i) to identify a more effective method in rapidly improving degraded cropland soil quality through fertilizer management; (ii) to analyze the bacterial and fungal community structure and their growth stage-related dynamics in degraded and restored soils.

Materials and Methods

Site description

The study was conducted at the Agricultural Ecosystem Research Station of Shandong Agricultural University in Jiang Jiazhuang Village, Pingyi County, Shandong Province of Eastern China (35°26′21″N, 117°50′11″E). The farm area is in a typical temperate and monsoonal climate, with a mean annual rainfall of 770.2 mm, and an average annual temperature of 13.2°C.

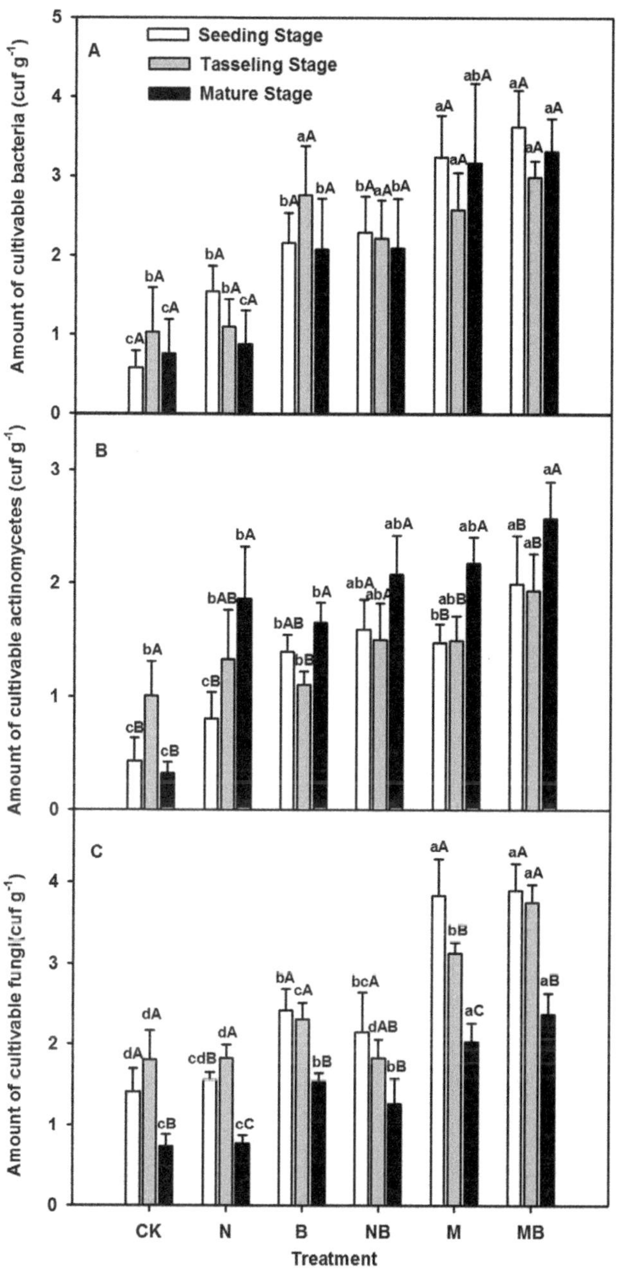

Figure 3. The clony-forming units of bacteria, fungi and actinomyces in various treatments during maize growth stages. CK: no fertilizer; N: N fertilizer; B: bacteria fertilizer; NB: N fertilizer + bacteria fertilizer; M: manure compost; MB: manure compost + bacteria fertilizer. Data are means ± SD (n = 3). Different lower case letters indicate significant differences (ANOVA, LSD test, $p<0.05$) among fertilizer treatments; different upper case letters indicate significant differences among maize growth stages.

Experimental design

Six treatments using different fertilizer applications were designed. Fertilizer treatments were established as follows: no fertilizer (CK), N fertilizer (N), N fertilizer + bacterial fertilizer (NB), manure compost (M), manure compost + bacterial fertilizer (MB), and bacterial fertilizer (B). The experimental pots (24 cm diameter, 45 cm height) were filled with soils collected from the 0–30 cm layer from a maize (*Zea mays* L.) cultivation area.

All the fertilizers were added as basal fertilization before planting maize. Seeds of maize were germinated on moist filter paper for 2 d, and seedlings were then planted in 90 pots (two plants per pot, 15 pots per treatment). The pots were located in a micro-region in the cropland at the study site. Urea (N = 46%) was chosen as the N treatment source. The Technology University of South China provided a bacterial fertilizer (HYSD001, Guangzhou Huayuan Biotechnology Company) that contained a variety of phosphorus-solubilizing and nitrogen-fixing bacteria as exogenous microorganisms. Cattle compost (M), which was fermented for three months (June to September) under high temperatures, was collected from the Hongyi Organic Farm in the village. According to the nitrogen fertilizer requirement (150 $kg \cdot hm^{-2}$) for maize growth, equal amounts of N in different treatments except for the B treatment were applied. Based on the 60% water content of cattle compost, the quantity of cattle compost (pH was 7.2; organic matter content was 440.43 $g \cdot kg^{-1}$; C/N was 26.7; nitrogen content was 1.6%) in the M and MB treatments was 3.75 $kg \cdot m^{-2}$, the quantity of urea (nitrogen content was 46%) in the N and NB treatments was 32.6 $g \cdot m^{-2}$, and the quantity of bacterial fertilizer in the B, MB, and NB treatments was 1.56 $g \cdot m^{-2}$. A conventional management scheme in China was conducted for pest and weed management during the experiment [21].

Sampling and processing

Soil samples were collected in triplicate at the maize seeding, tasseling, and mature stages from a 0–20 cm depth in each pot. Each soil sample was separated into two parts. One part was air-dried and stored at room temperature for determining soil chemical properties. The other part was passed through a 2-mm sieve, moistened to 60% of their water holding capacity, and immediately stored at 4°C for the measurement of soil microbial properties.

Soil microbial biomass and respiration

Microbial biomass C (MBC) and biomass N (MBN) contents were estimated by using chloroform fumigation extraction [22]. The soil samples were divided into two portions. One 15-g portion (in dry weight) rewetted to 60% was fumigated for 24 h at 25°C with chloroform (ethanol-free). After fumigant removal, the soils were extracted with 60 mL of 0.5 $mol \cdot L^{-1}$ K_2SO_4 and placed on a horizontal shaker at a speed of 300 $r \cdot min^{-1}$ for 30 min and filtered. The non-fumigated portion was extracted similarly as fumigation commenced. Total dissolved N and C in the K_2SO_4 extracts were determined by using an automated C analyzer (Shimadzu, TOC-VCPH, Japan). Regarding the incomplete extraction, two conversion factors were applied to calculate the biomass, as 0.54 for MBN (k_{EN}) and 0.45for MBC (k_{EC}) [23] in the following equation (1) and (2).

$$MBC = Ec/k_{EC} \quad (1)$$

$$MBN = E_N/k_{EN} \quad (2)$$

where E_C refers to the different organic carbon amount between fumigated and non-fumigated treatment, and E_N represents the difference in total nitrogen [24,25].

Basal soil respiration was determined by using the sealed jar incubation method, which employed a trap of 0.5 mol of NaOH alkali CO_2 [26]. At sampling, the jar lid was opened, the alkali trap was removed, and the solution was back-titrated with 0.5 mol

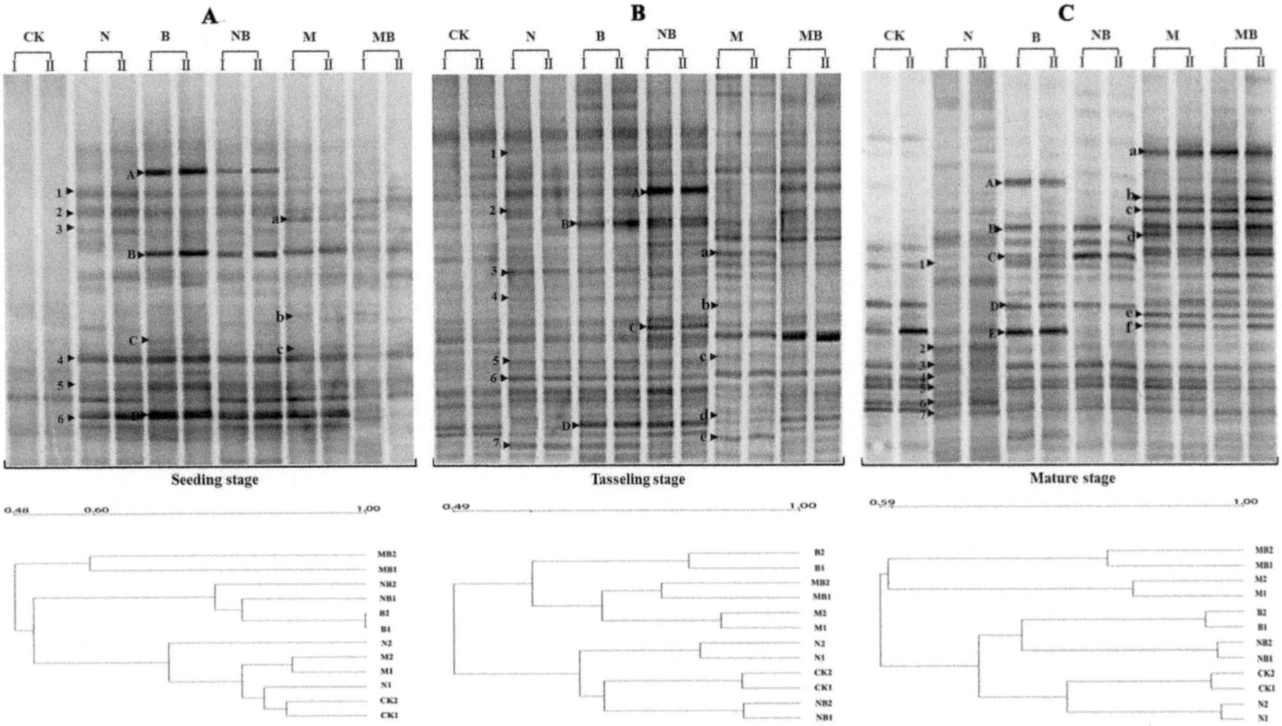

Figure 4. DGGE banding patterns of 16S bacteria fragment and clustering of DGGE profiles in various treatments during maize growth stages. CK: no fertilizer; N: N fertilizer; B: bacteria fertilizer; NB: N fertilizer + bacteria fertilizer; M: manure compost; MB: manure compost + bacteria fertilizer.

HCl in order to assess CO_2 release. The alkali trap was replaced during each measurement.

Plate counts of cultivable microorganisms

The total number of cultivable bacteria, fungi, and actinomyces were counted as colony forming units (CFUs) on agar plates using the dilution plate method. The media used for the enumeration of bacteria, fungi, and actinomyces were beef extract peptone medium, Czapek's medium, and Gause's No. 1 synthetic medium, respectively [27].

DNA extraction and DGGE

To determine soil microbial diversity, we used the molecular biology methods of DGGE. Briefly, soil DNA was extracted by using the Power Soil DNA Extraction Kit (Mobio Laboratories), following the manufacturer's instructions. For bacterial community analysis, the V3 region of 16S rRNA gene were amplified by touchdown polymerase chain reaction (PCR) using two different primer sets, 518R (5′-ATT ACC GCG GCT GCT GG) and GC-338F (5′-CGC CCG CCG CGC GCG GCG GGC GGG GCG GGG GCA CGG GGG GCC TAC GGG AGG CAG CAG), containing the GC clamp on the amplified 16S rDNA template. For fungal community analysis, the fragments of 18S rRNA gene were amplified by nested PCR using the primer sets GC-fung (5′-CGC CCG CCG CGC CCC GCG CCC GGC CCG CCG CCC CCG CCCCATTCCCCGTTACCCGTTG-3′) and NS1 (5′-GTAGTCATATGCTTGTCTC-3′).

PCR mixtures containing 25 μL of Premix Taq (Takara Biotechnologies), 1.5 μL of each primer, and 2 μL of the DNA template were made up to a volume of 50 μL with sterile Milli-Q water. The samples were amplified in a Peltier Thermal Cycler (PTC-200) (Bio-Rad Laboratories, Hercules, CA). The amplification conditions of bacteria and fungi samples were as follows: an initial denaturation of DNA for 5 min at 94°C, followed by 25 cycles of 30 s at 94°C, 30 s at 55°C, and 30 s at 72°C; 10 cycles of 30 s at 92°C, 30 s at 55°C, and 45 s at 72°C; and a final extension at 72°C for 10 min. Blank controls were used through all the steps.

DGGE was performed with a DCode universal mutation detection system (Bio-Rad Laboratories, Hercules, CA). Approximately 1 μg of the bacteria PCR product per lane were loaded onto 8% polyacrylamide (37.5:1 acrylamide: bisacrylamide) gels in a 1× TAE buffer with a bacteria denaturing gradient ranging from 40% to 60%, whereas that for fungi ranged from 15% to 40%. Gel electrophoresis was performed at 60°C for 13 h at 80 V. The gel was subsequently stained in a 1× TAE buffer containing 1:10000 dilution of SYBR Green I nucleic acid staining solution (GenScript, USA) for half an hour before being photographed on a Molecular Imager Gel Doc XR System (Bio-Rad Laboratories).

The soil microbial community diversity indexes were calculated in the following equation (3), (4) and (5):

$$\text{Shannon}-\text{Wiener index } (H): \ H = \sum_{i=1}^{S} Pi \ln Pi \qquad (3)$$

Where: Pi is the ratio of the activity on a particular substrate to the sum of activities on all substrates; S is the sum of band number in one sample.

Evenness index (E):

$$E = H/\ln S \qquad (4)$$

Table 1. Effect of different treatments on soil bacterial community structure diversity as evaluated by Shannon index (*H*), **Evenness** index (*E*) and Simpson index (*C*) in the DGGE-PCR.

Treatments	Shannon-Wiener (*H*)			Evenness index (*E*)			Simpson index (*C*)		
	Seeding	Tasseling	Mature	Seeding	Tasseling	Mature	Seeding	Tasseling	Mature
CK	2.16±0.25cB	2.62±0.01cA	2.53±0.06cA	0.96±0.01bA	0.98±0.01aA	0.95±0.01aA	0.07±0.03aA	0.05±0.01aA	0.06±0.01aA
N	2.84±0.03abA	2.69±0.04cA	2.40±0.05cB	0.96±0.01bA	0.98±0.01aA	0.98±0.01aA	0.05±0.01aA	0.05±0.01aA	0.05±0.01aA
B	2.78±0.02abA	2.82±0.01abA	2.76±0.21bA	0.97±0.01abA	0.98±0.01aA	0.97±0.01aA	0.06±0.01aA	0.05±0.01aA	0.06±0.01aA
NB	2.92±0.04aA	2.86±0.08abA	2.83±0.13bA	0.98±0.01abA	0.96±0.01aA	0.98±0.01aA	0.05±0.01aA	0.05±0.01aA	0.06±0.01aA
M	2.54±0.04bB	3.23±0.05aA	3.44±0.03aA	1.00±0.01aA	0.96±0.01aB	0.96±0.01aB	0.07±0.01aA	0.05±0.01aA	0.05±0.01aA
MB	2.60±0.12bB	3.31±0.06aA	3.41±0.11aA	0.96±0.01bA	0.97±0.01aA	0.96±0.01aA	0.09±0.01aA	0.06±0.01aA	0.05±0.01aA

Data analyzed by two-way ANOVA, LSD test, $p<0.05$. Different lower case letters indicate significant difference among fertilizer treatments; different upper case letters indicate significant difference among maize growth stages.

Where: H – Shannon index; S - band numbers of every substrate.

Simpson index (C):

$$C = \sum Pi^2 \quad (5)$$

Where: Pi is the ratio of the activity on a particular substrate to the sum of activities on all substrates.

Soil enzyme activity

Urease activity was determined by using the phenol-sodium hypochlorite colorimetric method [28], as indicated by mass (mg) of NH_3-N in 1 g soil incubated for 24 h (U). Catalase activity was analyzed by using the titration method [29], and its activity was indicated by depletion (mL) of $KMnO_4$ (0.1 mol L^{-1}) after 20 min of 1 g incubated soil. Invertase activity was determined by using the 3,5-dinitrosalicylate colorimetric method [24], which was indicated by the mass (mg) of glucose of 1 g soil after 24 h. Cellulase activity was measured by using the 3,5-dinitrosalicylate colorimetric method [29], which was indicated by the mass (mg) of decomposed generated glucose (mg) from cellulose of 1 g soil after 72 h.

Statistical analysis

All statistical analyses were carried out using the SPSS 16.0 software. A two-way ANOVA at the 0.05 level was conducted to determine the interactions between fertilizer treatments and maize stages. And one-way ANOVA was used for analysis of significant difference among fertilizer treatments and plant stages.

Results

Soil microbial biomass and soil respiration rate

Soil MBC and MBN over time in different treatments are presented in Fig. 1. For all sampling times, MBC significantly increased in the fertilization treatments against their controls (CK) ($p<0.01$) during the whole maize growth stage, with the largest increment of MBC occurring consistently in the MB treatment. Differences between the MB treatment and other treatments except for the M treatment reached significance at the tasseling stage ($p<0.01$). The ANOVA results showed that MBC in the M and MB treatments were significantly higher than those in the B ($p<0.05$), N ($p<0.01$) and NB treatments at the maize tasseling stage ($p<0.01$). However, no significant differences were noted between MB and M or between N and NB treatments (Fig. 1A). The fertilizers also affected soil MBN, which varied among different treatments. Compared with the CK treatment, all fertilizer treatments showed significantly higher MBN, and this was greater in the M and MB treatments than in the N and NB or B treatments ($p<0.01$). Nevertheless, no statistically significant differences ($p>0.05$) were observed among the B, N, and NB treatments.

Soils with different fertilizers showed higher respiration rates and greater cumulative CO_2 production compared to those observed in the CK treatment (Fig. 2). A significantly enhanced soil respiration rate was observed during the entire period. Compared with other fertilizer treatments, only M and MB treatments showed significant higher respiration rates at mature stage than seeding and tasseling stages. MB treatment exhibited drastically higher soil respiration rate than those of the NB and N treatments at the mature stage ($p<0.01$), whereas M treatment was

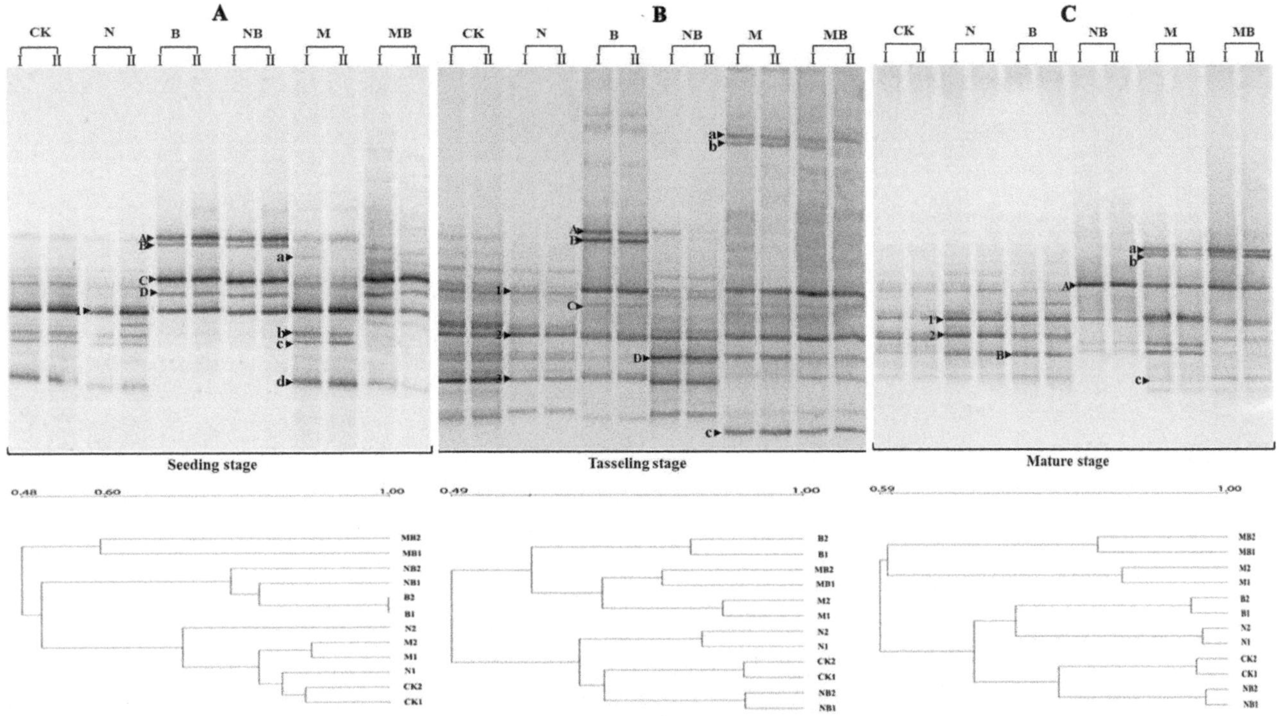

Figure 5. DGGE banding patterns of 18S rDNA fungi fragment and clustering of DGGE profiles in various treatments during maize growth stages. CK: no fertilizer; N: N fertilizer; B: bacteria fertilizer; NB: N fertilizer + bacteria fertilizer; M: manure compost; MB: manure compost + bacteria fertilizer.

only higher than NB treatment. However, soils with N and B treatments only showed higher respiration rates than those of the NB treatment, indicating that fertilizers could significantly enhance the microbial activities, and the highest number and greatest effects were observed from organic amendments.

Plate counts of cultivable microorganisms

There were extremely large variations in the number of bacteria, fungi, and actinomyces among various treatments (Fig. 3). The colony-forming units (CFUs) of bacteria in all fertilizer treatments (except N treatment) were significantly higher than those observed in the CK treatment ($p<0.01$) at the tasseling and mature stages (Fig. 3A). At the mature stage, the number of CFUs of bacteria in the M and MB treatments were significantly higher than that in the N treatment ($p<0.01$), with the bacterial numbers from the MB treatment considerably higher than those observed in the B and NB treatments ($p<0.05$) (Fig. 3A). However, the number of bacteria in the M treatment did not differ from that in the B and NB treatments ($p>0.05$). Fertilizers also largely influenced the amount of soil fungi. For instance, the amount of fungi in all fertilizer treatments was remarkably higher than that in the CK treatment ($p<0.01$) at the mature stage (Fig. 3B). At the same stage, the amount of fungi in the MB treatment was significantly higher than that in the B ($p<0.01$) and N ($p<0.05$) treatments, however this was not the case in the M and NB treatments. And the amount of fungi in all the fertilizer treatments (expect NB treatment) largely increased at mature stage. Similar to bacteria and fungi, the number of cultivable actinomyces was also affected. The amount of cultivable actinomyces in the M and MB treatments was considerably higher than that in the B and NB treatments, both were greatly higher than that observed in the CK and N treatments ($p<0.01$) (Fig. 3C). Meanwhile, the amount of cultivable actinomyces in the CK and N treatments initially increased, but eventually decreased at the end of growth. In the B, NB, M, and MB treatments, cultivable actinomyces decreased during the entire maize growth period.

Soil microbial community composition structure

Changes in the microbial community structure were determined by PCR-DGGE analysis, which targeted domains of bacteria and fungi. As shown in Fig. 4, different fertilizers significantly affected the community structure of soil bacteria and fungi at all sampling times. DGGE bands that could be enhanced by all fertilizers were marked with numbers. Bands enhanced by the N or NB treatments were marked with uppercase alphabet letters, whereas those enhanced by the M or MB treatments were marked with lower case alphabet letters (Fig. 4).

For bacterial community structure, all the fertilizers stimulated the generation of band numbers 1–6. N, B or NB treatments stimulated the occurrence of bands A–D, whereas M or MB enhanced the bands a–c at the seeding stage. The influences from N, B or NB treatment on bacterial community structure were lower than those from M or MB treatment (Fig. 4). The number of DGGE bands was the highest at the tasseling stage, indicating that the formed microbial community was the most diverse at this stage. We therefore confirmed that the largest influences of bacteria community structure occurred after the fertilizer treatments of M and MB. The results at the mature stage were similar to the tasseling stage, *e.g.*, the influences on bacterial community structure from the M or MB treatments were higher than that of the N or NB treatments (Fig. 4B and 4C). Furthermore, statistical analysis showed that manure compost and bacterial fertilizer could enhance the Shannon index, whereas the N and NB treatment decreased the Shannon index, especially from seeding stage to

Table 2. Effect of different treatments on soil fungi community structure diversity as evaluated by Shannon index (*H*), **Evenness** index (*E*) and Simpson index (*C*) in the DGGE-PCR.

Treatments	Shannon-Wiener (*H*)			Evenness index (*E*)			Simpson index (*C*)		
	Seeding	Tasseling	Mature	Seeding	Tasseling	Mature	Seeding	Tasseling	Mature
CK	1.86±0.18cA	2.08±0.01cA	1.04±0.01cB	0.96±0.01aA	0.96±0.01aA	0.95±0.01aA	0.12±0.03cB	0.09±0.01cB	0.18±0.01aA
N	2.03±0.11bA	1.81±0.12dA	0.69±0.01dB	0.93±0.01aA	0.96±0.01aA	0.98±0.01aA	0.15±0.01cA	0.17±0.02bA	0.19±0.01aA
B	1.83±0.16cB	2.55±0.04bA	1.06±0.01cC	0.95±0.02aA	0.98±0.01aA	0.97±0.01aA	0.24±0.02aB	0.29±0.01aA	0.17±0.01aC
NB	1.84±0.17cB	2.26±0.04cA	1.03±0.01cC	0.95±0.01aA	0.96±0.01aA	0.94±0.01aA	0.19±0.03bB	0.25±0.01aA	0.20±0.01aB
M	2.13±0.03abB	2.65±0.11abA	2.71±0.01aA	0.93±0.01aA	0.98±0.01aA	0.96±0.01aA	0.13±0.01cA	0.11±0.01cA	0.15±0.01bA
MB	2.20±0.02aB	2.79±0.08aA	2.55±0.20bA	0.96±0.01aA	0.98±0.02aA	0.96±0.034aA	0.12±0.01cA	0.08±0.005cB	0.14±0.06bA

Data analyzed by two-way ANOVA, LSD test, $p<0.05$. Different lower case letters indicate significant difference among fertilizer treatments; different upper case letters indicate significant difference among maize growth stages.

tasseling stage (Table 1). The Shannon indexes of the M or MB treatments were higher than those of the other treatments, which remained high during the entire experiment except seeding stage (Table 1). At tasseling stage the Shannon indexes of the N, NB, M and MB treatments significantly higher than CK and N treatments. The results at the mature stage were similar to the tasseling stage, the Shannon indexes of M and MB treatments were higher than other fertilizer treatments. Although the Shannon index in the N treatment was the lowest at tasseling and mature stage, a downtrend trend was observed. The evenness indexes (except seeding stage) and the Simpson indexes in all treatments were basically the same, and retained a smooth and steady trend during the entire growth period, and there was no significant difference among maize growth stages (Table 1).

Cluster analysis showed that bacterial community structures in treatments B and NB had a higher degree of similarity, while N, M and CK got closed and MB forms a separate category at the seeding stage (Fig. 4A). M and MB got together and B formed a separate category, whereas CK, NB and N are in another group at the tasseling stage (Fig. 4B). At the mature stage, M and MB had a higher degree of similarity, while other treatments classified to a similar category (Fig. 4C).

In terms of fungal community structure, the number of DGGE bands was highest at tasseling stage, and the largest influences on fungi community structure occurred with the M and MB treatments (Fig. 5). Statistical analysis showed that manure compost could enhance the Shannon index, whereas the N treatment decreased the Shannon index (Table 2). At seeding and tasseling stages, the Shannon index of the MB treatment appreciably higher than other treatments except M treatment, and the Shannon index of the M treatment was the highest at mature stage. However, the Shannon index of the N treatment rapidly decreased during the entire growth period, which was the lowest at tasseling and mature stages. The evenness index in all fertilizer treatments had no significant difference. The simpson index of M and MB treatments were notably lower than those of N, B and NB treatments during the entire experiment (Table 2). Cluster analysis of fungal community structure also showed that fungal community structure in treatments added bacterial fertilizer (B, NB and MB) were similar at the initial seeding stage, while other treatments (CK, N and M) formed another separate category (Fig. 5). M and MB become more and more similar with time, while CK, N and M treatments were in another group, and B forms a separate category at the tasseling and mature stages (Fig. 5).

The results of bacteria and fungi community structure analysis indicated that organic and chemical fertilizers could enhance the bacteria and fungi community by enhancing the microbial diversity at the tasseling and mature stages, while bacterial fertilizer significantly influenced the bacteria and fungi community in initial seeding stage (Table 2).

Soil enzyme activities

Urease activity in M and MB treatments were higher than that in the CK treatment ($p<0.01$), especially in the M treatment (Fig. 6). No significant differences were observed between the N and CK treatments at the seeding, tasseling stages and mature stages. However, urease activity in the N treatment rapidly decreased in the mature stage of maize (Fig. 6A). However, the catalase activities in the M and MB treatments were significantly higher than those observed in the CK treatment at tasseling and mature stages ($p<0.01$), and it was similar for B treatment at the mature stage. On the other hand, the activity of catalase in the N

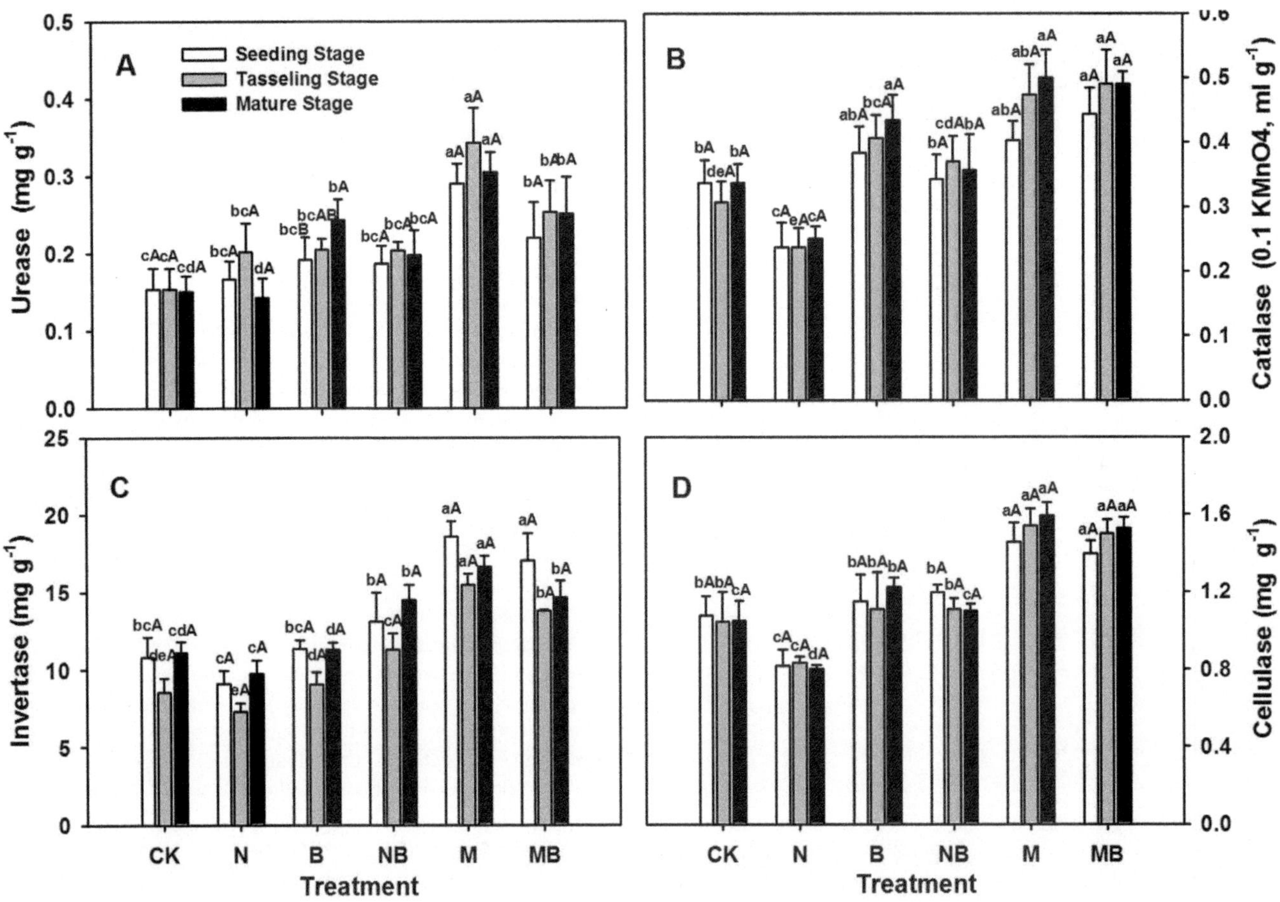

Figure 6. Soil enzyme activity in various treatments during maize growth stages. CK: no fertilizer; N: N fertilizer; B: bacteria fertilizer; NB: N fertilizer + bacteria fertilizer; M: manure compost; MB: manure compost + bacteria fertilizer. Data are means ± SD (n = 3). Different lower case letters indicate significant differences (ANOVA, LSD test, $p<0.05$) among fertilizer treatments; different upper case letters indicate significant differences among maize growth stages.

treatment was significantly lower than that in the CK treatment at all sampling times ($p<0.05$) (Fig. 6B).

Invertase activity in the M and MB treatments was appreciably higher than that in the CK treatment during the entire growth period ($p<0.01$), although no differences were observed between the B and CK treatments. N treatment showed the lowest invertase activity at all times, although this was not statistically different from that observed in the CK treatment (Fig. 6C). The M and MB treatments showed a significantly enhanced cellulase activity ($p<0.01$), whereas that in the N treatment decreased ($p<0.01$). No big differences in cellulase activity were observed among the NB, B, and CK treatments, indicating that the influence of bacterial fertilizer on cellulase activity was not significant (Fig. 6D). These results suggest that manure compost was better than bacteria fertilizer alone in enhancing soil enzyme activities, whereas chemical fertilizers degrade the activities of soil enzymes.

Soil physicochemical and chemical properties

The humus content in the M and MB treatments was remarkably higher than that in the B and CK treatments, whereas humus content in the N treatment was significantly lower than that in the CK treatment. Available N in the M and MB treatments were also considerably higher than that in the CK treatment, whereas there was no difference between the N and CK treatments. Fertilizers showed minimal impact on soil pH, total N, and available P during the entire maize growth period. Positive correlations were observed between soil organic matter (SOM) and microbial biomass (MBC and MBN), humus and microbial biomass (MBC and MBN), SOM and enzymes activities, humus and enzymes activities (urease, catalase, invertase, and cellulase) (Table 3), whereas the correlations among available N, microbial biomass, and enzymes activities were not significant. The addition of fertilizers enriched the soil microbial biomass and soil enzymes by enhancing the soil physicochemical properties of SOM and humus, especially through the addition of manure compost.

PCA combined with a hierarchical clustering analysis

Kaise-Meyer-Olkin Measure showed the KOM value of three phases is 0.798>0.7, and the F value is 0 with Bartlett's Test of Sphericity test. All data obey the normal distribution, and are suitable for principal component analysis (Fig.7).

At seeding stage, three factors were extracted as the main composition factor, and the cumulative variance contribution rate was 93.49 (Fig.7A). The first principal component contribution rate was the largest, accounting for 59.94, and the main relevant parameters were soil microbial indicators (bacteria, fungi, actinomycetes, MBN, MBC), enzyme activities (invertase urease and cellulase) and humus. The main relevant parameters of second and third principal component were SOM and Available N, accounting for 20.06 and 13.49, respectively. According to the selection

Table 3. Correlation of MBC, MBN, Available N, SOM, Humus and enzymes activities.

Variables	MBC	MBN	Available N	SOM	Humus	Urease	Catalase	Invertase
MBN	0.910**							
Available N	0.425	0.518*						
SOM	0.777**	0.820**	0.380					
Humus	0.685**	0.750**	0.621**	0.622**				
Urease	0.721**	0.694**	0.232	0.558*	0.826**			
Catalase	0.585*	0.620**	0.410	0.646**	0.871**	0.835**		
Invertase	0.582*	0.687**	0.238	0.580*	0.776**	0.732**	0.703**	
Cellulase	0.615**	0.663**	0.528*	0.614**	0.948**	0.838**	0.928**	0.790**

*, Pearson correlation is significant at the 0.05 level.
**, Pearson correlation is significant at the 0.01 level.

indexes of principal component, the ahierarchical clustering analysis was used and the result showed that treatments of N, B and NB form a separate category, while M and MB were in another group, and CK formed a separate category (Fig.7A). At tasseling stage, the cumulative variance contribution rate of three principal components factors was 90.12. The first principal component contribution rate was 67.68, with the main relevant parameters of soil microbial indicators (actinomycetes,Shannon, bacteri, MBN and MBC), enzyme activities (Cellulas, Catalase and Invertase) and humus (Fig.7B). The main relevant parameters of second and third principal component were SOM and simpson, accounting for 12.39 and 10.05, respectively. Cluster analysis showed that N, B, NB and CK had a higher degree of similarity, whereas M and MB formed a separate category, respectively (Fig.7B). At mature stage, the cumulative variance contribution rate was 94.45. The variance contribution of the first principal components was 64.154, and the mainly relevant parameters were soil microbial indicators (bacteria, actinomycetes, shannon and MBN), enzyme activities (cellulas and catalase) and humus. The variance contribution of the second principal components was SOM with 20.21. The variance contribution of the third principal components was simpson index, accounting for 10.093. Hierarchical clustering analysis showed that M and MB had a higher degree of similarity and treatments of B and NB forms a category, while CK and N are in another group (Fig.7C).

Discussion

Soil microbial community structure

In cropland soils, the application of fertilizers imparted a stronger effect on microbial communities, as shown by previous studies involving the analysis of PLFA, microbial biomass, and rRNA gene libraries [20]. DGGE profiles revealed that the band numbers in all fertilizer treatments significantly increased at various maize growth stages, with band numbers in M and MB treatments higher than those observed in the other fertilizer treatments (B, NB, and N) (Figs. 4 and 5). Our results suggest that cattle compost changes the soil bacteria and fungi community structure. Previous investigations have also demonstrated that animal compost increased bacterial and fungi diversity by increasing the carbon pool of the soil, thus improving the living conditions for indigenous microbial populations [30,31]. Our study showed that soil organic matter and humus in the M treatment were the highest. The DGGE profiles (Figs. 4C and Figs. 5B) and Shannon index data showed that the bacterial and fungi diversity of the M and MB treatments remained stable, whereas these declined in the N, B, and NB treatments at maize mature stage.

The M treatment resulted in an increase in exogenous microorganisms, which can directly influence the soil microorganism community structure [32]. The specific bands in the DGGE profiles showed that exogenous microorganisms emerged in the M and MB treatments, thus clearly supporting this statement (Figs. 4 and 5). The M treatment greatly improved microbial diversity, indicating that M is conducive to the establishment of a diverse microbial community structure (Tables 1 and 2).

The application of bacterial fertilizer provided a variety of nitrogen-fixing bacteria and phosphorus-solubilizing bacteria that could improve the soil microbial community structure (Figs. 4 and 5). New fungal species, which may act as indicators of exogenous microorganism, were thus introduced (Fig. 5). The Simpson index of soil fungi community in the MB treatment was lowest during the entire stages, and B and NB treatments was the highest

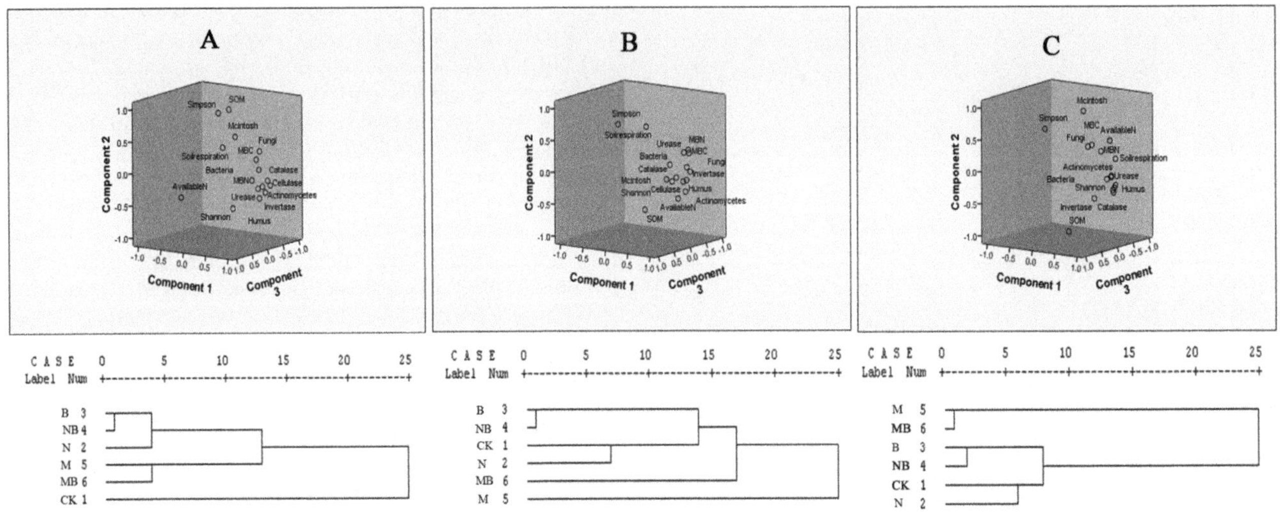

Figure 7. PCA of indexes combined with a hierarchical clustering analysis in various treatments during maize growth stages. CK: no fertilizer; N: N fertilizer; B: bacteria fertilizer; NB: N fertilizer + bacteria fertilizer; M: manure compost; MB: manure compost + bacteria fertilizer. A: Seeding stage; B: Tasseling stage; C: Mature stage.

respectively at seeding stage and last two stages (Table 2), suggesting that exogenous microorganisms played a dominant role in both stages, which combined with compost enhanced soil microbial community uniformity [33],while N fertilizer always displayed less or negative influences compared with the other treatments. For example, the Shannon index of soil bacterial and fungal communities showed that the soil microbial community diversity in the N and NB treatment gradually decreased compared to that in the B treatment, displaying an obvious negative influence from the administration of nitrogen fertilizers (Tables 1 and 2). This may be attributable to the absence of an external carbon source [34]. Nevertheless, only adding N fertilizers may inhibit the soil microorganism propagation rather than increase it, which may further degrade the bacteria and fungi community diversity [35].

Soil microbial biomass and soil activity

Soil microbial biomass is an important source of plant nutrients and is highly correlated with SOC (soil organic carbon) [34]. Enhancement of soil microbial activity was associated with high soil available N for plants [36]. The M and MB treatments significantly increased soil MBC, MBN, and respiration rates, which positively influenced microbial processes and development. These findings also showed that M significantly enhanced the density of soil bacteria, fungi, and actinomycetes. These indexes were significantly positively correlated with humus ($p<0.05$) (Table 3), indicating that the M treatment served as a sufficient carbon source that enhanced the microorganism biomass and activity.

Microbial carbon and nitrogen initially increased and then subsequently decreased during the maize growth period, achieving the highest value at the tasseling stage. This finding might be attributable to the competition of carbon and nitrogen source between soil microorganisms and crops. With the transformation of soil nutrients, the competitiveness of crops is enhanced, thus leading to a decrease in microbial carbon and nitrogen [37].

Effects on soil enzymes

Soil microbial enzymes are mainly driven by metabolic processes, largely reflecting the level of soil microbial activity and the intensity of biochemical reactions [38]. Our investigation demonstrated that different fertilizers have significantly affected soil enzyme activities ($p<0.01$) (Fig. 6). Four kinds of soil enzyme activities consistently displayed the highest levels in the M or MB treatments (Fig. 6). The M treatment can accelerate soil carbon and nitrogen circulation [14] and improve soil quality, with enzyme activities stemming from manure compost. As manure compost contains abundant organic matter, it can drastically increase SOM by providing a rich source of carbon and nutrients for enzyme production microorganisms.

However, the application of N fertilizer had less influence on soil urease activity, whereas it significantly decreased soil catalase, invertase, and cellulase activities ($p<0.01$) (Fig. 6). It is thus possible that the application of chemical fertilizers has inhibited enzyme production in microorganisms. Except for urease activity, the B and NB treatments showed no influence on the activities of the other three enzymes (Fig. 6), suggesting that NB can enhance the soil nitrogen transformation.

Improvement of soil fertility

The N treatment notably reduced microbial biomass and enzyme activities compared to those with the other treatments. This finding could be attributable to the acidifying effect of long-term N fertilizers, as showed by lower pH in previous studies [39]. However, no significant differences in soil pH were observed between most fertilizers and the CK treatment, which maintained a neutral pH during the maize growth period. It is possible that the acidifying effect of the N fertilizer may have not been reflected within a relatively short period. SOC is a pivotal component of soil fertility, and the level of organic matter is influenced by the application of fertilization [40]. In a healthy soil, the level of SOC is considered to be a functional indicator of the net input of organic residues [41]. In our study, manure compost enhanced the accumulation of SOC, which is consistent with the findings of several other previous studies from many countries [42,43]. The improved SOC may be due to a higher humectation rate and a constantly lower decay rate, since SOC is a nutrition pool that maintains microbial activity. SOC, especially humus, was significantly positively correlated with soil microbial biomass and soil enzyme activity (Table 3). However, there was no obvious change in the total N level under manure compost, which may partially be due to a slow release of N from manure compost and smaller losses of N as described by Bhandari et al. [42].

The results of PCA combined with a hierarchical clustering analysis showed that fertilizer treatments had been divided into three groups at the end stage (Fig.7). The application of manure compost treatments (M and MB) had a higher degree of similarity; the application of bacterial fertilizer bacterial fertilizer treatments (B and NB) formed another category, while treatments of CK and N are in another group. Compared to the application of bacterial fertilizer treatments, the distance of application of manure compost treatments was far away with the group of CK and N. These result exactly demonstrated that the application of manure compost has more influences on soil microbial community diversity and soil microenvironments than bacterial fertilizer.

Conclusions

Manure compost, especially manure compost + bacterial fertilizer, consistently resulted in higher levels of soil respiration rate, cultivable microorganisms, and soil enzyme activities, while N fertilizers showed no significant influence or negative results. The number of DGGE bands with bacteria and fungi also indicated that the Shannon index of manure compost treatment was the highest and remained high. Fertilizers, especially manure compost, significantly enhanced soil microbial properties in response to the increase in soil physicochemical properties of soil organic matter and humus. From a soil microbial point of views, manure compost application can be used as an environmentally friendly and rapid measure for restoring degraded cropland.

Author Contributions

Conceived and designed the experiments: GJ ZZ NW. Performed the experiments: ZZ HL. Analyzed the data: ZZ GJ HL NW. Contributed reagents/materials/analysis tools: LG JM ND GW. Wrote the paper: ZZ GJ.

References

1. Tilman D (1998) The greening of the green revolution. Nature 396: 211–212.
2. Leita L, De Nobili M, Mondini C, Muhlbachova G, Marchiol L, et al. (1999) Influence of inorganic and organic fertilization on soil microbial biomass, metabolic quotient and heavy metal bioavailability. Biol Fert Soils 28: 371–376.
3. Kaur T, Brar B, Dhillon N (2008) Soil organic matter dynamics as affected by long-term use of organic and inorganic fertilizers under maize–wheat cropping system. Nutr Cycl Agroecosys 81: 59–69.
4. Chaudhry AN, Jilani G, Khan MA, Iqbal T (2009) Improved processing of poultry litter to reduce nitrate leaching and enhance its fertilizer quality. Asian J Chem 21: 4997–5003.
5. Ahmad R, Jilani G, Arshad M, Zahir ZA, Khalid A (2007) Bio-conversion of organic wastes for their recycling in agriculture: an overview of perspectives and prospects. Ann Microbiol 57: 471–479.
6. Leite LFC, Oliveira FC, Araújo ASF, Galvão SRS, Lemos JO,et al. (2010) Soil organic carbon and biological indicators in an Acrisol under tillage systems and organic management in north-eastern Brazil. Soil Res 48: 258–265.
7. Francis I, Holsters M, Vereecke D (2010) The Gram-positive side of plant microbe interactions. Environ Microbiol 12: 1–12.
8. Wang W, Niu J, Zhou X, Wang Y (2011) Long-term change in land management from subtropical wetland to paddy field shifts soil microbial community structure as determined by PLFA and T-RFLP. Pol J Eco 59: 37–44.
9. Stark CH, Condron LM, O'Callaghan M, Stewart A, Di HJ (2008) Differences in soil enzyme activities, microbial community structure and short-term nitrogen mineralisation resulting from farm management history and organic matter amendments. Soil Biol Biochem 40: 1352–1363.
10. Hsu SF, Buckley DH (2008) Evidence for the functional significance of diazotroph community structure in soil. ISME J 3: 124–136.
11. Philippot L, Andert J, Jones CM, Bru D, Hallin S (2011) Importance of denitrifiers lacking the genes encoding the nitrous oxide reductase for N_2O emissions from soil. Global Change Biol 17: 1497–1504.
12. Hu J, Lin X, Wang J, Dai J, Chen R, et al. (2011) Microbial functional diversity, metabolic quotient, and invertase activity of a sandy loam soil as affected by long-term application of organic amendment and mineral fertilizer. J Soil Sediment 11: 271–280.
13. Zhang QC, Shamsi IH, Xu DT, Wang GH, Lin XY, et al. (2012) Chemical fertilizer and organic manure inputs in soil exhibit a vice versa pattern of microbial community structure. Appl Soil Ecol 57: 1–8.
14. Bending GD, Turner MK, Jones JE (2002) Interactions between crop residue and soil organic matter quality and the functional diversity of soil microbial communities. Soil Biol Biochem 34: 1073–1082.
15. Wardle DA, Yeates GW, Nicholson KS, Bonner KI, Watson RN (1999) Response of soil microbial biomass dynamics, activity and plant litter decomposition to agricultural intensification over a seven-year period. Soil Biol Biochem 31: 1707–1720.
16. Mäder P, Fliessbach A, Dubois D, Gunstl L, Fried P, et al. (2002) Soil fertility and biodiversity in organic farming. Science 296: 1694–1697.
17. Ge Y, Zhang J-b, Zhang L-m, Yang M, He J-z (2008) Long-term fertilization regimes affect bacterial community structure and diversity of an agricultural soil in northern China. J Soil Sediment 8: 43–50.
18. McCaig AE, Grayston SJ, Prosser JI, Glover LA (2001) Impact of cultivation on characterisation of species composition of soil bacterial communities. FEMS Microbiol Ecolo 35: 37–48.
19. Dinesh R, Srinivasan V, Hamza S, Manjusha A (2010) Short-term incorporation of organic manures and biofertilizers influences biochemical and microbial characteristics of soils under an annual crop. Bioresource technol 101: 4697–4702.
20. Qiu M, Zhang R, Xue C, Zhang S, Li S, et al. (2012) Application of bio-organic fertilizer can control Fusarium wilt of cucumber plants by regulating microbial community of rhizosphere soil. Biolo Fert Soil 48: 807–816.
21. Liu E, Yan C, Mei X, He W, Bing SH, et al. (2010) Long-term effect of chemical fertilizer, straw, and manure on soil chemical and biological properties in northwest China. Geoderma 158: 173–180.
22. Brookes P, Landman A, Pruden G, Jenkinson DS (1985) Chloroform fumigation and the release of soil nitrogen: a rapid direct extraction method to measure microbial biomass nitrogen in soil. Soil Biol Biochem 17: 837–842.
23. Jenkinson DS, Brookes PC, Powlson DS (2004) Measuring soil microbial biomass. Soil Biol Biochem 36: 5–7.
24. Wu J, Joergensen R, Pommerening B, Chaussod R, Brookes PC (1990) Measurement of soil microbial biomass C by fumigation-extraction—an automated procedure. Soil Biol Biochem 22: 1167–1169.
25. Joergensen RG, Mueller T (1996) The fumigation extraction method to estimate soil microbial biomass: Calibration of the value. Soil Biol Biochem 28: 33–37.
26. Anderson JP (1982) Soil respiration. Methods of soil analysis Part 2 Chemical and microbiological properties: 831–871.
27. Xu G, Zheng H (1986) Handbook of analysis methods of soil microbiology. Agricultural, Beijing, China.
28. Paul EA (2006) Soil microbiology, ecology and biochemistry: Academic press.
29. Guan S, Zhang D, Zhang Z (1986) Soil enzyme and its research methods. Agricultural, Beijing: 274–297.
30. Wu T, Chellemi DO, Graham JH, Martin KJ, Rosskopf EN (2008) Comparison of soil bacterial communities under diverse agricultural land management and crop production practices. Microb Ecol 55: 293–310.
31. Helgason B, Walley F, Germida J (2010) No-till soil management increases microbial biomass and alters community profiles in soil aggregates. Appl Soil Ecol 46: 390–397.
32. Li X, Rui J, Mao Y, Yannarell A, Mackie R (2014) Dynamics of the bacterial community structure in the rhizosphere of a maize cultivar. Soil Biol Biochem 68: 392–401.
33. Hedlund K (2002) Soil microbial community structure in relation to vegetation management on former agricultural land. Soil Biol Biochem 34: 1299–1307.
34. Dijkstra FA, Bader NE, Johnson DW, Cheng W (2009) Does accelerated soil organic matter decomposition in the presence of plants increase plant N availability? Soil Biol Biochem 41: 1080–1087.
35. Pengthamkeerati P, Motavalli P, Kremer R (2011) Soil microbial activity and functional diversity changed by compaction, poultry litter and cropping in a claypan soil. Appl Soil Ecol 48: 71–80.
36. Tu C, Ristaino JB, Hu S (2006) Soil microbial biomass and activity in organic tomato farming systems: Effects of organic inputs and straw mulching. Soil Biol Biochem 38: 247–255.
37. Inselsbacher E, Hinko-Najera Umana N, Stange FC, Gorfer M, Schüller E, et al. (2010) Short-term competition between crop plants and soil microbes for inorganic N fertilizer. Soil Biol Biochem 42: 360–372.
38. Mandal A, Patra AK, Singh D, Swarup A, Ebhin Masto R (2007) Effect of long-term application of manure and fertilizer on biological and biochemical activities in soil during crop development stages. Bioresource Technol 98: 3585–3592.
39. Saha S, Prakash V, Kundu S, Kumar N, Mina BL (2008) Soil enzymatic activity as affected by long term application of farm yard manure and mineral fertilizer under a rainfed soybean–wheat system in NW Himalaya. Eur J Soil Biol 44: 309–315.

40. Drenovsky R, Vo D, Graham K, Scow KM (2004) Soil water content and organic carbon availability are major determinants of soil microbial community composition. Microb Ecol 48: 424–430.
41. Gregorich E, Liang B, Ellert B, Drury CF (1996) Fertilization effects on soil organic matter turnover and corn residue C storage. Soil Sci Soc Am J 60: 472–476.
42. Bhandari A, Ladha J, Pathak H, Padre H, Dawe D, et al. (2002) Yield and soil nutrient changes in a long-term rice-wheat rotation in India. Soil Sci Soc Am J 66: 162–170.
43. Hao X, Liu S, Wu J, Hu R, Tong C, et al. (2008) Effect of long-term application of inorganic fertilizer and organic amendments on soil organic matter and microbial biomass in three subtropical paddy soils. Nutr Cycl Agroecosys81: 17–24.

10

Contribution for the Derivation of a Soil Screening Value (SSV) for Uranium, Using a Natural Reference Soil

Ana Luisa Caetano[1,2]*, Catarina R. Marques[1,2], Ana Gavina[6,2], Fernando Carvalho[3], Fernando Gonçalves[1,2], Eduardo Ferreira da Silva[4], Ruth Pereira[5,6]

1 Department of Biology, University of Aveiro, Campus Universitário de Santiago, Aveiro, Portugal, **2** CESAM, University of Aveiro, Campus Universitário de Santiago, Aveiro, Portugal, **3** Nuclear and Technological Institute (ITN) Department of Radiological Protection and Nuclear Safety, Sacavém, Portugal, **4** Department of Geosciences, University of Aveiro, GeoBioTec Research Center, Campus Universitário de Santiago, Aveiro, Portugal, **5** Department of Biology, Faculty of Sciences of the University of Porto, Porto, Portugal, **6** Interdisciplinary Centre of Marine and Environmental Research (CIIMAR/CIMAR), University of Porto, Porto, Portugal

Abstract

In order to regulate the management of contaminated land, many countries have been deriving soil screening values (SSV). However, the ecotoxicological data available for uranium is still insufficient and incapable to generate SSVs for European soils. In this sense, and so as to make up for this shortcoming, a battery of ecotoxicological assays focusing on soil functions and organisms, and a wide range of endpoints was carried out, using a natural soil artificially spiked with uranium. In terrestrial ecotoxicology, it is widely recognized that soils have different properties that can influence the bioavailability and the toxicity of chemicals. In this context, SSVs derived for artificial soils or for other types of natural soils, may lead to unfeasible environmental risk assessment. Hence, the use of natural regional representative soils is of great importance in the derivation of SSVs. A Portuguese natural reference soil PTRS1, from a granitic region, was thereby applied as test substrate. This study allowed the determination of NOEC, LOEC, EC_{20} and EC_{50} values for uranium. Dehydrogenase and urease enzymes displayed the lowest values (34.9 and <134.5 mg U Kg, respectively). *Eisenia andrei* and *Enchytraeus crypticus* revealed to be more sensitive to uranium than *Folsomia candida*. EC_{50} values of 631.00, 518.65 and 851.64 mg U Kg were recorded for the three species, respectively. Concerning plants, only *Lactuca sativa* was affected by U at concentrations up to 1000 mg U kg^1. The outcomes of the study may in part be constrained by physical and chemical characteristics of soils, hence contributinq to the discrepancy between the toxicity data generated in this study and that available in the literature. Following the assessment factor method, a predicted no effect concentration (PNEC) value of 15.5 mg $kg^{-1}{}_{dw}$ was obtained for U. This PNEC value is proposed as a SSV for soils similar to the PTRS1.

Editor: Stephen J. Johnson, University of Kansas, United States of America

Funding: ALC was supported by a PhD grant from Fundação para a Ciência e Tecnologia (FCT) (http://www.fct.pt/). The funders had no role in study design, data collection and analysis, decision to publish, or preparation of the manuscript.

Competing Interests: The authors have declared that no competing interests exist.

* Email: ana.caetano@ua.pt

Introduction

Uranium (U) is a natural soil component, being originated from rocks in the Earth's crust, where it mainly occurs in the form of oxides. Natural processes acting on rocks and soils, such as wind, water erosion, dissolution, precipitation and volcanic activity contribute for U dispersal in the environment [1]. The use of U as fuel in nuclear power plants has driven to its large-scale exploration worldwide. The U exploration became significantly important in the world during the Second World War, and later on during the Cold War, in both cases to supply military needs of the greatest potencies. Recently, the World Nuclear Association estimated worldwide reserves of U at 5.4 million tons in 2009, of which Australia had about 31%, followed by Kazakhstan (12%), Canada and Russia with 9% (http://www.world-nuclear.org/info/inf75.html). The remarkable energy crisis that is currently faced worldwide due to the exhaustion of carbon based energy resources is demanding further extraction of U, as nuclear energy arises as a potential solution. Hence, it is expected that the mining and milling of U will increase in the next decades, contributing for its widespread in the environment [2].

During the last century, Portugal has actively explored radioactive ores and was for some time ranked as one of the main U producers. The extraction of U ore in Portugal started in 1908, first driven by the interest in radium (being U a by-product) and then by the interest in its military applications, till 2001 [3,4]. Most of the old U mines were located in the granitic regions of the Iberian Meseta, in the centre-north of Portugal (Beiras), [5]. Nowadays, although the mining activities ceased, like in several other places in the world, the old U mines represent a serious environmental problem, due to waste accumulation (mainly tailings and sludge) and improper disposal of radioactive material, composed by U and its daughter radionuclides [1,5–16]. Soils and water are the two major environmental matrices affected by U contamination.

U has a long half-life, persisting in nature as different isotopes, with different chemical and radiological characteristics [17]. The toxic effects induced by this metal are caused by both properties.

However, since U isotopes mainly emit alpha particles, with little penetration capacity, the main radiation hazards only occur after ingestion or inhalation of these isotopes and daughter radionuclides [17]. Once in the soil, U interacts with all the components of this matrix, such as clay minerals, aluminum and iron oxides, organic matter and microorganism, in a very complex system, where pH and organic matter seem to have the major role in controlling U mobility (pH 6) and leaching (pH<6) [18]. The high mobility/availability of U will in turn increase the ecological risks posed to soil and water compartments [19–27].

The soil has been recognized as an important compartment that provides crucial ecosystem services (e.g., filtering of contaminants, reservoir of carbon and a bank of genes) and is the support of agro-sylvo-pastoral production [28,29] and of several other human activities. The soil compartment offers raw materials (e.g., peat, clay, ore) and contributes for climate regulation and biodiversity conservation, as well as other cultural services [30,31]. The recognition of the importance of maintaining the provision of such services has increased the necessity to create appropriate legal tools to correctly and effectively protect and manage this resource. In this sense, the Soil Framework Directive proposed by the Commission of the European Communities (CEC), aims to establish a common strategy for the protection and sustainable use of soils [32]. For that end, this proposal defines measures for the identification of the main problems faced by soils, the adoption of strategies to prevent their degradation, as well as for the rehabilitation of contaminated or degraded soils [33]. The Soil Framework Directive will fill in the gap regarding soil protection, since this compartment has never been a target of specific protection policies at the European Community level [32]. Many countries, committed in regulating the management of contaminated land, have adopted generic quality standards, the soil screening values (SSVs) [34]. SSVs are concentration thresholds above which, more site-specific evaluations are required to assess the risks posed by soil contamination [35]. The SSVs should provide a level of protection to terrestrial species and ecological functions of the soil [35–37]. SSVs are particularly useful for the first tier of Ecological Risk Assessment (ERA) processes applied to contaminated sites, supporting the decision-making at this initial stage of assessment [38], which at the end is aimed in setting priorities for remediation and risk reduction measures [39]. In the case of Portugal, SSVs for soils have never been established for metals or organics. Only threshold concentrations of metals on sewage sludge were legally established to regulate the application of this solid waste on agricultural soils [40]. However, they are not appropriate for soil ERA purposes.

The use of natural reference soils in ecotoxicological tests has been recommended by several authors [41–43]. This is because the properties of the OECD artificial soil are not representative of the great majority of natural soils [44]. Different levels of toxicity, for each contaminant, can be expected in soils with different properties [45–48], hence it is important each country derives their own SSVs using natural reference soils representing the main types of soils within their territories. In this context, the main aim of this work was to obtain ecotoxicological data for U, performing soil enzymes activity tests, invertebrates and plant tests, using for that a Portuguese natural reference soil (PTRS1), that represents one of the dominant types of soil from a granitic region (cambisol) of the country [49]. As a result, enough data are gatheredas to make the first proposal of a SSV for this metal.

Materials and Methods

The present study used a natural soil that was collected in a non-protected area, requiring no specific permission for its collection. Further, no work with endangered species was performed, and no vertebrate species were used in the ecotoxicological assays. Only tests with invertebrates and plants were performed. The invertebrates were obtained from laboratorial cultures maintained by the authors of this manuscript and plant seeds were obtained from a local supplier.

1. Test soil

The natural soil (PTRS1) used as test substrate in this study was collected in Ervas Tenras [Pinhel, Guarda, Portugal center; geographical coordinates: 40°44′4.27″N and 7°10′54.3″W), at 655 m altitude, in a granitic region.

A composite soil sample was collected and immediately brought to the laboratory where it was air dried. Another portion of the soil, was immediately sieved through a 2 mm mesh size and the sieved fraction (<2 mm) was stored in polyethylene bags, at −20°C, until further analysis of soil microbial parameters, which were performed within the period of one month. For the tests with soil organisms and plants, the soil was passed through a 4 mm mesh sieve and the sieved fraction (<4 mm) was defaunated through two freeze–thawing cycles (48 h at −20°C followed by 48 h at 25°C), before the beginning of the assays.

The physical and chemical properties (including total metal contents) of the PTRS1 soil were presented in a preliminary study by Caetano et al. [49], aimed in characterizing this soil as a reference substrate for ecotoxicological purposes. The main properties of the PTRS1 are also described in Tables 1 and 2. Briefly, soil-KCl 1 M and soil-deionized water suspensions (1:5 m/v) were used for pH (KCl, 1 M) and pH-H_2O measurements, respectively, according to ISO 17512–1 [50]- After 15 min of magnetic stirring and 1 h resting period, the pH of the suspension was measured using a WTW 330/SET-2 pH meter. A soil water suspension (1:5 w/v) was used for the measurement of soil conductivity [51] Ten grams of PTRS1 were mechanically shaken in polypropylene flasks with 50 ml with deionized water filtered in a Milli-Q equipment (hereinafter referred as deionized water), water for 15 min. The mixture was left to rest overnight for soil bulk settling [51]. The conductivity of the resulting suspension was measured using an LF 330/SET conductivity meter. Soil water content was determined from the loss of weight after drying at 105°C, for 24 h. Organic matter (OM) content was determined by loss of ignition of dried soil samples at 450°C during 8 h [52]. For determination of water holding capacity (WHC) polypropylene flasks were prepared with a filter paper-replaced bottom, which after being filled up with soil samples, were immersed in water for 3 h. After this period, samples were left for water drainage during 2 h and the WHC was determined accounting to the loss of weight after drying at 105°C until weight stabilization [50].

2. Test substance

For all the test organisms, the natural soil was spiked with a stock solution of uranyl nitrate 6-hydrate, $UO_2(NO_3)_2 6H_2O$ (98%, PANREAC) prepared with deionized water in order to obtain a range of concentrations, which were ascertained by range finding tests performed with the different test species.

For soil enzyme tests, the PTRS1 soil was spiked with the following concentrations: 0.0, 134.6, 161.5, 193.8, 232.5, 279.0, 334.8, 401.8, 482.2, 578.7, 694.4, 833.3, 1000 mg U $kg^{-1}{}_{dw}$. To obtain these concentrations, the stock solution of uranyl nitrate

Table 1. Physical and chemical properties of PTRS1 soil (retrieved from Caetano et al. [49]).

	pH (H_2O)	pH (KCl, 1 M)	Conductivity (mS cm^{-1})	OM (%)	WHC (%)	Particle-size distribution (%)				[U] (mg Cu kg^{-1} $soil_{dw}$)
						Clay (<4 µm)	Silt (4–63 µm)	Sand (63 µm–2 mm)	Gravel (>2 mm)	
PTRS1	5.9±0.09	4.3±0.02	4.8±0.23	6.5±0.004	23.9±1.84	3.3	22.8	46.9	23.9	9.0

pH (H_2O), pH (KCl, 1 M), OM (organic matter), and WHC (water holding capacity) are represented as average ± STDEV.

was diluted in the volume of deionized water required to adjust the soil moisture at 80% of its maximum water holding capacity (WHC_{max}).

The following U concentrations were used to expose the earthworms in the reproduction tests: 0.0, 113.1, 124.4, 136.9, 150.5, 165.6, 231.9, 324.6, 454.5, 500.0, 550.0, 605.0, 665.5 mg U $kg^{-1}{}_{dw}$. For potworms, collembolans and terrestrial plant assays the same range of concentrations was tested: 0.0, 167.4, 192.5, 221.4, 254.6, 292.7, 336.6, 420.8, 526.0, 657.5, 756.1, 869.6, 1000 mg U $kg^{-1}{}_{dw}$.

The volume of deionized water required to adjust the WHC of the soil to a given percentage of its maximum value was used to dilute the stock solution for these tests. After spiking the soil was left to rest for equilibration for 48 h before testing.

3. Ecotoxicological assessment

3.1 Soil microbial activity. For testing the effect of increasing concentrations of U on soil microbial parameters, a 30-day exposure was firstly conducted. Ten grams of sieved PTRS1 soil per replicate and concentration were spiked with different U concentrations, a total of three replicates were used per treatment. Six replicates with the same amount of soil only moistened with deionized water were also prepared for the control. The soil was incubated at 20±2°C and a photoperiod of 16 h^L:8 h^D. During the incubation period, the soil moisture was weekly monitored by weighing the pots, and whenever needed it was adjusted to 80% of its WHC_{max} by adding deionized water. At the end, 1 g of each replicate from the control and concentrations tested was stored in individual falcon tubes at −20°C for approximately one month. Thereby, a total of 9 sub-replicates were made for each concentration. The soil was thawed at 4°C before analysis.

The biochemical parameters analyzed were: the activity of arylsulphatase, dehydrogenase, urease, and cellulase enzymes and changes in the nitrogen mineralization (N mineralization) and potential nitrification.

For the determination of arylsulphatase activity, the method proposed by Tabatabai and Bremner [53] and Schinner et al. [54] was followed. After addition of 1 mL of p-nitrophenylsulfate (0.02 M), soil sub-samples were incubated for one hour, at 37°C. The nitrophenyl liberated by the activity of arylsulphatase was extracted and colored with a 4 mL of sodium hydroxide (0.5 M) and determined photometrically at 420 nm. The results were expressed as µg p-nitrophenylsulfate (p-NP) g^{-1} soil $_{dw}$ h^{-1}.

The method proposed by Öhlinger [55] was used to assess the dehydrogenase activity. The samples were suspended in 1 mL of trifeniltetrazol chloride (TTC) (3.5 g L^{-1}) and incubated at 40°C for 24 h. The triphenylformazan (TPF) produced was extracted with acetone and measured spectrophotometrically at 546 nm. The results were expressed as µg TPF g^{-1} $soil_{dw}$ h^{-1}.

The cellulase activity was tested according to the method proposed by Schinner et al. [54] and Schinner and von Mersi [56]. The reducing sugars produced during the incubation period, after addition of 1.5 mL of acetate buffer (2 M), caused the reduction of hexacyanoferrate (III) potassium to hexacyanoferrate (II) potassium in an alkaline solution. This last compound reacts with ferric ammonium sulfate in acid solution to form a ferric complex of hexacyanoferrate (II), of blue colour, which is colorimetrically measured at 690 nm and expressed as µg glucose g^{-1} $soil_{dw}$ 24 h^{-1}.

N mineralization activity was measured according to Schinner et al. [54]. For this purpose the soil samples were incubated for 7 days at 40°C. During this period, the organic forms of N were converted to inorganic forms (mainly ammonium ion, NH_4^+),

Table 2. Pseudo-total concentrations (mg/kg) of metals recorded in PTRS1 soil (average ± standard deviation) extracted with aqua régia, (retrieved from Caetano et al.[49]).

Metal	PTRS1
Ag	0.1±0.0
Al	25628.5±5130.0
B	2.2±0.8
Ba	45.8±8.0
Be	1.2±0.2
Cd	0.1±0.1
Co	5.6±1.1
Cr	10.8±2.1
Cu	9.0±1.8
Fe	24921.4±4534.4
Li	124.4±22.9
Hg	5253.5±1025.5
Mn	386.8±77.9
Mo	0.9±0.2
Na	78.1±14.9
Ni	4.6±0.9
Pb	12.5±2.2
Sb	0.2±0.0
Sn	10.4±1.9
U	7.8±1.7
V	37.8±14.1
Zn	57.1±8.9

which were determined by a modification of the Berthelot reaction, after extraction with 3 mL of potassium chloride (2 M). The reaction of ammonia with sodium salicylate in the presence of sodium dichloroisocyanurate formed a green colored complex in alkaline pH that was measured at 690 nm and expressed as μg N g^{-1} $soil_{dw}$ d^{-1}.

The urease activity was assayed according to the method proposed by Kandeler and Gerber [57] and, Schinner et al. [54]. The samples were incubated for 2 h at 37°C after the addition of 4 mL of a buffered urea solution (720 mM). The ammonia released was extracted with 6 mL of potassium chloride (2 M) and determined by the modified Berthelot reaction. The quantification was based on the reaction of sodium salicylate with ammonia in the presence of chlorinated water. UR was detected at 690 nm and expressed as μg N g^{-1} $soil_{dw}$ 2 h^{-1}.

The quantification of potential nitrification was determined by the method of Kandeler [58], which is a modification of the technique proposed by Berg and Rosswall [59]. The ammonium sulphate (4 mL, 10 mM) was used as substrate, and soil samples were incubated for 5 h, at 25°C. Nitrate released during the incubation period was extracted with 1 mL of potassium chloride (2 mM) and determined colorimetrically at 520 nm. This reaction was expressed as μg nitrite (N) g^{-1} $soil_{dw}$ h^{-1}.

3.2. Invertebrate and plant tests. Test organisms and culture conditions: The earthworm *Eisenia andrei* (Oligochaeta: Lumbricidae), the potworm *Enchytraeus crypticus* (Oligochaeta: Enchytraeidae) and the springtail *Folsomia candida* (Collembola: Isotomidae) were used as invertebrate test organisms. All organisms were obtained from laboratorial cultures, kept under controlled environmental conditions (temperature: 20±2°C; photoperiod: 16 h^L:8 h^D). The earthworms (*E. andrei*) are maintained in plastic boxes (10 to 50 L) containing a substrate composed by peat, dry and defaunated horse manure (through two freeze–thawing cycles (48 h at −20°C followed by 48 h at 65°C), and deionized water. The pH of the culture medium is adjusted to 6.0–7.0 with $CaCO_3$. The organisms are fed, every 2 weeks, with six table spoons of oatmeal previously hydrated with deionized water and cooked for 5 min. The potworms (*E. crypticus*) are cultured in plastic containers (25.5 cm length; 17.4 cm width; 6.5 cm height), which are filled with pot soil moistened to the nearest 60% of its WHC_{max} and with pH adjusted to 6.0±0.5. The organisms are fed twice a week with a tea spoon of macerated oat. The collembolans (*F. candida*) are maintained in plastic containers filled with culture medium composed by moistened Plaster of Paris mixed with activated charcoal 8:1 (w:w). They are fed with half of a tea spoon of granulated dry yeast, twice a week. The food is added in small amounts to avoid spoilage by fungi.

Seeds from four plant species (two dicotyledonous and two monocotyledonous), purchased from a local supplier, were used for seed germination and growth tests: *Avena sativa, Zea mays, Lacuta sativa* and *Lycopersicon esculentum*.

Reproduction tests with invertebrates: Previous studies from our team, at least with earthworms from the same laboratorial cultures, have proved that these organisms were not exposed to meaningful levels of metals (especially U, in laboratorial culture conditions) [60]. The accomplishment of validity criteria, by all the controls of the assays (herein described) with the three invertebrate species, also confirmed that the test animals were not previously exposed to toxic levels of metals through test containers, substrates or food. The reproduction tests with *E. andrei, E. albidus* and *F. candida* were carried out according to the ISO guidelines 11268-2 [61], 16387 [62] and 11267 [63], respectively. Each replicate of the invertebrate tests contained 10 individuals in a certain developmental stage: the earthworms had a fully developed clitellum and an individual fresh weight between 250 and 600 mg; the potworms were 12-mm size; and the springtails were 10–12 days old. Five hundred grams of dry soil were weighted per test vessel for earthworms. For the tests with potworms and collembolans 20 g and 30 g of soil were weighted per replicate, respectively. Following an ECx sampling design, which considers more concentrations and less number of replicates, two replicates per concentration and five replicates for the control were prepared in the reproduction tests with *E. andrei*. Adult earthworms were removed from the test containers after 28 days. The produced cocoons persisted in the soil until 56 days have been completed. After this period, the juveniles from each test container were counted. During the test, organisms were fed once a week, with 5 g per box of defaunated horse manure (using the same procedure above described), and the soil moisture content was weekly monitored (following the procedures outlined in ISO guideline 11268-2 [61]).

The *E. albidus* reproduction test was held for 28 days and the adults were left in the vessels until the end of the test. About 2 mg of rolled oats were placed on the soil surface, weekly to feed the animals. At the end of the test, the potworms were killed with alcohol, colored with Bengal red and counted according to the Ludox Flotation Method, as described in ISO 16387 [62]. The reproduction tests with *F. candida* took four weeks to be completed. The collembolans were fed with granulated dry yeast, obtained from a commercial supplier, being weekly added (about 2 mg of yeast per test vessel) to the soil surface. At the end of the test, the containers were filled with water and the juveniles were

Table 3. Toxicity data obtained for copper (mg U kg^{-1} $soil_{dw}$) in PTRS1 soil on soil microbial processes, invertebrates and plants.

Biota	Endpoint	NOEC	LOEC	EC_{20}	EC_{50}
Microbial parameters					
Arylsulphatase		232.5	279	155.3 (84.76–255.87)	295.6 (216.09–375.17)
Dehydrogenase		<134.5	≤134.5	34.9 (20.52–59.35)	110.3 (83.25–137.47)
Nitrogen mineralization	Enzyme activity	694.4	833.3	152.2 (46.66–257.79)	347.0 (211.25–482.91)
Celulase		≤134.5	≥134.5	n.d.	n.d.
Urease		<134.5	≤134.5	<134.5	<134.5
Potencial nitrification		<134.5	≤134.5	429.5 (229.53–629.46)	610.0 (459.07–761.11)
Invertebrates					
Eisenia andrei	Rep. (56 days)	500.0	550.0	474.8 (391.47–558.04)	631.0 (532.78–699.21)
Enchytraeus crypticus	Rep. (28 days)	420.8	526.0	469.7 (355.47–584.04)	518.6 (480.40–556.90)
Folsomia candida	Rep. (28 days)	675.5	756.1	343.4 (172.23–514.60)	851.64 (606.10–1097.18)
Plants					
Avena sativa	Germination	≥1000	>1000	n.d.	n.d.
Zea mays	Germination	≥1000	>1000	n.d.	n.d.
Lactuca sativa	Germination	≥1000	>1000	n.d.	n.d.
Lycopersicon esculentum	Germination	≥1000	>1000	n.d.	n.d.
Avena sativa	Fresh mass	≥1000	>1000	n.d	n.d.
Zea mays	Fresh mass	≥1000	>1000	n.d	n.d.
Lactuca sativa	Fresh mass	≥1000	>1000	n.d	n.d.
Lycopersicon esculentum	Fresh mass	≥1000	>1000	n.d	n.d.
Avena sativa	Dry mass	≥1000	>1000	n.d.	n.d.
Zea mays	Dry mass	>1000	>1000	n.d.	n.d.
Lactuca sativa	Dry mass	<167.4	≤167.4	n.d.	n.d.
Lycopersicon esculentum	Dry mass	≥1000	>1000	n.d.	n.d.

For ECx point estimates the 95% confidence limits are presented in brackets. n.d.- not determined; Rep. – reproduction.

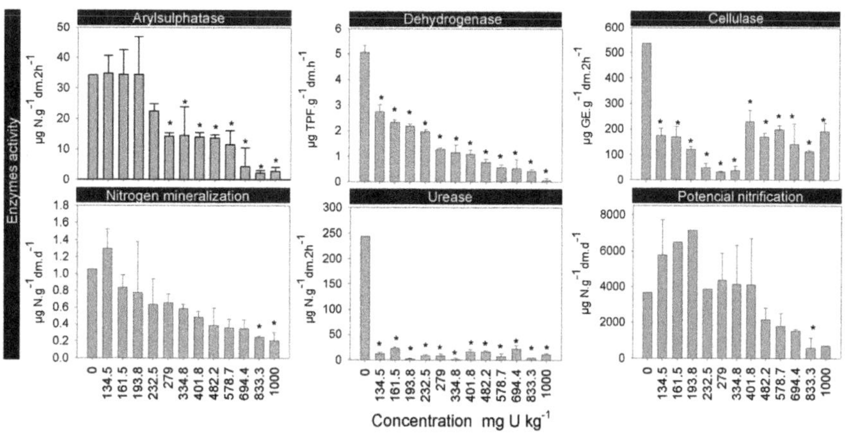

Figure 1. Soil enzyme activities, N mineralization and potential nitrification. Response of the arylsulphatase, dehydrogenase, cellulase urease, activity, N mineralization and potential nitrification to soils spiked with a range of uranium concentrations. The error bars indicate the standard deviation. The asterisks point out significantly differences from the control ($P<0.05$).

Table 4. Toxicity of copper (mg U $kg^{-1}soil_{dw}$) reported in the literature for the reproduction of soil invertebrates using different soil types with different physical and chemical characteristics.

Species	Endpoint	Soil type	Physical-chemical parameters			Point estimates (mg U $kg^{-1}soil_{dw}$)				Reference
			pH	OM (%)	Clay (%)	NOEC	LOEC	EC_{20}	EC_{50}	
		Canadian soil	6.2	1.0	2.0	n.d.	n.d.	>1000	n.d.	Sheppard and Stephenson [86]
Eisenia fetida	Rep. (56 days)	Canadian soil	6.2	1.0	2.0	n.d.	n.d.	>1120	n.d.	Sheppard and Stephenson [86]
		Canadian soil	7.5	2.2		>838	n.d.	n.d.	n.d.	Sheppard and sheppard [98]
		Canadian soil	7.5	18.4		>994	n.d.	n.d.	n.d.	Sheppard and sheppard [98]
Folsomia Candida	Rep. (28 days)	Canadian soil	7.5	2.2	24.0	n.d.	n.d.	840.0	n.d.	Sheppard and Stephenson [86]
		Canadian soil	7.5	n.d.	n.d.	n.d.	n.d.	>720	n.d.	Sheppard and Sheppard [98]
Elymus lanceolatus	Germination	Canadian soil	6.2	1	2	n.d.	>1000	n.d.	n.d.	Sheppard and Stephenson [86]
Elymus lanceolatus	Germination	Canadian soil	7.5	2.2	24	n.d.	>1001	n.d.	n.d.	Sheppard and Stephenson [86]
Zea mays	Dry mass	European soil	5.2	2.5	n.d.	n.d.	>100	n.d.	n.d.	Stojanović et al., [108]

OM - organic matter, Rep. - reproduction, n.d. - not determined., germ.- germination.

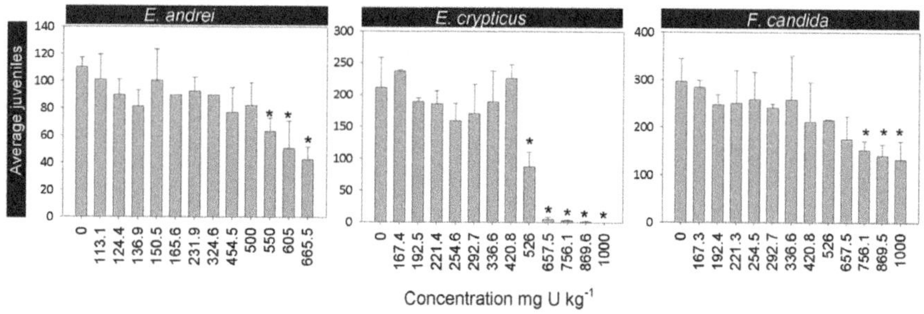

Figure 2. Reproduction of invertebrates. Results obtained exposing *Eisena andrei, Enchytraeus crypticus* and *Folsomia candida,* to natural PTRS1 soil, contaminated with different concentrations of U. The error bars indicate the standard deviation. The asterisks point out significantly differences from the control (P<0.05).

counted after flotation. The addition of a few dark ink drops provided a higher contrast between the white individuals and the black background. The organisms were then counted through the use of the ImageJ software (online available: http://rsb.info.nih.gov/ij/download.html). The exposure was carried out at 20±2°C and a photoperiod of $16^{L}:8^{D}$. For both species five replicates of uncontaminated natural PTRS1 soil were prepared for the control. The same ECx sampling design applied for earthworms was followed. However, in order to reduce the variability of the results, three replicates were prepared per test concentration (instead of two for the earthworms).

Seed germination and plant growth tests: Germination and growth tests with terrestrial plants were performed following standard procedures described by the ISO guideline 11269-2 [64]. For this purpose, 200 g_{dw} of the spiked soil with the concentrations described above was placed in plastic pots (11.7 cm diameter, 6.2 cm height) and tested. In this case, the amount of water required to adjust the WHC_{max} of the soil to 45% was used to dilute the stock solution and to moist the soil at the beginning of the test. The soil was placed in the plastic pots (11.7 cm diameter, 6.2 cm height). In the bottom of each plastic pot a hole was previously made to let a rope passing through, hence allowing communication with the pot below that was filled with deionized water. After soil spiking and soil saturation with water twenty seeds were added to each pot and gently covered with the spiked soil. The level of water in the lower recipient was adjusted whenever needed, as to guarantee the necessary conditions of moisture according to, the recommendations specified in [64]. Five replicates of uncontaminated natural PTRS1 soil were prepared for the control, while three replicates were tested per

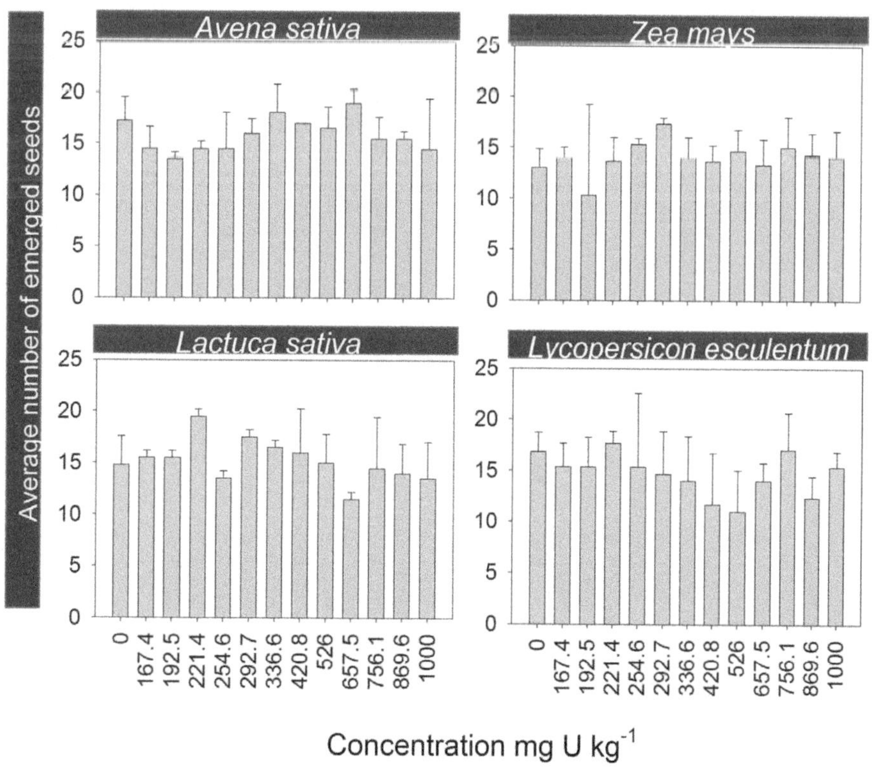

Figure 3. Seed germination of plants. Average number of emerged seeds in monocotyledonous, *Avena sativa and Zea mays* and in dicotyledonous species, *Lycopersicon esculentum and Lactuca sativa,* grown in PTRS1 soil contaminated with U. The error bars indicate the standard deviation.

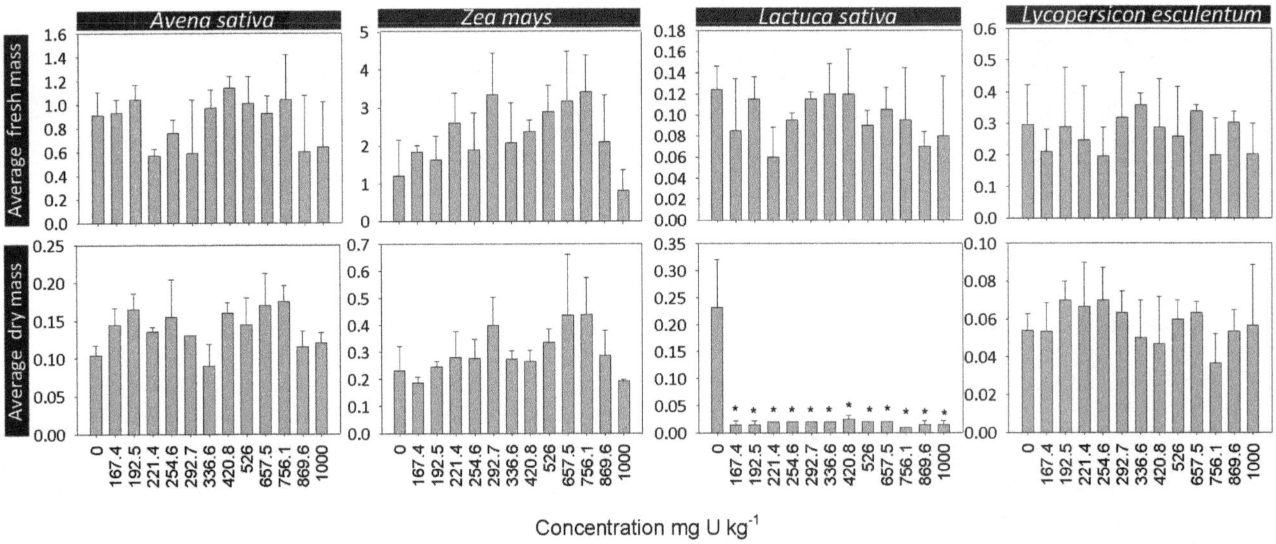

Figure 4. Growth of plants. Average values of fresh mass and dry mass in monocotyledonous, *Avena sativa and Zea mays* and in dicotyledonous species, *Lycopersicon esculentum and Lactuca sativa* grown in PTRS1 soil, contaminated with U. The error bars indicate the standard deviation. The asterisks point out significantly differences from the control (P<0.05).

concentration, in order to minimize the variability of the results, and to follow the ECx sampling design, similarly used for the invertebrate tests.

At the beginning of the test, nutrients (Substral - Plants fertilizer using 1 bottle cap for 2 L of water proportion according to the manufacturer recommendation; Fertilizer NPK: 6-3-6; nitrogen (N): 6%; phosphate (P_2O_5): 3%; potassium (K_2O): 6%; iron (Fe): 0,03%; trace elements: Cu, Mn, Mo and Zn), were added in each lower recipient containing the water. Pots were maintained at constant conditions of temperature (20±2°C), photoperiod (16 hL: 8 hD) and light intensity (25.000 lux). Daily observations were carried out to record the number of emerged seeds. Only the first five emerged seeds were left to grow, the remaining ones were counted and harvested. Fourteen days later, the assay was finished and the fresh and dry biomass above soil was assessed for each test species at the end of the exposure period.

The endpoints seed germination, and fresh and dry biomass, above soil, were assessed for each species at the end of the exposures according to the methods outlined in ISO guideline 11269-2 [64].

For this work, a battery of enzymes involved in different biogeochemical cycles S (sulfur cycle), N (Nitrogen cycle), C (Carbon cycle), as well as enzymes more indicative of the good physiological conditions of the whole microbial community (e.g. dehydrogenases) were selected. The species of invertebrates and plants were selected based on the availability of standard protocols. Since we aimed to obtain data for the derivation of SSVs, for regulatory purposes, this procedure is recommended.

Statistical Analysis

A one-way analysis of variance (one-way ANOVA) was performed to test significant differences between the uranium concentrations tested for each endpoint analyzed: the activity of enzymes, the number of juveniles produced by potworms and collembolans, the number of emerged seeds, and the fresh and dry mass of the plants. The Kolmogorov-Smirnov test was applied to check data normality, whereas homoscedasticity of variances was checked by the Levene's test. When these two assumptions of the one-way ANOVAs were not met, a Kruskal-Wallis analysis was performed. The statistical analysis was run in the SigmaPlot 11.0 software for Windows. When statistical significant differences were recorded, the Dunnett's (for parametric one-way ANOVA) or the Dunn's test (for non-parametric ANOVA) was carried out to perceive which concentrations were significantly different from the respective control. Based on the outcomes of the multiple comparison tests the NOEC (no-observed-effect-concentration) and LOEC (low-observed-effect-concentration) values were deter-

Table 5. Soil quality guideline values derived for copper in Portugal, USA and Canada (mg U Kg^{-1} soildw).

Portugal				Canada	Other reference
Backgound concentrations	PNEC		Proposed SSV[b]	SQG_E[c]	
	NOEC	EC_{20}			
7.8[a]	23.3	15.5	15.5	23	100[d]

[a]Caetano et al.[49];
[b]SSV - soil screening value;
[c]Canadian Soil Quality Guidelines for environmental health (SQGe), Scott-Fordsmand and Pedersen [116].;
[d]Sheppard and Sheppard [98].

mined. The EC_{20} and EC_{50} values for each endpoint were calculated whenever possible, after fitting the data to a log-logistic model using the STATISTICA 7.0 software.

PNEC-Based SSV Derivation

Following the approach suggested by the Technical Guidance Document published by the European Commission [65], a predicted no effect concentration (PNEC) for U in the PTRS1 soil was determined, based on the assessment factor method For that, it was by used the lowest point estimate (i.e., NOEC and EC_{20} values) and applied the appropriate assessment factor based on the criteria of the Guidance Document [65]. The lowest point estimate calculated was for arylsulphatase activity. Considering that more than three NOEC values were obtained in this study, for at least three species, an assessment factor of 10 was applied, The PNEC value was calculated through the application of the following equation:

$$PNEC = \frac{\text{lowest point estimate}}{10}$$

Results and Discussion

1. Soil microbial activity

As far as authors are aware, this study is one of the few studies gathering data regarding the ecotoxicity of spiked soils with U on soil microbial parameters. Only a study from Sheppard et al. [66] has analyzed the effect of U on soil phosphatase activity in eleven different Canadian soils (including an agricultural, a boreal forest and a garden soil). This study recorded a significantly depressed activity only at the highest concentration tested (1000 mg U kg $soil_{dw}^{-1}$) for all the soils. These results suggested that probably, soil phosphatase activity was one of the less sensitive soil microbial parameters to U. In fact, Pereira et al. [7] also reported the low sensitivity of this parameter in mine soils contaminated with metals.

The variation in soil enzyme activities, N mineralization and potential nitrification in the PTRS1 soil, spiked with different U concentrations, is shown in Figure 1, and the Table 3 summarizes the toxicity values obtained for each biochemical parameter.

U had a clear inhibitory effect in almost all functional parameters tested. Overall, dehydrogenase and urease were the most affected soil enzymes by U, being their activity significantly inhibited at concentrations equal or lower than 134.5 mg U kg $soil_{dw}^{-1}$ (Table 3). Dehydrogenases have a relevant role in the oxidation of soil organic matter (SOM), being a good indicator of the active microbial biomass in the soil compartment [67]. As such, U (in the form of uranyl) strongly affected the normal microbial activity in PTRS1 soil. Meyer et al. [68] also observed a significant reduction in respiration rates of a soil exposed to depleted uranium (DU), but only for concentration equal and higher than 500 mg U kg $soil_{dw}^{-1}$. Indeed, the inhibition of urease activities indicates that U had a deleterious effect on soil N-cycle (Figure 1, Table 3). The reduction in the activity of this enzyme may have been caused by a negative effect of U on the overall microbial biomass, which in turn was translated into a reduction in the oxidation rate of organic N into ammonium [58,69]. Arylsulphatase is regularly involved in the S-cycle by catalyzing hydrolysis reactions in the biogeochemical transformation of S [67]. This parameter was significantly affected by U at a LOEC of 279.0 mg U kg $soil_{dw}^{-1}$. On its turn, the cellulase activity was significantly inhibited at intermediate U concentrations. However in the highest concentrations the tendency was reversed and the activity increased, but not for levels significantly different from the control (Figure 1). Thereby, we can conclude that the C-metabolism associated with the degradation of SOM and catalyzed by these extracellular enzymes [70] was constrained by U. N mineralization and potential nitrification are indicators of the functioning of the N-cycle, hence providing an overview of the activity of specific microbial groups (nitrifying bacteria) directly involved in both processes [71]. The general pattern of response observed for these two parameters corresponded to stimulation at the lower U concentrations and inhibition under the highest ones (Figure 1), leading to EC_{50} values of 347.0 and 610.0 mg U kg $soil_{dw}^{-1}$ (Table 3), respectively. It has been stated that N mineralization is normally less sensitive than potential nitrification, since the former is carried out by a wider diversity of microorganisms [71]. However, our data showed the opposite (Figure 1). Meyer et al [68] did not observe effects on nitrogen mineralization of the test soil for U concentrations up to 25000 mg kg $soil_{dw}^{-1}$, however the form of U tested by these authors (schoepite $UO2(OH_2).H_2O$) was less soluble than the one tested in this soil.

The sensitivity of soil microbial parameters to metals has already been demonstrated by several authors, either in metal-polluted or in artificially spiked soils (e.g.,[4,72–77]). Dehydrogenase and urease had generally been referred as the most affected enzymes for different metals (e.g., Cu, Pb, Zn, Cd, Fe, Cr, Ni), (e.g.,[72,73,78,79]). Arylsulphatase and cellulase, however, have shown contradictory responses in different studies. Some authors observed negative correlations between arylsulphatase and cellulase activities and Zn [75] and Cu concentrations, respectively [80,81]; while others observed positive correlations between arylsulphatase and Cd [81], and no changes on cellulase activities in the presence of metals in urban soils was observed [82]. Usually, potential nitrification is negatively influenced by the presence of metals and metalloids such as Pb, Cu and As [7,81]; and the inhibitory effect of some metals (like Zn, Cd and Pb) on N mineralization was also reported by Dai et al. [83]. Antunes et al [81] found negative correlations (based on the Spearman coefficient) between U levels in soils from an abandoned U mine (presenting a mixture of metals) and the activities of urease and cellulase enzymes. For dehydrogenase, potential nitrification and arylsulphatase no significant correlations were detected. Nevertheless, this study analyzed mine contaminated soils, where the mixture of metals, may cause either synergistic or antagonistic effects, and where a well adapted and functional microbial community was likely established.

The inhibition of soil enzyme activities recorded could have been caused by toxicological effects of metals on soil microorganisms with subsequent decrease in their abundance and/or biomass; and/or by the direct inactivation of extracellular enzymes by metals [84]. Notwithstanding, the levels of metals may be not the sole effect on soil microbial activity. Soil properties (e.g., pH, organic matter content, nutrients and soil texture) may also interfere and modulate the bioavailability and toxicityof metals on soil enzymes [74,85]. Clays can retain and protect extracellular hydrolases, namely urease [73]. But the low clay content of PTRS1 soil (3.32%) (Table 1) might have increased U bioavailability, leading to the impairment of soil microbial community through cytotoxic effects, hence reducing their metabolic activity [81]. Additionally, the low pH of PTRS1 soil (Table 1) might have contributed for U availability and impacts on enzyme processes, potential nitrification and N mineralization, particularly at higher

U concentrations, as previously observed by Coppolecchia et al. [75] for arylsulphatase in the presence of Zn and low pH.

The above results illustrated well the effects of U in the performance of soil enzymes, reinforcing the importance of these parameters as bioindicators of soil quality. Indeed, the EC_{20} values calculated for dehydrogenase (34.9 mg Ukg $soil_{dw}^{-1}$), urease (< 135.5 mg Ukg $soil_{dw}^{-1}$), N mineralization (152.2 mg Ukg $soil_{dw}^{-1}$) and arylsulphatase (155.3 mg Ukg $soil_{dw}^{-1}$) are within the environmental concentrations quantified in soils from an abandoned U mine, following extractions with *aqua regia* or with rainwater [8]. In this sense, the data herein generated represent a great asset for the derivation of SSVs, since they have a great ecological representativeness.

2. Uranium toxicity to the reproduction of soil invertebrates

The reproduction tests with the three invertebrate species revealed that *E. andrei, E. crypticus and F. candida* were quite sensitive to U in the PTRS1 soil. Tests fulfilled the validity criteria established by the standard guidelines for control replicates [61–63]. The resulting NOEC, LOEC, EC_{20} and EC_{50} values obtained in this study and toxicity data available in the literature are summarized in the Table 3.

The effects of U in the reproduction of *E. andrei* were evident, since statistical significant differences were found between the control and the highest tested concentrations of U for this organism ($F = 5.218$, d.f. $= 23$, $p = 0.002$) (Figure 2). The tested metal did not significantly affect the reproduction of *E. andrei* at concentrations up to 500.0 mg U kg $soil_{dw}^{-1}$ (NOEC) but compromised this endpoint for concentrations above 550.0 mg U kg $soil_{dw}^{-1}$ (LOEC). EC_{20} and EC_{50} values of U for *E. andrei* reproduction were 474.83 mg U kg $soil_{dw}^{-1}$ and 631.00 mg U kg $soil_{dw}^{-1}$, respectively (Table 3). The results obtained in our study, did not support those of Sheppard and Stephenson [86] (Table 4) that did not record toxic effects for *E. andrei* below 1000 mg U kg $soil_{dw}^{-1}$ (soils (carbonated): pH 7.5, 18% organic matter, 18% clay). However, they found an inhibition of juveniles production in two soils spiked with U, presenting low organic matter (2.2% and 1%) and a pH of 7.5 and 6.2, respectively (Table 4). According to the literature, the adsorption of metals to soil components is dependent on its physical and chemical properties, therefore influencing their toxicity to soil organisms [41,47,48]. Chelinho et al. [87], observed that soils with an organic matter content below 4% reduced or completely inhibited earthworms reproduction. However, the PTRS1 natural soil, had a high organic matter content, 6.2% (according to the classification provided by Murphy et al. [88]). Besides, as previously checked, the intrinsic properties of this soil did not compromise the performance of earthworms [49]. A high organic matter content of soils is usually related with a decrease in the toxicity of the contaminants for the organisms [41,43,89]. However, this was not the case in the study. In fact, Lourenço [60,90] exposed *E. andrei* to a U mine soil with 215.72±8.50 mg U kg $soil_{dw}^{-1}$, a pH of 7.79±0.01, and 7.71±0.60% of organic matter and observed that the bioaccumulation of U and daughter radionuclides was in tandem with loss of DNA integrity of coelomocyte cells, changes in the frequency of cells of immune system and also with histopathological changes (especially of the epidermis and chloragogenous tissue and intestinal epithelium). In fact, some other authors [91] had also suggested that the direct dermal exposure of earthworms to metals in the soil pore water, the ingestion of water, and/or soil particles may strongly favor the bioaccumulation of metals. Since pH is variable in the different compartments of gastrointestinal tract of earthworms, it can increase the mobilization of contaminants from soil after its ingestion [92,93].

Although, other metals were present in the contaminated soil tested by Lourenço et al. [60,90] U likely had a crucial role in the toxic effects observed, because it persisted in the whole body till 56 days. These authors suggested that the changes observed in DNA integrity were likely early warning indicators of effects on the growth and reproduction of the organisms. And in fact, effects on reproduction were observed in our study. Further, Giovanetti et al. [94]. exposed *E. fetida* natural U- and DU-contaminated soil (no information on soil type) for 7 and 28 days. Regarding natural U, no mortality or significant changes in weight were observed for both exposure periods up to 600 mg U $kg^{-1}{}_{dw}$. The chloragogeneous tissue, the main storage tissue of U, presented meaningful changes after 7 days of exposure for 300 mg U Kg^{-1}, while DNA strand breaks increased in a dose dependent manner above 150 mg U kg $soil_{dw}^{-1}$.

Regarding *E. crypticus* reproduction, it was significantly not reduced above 526.0 mg U kg $soil_{dw}^{-1}$ (LOEC) (Table 3) ($F = 31.05$, d.f. $= 12$, $p<0.05$). The EC_{20} and EC_{50} values estimated were respectively 469.7 and 518.6 mg U kg $soil_{dw}^{-1}$. Although no toxicity values are reported for the lowest concentrations tested, enchytraeids showed considerable sensitivity to U, since the number of juveniles was minimal or no juveniles were produced by *E. crypticus* at concentrations above 657.5 mg U kg $soil_{dw}^{-1}$ (Figure 2). Despite enchytraeids are commonly used in standardized toxicity tests, to the best of our knowledge, no data are available in the literature regarding the effects of U on the reproduction of this species. The available information concerns only the toxic effects caused by other metals or by natural soil properties in the reproduction of this species [43,48,95–97]. Thus, taking into account this literature review, pH and CEC were the most important parameters controlling the high sensitivity of enchytraeids to metals. Additionally, and according to Kuperman et al. [43], adults survival and juveniles production by *E. crypticus* can be maximized in natural soils with properties within the following ranges: 4.4–8.2 pH; 1.2–42% OM; 1–29% clay. The PTRS1 natural soil used as test substrate fell into in these ranges (Table 3), and similarly to *E. andrei,* the reproduction of this species was not compromised during the validation of the PTRS1 natural soil as a reference soil [49], meaning that the soil properties did not limit the performance of *E. crypticus*.

Concerning *F. candida*, U affected the production of juveniles, as shown by a significant decrease of this endpoint along the concentrations tested ($F = 11.6$, d.f. $= 12$, $p<0.05$) (Figure 2). The number of juveniles was not significantly affected up to 675.50 mg U kg $soil_{dw}^{-1}$ (NOEC), but it was significantly decreased for U concentrations equal to or greater than 756.10 mg U Kg^{-1} (LOEC) (Table 3). The EC_{20} value estimated for reproduction was 343.41 mg U kg $soil_{dw}^{-1}$ which is considerably lower than the toxicity data reported by Shepard et al. [98], EC_{20}>710 mg U kg $soil_{dw}^{-1}$, in two loam soils with pH 7.5 (Table 4). The low sensitivity of *F. candida* to U was also observed by Sheppard and Stephenson [86] which tested 3 soils amended with a range of U concentrations and aged for 10 years before testing. In this study, the lowest EC_{20} value obtained was 840 mg U kg $soil_{dw}^{-1}$ in a loam soil (pH 7.5, 24% clay, 2.2% OM) (Table 4). Despite this, *F. candida* was more sensitive in the study of Sheppard and Stephenson (since their EC_{20} value was similar to the EC_{50} recorded in our study 851.64 mg U kg $soil_{dw}^{-1}$). When considering the number of juveniles produced, U was less toxic to *F. candida* comparatively to *E. andrei* and *E. crypticus*. The lower sensitivity of *F. candida* is also consistent with other studies, when the effects of other metals in the reproduction of the three species was investigated [97,99], or even when other species of collembolans were analyzed [86]. The exposure of *F. candida* to chemicals in soil is apparently lower than for earthworms, which are exposed both by ingestion of contaminated soil (mineral particles, organic matter and chemicals in the soil solution) and also through direct dermal contact [100]. Despite the widely known influence of soil parameters on the bioavailability of chemicals and their influence on the reproduction of soil organisms, less is known about the intrinsic effects of physico-chemical parameters of the soils in the reproduction of *F. candida*. In general, several authors had reported a high tolerance of *F. candida* reproduction to a wide range of soil textural classes, organic matter contents and soil pH [48,101,102]. Once again the performance of this species was not compromised by the intrinsic properties of the PTRS1 soil. Hence, the effects observed can undoubtedly be attributed to U exposure.

3. Phytotoxicity of uranium

Relatively to terrestrial plants tests, all the validity criteria as described by the standard guidelines were attained [64]. Data obtained showed no significant effects on seeds emergence for all species tested ($p>0.05$). In fact, it was observed a relatively high rate of germination, either in monocotyledonous and dicotyledonous species (Figure 3). This outcome was somewhat expected, based on previous studies (e.g.,[22]). Seed coats form a barrier which protects embryos from a wide range of contaminants, especially metals. Thus, the germination relies almost exclusively on the seed reserves making it a less sensitive endpoint to the toxicity of soil pollutants [103].

An apparent hormetic effect was recorded for the other endpoints measured for almost all plant species. Such occurrence was recorded by other authors and it was attributed to the use of U as uranyl nitrate, which corresponds to a supplementary dose of N given to plants [98].

With regard to production of fresh- and dry-mass, it was possible to perceive that the tested plants displayed different sensitivities to this metal. However, no significant differences were generally observed comparatively to the control, exception for *L. sativa* dry mass ($H=22.8$, d.f. $=12$, $p=0.029$). Thus, and according to Figure 4, *L. sativa* was the most sensitive terrestrial plant to U. The high sensitivity of *L. sativa* was also found by Hubálek et al. [104] and Soudek [105]. This was probably caused by the high capacity of this species to bioaccumulate high concentrations of metals, including U [22].

The exposure of plants to metals, was already extensively studied, showing that these contaminants can induce biological effects on germination, growth and development, as well as, alterations in the nutrient profile of plants [22,106]. However, only some studies (e.g., [66,107] and others reviewed [98]) have assessed the ecotoxicological effects of U on terrestrial plant species.

Based on our study, once again was proved the diverse ecotoxicological outcomes for U effects on plant species, since no effects were observed, in the range of tested concentrations for the three evaluated endpoints (in three out of four species), in PTRS1 soil. Similar results were obtained by Sheppard and Sheppard [86] in acidic soils (Table 4), when testing the emergence and growth of wheatgrass *Elymus lanceolatus*. Like in our study, these authors did not observe any effect on this species (up to 1000 mg U kg $soil_{dw}^{-1}$). In opposition, Sheppard and Sheppard [81] revised data on U toxicity to terrestrial plants and reported EC_{25} values ranging from 300 to 500 mg U kg $soil_{dw}^{-1}$, considering only the most reliable studies. Stojanović et al. [108] also reported phytotoxic effects of U on *Zea mays* exposed, in different soil types, to 250, 500 and 1000 mg U kg $soil_{dw}^{-1}$, but especially at the highest concentration tested and in the most acidic soil. However, no statistical analysis of the data was performed in this study.

Soil properties are also the factors that most strongly affect U uptake and phytotoxic effects, [18,109–111]. The bivalent uranyl ion (UO_2^{2+}) is sorbed to the negatively charged surfaces of clay minerals and organic compounds. In acidic soils subjected to pH increase, more negatively charged binding sites are available on mineral surfaces due to the progressive reduction of protons occupying these sites. However, pH values close to 6, like the one of PTRS1, favors U availability, since the concentrations of carbonates tends to increase, and U is released to the soil solution in the form of U-carbonate complexes [18]. The natural soil PTRS1, besides being acidic, has a lower clay content, which means lower binding sites for the bivalent uranyl ion (UO_2^{2+}), hence constraining U bioavailability. other soil properties and plant mechanisms may explain the reduced sensitivity of the plants in comparison with soil microbial parameters and invertebrates. Viehweger and Geipel [112] reported an increased U absorption by *Arabidopsis halleri* attributed to Fe deficiency in the medium of hydroponically grown plants. With respect to this metal, in the natural PTRS1 soil, the analyses done by Caetano et al. [49] showed that Fe surpassed the soil benchmark values proposed by two EPA regions (http://rais.ornl.gov/tools/eco_search.php http://rais.ornl.gov/tools/eco_search.php). In this sense, it is hypothesized that the high Fe content of the PTRS1 soil, may have also contributed for reducing the absorption of U by plants. As far as plant mechanisms are considered, in several studies reviewed by Mitchell et al. [113] the transport of U within plants was reduced and higher concentrations were consistently found in the roots. Using X-ray absorption spectroscopy (XAS) and transmission electron microscopy (TEM), Laurette et al. [114] observed that when plants are exposed to U and phosphates, needle-like U-phosphates are formed and precipitate, both outside and inside the cells, or persist in the subsurface of root tissues. The precipitation of U-phosphate complexes acts as a protective mechanism preventing U translocation to the shoots and leaves. This can also occur when the culture medium of the plants has no phosphate, since some plants are able to exudate phosphates. Further, U may be also absorbed like UO_2^{2+} and linked to endogenous organophospate groups [114]. In opposition, when translocation occurs within plants, U has mainly formed U-carboxylated complexes. Plants can also exudate organic acids to the rhizosphere environment or UO_2^{2+} may form complexes with endogenous compounds like malic, citric, oxalic and acetic acid [114]. In summary, the different resistance mechanisms described above could explain the lack of toxic effects observed for *A. sativa*, *Z. mays* and *L. esculentum*, in opposition to *L. sativa*. Most concerning is the fact that the majority of studies testing the phytotoxicity of U, including those performed by us, were made with the addition of nutrients solution, which increased the availability of phosphates to the test soil, likely decreasing the sensitivity of plants to U. Hence, to enhance the protection level of SSVs derived for plants, more assays with different plant species should be performed and the addition of nutrients should be prevented, or at least the tests may include replicates with and without nutrients.

Derivation of a Soil Screening Value (SSV) for Uranium Applying Assessment Factors

The PNEC values obtained for U were based in EC_{20} and NOEC values varied between 15.5 and 23.3 mg kg $soil_{dw}^{-1}$, respectively (Table 5). These values were six to four times lower than the PNEC value suggested by Sheppard and Sheppard [86] (Table 5), which was 100 mg Ukg $soil_{dw}^{-1}$. In opposition, they are close to the lowest Canadian Soil Quality Guideline for both environment and human health (23 mg U kg $soil_{dw}^{-1}$). Thereby, while more ecotoxicological data is being obtained or other methods are being applied to derive soil screening values (SSVs) we prefer to be precautious by proposing a PNEC of 15.5 mg Kg^{-1} $soil_{dw}$ as a SSV for U, in soils similar to the PTRS1. This SSV value is near the background value found in non-contaminated soils [8,48], but not in some areas with naturally occurring U anomalies in soils, where concentrations ranging between 13–724 mg U kg $soil_{dw}^{-1}$ can be found [115].

Conclusion

With the present study it was possible to generate a set of important ecotoxicological data for the derivation of a SSV for U, using a Portuguese natural soil representative of a granitic region, where this type of mine exploration occurred.

Soil enzyme activities were clearly inhibited by U. The obtained results depended not only on the concentrations of U but also on the properties of soil, which were likely responsible for the bioavailability of U and subsequent impairments on soil microbial population and, consequently, in their activity. Dehydrogenase and urease were particularly sensitive to U. Further, and comparatively to the remaining effect concentrations obtained/estimated for invertebrates and plants, the soil microbial parameters were more affected by U contamination[1].

The toxic effects of U in soil invertebrates were also confirmed, but the tested species showed a variable sensitivity to this metal. The increasing order of species sensitivity to U based on EC_{50} values for reproduction was *E. crypticus* > *E. andrei* > *F. candida*. However, if EC_{20} values are considered *F. candida* is the most sensitive invertebrate, since its EC_{20} value was 343.41 mg U kg $soil_{dw}^{-1}$, compared to 474.83 mg U kg $soil_{dw}^{-1}$ and 469.76 mg U kg $soil_{dw}^{-1}$ EC_{20} values estimated for *E. andrei* and *E. crypticus*, respectively. The EC_{20} values estimated were lower than the NOEC values for *E. andrei* and *F. candida*. Thus, the EC_{20} point estimate should be selected for the derivation of more protective SSVs. Relatively to the plants, the tested species showed no adverse effects caused by U in PTRS1, with the exception of *L. sativa* dry mass yield. Considering the results obtained, it was possible to verify a great variability between the EC_x values estimated in this study and those reported in the scientific literature. Multiple factors can contribute to this discordance, but probably at least for some species, soils physical and chemical properties were the main factors responsible for such differences. Although, this reinforces, at least in part, the importance of using natural soils representatives of the main types of soil from each region in ecotoxicological evaluations and for the derivation of SSVs, the data generated suggests that the SSV (15.5 mg Kg^{-1} $soil_{dw}$) derived for U, was six times lower than the PNEC value proposed by other authors. Nevertheless, as mentioned previously, more data should be obtained following standard protocols.

Author Contributions

Conceived and designed the experiments: ALC CRM RP. Performed the experiments: ALC AG. Analyzed the data: ALC CRM RP. Contributed reagents/materials/analysis tools: FC FG EFS. Contributed to the writing of the manuscript: ALC CRM RP.

References

1. Gavrilescu M, Pavel L, Cretescu I (2009) Characterization and remediation of soils contaminated with uranium. J Hazard Mater 163: 475–510. doi:10.1016/j.jhazmat.2008.07.103.
2. Malyshkina N, Niemeier D (2010) Future sustainability forecasting by exchange markets: basic theory and an application. Environ Sci Technol 44: 9134–9142. Available: http://www.ncbi.nlm.nih.gov/pubmed/21058697.
3. Carvalho F, Oliveira J, Faria I (2009) Alpha emitting radionuclides in drainage from Quinta do Bispo and Cunha Baixa uranium mines (Portugal) and associated radiotoxicological risk. Bull Environ Contam Toxicol 83: 668–673. Available: http://www.ncbi.nlm.nih.gov/pubmed/19590808. Accessed 19 November 2012.
4. Pereira R, Barbosa S, Carvalho FP (2014) Uranium mining in Portugal: a review of the environmental legacies of the largest mines and environmental and human health impacts. Environ Geochem Health 36: 285–301. Available: http://www.ncbi.nlm.nih.gov/pubmed/24030454. Accessed 7 March 2014.
5. Carvalho F, Madruga M, Reis M, Alves J, Oliveira J, et al. (2007) Radioactivity in the environment around past radium and uranium mining sites of Portugal. J Environ Radioact 96: 39–46. Available: http://www.ncbi.nlm.nih.gov/pubmed/17433852. Accessed 25 October 2012.
6. Vandenhove H, Sweeck L, Mallants D, Vanmarcke H, Aitkulov A, et al. (2006) Assessment of radiation exposure in the uranium mining and milling area of Mailuu Suu, Kyrgyzstan. J Environ Radioact 88: 118–139. Available: http://www.ncbi.nlm.nih.gov/pubmed/16581165.
7. Pereira R, Sousa J, Ribeiro R, Gonçalves F (2006) Microbial Indicators in Mine Soils (S. Domingos Mine, Portugal). Soil Sediment Contam 15: 147–167. Available: http://www.informaworld.com/openurl?genre=article&doi=10%2e1080%2f15320380500506813&magic=crossref%7c%7cD404A21C5BB053405B1A640AFFD44AE3. Accessed 5 October 2013.
8. Pereira R, Antunes S, Marques S, Gonçalves F (2008) Contribution for tier 1 of the ecological risk assessment of Cunha Baixa uranium mine (Central Portugal): I soil chemical characterization. Sci Total Environ 390: 377–386. Available: http://www.ncbi.nlm.nih.gov/pubmed/17919686. Accessed 6 September 2011.
9. Arogunjo A, Höllriegl V, Giussani A, Leopold K, Gerstmann U, et al. (2009) Uranium and thorium in soils, mineral sands, water and food samples in a tin mining area in Nigeria with elevated activity. J Environ Radioact 100: 232–240. Available: http://www.ncbi.nlm.nih.gov/pubmed/19147259. Accessed 14 October 2013.
10. Momčilović M, Kovačević J, Dragović S (2010) Population doses from terrestrial exposure in the vicinity of abandoned uranium mines in Serbia. Radiat Meas 45: 225–230. Available: http://linkinghub.elsevier.com/retrieve/pii/S1350448710000363. Accessed 25 October 2012.
11. Figueiredo M, Silva T, Batista M, Leote J, Ferreira M, et al. (2011) Uranium in surface soils: An easy-and-quick assay combining X-ray diffraction and X-ray fluorescence qualitative data. J Geochemical Explor 109: 134–138. Available: http://linkinghub.elsevier.com/retrieve/pii/S0375674210001366. Accessed 23 October 2012.
12. Patra A, Sumesh C, Mohapatra S, Sahoo S, Tripathi R, et al. (2011) Long-term leaching of uranium from different waste matrices. J Environ Manage 92: 919–925. Available: http://www.ncbi.nlm.nih.gov/pubmed/21084148. Accessed 14 October 2013.
13. Scheele F (2011) Uranium from Africa: Mitigation of Uranium Mining Impacts on Society and Environment by Industry and Governments. SSRN Electron J. Available: http://papers.ssrn.com/abstract=1892775. Accessed 14 October 2013.
14. Niemeyer J, Moreira-Santos M, Nogueira M, Carvalho G, Ribeiro R, et al. (2010) Environmental risk assessment of a metal-contaminated area in the Tropics. Tier I: screening phase. J Soils Sediments 10: 1557–1571. Available: http://www.springerlink.com/index/10.1007/s11368-010-0255-x. Accessed 29 October 2012.
15. Carvalho F (2011) Environmental Radioactive Impact Associated to Uranium Production Nuclear and Technological Institute Department of Radiological Protection and Nuclear Safety,. Journal, Am Sci Environ Publ Sci 7: 547–553.
16. Wang J, Lu A, Ding A (2007) Effect of cadmium alone and in combination with butachlor on soil enzymes. Heal (San Fr: 395–403. doi:10.1007/s10653-007-9084-2.
17. ASTDR (2011) Agency for Toxic Substances & Disease Registry-Toxicological profile for uranium. Draft. U.S. Department of Health and Human Services. Public Health Service and Agency for Toxic Substances and Disease Registry. 452 + annexes.
18. Vandenhove H, Van Hees M, Wouters K, Wannijn J (2007) Can we predict uranium bioavailability based on soil parameters? Part 1: effect of soil parameters on soil solution uranium concentration. Environ Pollut 145: 587–595. Available: http://www.ncbi.nlm.nih.gov/pubmed/16781802. Accessed 19 November 2012.
19. Gongalsky K (2003) Impact of pollution caused by uranium production on soil macrofauna. Environ Monit Assess 89: 197–219. Available: http://www.ncbi.nlm.nih.gov/pubmed/14632090. Accessed 15 October 2013.
20. Geras'kin S, Evseeva T, Belykh E, Majstrenko T, Michalik B, et al. (2007) Effects on non-human species inhabiting areas with enhanced level of natural

radioactivity in the north of Russia: a review. J Environ Radioact 94: 151–182. Available: http://www.ncbi.nlm.nih.gov/pubmed/17360083. Accessed 26 October 2012.

21. Joner E, Munier-Lamy C, Gouget B (2007) Bioavailability and microbial adaptation to elevated levels of uranium in an acid, organic topsoil forming on an old mine spoil. Environ Toxicol Chem 26: 1644–1648. Available: http://www.ncbi.nlm.nih.gov/pubmed/17702337.
22. Pereira R, Marques CR, Ferreira MJS, Neves MFJV, Caetano AL, et al. (2009) Phytotoxicity and genotoxicity of soils from an abandoned uranium mine area. Appl Soil Ecol 42: 209–220. Available: http://linkinghub.elsevier.com/retrieve/pii/S0929139309000778. Accessed 6 September 2011.
23. Kenarova A, Radeva G, Danova I, Boteva S, Dimitrova I (2010) Soil bacterial abundance and diversity of uranium impacted. Second Balkan conference on Biology: 5–9.
24. Islam E, Sar P (2011) Molecular assessment on impact of uranium ore contamination in soil bacterial diversity. Int Biodeterior Biodegradation 65: 1043–1051. Available: http://linkinghub.elsevier.com/retrieve/pii/S0964830511001697. Accessed 29 October 2012.
25. Geng F, Hu N, Zheng J-F, Wang C-L, Chen X, et al. (2011) Evaluation of the toxic effect on zebrafish (*Danio rerio*) exposed to uranium mill tailings leaching solution. J Radioanal Nucl Chem 292: 453–463. Available: http://link.springer.com/10.1007/s10967-011-1451-x. Accessed 15 October 2013.
26. Lourenço J, Pereira R, Silva A, Carvalho F, Oliveira J, et al. (2012) Evaluation of the sensitivity of genotoxicity and cytotoxicity endpoints in earthworms exposed in situ to uranium mining wastes. Ecotoxicol Environ Saf 75: 46–54. Available: http://www.sciencedirect.com/science/article/pii/S0147651311002685. Accessed 14 October 2013.
27. Islam E, Sar P (2011) Molecular assessment on impact of uranium ore contamination in soil bacterial diversity. Int Biodeterior Biodegradation 65: 1043–1051. Available: http://www.sciencedirect.com/science/article/pii/S0964830511001697. Accessed 25 October 2013.
28. Lavelle P, Decaens T, Aubert M, Barot S, Blouin M, et al. (2006) Soil invertebrates and ecosystem services. Eur J Soil Biol 42: S3–S15. Available: http://linkinghub.elsevier.com/retrieve/pii/S1164556306001038. Accessed 19 July 2011.
29. O'Halloran K (2006) Toxicological Considerations of Contaminants in the Terrestrial Environment for Ecological Risk Assessment. Hum Ecol Risk Assess An Int J 12: 74–83. Available: http://www.tandfonline.com/doi/abs/10.1080/10807030500428603. Accessed 7 September 2011.
30. Barrios E (2007) Soil biota, ecosystem services and land productivity. Ecol Econ 64: 269–285. Available: http://linkinghub.elsevier.com/retrieve/pii/S0921800907001693. Accessed 4 October 2012.
31. Dominati E, Patterson M, Mackay A (2010) A framework for classifying and quantifying the natural capital and ecosystem services of soils. Ecol Econ 69: 1858–1868. Available: http://www.sciencedirect.com/science/article/pii/S0921800910001928. Accessed 14 October 2013.
32. CEC (2006) Comission of the European Cominities, Directive 2004/35/EC. Commissiona of the European Communities, Brussels.
33. Bone J, Head M, Barraclough D, Archer M, Scheib C, et al. (2010) Soil quality assessment under emerging regulatory requirements. Environ Int 36: 609–622. Available: http://www.ncbi.nlm.nih.gov/pubmed/20483160. Accessed 6 October 2012.
34. Jensen J, Mesman M (2006) Ecological Risk Assessment of Contaminated Land. Decision support for site specific investigations. ISBN 90-6960-138-9 978-90-6960-138-0.
35. Fishwick S (2004) Soil screening values for use in UK ecological risk assessment. Soil Quality & Protection, Air Land and Water Group. Science Environment Agency's Project Manager.
36. USEPA (2003) United States Environmental Protection Agency. Guidance for Developing Ecological Soil Screening Levels (Eco-SSLs). Attachment 1–2: Assessment of Whether to Develop Ecological Soil Screening Levels for Microbes and Microbial Processes, OSWER Directi. Environ Prot.
37. Carlon C (2007) Derivation methods of soil screening values in europe. a review and evaluation of national procedures towards harmonisation. European Commission, Joint Research Centre, Ispra, EUR 22805-EN, 306.
38. Provoost J, Reijnders L, Swartjes F, Bronders J, Carlon C, et al. (2008) Parameters causing variation between soil screening values and the effect of harmonization. J Soils Sediments 8: 298–311. Available: http://www.springerlink.com/index/10.1007/s11368-008-0026-0. Accessed 29 October 2012.
39. Van Gestel C (2012) Soil ecotoxicology: state of the art and future directions. Zookeys 296: 275–296. Available: http://www.pubmedcentral.nih.gov/articlerender.fcgi?artid=3335420&tool=pmcentrez&rendertype=abstract. Accessed 29 October 2012.
40. MAOTDR (2006) Ministerio do Ambiente do Ordenamento do Territorio e do Desenvolvimento Regional. Diário da República, 1. a série – N. o 208–27 de Outubro.
41. Römbke J, Jänsch S, Junker T, Pohl B, Scheffczyk A, et al. (2006) Improvement of the applicability of ecotoxicological tests with earthworms, springtails, and plants for the assessment of metals in natural soils. Environ Toxicol Chem 25: 776–787. Available: http://www.ncbi.nlm.nih.gov/pubmed/16566163.
42. Van Assche F, Alonso JL, Kapustka L, Petrie R, Stephenson GL, et al. (2002) Terrestrial plant toxicity tests. In: Fairbrother A., Glazebrock P.W., Van Straalen N.M., Tarazona J.V. (eds) Test methods to determine hazards of sparingly soluble metal compounds in soils. SETAC, Pensacola, Forida, 37–57. Available: https://www.setac.org/store/view_product.asp?id=1038018.
43. Kuperman R, Amorim M, Römbke J, Lanno R, Checkai R, et al. (2006) Adaptation of the enchytraeid toxicity test for use with natural soil types. Eur J Soil Biol 42: S234–S243. Available: http://www.sciencedirect.com/science/article/pii/S1164556306000719. Accessed 14 October 2013.
44. Hofman J, Hovorková I, Machát J (2009) Comparison and Characterization of OECD Artificial Soils. Ecotoxicological Characterization of Waste 2009, 223–229. Available: http://www.rivm.nl/bibliotheek/rapporten/711701047.html.
45. Song J, Zhao F, McGrath S, Luo Y (2006) Influence of soil properties and aging on arsenic phytotoxicity. Environ Toxicol Chem 25: 1663–1670. Available: http://www.ncbi.nlm.nih.gov/pubmed/16764487. Accessed 14 October 2013.
46. Rooney C, Zhao F, McGrath S (2007) Phytotoxicity of nickel in a range of European soils: influence of soil properties, Ni solubility and speciation. Environ Pollut 145: 596–605. Available: http://www.ncbi.nlm.nih.gov/pubmed/16733077. Accessed 14 October 2013.
47. Van Gestel C, Borgman E, Verweij R, Ortiz M (2011) The influence of soil properties on the toxicity of molybdenum to three species of soil invertebrates. Ecotoxicol Environ Saf 74: 1–9. Available: http://www.ncbi.nlm.nih.gov/pubmed/20951431. Accessed 14 October 2013.
48. Domene X, Chelinho S, Campana P, Natal-da-Luz T, Alcañiz JM, et al. (2011) Influence of soil properties on the performance of Folsomia candida: implications for its use in soil ecotoxicology testing. Environ Toxicol Chem 30: 1497–1505. Available: http://www.ncbi.nlm.nih.gov/pubmed/21437938. Accessed 13 March 2013.
49. Caetano A, Gonçalves F, Sousa J, Cachada A, Pereira E, et al. (2012) Characterization and validation of a Portuguese natural reference soil to be used as substrate for ecotoxicological purposes. J Environ Monit 14: 925–936. Available: http://www.ncbi.nlm.nih.gov/pubmed/22297688. Accessed 31 May 2013.
50. ISO 17512-1 (2008) International Organization for Standardization 17512-1: 2008.Soil Quality: Avoidance Test for Testing the Quality of Soils and the Toxicity of Chemicals-Test with Earthworms (*Eisenia Fetida*). Geneva, Switzerland. International Organization for Standard. Geneva, Switzerland.
51. FAOUN (1984) Food and agriculture organization of the United Nations – physical and chemical methods of soil and water analysis. Soils Bull. 10, 1–275.
52. SPAC (2000) Soil and Plant Analysis Council – Handbook of Reference Methods. CRC Press, Boca Raton, Florida.
53. Tabatabai M, Bremner J (1970) Arylsulfatase activity in soils. Soil Science Society of America, 34: 225–9.
54. Schinner F, Kandeler E, Öhlinger R, Margesin R (1996) Methods in soil biology. Germany: Springer-Verlag.
55. Öhlinger R (1996) Soil sampling and sample preparation. In: Schinner F, Öhlinger R, Kandeler E, Margesin R, editors. Methods in Soil Biology. Springer Verlag.
56. Schinner F, von Mersi W (1990) Xylanase, CM-cellulase and invertase activity in soil, an improved method. Soil Biol Biochem 22: 511–5.
57. Kandeler E, Gerber H (1988) Short-term assay of soil urease activity using colorimetric determination of ammonium. biol Fert Soils 6: 68–72.
58. Kandeler E (1996) Potential nitrification. In: Schinner F, Öhlinger R, Kandeler E, Margesin R, editors. Methods in soil biology. Germany: Springer-Verlag-Berlin-Heidelberg, 146–9.
59. Berg P, Rosswall T (1985) Ammonium oxidizer numbers, potential and actual oxidation rates en two Swedish arable soils. Biology Fertility Soils 1: 131–40.
60. Lourenço JI, Pereira RO, Silva AC, Morgado JM, Carvalho FP, et al. (2011) a) Genotoxic endpoints in the earthworms sub-lethal assay to evaluate natural soils contaminated by metals and radionuclides. J Hazard Mater 186: 788–795. Available: http://www.ncbi.nlm.nih.gov/pubmed/21146299. Accessed 21 September 2013.
61. ISO 11268-2 (1998) International Organization for Standardization ISO 11268-2. Soil quality: effects of pollutants on earthworms (*Eisenia fetida*) - Part 2: Determination of effects on reproduction. ISO 11268-2. Geneva, Switzerland: International Organization for Standardiza. Geneva, Switz.
62. ISO 16387 (2004) International Organization for Standardization ISO 16387. Soil quality: effects of pollutants on Enchytraeidae (*Enchytraeus* sp.)- Determination of effects on reproduction and survival. ISO16387. Geneva, Switzerland: International Organization for Standard. Geneva, Switz.
63. ISO 11267 (1999) International Organization for Standardization ISO 11267. Soil quality: inhibition of reproduction of Collembola (*Folsomia candida*) by soil pollutants. ISO 11267. Geneva, Switzerland: International Organization for Standardization. Geneva, Switz.
64. ISO 11269-2 (2005) International Organization for Standardization ISO 11269-2. Soil quality: determination of the effects of pollutants on soil flora - Part 2: Effects of chemicals on the emergence and growth of higher plants. ISO 11269-2. Geneve, Switzerlan: International O. Geneva, Switz.
65. EC. Commission European (2003) Commission European. Technical Guidance Document on Risk Assessment. in support of Commission Directive 93/67/EEC on Risk Assessment for new notified substances Commission Regulation (EC) No 1488/94 on Risk Assessment for existing substances Directive 98/.
66. Sheppard S, Evenden W (1992) Bioavailability Indices for Uranium: Effect of Concentration in Eleven Soils. Arch Environ Contain Toxicol 23, 117–124.

67. Taylor J, Wilson B, Mills M, Burns R (2002) Comparison of microbial numbers and enzymatic activities in surface soils and subsoils using various techniques. 34.
68. Meyer MC, Paschke MW, McLendon T, Price D (1998) Decreases in Soil Microbial Function and Functional Diversity in Response to Depleted Uranium. J Environ Qual 27: 1306. Available: https://www.agronomy.org/publications/jeq/abstracts/27/6/JEQ0270061306. Accessed 31 March 2014.
69. Wang M, Markert B, Shen W, Chen W, Peng C, et al. (2011) Microbial biomass carbon and enzyme activities of urban soils in Beijing. Environ Sci Pollut Res Int 18: 958–967. Available: http://www.ncbi.nlm.nih.gov/pubmed/21287285. Accessed 25 September 2013.
70. Alvarenga P, Palma P, Gonçalves AP, Baião N, Fernandes RM, et al. (2008) Assessment of chemical, biochemical and ecotoxicological aspects in a mine soil amended with sludge of either urban or industrial origin. Chemosphere 72: 1774–1781. Available: http://www.ncbi.nlm.nih.gov/pubmed/18547605. Accessed 13 May 2014.
71. Winding A, Hund-Rinke K, Rutgers M (2005) The use of microorganisms in ecological soil classification and assessment concepts. Ecotoxicol Environ Saf 62: 230–248. Available: http://www.ncbi.nlm.nih.gov/pubmed/15925407.
72. Khan S, Cao Q, Hesham AE-L, Xia Y, He J-Z (2007) Soil enzymatic activities and microbial community structure with different application rates of Cd and Pb. J Environ Sci (China) 19: 834–840. Available: http://www.ncbi.nlm.nih.gov/pubmed/17966871. Accessed 15 October 2013.
73. Lee S, Kim E, Hyun S, Kim J (2009) Metal availability in heavy metal-contaminated open burning and open detonation soil: assessment using soil enzymes, earthworms, and chemical extractions. J Hazard Mater 170: 382–388. Available: http://www.ncbi.nlm.nih.gov/pubmed/19540045. Accessed 5 April 2013.
74. Papa S, Bartoli G, Pellegrino A, Fioretto A (2010) Microbial activities and trace element contents in an urban soil. Environ Monit Assess 165: 193–203. Available: http://www.ncbi.nlm.nih.gov/pubmed/19444636. Accessed 13 May 2013.
75. Coppolecchia D, Puglisi E, Vasileiadis S, Suciu N, Hamon R, et al. (2011) Soil Biology & Biochemistry Relative sensitivity of different soil biological properties to zinc. Soil Biol Biochem 43: 1798–1807. Available: http://dx.doi.org/10.1016/j.soilbio.2010.06.018.
76. Lee S-H, Park H, Koo N, Hyun S, Hwang A (2011) Evaluation of the effectiveness of various amendments on trace metals stabilization by chemical and biological methods. J Hazard Mater 188: 44–51. Available: http://www.ncbi.nlm.nih.gov/pubmed/21333442. Accessed 17 August 2013.
77. Hu B, Liang D, Liu J, Xie J (2013) Ecotoxicological effects of copper and selenium combined pollution on soil enzyme activities in planted and unplanted soils. Environ Toxicol Chem 32: 1109–1116. Available: http://www.ncbi.nlm.nih.gov/pubmed/23401089. Accessed 8 October 2013.
78. Gülser F, Erdoğan E (2008) The effects of heavy metal pollution on enzyme activities and basal soil respiration of roadside soils. Environ Monit Assess 145: 127–133. Available: http://www.ncbi.nlm.nih.gov/pubmed/18027096. Accessed 17 August 2013.
79. Thavamani P, Malik S, Beer M, Megharaj M, Naidu R (2012) Microbial activity and diversity in long-term mixed contaminated soils with respect to polyaromatic hydrocarbons and heavy metals. J Environ Manage 99: 10–17. Available: http://dx.doi.org/10.1016/j.jenvman.2011.12.030.
80. Alvarenga P, Palma P, de Varennes A, Cunha-Queda AC (2012) A contribution towards the risk assessment of soils from the São Domingos Mine (Portugal): chemical, microbial and ecotoxicological indicators. Environ Pollut 161: 50–56. Available: http://www.ncbi.nlm.nih.gov/pubmed/22230067. Accessed 5 October 2013.
81. Antunes S, Pereira R, Marques S, Castro B, Gonçalves F (2011) Impaired microbial activity caused by metal pollution A field study in a deactivated. Sci Total Environmen 410–411.
82. Sivakumar S, Nityanandi D, Barathi S, Prabha D, Rajeshwari S, et al. (2012) Selected enzyme activities of urban heavy metal-polluted soils in the presence and absence of an oligochaete, *Lampito mauritii* (Kinberg). J Hazard Mater 227–228: 179–184. Available: http://www.ncbi.nlm.nih.gov/pubmed/22658212. Accessed 17 August 2013.
83. Dai J, Becquer T, Rouiller JH, Reversat G, Bernhard-Reversat F, et al. (2004) Influence of heavy metals on C and N mineralisation and microbial biomass in Zn-, Pb-, Cu-, and Cd-contaminated soils. Appl Soil Ecol 25: 99–109. Available: http://www.sciencedirect.com/science/article/pii/S0929139303001355. Accessed 15 October 2013.
84. Kızılkaya R, Bayraklı B (2005) Effects of N-enriched sewage sludge on soil enzyme activities. Appl Soil Ecol 30: 192–202. Available: http://www.sciencedirect.com/science/article/pii/S0929139305000594. Accessed 15 October 2013.
85. Turner BL, Hopkins DW, Haygarth PM, Ostle N (2002) β-Glucosidase activity in pasture soils. Available: http://www.sciencedirect.com/science/article/pii/S0929139302000203. Accessed 15 October 2013.
86. Sheppard SC, Stephenson GL (2012) Ecotoxicity of aged uranium in soil using plant, earthworm and microarthropod toxicity tests. Bull Env Contam Toxicol 8843–47: 43–47. doi:10.1007/s00128-011-0442-5.
87. Chelinho S, Domene X, Campana P, Natal-da-Luz T, Scheffczyk A, et al. (2011) Improving ecological risk assessment in the Mediterranean area: selection of reference soils and evaluating the influence of soil properties on avoidance and reproduction of two oligochaete species. Environ Toxicol Chem 30: 1050–1058. Available: http://www.ncbi.nlm.nih.gov/pubmed/21305581. Accessed 17 October 2013.
88. Murphy S, Giménez D, Muldowney LS, Heckman JR (2012) Soil Organic Matter Level and Interpretation How is the Organic Matter Content of Soils Determined? How Is Organic Matter Level: 1–3.
89. Natal-da-Luz T, Ojeda G, Pratas J, Van Gestel CA, Sousa JP (2011) Toxicity to *Eisenia andrei* and *Folsomia candida* of a metal mixture applied to soil directly or via an organic matrix. Ecotoxicol Environ Saf 74: 1715–1720. Available: http://www.ncbi.nlm.nih.gov/pubmed/21683441. Accessed 13 March 2013.
90. Lourenço J, Silva A, Carvalho F, Oliveira J, Malta M, et al. (2011) b) Histopathological changes in the earthworm Eisenia andrei associated with the exposure to metals and radionuclides. Chemosphere 85: 1630–1634. Available: http://www.ncbi.nlm.nih.gov/pubmed/21911243. Accessed 4 October 2013.
91. Hobbelen PHF, Koolhaas JE, van Gestel CAM (2006) Bioaccumulation of heavy metals in the earthworms *Lumbricus rubellus* and *Aporrectodea caliginosa* in relation to total and available metal concentrations in field soils. Environ Pollut 144: 639–646. Available: http://www.ncbi.nlm.nih.gov/pubmed/16530310. Accessed 15 October 2013.
92. Li L, Wu J, Tian G, Xu Z (2009) Effect of the transit through the gut of earthworm (*Eisenia fetida*) on fractionation of Cu and Zn in pig manure. J Hazard Mater 167: 634–640. Available: http://www.ncbi.nlm.nih.gov/pubmed/19232822. Accessed 19 September 2013.
93. Peijnenburg W, Jager T (2003) Monitoring approaches to assess bioaccessibility and bioavailability of metals: matrix issues. Ecotoxicol Environ Saf 56: 63–77. Available: http://www.ncbi.nlm.nih.gov/pubmed/12915141. Accessed 15 October 2013.
94. Giovanetti A, Fesenko S., Cozzella ML, Asencio LD, Sansone U (2010) Bioaccumulation and biological effects in the earthworm *Eisenia fetida* exposed to natural and depleted uranium. J Environ Radioact 101: 509–516. Available: http://www.ncbi.nlm.nih.gov/pubmed/20362371. Accessed 15 October 2013.
95. Peijnenburg W, Baerselman R, de Groot C, Jager T, Posthuma L, et al. (1999) Relating environmental availability to bioavailability: soil-type-dependent metal accumulation in the oligochaete *Eisenia andrei*. Ecotoxicol Environ Saf 44: 294–310. Available: http://www.ncbi.nlm.nih.gov/pubmed/10581124.
96. Amorim M, Römbke J, Scheffczyk A, Soares AM (2005) Effect of different soil types on the enchytraeids *Enchytraeus albidus and Enchytraeus luxuriosus* using the herbicide Phenmedipham. Chemosphere 61: 1102–1114. Available: http://www.ncbi.nlm.nih.gov/pubmed/16263380.
97. Kuperman R (2004) Manganese toxicity in soil for *Eisenia fetida, Enchytraeus crypticus (Oligochaeta),* and *Folsomia candida* (Collembola). Ecotoxicol Environ Saf 57: 48–53. Available: http://linkinghub.elsevier.com/retrieve/pii/S0147651303001544. Accessed 22 June 2011.
98. Sheppard S, Sheppard M (2005) Derivation of ecotoxicity thresholds for uranium. J Environ 79: 55–83. Available: http://www.ncbi.nlm.nih.gov/pubmed/15571876. Accessed 19 November 2012.
99. Lock K, Janssen CR (2001) Cadmium Toxicity for Terrestrial Invertebrates: Taking Soil Parameters Affecting Bioavailability into Account. Environ Toxicol: 315–322.
100. Layinka T, Idowu B, Dedeke A, Akinloye A, Ademolu O, et al. (2011) Earthworm as bio-indicator of heavy metal pollution around Lafarge, Wapco Cement Factory, Ewekoro, Nigeria. Proceedings of the Environmental Management Conference, Federal University of Agriculture, Abeokuta, Nigeria: 489–496.
101. Amorim M, Rçmbke J, Scheffczyk A, Nogueira A, Soares A (2005) Effects of Different Soil Types on the Collembolans *Folsomia candida* and *Hypogastrura assimilis* using the Herbicide Phenmedipham. 352: 343–352. doi:10.1007/s00244-004-0220-z.
102. Jänsch S, Amorim MJ, Römbke J (2011) Identification of the ecological requirements of important terrestrial ecotoxicological test species. Environmental Reviews, 13(2): 51–83. Available: http://www.nrcresearchpress.com/doi/abs/10.1139/a05-007#.Ul0AxlDENPB. Accessed 15 October 2013.
103. Liu X, Zhang S, Shan X-Q, Christie P (2007) Combined toxicity of cadmium and arsenate to wheat seedlings and plant uptake and antioxidative enzyme responses to cadmium and arsenate co-contamination. Ecotoxicol Environ Saf 68: 305–313. Available: http://www.ncbi.nlm.nih.gov/pubmed/17239437. Accessed 15 October 2013.
104. Hubálek T, Vosáhlová S, Matejů V, Kovácová N, Novotný C (2007) Ecotoxicity monitoring of hydrocarbon-contaminated soil during bioremediation: a case study. Arch Environ Contam Toxicol 52: 1–7. Available: http://www.ncbi.nlm.nih.gov/pubmed/17106791. Accessed 13 August 2013.
105. Soudek P, Petrová S, Benešová D, Dvořáková M, Vaněk T (2011) Uranium uptake by hydroponically cultivated crop plants. J Environ Radioact 102: 598–604. Available: http://www.ncbi.nlm.nih.gov/pubmed/21486682. Accessed 13 September 2013.
106. Gopal R, Rizvi A (2008) Excess lead alters growth, metabolism and translocation of certain nutrients in radish. Chemosphere 70: 1539–1544. Available: http://www.ncbi.nlm.nih.gov/pubmed/17923149. Accessed 15 October 2013.
107. Sheppard S, Evenden W, Anderson A (1992) Multiple assays of uranium toxicity in soil. Environ. Toxicol. Water Qual., 7: 275–294.
108. Stojanović M, Stevanović D, Milojković J, Grubišić M, Ileš D (2009) Phytotoxic Effect of the Uranium on the Growing Up and Development the Plant of Corn.

Water, Air, Soil Pollut 209: 401–410. Available: http://www.springerlink.com/index/10.1007/s11270-009-0208-4. Accessed 25 September 2011.

109. Bednar A, Medina V, Ulmer-Scholle D, Frey B, Johnson B, et al. (2007) Effects of organic matter on the distribution of uranium in soil and plant matrices. Chemosphere 70: 237–247. Available: http://www.sciencedirect.com/science/article/pii/S0045653507008168. Accessed 15 October 2013.

110. Tunney H, Stojanovic M, Mrdakovic Popic J, McGrath D, Zhang C (2009) Relationship of soil phosphorus with uranium in grassland mineral soils in Ireland using soils from a long-term phosphorus experiment and a National Soil Database. J Plant Nutr Soil Sci 172: 346–352. Available: http://doi.wiley.com/10.1002/jpln.200800069. Accessed 13 September 2013.

111. Soudek P, Petrová Š, Benešová D, Vaněk T (2011) Uranium uptake and stress responses of in vitro cultivated hairy root culture of Armoracia rusticana. Agrochimical 1: 15–28.

112. Viehweger K, Geipel G (2010) Uranium accumulation and tolerance in *Arabidopsis halleri* under native versus hydroponic conditions. Environ Exp Bot 69: 39–46. Available: http://dx.doi.org/10.1016/j.envexpbot.2010.03.001.

113. Mitchell N, Pérez-Sánchez D, Thorne MC (2013) A review of the behaviour of U-238 series radionuclides in soils and plants. J Radiol Prot 33: R17–48. Available: http://www.ncbi.nlm.nih.gov/pubmed/23612607. Accessed 18 September 2013.

114. Laurette J, Larue C, Llorens I, Jaillard D, Jouneau P-H, et al. (2012) Speciation of uranium in plants upon root accumulation and root-to-shoot translocation: A XAS and TEM study. Environ Exp Bot 77: 87–95. Available: http://www.sciencedirect.com/science/article/pii/S0098847211002814. Accessed 15 October 2013.

115. Pereira A, Neves L (2012) Estimation of the radiological background and dose assessment in areas with naturally occurring uranium geochemical anomalies–a case study in the Iberian Massif (Central Portugal). J Environ Radioact 112: 96–107. Available: http://www.ncbi.nlm.nih.gov/pubmed/22694913. Accessed 14 September 2013.

116. Scott-Fordsmand JJ, Pedersen MB (1995) Soil quality criteria for selected inorganic compounds. Danish Environmental Protection Agency, Working Report No. 48, 200.

11

Decomposition of Organic Carbon in Fine Soil Particles Is Likely More Sensitive to Warming than in Coarse Particles: An Incubation Study with Temperate Grassland and Forest Soils in Northern China

Fan Ding[1,2], Yao Huang[1], Wenjuan Sun[1]*, Guangfu Jiang[1,2], Yue Chen[1,2]

1 State Key Laboratory of Vegetation and Environmental Change, Institute of Botany, Chinese Academy of Sciences, Beijing, China, **2** University of Chinese Academy of Sciences, Beijing, China

Abstract

It is widely recognized that global warming promotes soil organic carbon (SOC) decomposition, and soils thus emit more CO_2 into the atmosphere because of the warming; however, the response of SOC decomposition to this warming in different soil textures is unclear. This lack of knowledge limits our projection of SOC turnover and CO_2 emission from soils after future warming. To investigate the CO_2 emission from soils with different textures, we conducted a 107-day incubation experiment. The soils were sampled from temperate forest and grassland in northern China. The incubation was conducted over three short-term cycles of changing temperature from 5°C to 30°C, with an interval of 5°C. Our results indicated that CO_2 emissions from sand (>50 μm), silt (2–50 μm), and clay (<2 μm) particles increased exponentially with increasing temperature. The sand fractions emitted more CO_2 (CO_2-C per unit fraction-C) than the silt and clay fractions in both forest and grassland soils. The temperature sensitivity of the CO_2 emission from soil particles, which is expressed as Q_{10}, decreased in the order clay>silt>sand. Our study also found that nitrogen availability in the soil facilitated the temperature dependence of SOC decomposition. A further analysis of the incubation data indicated a power-law decrease of Q_{10} with increasing temperature. Our results suggested that the decomposition of organic carbon in fine-textured soils that are rich in clay or silt could be more sensitive to warming than those in coarse sandy soils and that SOC might be more vulnerable in boreal and temperate regions than in subtropical and tropical regions under future warming.

Editor: Dafeng Hui, Tennessee State University, United States of America

Funding: This work was jointly supported by the National Natural Science Foundation of China (Grant No. 41075108), the Ministry of Science and Technology of China (2010CB950604) and the CAS Strategic Priority Research Program (Grant No. XDA05050507). The funders had no role in study design, data collection and analysis, decision to publish, or preparation of the manuscript.

Competing Interests: The authors have declared that no competing interests exist.

* E-mail: sunwj@ibcas.ac.cn

Introduction

As the largest terrestrial carbon pool, soil plays a vital role in the global carbon cycle [1]. A small change in the magnitude of soil carbon stocks could result in a large impact on the atmospheric enrichment of CO_2 [2]. Soil organic carbon (SOC) loss via soil heterotrophic respiration (i.e., SOC decomposition) is one of the largest components of global CO_2 flux from the terrestrial ecosystem to the atmosphere [3,4]. It is widely recognized that temperature has a profound influence on SOC decomposition [5] and that soils are expected to release CO_2 at faster rates with rising temperature [3]. Therefore, understanding the temperature sensitivity of SOC decomposition is critical for predicting changes in global soil respiration, changes in soil carbon stocks to future warming, and especially, the feedback of soils on climate change [2,6–8].

The temperature sensitivity of SOC decomposition, which is commonly expressed as Q_{10}, refers to the change in decomposition rate with a 10°C increase in temperature. A recent study constructed a database of worldwide soil respiration observations matched with high-resolution historical climate data and estimated that global soil respiration had a Q_{10} of 1.5 [3]. Nevertheless, soil respiration experiments have unanimously indicated that Q_{10} is spatially heterogeneous on a global scale [9,10]. Thus, to accurately predict changes in global soil carbon stocks to future warming, knowing which factors determine the temperature dependence of SOC decomposition on large scales is critical [6]. It is known that the Q_{10} of soil heterotrophic respiration varies with temperature [5,11], soil moisture [12,13], and the quantity and quality of SOC [14,15]. However, it is not clear whether SOC decomposition has different warming responses in different soil textures. It is most likely so because the SOC stock in fine-textured soils is less vulnerable to disturbances, e.g., tillage, than in coarse-textured soils [16,17].

Such different vulnerabilities are assumed to result from fine-textured soils having greater fractions of silt-sized or clay-sized particles, whereas coarse-textured soils have more sand-sized particles; these primary particles contain different mineral compositions and provide various affinities to organic matter [18]. Clay and silt particles (e.g., sesquioxides and layer silicates) provide large specific surface areas and numerous reactive sites at which SOC can be sorbed by strong ligand exchange and polyvalent cation bridges, whereas the sand particles, which is dominated by quartz particles, exhibits only weak bonding affinities to SOC [19,20]. The bonding of SOC to minerals is recognized to prevent decomposition [21,22] and may therefore decrease the temperature sensitivity of decomposition [23]. Thus, the decomposition of organic carbon associated with sand, silt, and clay fractions likely respond to warming differently. The temperature sensitivity of SOC decomposition from different particle size fractions is intimately linked to the response of CO_2 emission from different soil textures to future warming and the emission's feedback on climate change.

Different soil particle size fractions possess different sizes of carbon pools [20]. Clay and silt have large specific surface areas and can sorb abundant SOC [24]. For example, in temperate arable soils, 50–75% of the total SOC is associated with clay, whereas silt accounts for 20–40% and sand accounts for less than 10% [18]. Additionally, the SOC associated with the clay and silt fractions is assumed to be stable relative to that associated with the sand fraction [21]. In this context, the long-term stabilization of soil carbon pools under climate warming mainly depends on the warming response of the carbon that is associated with the clay and silt fractions.

However, to our knowledge, there were few studies who explored difference in the warming responses of organic carbon associated with these primary particles. In this study, we conducted a 107-day incubation experiment to investigate the temperature sensitivity of SOC decomposition in sand (>50 μm), silt (2–50 μm), and clay (<2 μm) particles from grassland and forest soils. This study aims to answer the following questions: 1) Is there a difference in the Q_{10} of SOC decomposition among different size soil particles and between the two different ecosystems? 2) Which factors determine the Q_{10}? A further objective is to better understand the features of CO_2 emission and their future trends with global warming for different textured soils.

Materials and Methods

Sites and sampling

Soil samples were collected from two sites at the typical steppe grassland and temperate forest ecosystems in northern China. The grassland site was at the Duolun Restoration Ecology Experimentation and Demonstration Station (42°02′N, 116°17′E, 1324 m a.s.l.) and the forest site was at the Beijing Forest Ecosystem Research Station (39°58′N, 115°26′E, 1252 m a.s.l.) at the Institute of Botany of the Chinese Academy of Sciences. The grassland is permanent grassland situated in the temperate continental semi-arid climate zone. The dominant species are *Stipa baicalensis*, *Thalictrum petaloideum*, and *Artemisia frigida*. The mean annual temperature (MAT) is 1.6°C, and the mean annual precipitation (MAP) is 386 mm, with 100 frost-free days. The soil is sandy Haplic Calcisols (FAO) composed of 61.2% sand, 23.0% silt, and 15.8% clay, and the pH is 7.0. The forest is an artificial forest (with age >40 years) with a dominant *pinus tabuliformis Carrière* specie and the climate is warm temperate humid monsoon climate zone. The MAT and MAP are 5°C and 650 mm, respectively, with 160 frost-free days. The soil is a clay loam that belongs to Eutric Cambisol (FAO) and is composed of 41.0% sand, 42.5% silt, and 16.5% clay, and the pH is 7.0.

At each site, 40 cores of mineral soils at 0–20 cm depths were randomly collected using an auger with 4 cm in diameter and were composited into one soil sample. A small fraction of the soils (approximately 1 kg) was immediately placed in a sterile plastic bag on ice when transported to the laboratory and stored at 4°C for inoculum preparation. The remaining soils were placed into cloth bags and air-dried at room temperature. The dried soils were finely crumbed and sieved through a 2-mm mesh to remove visibly identifiable plant debris and gravel. Fine roots were removed through static electricity sorption.

Particle-size fractionation

Twenty grams of dried soil were dispersed in 100 ml of deionized water by the ultrasonic probe of a JY92-II Ultrasonic Homogenizer (6 mm diameter probe, 300 W). The ultrasonic time was chosen as 3 min and 8.5 min for grassland and forest soil, respectively, after calibrating against conventional soil texture analyses [25]. The dispersed suspensions were separated into three size fractions: sand (>50 μm), silt (2–50 μm), and clay (<2 μm). The sand fraction was separated using distilled water to wash all of the fine particles in the soil suspension through a 50-μm sieve until the elutriate solution that exited the sieve was clear. The elutriate was divided into silt and clay fractions by repeated centrifugation using an Avanti J-25 Beckman Centrifuge [26]. The clay fraction was concentrated from the large volumes of suspensions by flocculation with $CaCl_2$ and centrifugation [27]. The three size fractions were oven-dried at 40°C [28], gently crushed, weighed, and then stored for incubation and future analysis.

Laboratory incubation

For each soil particle, a 10-g dried sample was placed into a 150-ml brown flask. Then, 2 ml of inoculum were added to each sample, and the inoculum solution was prepared by shaking 100 g of fresh soil with 1000 ml of deionized water and leaving to stand overnight [28]. To improve soil aeration and make the inoculum disperse uniformly, the sample plus inoculum was loosened, stirred, and leveled slightly. The soil moisture was adjusted to 60% water-filled pore space (WFPS) [29]. The incubation lasted 107 days (Figure 1). The soil samples were incubated at room temperature (ca. 20°C) when there was no measurement. The samples were regularly corrected for water loss by adding deionized water. Four replicates were conducted in the experiment. Empty flasks were set as blank. The volume of head space (flask volume minus soil solid) was accurately measured after complementing the experiment by the method of water replacement [30].

Following Fang et al. [31] and Xu et al. [32], soil respiration rates under changing temperature were measured in sequence. Three temperature-change cycles were started at the 9th, 47th, and 100th days after the beginning of the incubation, and each cycle lasted 8 days (Figure 1). The first temperature-change cycle began after 8 days of incubation to stabilize the microbial activity to avoid an undesired microbial peak [33]. Within each cycle, the temperature was continuously increased from 5°C to 30°C in intervals of 5°C and then gradually decreased from 30°C to 5°C with the same interval. After each temperature was changed to the next, the incubation flasks were kept open for 3 hours to allow the soil samples to adapt to the changed temperature [30]; the flasks were then sealed using butyl rubber septa and flushed with ambient air for one minute using a pump with an "inlet-outlet" needle inserted through the septa to produce the same initial gas conditions among all flasks. After specific times of enclosure (20, 15, 9, 6, 4, and 3 hours for 5, 10, 15, 20, 25, and 30°C, respectively), 5 ml of gas was taken through a gas-tight injection syringe from the headspace of the flask. The CO_2 concentrations were analyzed with a gas chromatograph (Agilent 7890A, USA).

Determination of soil physicochemical properties

The soil texture was measured using the hydrometer method [34]. The SOC and soil total nitrogen (STN) concentrations were

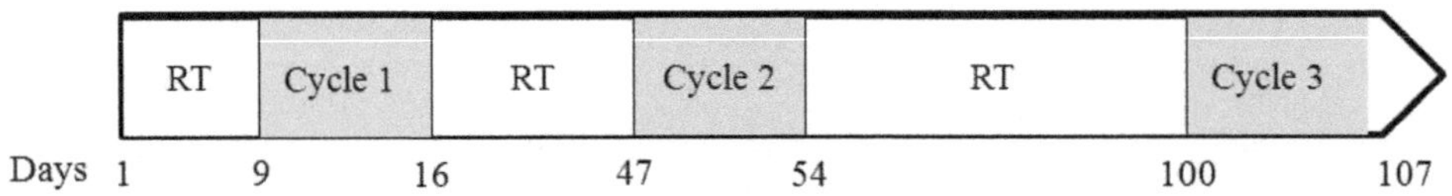

Figure 1. Schematic diagram of the time sequence for the three temperature-change cycles. RT signifies that the soil samples were incubated at room temperature.

determined using the $K_2Cr_2O_7$-H_2SO_4 wet oxidation method [35] and the Kjeldahl method [36], respectively. The microbial biomass carbon (MBC) and microbial biomass nitrogen (MBN) in the soil samples at the end of incubation were measured using chloroform fumigation-extraction (K_2SO_4) [37] with a Multi N/C 3100-TOC/TN Analyzer (Analytik Jena, Germany). The K_2SO_4-extractable organic carbon and nitrogen from the non-fumigated soil samples were considered as the dissolved organic carbon (DOC) and total dissolved nitrogen (TDN) [38,39].

Calculations and statistical analysis

Similar to Lang et al. [40], the CO_2 emission rate was calculated using the following equation:

$$R = V \times \frac{273}{(273+T)} \times (C_s - C_b) \times \rho \times \frac{12}{44} / (W \times c)/t \quad (1)$$

where R is the CO_2 emission rate (μg CO_2-C g^{-1} OC h^{-1}); V is the flask headspace volume (L); T is the incubation temperature (°C); C_s and C_b are the CO_2 concentrations in the sample and blank flasks (ppm), respectively; ρ is the standard state CO_2 density (g L^{-1}); W is the dry weight of the soil particles (kg); c is the SOC concentration of the soil particles (g kg^{-1}); and t is the time enclosed (h). The final CO_2 emission rate at a given temperature was calculated as the average of the values between the increasing temperature and decreasing temperature periods [31,32]. To illustrate the temperature sensitivity of CO_2 emission from soil particles, we used two methods to calculate Q_{10}. One method involves the commonly used first-order exponential model:

$$R = A \cdot e^{BT} \quad (2)$$

where R is the CO_2 emission rate (μg CO_2-C g^{-1} SOC h^{-1}), T is the temperature (°C), and A and B are model parameters that are to be determined. The parameter A is a carbon quality index, which refers to the availability and lability of the carbon substrates [15,41]. The Q_{10} value was then computed as following:

$$Q_{10} = e^{10 \times B} \quad (3)$$

The Q_{10} calculated by Eqn. 3 was a constant value in temperature range of 5–30°C. To examine the Q_{10} variation with temperature, the Q_{10} values were also calculated separately for each temperature interval of 5 degrees (e.g., 5–10°C) and 10 degrees (e.g., 5–15°C) based on the following equation [11]:

$$Q_{10} = (R_2 - R_1)^{10/(T_2 - T_1)} \quad (4)$$

where R_2 and R_1 are the CO_2 emission rates (μg CO_2-C g^{-1} SOC h^{-1}) measured at temperatures T_2 (high temperature) and T_1 (low temperature), respectively.

All statistical analysis was conducted using SPSS 13.0 (SPSS Inc., 2004). The means were compared using the Duncan's multiple range test (DMRT).

Results

Physicochemical characteristics of soils

The SOC and STN concentrations of the bulk soil were very similar between the grassland and forest (Table 1). The organic carbon concentrations in grassland and forest soil particles both decreased with increasing particle size, from 55 g kg^{-1} and 73 g kg^{-1} in the clay particle (<2 μm) to 9 and 13 g kg^{-1} in the sand particles (>50 μm), respectively. Similar trends were observed for the total nitrogen concentrations. The C:N ratios were similar between the clay and silt particles, which were lower than those of the bulk soil and the sand particles. The total percentages of the carbon masses associated with clay and silt particles in the total SOC were more than 60% for both grassland and forest soils, and the percentages of their nitrogen masses in the total STN were both approximately 75%.

Relationship between soil organic carbon decomposition rate and temperature

The SOC decomposition rate from soil particle size fractions, which is expressed as the CO_2-C emission rate per unit of fraction-C, had an exponential relationship with the incubation temperature in all measurements (Figure 2). The sand fractions, which had a lower organic carbon content (Table 1), exhibited greater SOC decomposition rates than the silt and clay fractions, which latter two were similar to one another. The decomposition rates in all the soil fractions decreased continuously as incubation time progressed (from an earlier temperature-change cycle to a later one), except for the sand fraction from the second to the third cycle. As a whole, the decomposition rates from grassland soil particles were greater than those from the corresponding forest soil particles under same condition (Figure 2).

The parameter A in the fitted equation (Eqn. 2) for the relationship between the decomposition rate and temperature is an index for carbon availability and lability. The value of parameter A was greater for the sand fraction than for the silt and clay fractions in both the grassland and forest soils (Figure 2). Furthermore, the values of parameter A were approximately equal between the two land uses for the clay and silt fractions during each temperature change cycle but were greater in the grassland than in the forest for the sand fraction (Figure 2).

Temperature sensitivity of soil organic carbon decomposition for different soil particle-size fractions

The Q_{10} values calculated using the exponential model (Eqn. 3) were significantly different among the soil particle size fractions during each temperature-change cycle; the clay fraction exhibited the greatest value, followed by the silt and sand fractions (Table 2). The calculated mean Q_{10} values of the three cycles were 2.68±0.16 and 2.43±0.16 for the clay fraction, 2.45±0.06 and

Table 1. Physicochemical composition of the soils.

Land uses and particle size fractions	Fraction [a] (%)	Organic carbon		Total nitrogen		C:N ratio
		Concentration (g kg^{-1})	Share of the total (%)	Concentration (g kg^{-1})	Share of the total (%)	
Grassland						
Clay (<2 μm)	11.2 (1.6)	55.3	28.5	5.8	31.9	9.5
Silt (2–50 μm)	28.0 (1.3)	30.9	40.0	3.1	43.2	9.9
Sand (>50 μm)	57.4 (1.0)	8.9	23.6	0.7	18.7	13.4
Bulk soil		21.7		2.0		10.7
Recovery (%)	96.6 (1.1)		92.1		93.9	
Forest						
Clay (<2 μm)	18.1 (2.8)	72.6	44.6	6.1	51.2	12.0
Silt (2–50 μm)	35.1 (2.7)	17.4	20.8	1.5	24.2	11.8
Sand (>50 μm)	43.9 (1.2)	13.0	19.4	0.7	14.1	18.9
Bulk soil		29.5		2.1		13.8
Recovery (%)	97.1 (1.0)		84.8		89.5	

[a]Values in the parentheses represent SD.

2.22±0.10 for the silt, and 2.33±0.08 and 2.09±0.09 for the sand, in the grassland and forest soils, respectively. The mean Q_{10} values of the grassland soil particles were significantly greater than those that of the corresponding forest soil particles (Table 2). As the incubation time progressed (from an earlier temperature-change cycle to a later one), Q_{10} increased for the grassland clay and silt fractions (Table 2).

The Q_{10} values decreased in power-law with increasing temperature for all soil particles when calculating using Eqn. 4 during each individual temperature interval (Figure 3). The average Q_{10} values of all soil particle fractions were 4.22±0.14 during the minimum temperature interval and 1.81±0.17 during the maximum temperature interval (Figure 3). The grassland soil particles had greater Q_{10} than the corresponding forest soil particles during the temperature interval of 5–15°C, whereas the Q_{10} values for the two land uses tended to approach one other during 15–30°C (Figure 3). Nevertheless, the grassland soil particles exhibited a more rapid decline in Q_{10} with increasing temperature (i.e., greater absolute values of the power-law exponents in Figure 3) than did the forest soil particles. Furthermore, the Q_{10} values calculated using Eqn. 4 followed as clay>silt>sand (Figure 3), which was consistent with the pattern in Q_{10} calculated using the exponential model (Eqn. 3) (Table 2).

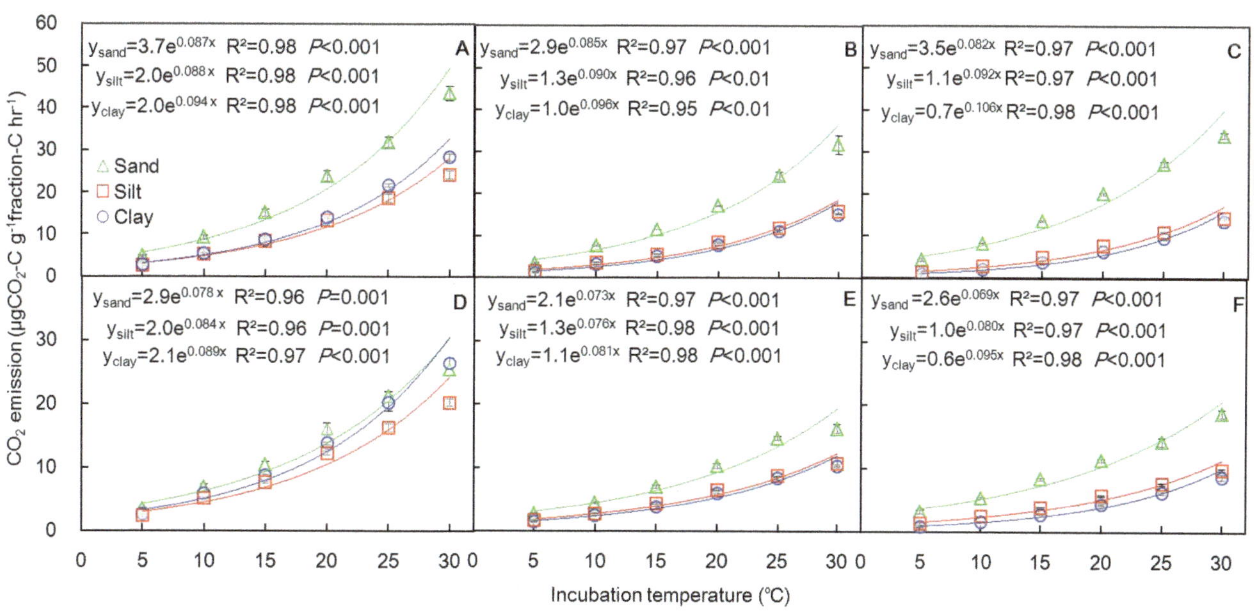

Figure 2. Relationship between CO2 emission rates from particle-size fractions and incubation temperature for grassland (A, B, C) and forest soils (D, E, F). The panels from left to right represent temperature-change cycles in the order 1, 2, 3. The vertical bars denote the standard deviations (n = 4).

Table 2. The Q_{10}^* values (±SD) of SOC decomposition for different particle-size fractions in grassland and forest soils (n = 4).

Land uses	Temperature-change cycles	Particle size fractions		
		Clay (<2 μm)	Silt (2–50 μm)	Sand (>50 μm)
Grassland	1	2.56±0.07a¶	2.41±0.05b	2.39±0.04b
	2	2.60±0.04a	2.45±0.06b	2.34±0.09c
	3	2.88±0.04a	2.51±0.02b	2.27±0.07c
	Mean§	2.68±0.16aA#	2.45±0.06bA	2.33±0.08cA
Forest	1	2.45±0.07a	2.32±0.08b	2.19±0.04c
	2	2.24±0.01a	2.13±0.08b	2.08±0.03b
	3	2.59±0.04a	2.23±0.05b	2.00±0.05c
	Mean§	2.43±0.16aB	2.22±0.10bB	2.09±0.09cB

*Q_{10} value was calculated using Eqn. 3 with the parameters for the fitted curves in Figure 2.
§The mean was calculated including each replicate among the three temperature-change cycles (n = 12).
¶Different lowercase letters in the same row indicate significance differences ($p<0.05$) among soil fractions.
#Different capital letters in the same column indicate significance differences ($p<0.05$) between two land uses.

Dependence of Q_{10} on soil biochemical properties

Generally, there were large differences among various particle-size fractions in most soil biological and chemical properties at the end of incubation (Table 3). Specifically, MBC, MBN, DOC, and TDN were greatest in the clay fraction and were generally of the same order of magnitude as Q_{10} (Table 2). However, the ratios of MBC/OC and MBN/TN shifted to be greatest in the sand fraction. The microbial C/N ratios were not significantly different among the soil particles, but they were generally greater than the soil particles' C/N ratios (except for the forest sand particles; see Table 1). Additionally, the grassland soil particles had greater MBC/OC and less qCO_2 values than the corresponding forest soil particles.

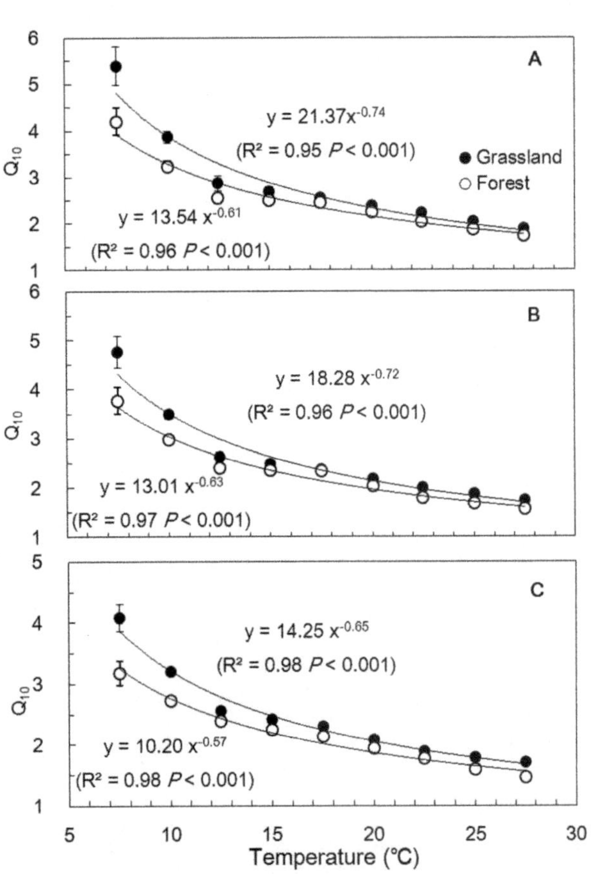

Figure 3. Dependence of Q10 on incubation temperature for (A) clay, (B) silt, and (C) sand. The Q10 for each temperature interval of 5 degrees (e.g., 5–10°C) and 10 degrees (e.g., 5–15°C) was calculated using Eqn. 4. The average temperature for each interval was taken as representative temperature. Data from all the three temperature-change cycles were included. Vertical bars denote the standard errors (n = 12).

The soil biological and chemical properties at the end of incubation represented the soil conditions during the 3rd temperature-change cycle; thus we analyzed the relationship between the Q_{10} in the 3rd cycle and these soil biological and chemical properties. Table 4 presents the Pearson correlation coefficients of Q_{10} during an intermediate temperature range of 15–25°C with various soil biochemical properties. Clearly, the Q_{10} values were highly correlated with MBC and TDN. The Q_{10} had significant linear relationships with MBC (Figure 4A) and TDN (Figure 4B) for the 5–15°C and 20–30°C temperature intervals and also for other temperature intervals (not shown). It should be noted that the slopes of regression line were different between the temperature ranges of 5–15°C and 20–30°C. Over an increasing temperature interval of 10°C, the slopes with low baseline levels of 5°C were steeper than those with high baseline levels of 20°C. Note that the datasets were well-mixed for the grassland and forest, thereby suggesting that these two land use soils had similar features in terms of the dependence of Q_{10} on MBC and TDN.

Discussion

Carbon stability and microbial activity among different size fractions

The carbon associated with different particle-size fractions has different availability and lability [42]. The sand fraction had greater organic carbon availability and lability (greater parameter A), and thus lower carbon stability, than did the silt and clay fractions (Fig. 2). This pattern of SOC stability among the different size fractions was consistent with previous studies [43,44]. The difference in SOC stability among soil primary particles is the reason why coarse sandy soils that are rich in sand particles are less

Table 3. Soil biological and chemical properties in different particle-size fractions at the end of incubation (mean±SE, n = 4).

Land uses and soil fractions	MBC (mg/kg)	MBN (mg/kg)	DOC (mg/kg)	TDN (mg/kg)	MBC/OC* (‰)	MBN/TN* (‰)	Micro C/N (kg/kg)	qCO_2[§] (mg CO_2-C kg^{-1} MBC h^{-1})
Grassland								
Clay (<2 µm)	1096±43a[¶]	35±7a	236±21a	552±3a	20±1c	6±1b	36±7a	480±25a
Silt (2–50 µm)	831±22b	29±4a	65±4b	179±1b	27±1b	9±1b	29±3a	400±14b
Sand (>50 µm)	457±16c	22±1a	38±4b	39±1c	51±2a	33±2a	21±2a	531±14a
Forest								
Clay (<2 µm)	515±46a	61±24a	262±17a	372±21a	7±1b	10±4a	17±8a	896±72b
Silt (2–50 µm)	188±7b	11±0b	114±9b	65±1b	11±0a	8±0a	17±1a	708±32c
Sand (>50 µm)	169±15b	13±3b	122±6b	31±2b	13±1a	19±5a	16±3a	1117±93a

*MBC/OC and MBN/TN refer to the ratio of MBC to fraction-OC and of MBN to fraction-TN, respectively.
§qCO_2 is the metabolic quotient calculated by the CO_2 emission rate at 25°C in the third incubation cycle divided by MBC.
¶Different lowercase letters in the same column for the same land use indicate significance differences ($p<0.05$) among soil fractions.

stable and exhibit faster SOC turnover rates than fine-textured soils that are rich in clay and silt particles [45].

Land-use type can affect the SOC stability by inputting carbon into the soil via litter or roots with different chemical recalcitrance [18,46]. The sand fraction had greater carbon stability (smaller parameter A) in the forest than in the grassland soil (Fig. 2). A possible reason is that the organic matter in the sand fraction is primarily composed of partially decomposed plant litter [18]; forest litter has higher lignin concentration and higher C:N ratio than grassland litter and is thus usually more recalcitrant [47]. In contrast, the carbon stability (parameter A) in clay or silt fractions was similar between grassland and forest during each temperature-change cycle (Fig. 2), thereby suggesting that the organic carbon stabilities in clay and silt fractions were not affected by land-cover type. The carbon stability in these fine fractions is primarily related to organo-mineral association (i.e., chemical protection) [48]. Overall, the SOC stability in the sand fraction is primarily determined by the chemical recalcitrance of organic matter but is more strongly determined by the association of SOC with minerals in the clay or silt fractions. This pattern was consistent with different determinants of organic matter-derived nitrogen preservation between the sandy soils and clay soils [49].

Microbial activities and distribution may differ among different size fractions because these fractions provide distinct microhabitats and various organic carbon qualities and quantities [50–52]. Clay fraction had the greatest organic carbon concentration (Table 1) and had the maximum amount of MBC (Table 3), thereby indicating that the greatest biological activity was in clay fractions [18]. However, the clay fraction had the least MBC/OC ratio (Table 3), an eco-physiological parameter that refers to the microbial efficiency in using carbon substrate [53]. The microbes in the clay fraction had the lowest carbon use efficiency, which is likely related to low carbon availability due to strong organo-mineral association. Moreover, the grassland soil particles had greater MBC/OC and lower qCO_2, i.e., microbial community respiration per unit microbial biomass [53], than the corresponding forest particles (Table 3), thereby suggesting that the grassland soil contains a microbial community that is more efficient at using carbon energy.

Effect of climatic and soil parameters on Q_{10}

First, temperature can determine the Q_{10} of SOC decomposition. The lower Q_{10} of soil respiration at higher temperatures was not only assumed by the Arrhenius equation [6] but also confirmed by two comprehensive analyses of incubation studies of soils from all types of terrestrial ecosystems [11,54]. Similarly, we found that the Q_{10} values were negatively correlated with the incubation temperature (Figure 3). The relationship was fitted by a power function: $y=\alpha x^{\beta}$. Clearly, higher values of α produce higher values of Q_{10}. Otherwise, with larger absolute values of β, Q_{10} decreases more rapidly with increasing temperature. The parameters α and β in the Q_{10}-T curves were found to be significantly correlated with the MBC and soil C:N ratio (Figure 5). An increase in the MBC concentration would lead to α and Q_{10} rising (Figure 5A), which is consistent with Figure 4A. Significant inverse relationships were found between the parameter α and the soil C:N ratio (Figure 5B), thereby suggesting that higher soil C:N ratios was linked with higher tolerance to increasing temperature. A more rapid rate of Q_{10} decrease with increasing temperature (greater absolute value of β) in grassland versus forest soil particles could be related to lower C:N ratios (Table 1) or greater MBC in grassland soil particles (Table 3). It should be mentioned that no significant relationship was found between the values of α and β and the soil properties of MBN, DOC, and DTN.

Table 4. Pearson correlation coefficient for Q_{10} (15–25°C) in the third temperature-change cycle and soil biological and chemical properties.

	Q_{10} (15–25°C) [a]	MBC	MBN	DOC	TDN	Micro C/N	DOC/TDN	MBC/OC	MBN/TN	qCO_2
Q_{10} (15–25°C)	1									
MBC	0.87***	1								
MBN	0.52*	0.44*	1							
DOC	0.58**	0.32	0.38	1						
TDN	0.90***	0.80***	0.45*	0.81***	1					
Micro C/N	0.51*	0.59**	−0.29	0.25	0.54**	1				
DOC/TDN	−0.85***	−0.73***	−0.42	−0.20	−0.59**	−0.45*	1			
MBC/OC	−0.02	0.19	−0.09	0.65**	0.29	0.13	0.19	1		
MBN/TN	−0.43*	−0.29	0.04	−0.59**	−0.54**	−0.44*	0.19	0.71***	1	
qCO_2	−0.66***	−0.70***	−0.24	0.18	−0.35	−0.37	0.85***	−0.54*	−0.05	1

[a] Q_{10} (15−25°C) was calculated using Eqn. 4 for each replicate.
*, **, ***Significant at probability levels of 0.05, 0.01, and 0.001, respectively.

The negative relationship between Q_{10} of SOC decomposition and temperature has two implications. First, in a given region, SOC decomposition should be more sensitive to warming during winter than warming during summer. This trend was observed in the Alaskan tundra, with an increase of greater than 50% in CO_2 fluxes during the snow-covered period by winter warming and a 20% increase during the growing season by summer warming [55]. Nevertheless, considering that the initial soil respiration rate is much less in winter than in summer, the absolute magnitude of the CO_2 emission increase from warming could be still less in winter than in summer. Second, the SOC stock in boreal and temperate regions (low MAT), especially peat lands and permafrost, should be more vulnerable to warming than that in

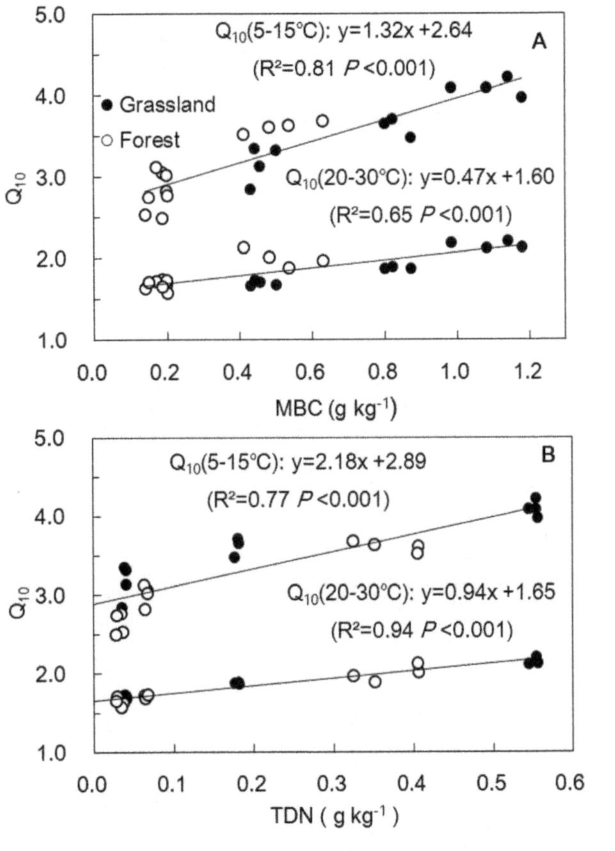

Figure 4. Dependence of Q10 in the third temperature-change cycle on microbial biomass carbon (A) and total dissolved nitrogen (B). The Q10 (5–15°C) and Q10 (20–30°C) were calculated using Eqn. 4 for each replicate.

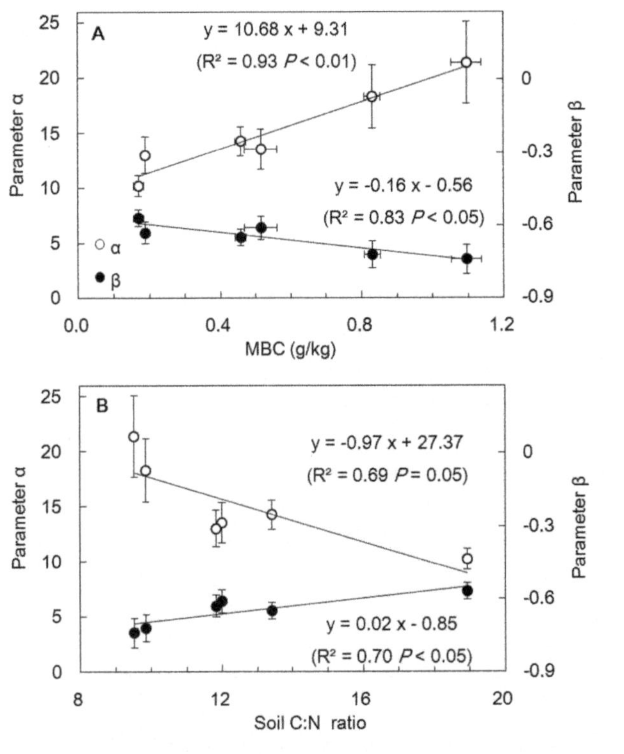

Figure 5. Dependence of parameters in the Q10−T curve on the soil microbial biomass carbon (A) and soil C: N ratio (B). Vertical bars denote the 95% CI of the estimated parameters in Fig. 3. For each parameter α and β, six points represent the three particles of clay, silt, and sand from the grassland and forest soils. Horizontal bars represent the standard errors (n = 4).

subtropical and tropical regions (high MAT) [6]. Moreover, peat lands and permafrost contain more than 50% of global soil carbon stock [56] and have been experiencing remarkable warming [57]. Consequently, peat lands and permafrost are expected to experience largest potential soil carbon loss by 2100 due to global warming [6].

Second, soil texture could have an impact on Q_{10} of SOC decomposition. The Q_{10} decreased in the order clay>silt>sand seemingly regardless of land-cover types, soil types, and climatic factors (Table 2). Consequently, SOC decomposition in fine-textured soils (rich in clay and silt content) could be more sensitive to global warming than that in coarse-textured soils (rich in sand content). Moreover, fine-textured soils generally have greater capacities to preserve SOC [58] and thus contain more SOC than do coarse-textured soils [1,59]. Therefore, current global model predictions that use constant Q_{10} values for different soil textures may predict an underestimated response of the soil carbon cycle to climate warming.

Third, the soil nitrogen availability, indicated by TDN, could also affect the Q_{10} values, and greater Q_{10} values were related to higher TDN content regardless of land use type (Figure 4B). This result agrees with other incubation studies: Chang et al. [60] demonstrated the existence of a positive relationship between Q_{10} and the soil total nitrogen content, and Cusack et al. [61] observed that nitrogen fertilization significantly increased the Q_{10} of slow SOC pools. This relationship between Q_{10} and nitrogen availability could result from the nitrogen availability driving the activities of the soil enzymes that acquire carbon [62]. Low nitrogen availability may inhibit enzyme activity and thus constrain the warming response of SOC decomposition. This close relation of Q_{10} with soil nitrogen availability suggests that atmospheric nitrogen deposition could stimulate the positive feedback of soil respiration to climate warming.

Q_{10} for labile carbon pools vs. Q_{10} for recalcitrant carbon pools

The question of whether the decomposition of recalcitrant and labile carbon has different temperature dependencies is important because most experimental work has been done on the temperature sensitivity of the decomposition of more labile fractions, yet the long-term effect of warming on soil carbon storage should be primarily determined by the temperature dependence of the more recalcitrant carbon pool [7]. Kinetic theory and the Arrhenius equation both suppose that recalcitrant carbon pools are more warming-sensitive than labile carbon pools [6,63]. Many laboratory incubation studies have supported this supposition [32,64,65]. Our study further affirmed this result by showing that the decomposition of the carbon associated with the clay and silt fractions (recalcitrant carbon pool), with slower turnover rates (Figure 2), yet had higher Q_{10} than that with the sand fraction (labile carbon pool) (Table 2). Furthermore, the organic carbon associated with the clay and silt fractions usually accounts for a large portion of the total SOC, i.e., more than 60% in this study (Table 1). Therefore, long-term response of the SOC stock to climate warming would be more severe than the current predictions that assume the same Q_{10} for labile and recalcitrant carbon pools.

However, the supposition that recalcitrant carbon pools have greater warming dependence has not obtained wide acceptance [31,66]. Based on this supposition, Q_{10} in the incubation studies is expected to rise with prolonged incubation time because SOC is generally composed of a large proportion of labile carbon during the early incubation period, yet of a small proportion of labile carbon during the late incubation period due to depletion of labile carbon [67]. And yet, Fang et al. [31] and Reichstein et al. [66] found that Q_{10} was stable and independent of the incubation time. Similarly, our study only observed increasing trends in Q_{10} for the grassland clay and silt fractions from the 1st temperature-change cycle to the 2nd and the 3rd (Table 2) and did not found such trends in Q_{10} for the forest soil fractions or the grassland sand fractions. The inconsistency between these observations and the expectation derived from the above supposition could be explained by that Q_{10} having a significant positive relation with MBC and lower MBC values resulting in a smaller Q_{10} (Figure 4A); MBC generally declines with prolonged incubation because of the depletion of labile carbon [31,68]. Consequently, as incubation is prolonged, increasing trend of Q_{10} due to increased recalcitrant carbon proportion could be counteracted by decreasing trend of Q_{10} as a result of declined MBC [67]. If the MBC did not decrease sharply from early to late incubation stage, the Q_{10} would have increased as incubation progressed in Fang et al. [31] and Reichstein et al. [66], as well as in our study. However, the mechanism for how Q_{10} is affected by MBC is not clear. Assuming that the microbial community composition does not change and that every single microbe maintains the same warming response with prolonged incubation time, a decline in MBC alone should not induce lower Q_{10}. Possible mechanisms could be either microbial community shift to species with low thermal response [69] or microbe acclimatization to temperature increase [70]. On the other hand, soil carbon recalcitrance did not certainly increase continuously with prolonged incubation. In other words, carbon availability and lability did not certainly decease continuously. Such as, we can see that the parameter A for the sand fraction increased from the second to the third cycle both for grassland and forest soils (Figure 2). Tian et al. [71] also observed the phenomenon and affirmed that a pulse-dynamic pattern of soil respiration and increased carbon availability during a long incubation could exist. Overall, these results suggest that it is unreliable to judge whether the decomposition of recalcitrant and labile carbon have different temperature dependencies by comparing Q_{10} values after different incubation time.

Conclusions

This study presented three implications for understanding the soil-carbon response to climate warming. First, the Q_{10} of SOC decomposition in soil particles decreased with increasing particle size (clay>silt>sand) for grassland and forest soils. This result suggested that SOC decomposition from fine-textured soils rich in clay or silt content could be more sensitive to warming than coarse sandy soils. Second, Q_{10} decreased with increasing temperature, thereby implicating that the SOC might be more vulnerable in low MAT regions than in high MAT regions under future warming. Third, the soil nitrogen availability facilitated a temperature dependence of SOC decomposition, suggesting that atmospheric nitrogen deposition could stimulate positive feedback of soil respiration on climate warming.

Author Contributions

Conceived and designed the experiments: YH WS. Performed the experiments: FD GJ YC. Analyzed the data: YH FD WS. Wrote the paper: FD WS.

References

1. Jobbagy E, Jackson R (2000) The vertical distribution of soil organic carbon and its relation to climate and vegtation. Ecol Appl 10: 423–436.
2. Cox PM, Betts RA, Jones CD, Spall SA, Totterdell IJ (2000) Acceleration of global warming due to carbon-cycle feedbacks in a coupled climate model. Nature 408: 184–187.
3. Bond-Lamberty B, Thomson A (2010) Temperature-associated increases in the global soil respiration record. Nature 464: 579–582.
4. Schimel DS, Braswell BH, Holland EA, McKeown R, Ojima DS, et al. (1994) Climatic, edaphic, and biotic controls over storage and turnover of carbon in soils. Global Biogeochem Cy 8: 279–293.
5. Lloyd J, Taylor JA (1994) On the temperature dependence of soil respiration. Funct Ecol 8: 315–323.
6. Davidson EA, Janssens IA (2006) Temperature sensitivity of soil carbon decomposition and feedbacks to climate change. Nature 440: 165–173.
7. Kirschbaum MUF (2006) The temperature dependence of organic-matter decomposition—still a topic of debate. Soil Biol Biochem 38: 2510–2518.
8. Friedlingstein P, Cox P, Betts R, Bopp L, Von Bloh W, et al. (2006) Climate-carbon cycle feedback analysis: Results from the C(4)MIP model intercomparison. J Climate 19: 3337–3353.
9. Zhou T, Shi P, Hui D, Luo Y (2009) Global pattern of temperature sensitivity of soil heterotrophic respiration (Q_{10}) and its implications for carbon-climate feedback. J Geophys Res 114, G02016.
10. Raich JW, Schlesinger WH (1992) The global carbon-dioxide flux in soil respiration and its relationship to vegetation and climate. Tellus B 44: 81–99.
11. Kirschbaum MUF (1995) The temperature-dependence of soil organic-matter decomposition, and the effect of global warming on soil organic-C storage. Soil Biol Biochem 27: 753–760.
12. Davidson EA, Belk E, Boone RD (1998) Soil water content and temperature as independent or confounded factors controlling soil respiration in a temperate mixed hardwood forest. Global Change Biol 4: 217–227.
13. Suseela V, Conant RT, Wallenstein MD, Dukes JS (2012) Effects of soil moisture on the temperature sensitivity of heterotrophic respiration vary seasonally in an old-field climate change experiment. Global Change Biol 18: 336–348.
14. Wan S, Luo Y (2003) Substrate regulation of soil respiration in a tallgrass prairie: results of a clipping and shading experiment. Global Biogeochem Cy 17, 1054.
15. Fierer N, Colman BP, Schimel JP, Jackson RB (2006) Predicting the temperature dependence of microbial respiration in soil: a continental-scale analysis. Global Biogeochem Cy 20, GB3026.
16. Dieckow J, Bayer C, Conceicao PC, Zanatta JA, Martin-Neto L, et al. (2009) Land use, tillage, texture and organic matter stock and composition in tropical and subtropical Brazilian soils. Eur J Soil Sci 60: 240–249.
17. Zinn YL, Lal R, Resck DVS (2005) Changes in soil organic carbon stocks under agriculture in Brazil. Soil Till Res 84: 28–40.
18. Christensen BT (2001) Physical fractionation of soil and structural and functional complexity in organic matter turnover. Eur J Soil Sci 52: 345–353.
19. Sposito G, Skipper NT, Sutton R, Park S-h, Soper AK, et al. (1999) Surface geochemistry of the clay minerals. P Natl Acad Sci USA 96: 3358–3364.
20. von Lützow M, Kögel-Knabner I, Ekschmitt K, Flessa H, Guggenberger G, et al. (2007) SOM fractionation methods: Relevance to functional pools and to stabilization mechanisms. Soil Biol Biochem 39: 2183–2207.
21. Six J, Conant RT, Paul EA, Paustian K (2002) Stabilization mechanisms of soil organic matter: Implications for C-saturation of soils. Plant Soil 241: 155–176.
22. Baldock JA, Skjemstad JO (2000) Role of the soil matrix and minerals in protecting natural organic materials against biological attack. Org Geochem 31: 697–710.
23. Arevalo CBM, Chang SX, Bhatti JS, Sidders D (2012) Mineralization potential and temperature sensitivity of soil organic carbon under different land uses in the parkland region of Alberta, Canada. Soil Sci Soc Am J 76: 241–251.
24. Feng W, Plante AF, Six J (2013) Improving estimates of maximal organic carbon stabilization by fine soil particles. Biogeochem 112: 1–13.
25. Yang XM, Drury CF, Reynolds WD, MacTavish DC (2009) Use of sonication to determine the size distributions of soil particles and organic matter. Can J Soil Sci 89: 413–419.
26. Jackson ML (1969) Soil chemical analysis-advanced course. Wisconsin: University of Wisconsin. pp. 20–65.
27. Tiessen H, Stewart JWB (1983) Particle-size fractions and their use in studies of soil organic-matter. 2. cultivation effects on organic-matter composition in size fractions. Soil Sci Soc Am J 47: 509–514.
28. Christensen BT (1987) Decomposability of organic-matter in particle-size fractions from field soils with straw incorporation. Soil Biol Biochem 19: 429–435.
29. Linn DM, Doran JW (1984) Effect of water-filled pore-space on carbon-dioxide and nitrous-oxide production in tilled and nontilled soils. Soil Sci Soc Am J 48: 1267–1272.
30. Chen X, Tang J, Jiang L, Li B, Chen J, et al. (2010) Evaluating the impacts of incubation procedures on estimated Q_{10} values of soil respiration. Soil Biol Biochem 42: 2282–2288.
31. Fang CM, Smith P, Moncrieff JB, Smith JU (2005) Similar response of labile and resistant soil organic matter pools to changes in temperature. Nature 433: 57–59.
32. Xu X, Luo YQ, Zhou JZ (2012) Carbon quality and the temperature sensitivity of soil organic carbon decomposition in a tallgrass prairie. Soil Biol Biochem 50: 142–148.
33. Huang Y, Zou JW, Zheng XH, Wang YS, Xu XK (2004) Nitrous oxide emissions as influenced by amendment of plant residues with different C:N ratios. Soil Biol Biochem 36: 973–981.
34. Bouyoucos GJ (1962) Hydrometer method improved for making particle size analyses of soils. Agron J 54: 464–465.
35. Nelson DW, Sommers LE (1996) Total carbon, organic carbon, and organic matter. In: Sparks DL, Page AL, Helmke PA, Loeppert RH, Soltanpour PN, et al., editors. Methods of soil analysis Part 3, chemical methods. Wisconsin: ASA, CSSA and SSSA. pp. 961–1010.
36. Bremner J (1996) Nitrogen-total. In: Sparks DL, Page AL, Helmke PA, Loeppert RH, Soltanpour PN, et al., editors. Methods of soil analysis Part 3, chemical methods. Wisconsin: ASA, CSSA and SSSA. pp. 1085–1121.
37. Vance ED, Brookes PC, Jenkinson DS (1987) An extraction method for measuring soil microbial biomass-C. Soil Biol Biochem 19: 703–707.
38. Jones DL, Willett VB (2006) Experimental evaluation of methods to quantify dissolved organic nitrogen (DON) and dissolved organic carbon (DOC) in soil. Soil Biol Biochem 38: 991–999.
39. Li LJ, Zeng DH, Yu ZA, Fan ZP, Mao R (2010) Soil microbial properties under N and P additions in a semi-arid, sandy grassland. Biol Fert Soils 46: 653–658.
40. Lang M, Cai Z, Chang SX (2011) Effects of land use type and incubation temperature on greenhouse gas emissions from Chinese and Canadian soils. J Soil Sediment 11: 15–24.
41. Mikan CJ, Schimel JP, Doyle AP (2002) Temperature controls of microbial respiration in arctic tundra soils above and below freezing. Soil Biol Biochem 34: 1785–1795.
42. Balesdent J, Mariotti A, Guillet B (1987) Natural ^{13}C abundance as a tracer for studies of soil organic matter dynamics. Soil Biol Biochem 19: 25–30.
43. Balesdent J, Mariotti A, (1996) Measurement of soil organic matter turnover using ^{13}C natural abundance. In: Boutton T, Yamasaki S, editors. Mass spectrometry of soils. New York: Marcel Dekker. pp. 83–111.
44. Ludwig B, John B, Ellerbrock R, Kaiser M, Flessa H (2003) Stabilization of carbon from maize in a sandy soil in a long-term experiment. Eur J Soil Sci 54: 117–126.
45. Franzluebbers A (1999) Potential C and N mineralization and microbial biomass from intact and increasingly disturbed soils of varying texture. Soil Biol Biochem 31: 1083–1090.
46. Chapman SK, Palanivel RU, Langley JA (2012) Soil carbon stability responds to land-use and groundcover management in southern appalachian agroecosystems. Soil Sci Soc Am J 76: 2221–2229.
47. Kirschbaum MU, Guo LB, Gifford RM (2008) Why does rainfall affect the trend in soil carbon after converting pastures to forests?: A possible explanation based on nitrogen dynamics. Forest Ecol Manag 255: 2990–3000.
48. Manjaiah KM, Kumar S, Sachdev MS, Sachdev P, Datta SC (2010) Study of clay-organic complexes. Curr Sci 98: 915–921.
49. Chivenge P, Vanlauwe B, Gentile R, Six J (2011) Comparison of organic versus mineral resource effects on short-term aggregate carbon and nitrogen dynamics in a sandy soil versus a fine textured soil. Agr Ecosyst Environ 140: 361–371.
50. Sessitsch A, Weilharter A, Gerzabek MH, Kirchmann H, Kandeler E (2001) Microbial population structures in soil particle size fractions of a long-term fertilizer field experiment. Appl Environ Microb 67: 4215–4224.
51. Lagomarsino A, Grego S, Kandeler E (2012) Soil organic carbon distribution drives microbial activity and functional diversity in particle and aggregate-size fractions. Pedobiologia 55: 101–110.
52. Kogel-Knabner I, Guggenberger G, Kleber M, Kandeler E, Kalbitz K, et al. (2008) Organo-mineral associations in temperate soils: Integrating biology, mineralogy, and organic matter chemistry. J Plant Nutr Soil Sci 171: 61–82.
53. Anderson TH, Domsch KH (2010) Soil microbial biomass: The eco-physiological approach. Soil Biol Biochem 42: 2039–2043.
54. Hamdi S, Moyano F, Sall S, Bernoux M, Chevallier T (2013) Synthesis analysis of the temperature sensitivity of soil respiration from laboratory studies in relation to incubation methods and soil conditions. Soil Biol Biochem 58: 115–126.
55. Natali SM, Schuur EAG, Trucco C, Pries CEH, Crummer KG, et al. (2011) Effects of experimental warming of air, soil and permafrost on carbon balance in Alaskan tundra. Global Change Biol 17: 1394–1407.
56. Tarnocai C, Canadell JG, Schuur EAG, Kuhry P, Mazhitova G, et al. (2009) Soil organic carbon pools in the northern circumpolar permafrost region. Global Biogeochem Cy 23,GB2023.
57. Chapin FS, Sturm M, Serreze MC, McFadden JP, Key JR, et al. (2005) Role of land-surface changes in Arctic summer warming. Science 310: 657–660.
58. Hassink J (1994) Effects of soil texture and grassland management on soil organic C and N and rates of C and N mineralization. Soil Biol Biochem 26: 1221–1231.
59. Burke I, Yonker C, Parton W, Cole C, Schimel D, et al. (1989) Texture, climate, and cultivation effects on soil organic matter content in US grassland soils. Soil Sci Soc Am J 53: 800–805.
60. Chang XF, Zhu XX, Wang SP, Luo CY, Zhang ZH, et al. (2012) Temperature and moisture effects on soil respiration in alpine grasslands. Soil Sci 177: 554–560.

61. Cusack DF, Torn MS, McDowell WH, Silver WL (2010) The response of heterotrophic activity and carbon cycling to nitrogen additions and warming in two tropical soils. Global Change Biol 16: 2555–2572.
62. Cusack DF (2013) Soil nitrogen levels are linked to decomposition enzyme activities along an urban-remote tropical forest gradient. Soil Biol Biochem 57: 192–203.
63. Bosatta E, Agren GI (1999) Soil organic matter quality interpreted thermodynamically. Soil Biol Biochem 31: 1889–1891.
64. Hartley IP, Ineson P (2008) Substrate quality and the temperature sensitivity of soil organic matter decomposition. Soil Biol Biochem 40: 1567–1574.
65. Leifeld J, Fuhrer J (2005) The temperature response of CO_2 production from bulk soils and soil fractions is related to soil organic matter quality. Biogeochem 75: 433–453.
66. Reichstein M, Subke JA, Angeli AC, Tenhunen JD (2005) Does the temperature sensitivity of decomposition of soil organic matter depend upon water content, soil horizon, or incubation time? Global Change Biol 11: 1754–1767.
67. Conant RT, Steinweg JM, Haddix ML, Paul EA, Plante AF, et al. (2008) Experimental warming shows that decomposition temperature sensitivity increases with soil organic matter recalcitrance. Ecology 89: 2384–2391.
68. Follett RF, Paul EA, Pruessner EG (2007) Soil carbon dynamics during a long-term incubation study involving C-13 and C-14 measurements. Soil Sci 172: 189–208.
69. Zogg GP, Zak DR, Ringelberg DB, MacDonald NW, Pregitzer KS, et al. (1997) Compositional and functional shifts in microbial communities due to soil warming. Soil Sci Soc Am J 61: 475–481.
70. Luo YQ, Wan SQ, Hui DF, Wallace LL (2001) Acclimatization of soil respiration to warming in a tall grass prairie. Nature 413: 622–625.
71. Tian QX, He HB, Cheng WX, Zhang XD (2014) Pulse-dynamic and monotonic decline patterns of soil respiration in long term laboratory microcosms. Soil Biol Biochem 68: 329–336.

12

Combining Forces - The Use of Landsat TM Satellite Imagery, Soil Parameter Information, and Multiplex PCR to Detect *Coccidioides immitis* Growth Sites in Kern County, California

Antje Lauer[1]*, Jorge Talamantes[2], Laura Rosío Castañón Olivares[3], Luis Jaime Medina[2], Joe Daryl Hugo Baal[1], Kayla Casimiro[1], Natasha Shroff[1], Kirt W. Emery[4]

1 Department of Biology, California State University, Bakersfield, California, United States of America, **2** Department of Physics & Engineering, California State University, Bakersfield, California, United States of America, **3** Laboratorio de Micología Médica, Facultad de Medicina, Universidad Nacional Autónoma de México, Mexico City, Mexico, **4** County of Kern Public Health Services Department, Bakersfield, California, United States of America

Abstract

Coccidioidomycosis is a fungal disease acquired through the inhalation of spores of *Coccidioides* spp., which afflicts primarily humans and other mammals. It is endemic to areas in the southwestern United States, including the San Joaquin Valley portion of Kern County, California, our region of interest (ROI). Recently, incidence of coccidioidomycosis, also known as valley fever, has increased significantly, and several factors including climate change have been suggested as possible drivers for this observation. Up to date details about the ecological niche of *C. immitis* have escaped full characterization. In our project, we chose a three-step approach to investigate this niche: 1) We examined Landsat-5-Thematic-Mapper multispectral images of our ROI by using training pixels at a 750 m×750 m section of Sharktooth Hill, a site confirmed to be a *C. immitis* growth site, to implement a Maximum Likelihood Classification scheme to map out the locations that could be suitable to support the growth of the pathogen; 2) We used the websoilsurvey database of the US Department of Agriculture to obtain soil parameter data; and 3) We investigated soil samples from 23 sites around Bakersfield, California using a multiplex Polymerase Chain Reaction (PCR) based method to detect the pathogen. Our results indicated that a combination of satellite imagery, soil type information, and multiplex PCR are powerful tools to predict and identify growth sites of *C. immitis*. This approach can be used as a basis for systematic sampling and investigation of soils to detect *Coccidioides* spp.

Editor: Vishnu Chaturvedi, California Department of Public Health, United States of America

Funding: The work was supported by National Science Foundation grant HRD-0331537, and support from Chevron (Research Experience Vitalizing Science Undergraduate Program (REVS-UP) at CSUB. The funders had no role in study design, data collection and analysis, decision to publish, or preparation of the manuscript.

Competing Interests: The authors have the following interest: This study was partly funded by Chevron (Research Experience Vitalizing Science Undergraduate Program [REVS-UP]) at CSUB. There are no patents, products in development or marketed products to declare.

* Email: alauer@csub.edu

Introduction

Valley fever research has predominantly focused on the medical and epidemiological aspects of *Coccidioides immitis* and *Coccidioides posadasii*, the fungi that cause coccidioidomycosis ([1,2], and references therein). *Coccidioides* spp. can have a complete life cycle as soil dwelling organisms but if the soil is disturbed, their arthroconidia can become air-borne and are able to infect a host via the respiratory tract. About 60% of infected patients report no symptoms [3]; about 25% exhibit severe flu-like symptoms, such as cough, sputum, fever, and muscle aches; the remaining 15% become very ill with pneumonia-like symptoms (e.g. pleurisy and heavier sputum) requiring medication and bed rest. In a small number of cases (about 0.5–1%), the disease disseminates beyond the lungs to e.g. the skin, bones, and/or meninges of the brain, and the disease can be fatal. Certain sectors of the population seem to be more susceptible to infection, such as the very young, persons newly arrived to the endemic areas (since immunity develops with infection), field-, and construction workers, and those with impaired immune systems [4].

Coccidioides spp. are endemic in the southern part of the San Joaquin Valley in California, southern California, the southern part of Arizona, New Mexico and Texas, most of northern Mexico, and some areas in Guatemala, Honduras, Venezuela, northeastern Brazil, Argentina, and Paraguay [5,6]. Given its geographic distribution, it is evident that *C. posadasii* is able to flourish in desert regions of the Americas (besides California), in contrast to its close relative *C. immitis* which seems to be restricted to areas in California. Of the two fungal species, it is *C. immitis* which afflicts the San Joaquin Valley portion of Kern County, California [7,8] which is the Region of Interest (ROI) of this study. However, population genomic sequencing of *Coccidioides* spp.

revealed recent hybridization between both species [9], and nothing is known about the distribution and ecology of these hybrids.

It is reasonable to expect that climatic fluctuations might affect the rate at which humans become infected [4]. For example, an extended drought might decimate less heat tolerant, non-sporeforming soil microorganisms that had acted as natural antagonists to the pathogen in its natural environment. A wetter than-normal rainy season could help *Coccidioides* spp. bloom, and windy spells might facilitate the dispersal of its arthroconidia. The "grow and blow" hypothesis has first been introduced by Comrie and Glueck [10]. It has long been surmised that *Coccidioides* spp. are generally poor competitors [11], but that they are more heat-resistant than competing microorganisms – thus, it can be expected that hot summers might favor its presence or dominance. Indeed, anecdotal evidence to these effects is well documented in the literature [12–20]. There have been a number of attempts at demonstrating this connection quantitatively with various degrees of success [10,11,21–25]. Yet, despite extensive study, there is currently no ecologically consistent link identified between the environment and coccidioidomycosis rates [26]. The predicted warming of the climate in California will add another piece of the puzzle in the already complicated interrelationships of environmental factors that might support or suppress the growth of pathogens with environmental reservoirs [27,28]. However, occasionally, the pathogen was detected in other regions, such as the recent detection of *Coccidioides immitis* in soils of Eastern WA [29].

There also have been several attempts to characterize the ecological niche of *Coccidioides* spp. in more detail [4,31–33], but we still do not have a complete description of this niche. To date, Fisher et al. [31] present the most comprehensive review of this subject. We need to direct attention to a few fundamental points about what is known in regards to this niche. First, it is important to realize that *C. immitis* and *C. posadasii* do not grow in disturbed soils [4,30,31] such as cultivated fields, gardens, etc. Second, whereas it was initially thought that *Coccidioides* spp. ecological niche corresponds to the Lower Sonoran Life Zone (as defined and described by Merriam [34]), or similar environments [14,35,36]. Later research [30] showed that this is not quite correct, and indeed more recent works [31,32,37,38] suggested that the fungus's niche corresponds more closely with thermic and hyperthermic soils in which temperatures can reach or exceed 22°C in 50 cm depth. Fisher et al. [31] described sites where *Coccidioides* spp. were suspected to have been present because humans or animals were reported to have been infected at these sites. Fisher et al. [31] also made the general observations that the vegetation at those sites ranged from sparse to relatively thick cover in lower Sonoran Deserts, Chaparral-upper Sonoran brush and grasslands, as well as Mediterranean savannas and forested foothills. Furthermore, they stated that the temperature regimes, climate conditions in general, and soil textures are the only indicative variables of the presence of *Coccidioides* spp. Microbial diversity in soils is highly influenced by the habitat's chemical and physical parameters. But biotic soil factors such as plant and microeukaryote diversity influence fungal and bacterial soil communities as well through root exudation (additional available nutrients), microbial antagonism (antibiotic production) and synergism, as well as through selective grazing by microeukaryotes [32,39–40]. It is currently being discussed that the pathogen is in fact not very competitive as a soil saprophyte because it has lost the ability to produce a variety of enzymes that are involved in important biodegradation processes of soil organic matter, which might explain the difficulty to detect it in bulk soil [41].

There are few published data available about the distribution of *C. immitis* growth sites in Kern County, California [42] most probably because it has been very difficult in the past to isolate and identify *Coccidioides* spp. from soil and dust samples [4,14,30]. Recently, first attempts using molecular biological techniques to identify *C. immitis* in bulk soil samples from Kern County, predominantly around Bakersfield, have been performed [32]. Based on that study, it appeared that *C. immitis* is likely to be found in the Bakersfield area at locations that are non-agricultural and have about equal parts of sand, clay, and silt (clay loam), a pH between 7.8 and 8.5, an available water capacity of about 0.15–0.2 cm/cm, a water content of about 30% (1/3 bar), an available water supply (0–25 cm) of 4–5 cm, and a Cation Exchange Capacity (CEC7) of over 20 milliequivalents per 100 grams.

The idea of using remote sensing (RS) techniques to piece together environmental characteristics, environmental change, and their relationship to disease transmission has been used extensively in connection with other diseases such as malaria [43], cholera [44], and African trypanosomiases [45]. Even though the ecological niche of *C. immitis* is not well characterized, we present here a RS technique that allows the mapping of sites around Bakersfield, California, where the pathogen is suspected to grow based on data obtained in a previous study by Lauer et al. [32]. Our method utilized a location well-known for being a *C. immitis* growth site (Sharktooth hill [STH], Bakersfield, California) as a basis, and then examined satellite images of the ROI to find all locations with similar spectral signatures. This is similar to characterizing the growth sites by the vegetation that tends to grow in the same environment as *C. immitis*, using the vegetation type as a marker. This is reasonable because the vegetation type closely reflects the co-variation of the relevant physical and chemical parameters such as clay and sand content, temperature, pH, nutrients, water content, etc., and also affects the development of the microbial diversity in the soils [46].

To validate our approach, we investigated if a combination of remote sensing and soil parameter information can predict locations which might be suitable to support the growth of *C. immitis*, followed by a molecular biological approach to detect the fungus in these soils with a culture independent polymerase chain reaction (PCR) based method [32,47].

Material and Methods

No specific permissions were required for the soil sampling. Our field study did also not involve endangered or protected species.

Multispectral image analysis

Landsat-5 Thematic Mapper (TM) L1G corrected multispectral images were downloaded from the United States Geological Survey archive (http://EarthExplorer.usgs.gov). The satellite relayed a continuous data stream which was then framed into individual scenes each 23.92 sec (see, e.g., http://landsat.gsfc.nasa.gov/about/wrs.html). The images for path 42, rows 35 and 36: Worldwide Reference System to cover our ROI were downloaded, and then the two images were mosaicked. Most of the analysis that is presented here was performed on a spatial subset of this mosaicked image. This subset corresponds to an area approximately one million hectares that covers the San Joaquin Valley portion of Kern County. Our work mainly focused on a multispectral image taken on April 20, 2008 at 10:23 PM local time. This image was chosen because it was obtained at a date (during spring) were microbial activity and biomass in the soil is generally considered high, because of supportive environmental parameters, such as moderate temperatures and increased water

Table 1. Location and description of sampling sites used as test data for the remote sensing approach.

sampling sites and year sampled	coordinates	soil type (map unit symbol)	GS or AS of the pathogen	rodent activity*
Bakersfield city				
1. CSUB Children Center ('08, '09)	119° 06′ 29.0′′ W, 35° 20′ 57.0′′ N	Wasco sandy loam (243)	AS	yes
2. Belle Terrace/P Str. ('11)	119° 00′ 37.2′′ W, 35° 20′ 49.8′′′ N	Kimberlina Urban land, Cajon-complex (180)	GS	no
3. Belle Terrace/Gay Str. ('11)	118° 59′ 22.7′′ W, 35° 20′ 40.0′′′ N	Kimberlina Urban land, Cajon-complex (180)	GS	yes
4. Marella Way ('11)	118° 63′ 15.0′′ W, 35° 21′ 40.0′′′ N	Kimberlina Urban land, Cajon-complex (180)	NS	no
5. Flood Plain CSUB ('08, '09)	119° 06′ 05.0′′ W, 35° 21′ 16.0′′ N	River Wash (229)	AS	no
SW Bakersfield				
6. Bike Path West ('08, '09)	119° 15′ 06.0′′ W, 35° 18′ 20.0′′ N	Cajon sandy loam (125)	NS	yes
7. Lake Webb ('08, '09)	119° 16′ 27.0′′ W, 35° 13′ 53.0′′ N	Zalvidea sandy loam (240)	AS	no
8. Cole's Levee Rd. I ('08, '09, '11)	119° 13′ 60.0′′ W, 35° 14′ 08.0′′ N	Garces loam (180)	GS	yes
9. Cole's Levee Rd. II ('11)	119° 13′ 65.3′′ W, 35° 14′ 09.7′′ N	Garces loam (180)	GS	yes
10. Olen Avenue ('11)	119° 14′ 50.0′′ W, 35° 14′ 72.0′′ N	Garces loam (180)	GS	yes
11. Valley Street Field ('08, '09)	118° 52′ 18.0′′ W, 35° 24′ 29.0′′ N	Delano sandy loam (139)	AS	no
NE Bakersfield				
12. Across CALM ('11)	118° 53′ 14.1′′ W, 35° 25′ 50.3′′ N	Chanac Clay Loam (130)	GS	yes
13. Ant Hill Oil Field ('08, '09, '11)	118° 51′ 25.0′′ W, 35° 23′ 50.0′′ N	Chanac Clay Loam (131)	GS	yes
14. Round Mt. Rd. I ('08, '09)	118° 52′ 20.0′′ W, 35° 27′ 10.0′′ N	Xeric Torriorthents-Calcic Haploxerept association (174)	AS	yes
15. Round Mt. Rd. II ('08, '09)	118° 53′ 30.0′′ W, 35° 28′ 42.0′′ N	Xeric Torriorthents-Calcic Haploxerept association (174)	NS	yes
16. Sharktooth hill I	118° 55′ 03.4′′ W, 35° 27′ 44.5′′ N	Chanac Pleito Premier Association (305)	nd	yes
17. Sharktooth hill 2	118° 54′ 37.0′′ W, 35° 28′ 20.0′′ N	Pleito Trigo Chanac Complex (205)	GS	yes
18. Sharktooth hill 3	118° 54′ 33.0′′ W, 35° 28′ 21.3′ N	Pleito Trigo Chanac Complex (205)	GS	yes
NW Bakersfield				
19. Acari Rd. ('11)	119° 15′ 26.8′′ W, 35° 23′ 16.1′′ N	Garces silt loam (156)	NS	no
20. Elementary Lne. ('11)	119° 15′ 16.1′′ W, 35° 25′ 20.5′′ N	Panoche clay loam (211)	GS	no
21. Beech Str. ('11)	119° 15′ 43.5′′ W, 35° 26′ 39.6′′ N	Garces silt loam (156)	GS	yes
Wasco				
22. Gun Club Rd.('11)	119° 29′ 54.0′′ W, 35° 39′ 34.9′′ N	Garces silt loam (156)	NS	yes
23. McCoy Rd. ('11)	119° 31′ 34.3′′ W, 35° 37′ 24.8′′ N	Garces silt loam (156)	NS	yes
Arvin				
24. Di Giorgio Rd. ('11)	118° 57′ 28.7′′ W, 35° 15′ 06.6′′ N	Garces loam (180)	GS	yes
25. Bear Mt. Rd. ('11)	118° 57′ 05.9′′ W, 35° 12′ 30.0′′ N	Garces loam (180)	GS	yes

Growth sites (GS), accumulation sites (AS) and negative sites (NS) were determined by multiplex PCR results, nd: not determined.
* Proof of rodent activity was observed in the immediate neighborhood of the sampling site. Soil disturbing activity was also observed by burrowing owls, coyotes, kit foxes, spiders or large ants at some locations. The dominant rodents observed were ground squirrels, kangaroo rats and hares.

content. Furthermore, this image had 0% cloud cover. Our analysis started by defining a 25 pixel ×25 pixel area centered at latitude 35° 28′ 20.29′′ N, and longitude 118° 54′ 37.04′′ W. This location is at STH, an area where *C. immitis* has been repeatedly detected ([30,31,48], this study). These 625 pixels were used to train the algorithm, and thus, they define a spectral class which is referred to in what follows as the "STH-vegetation class". To implement the Maximum Likelihood Classification (MLC) method distributed (Richards & Jia, 2006), TM bands 1 (0.45–0.52 μm, blue-green), 2 (0.52–0.60 μm, green), 3 (0.63–0.69 μm, red), 4 (0.76–0.90 μm, near infrared), 5 (1.55–1.75 μm, mid infrared), and 7 (2.08–2.35 μm, mid infrared) were used. Band 6 (10.40–12.50 μm, thermal infrared) was not used in our MLC scheme because the resolution was 60 m instead of the 30 m (as it is for the other bands). However, this band was used to compute surface temperatures as described in more detail below. Our MLC scheme then entailed computing, for each of the pixels in the ROI, the probability that it belonged to the STH-vegetation class. This probability was assumed to be normally distributed [49], and thus is given by

$$p(x)=\frac{1}{(2\pi)^{N/2}|\Sigma|^{1/2}}e^{-\left[\frac{1}{2}(x-m)^{T}\sum^{-1}((x-m))\right]}$$

where $\boldsymbol{x}$ is a vector location in pixel space, $N=6$ is the dimensionality of pixel space, $\boldsymbol{\Sigma}$ the covariance matrix of the distribution, and $\boldsymbol{m}$ is the mean position of the spectral class. ($\boldsymbol{m}$ and $\boldsymbol{\Sigma}$ are computed from the training pixels). A threshold value was set at p_0, meaning that if a pixel had a probability $p \geq p_0$ of being in the STH-vegetation class, then the pixel was put into this

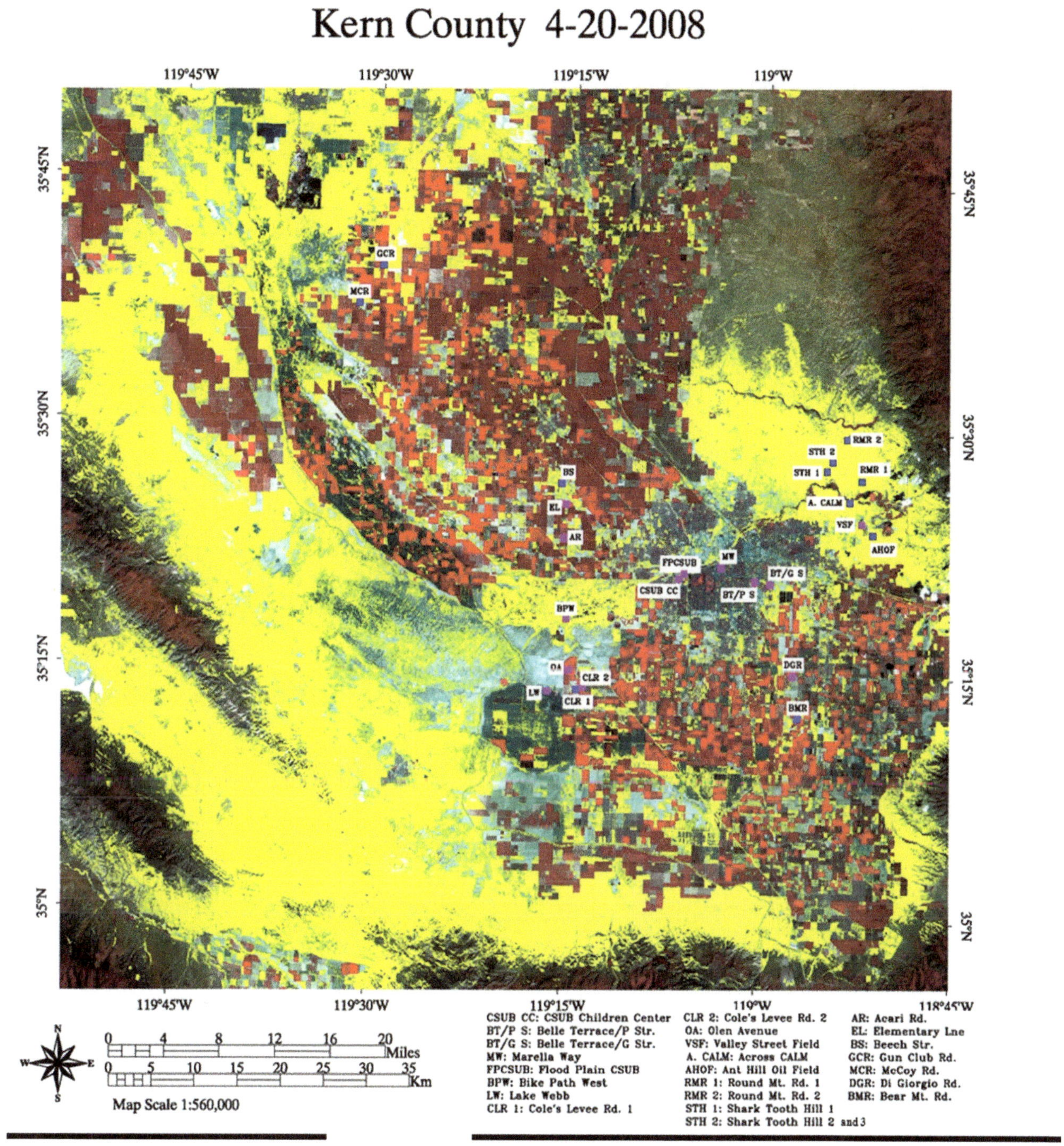

Figure 1. False color image of the ROI on April 20, 2008. Yellow pixels indicate locations in the STH-vegetation class, $p_0 = 0.95$ and $f = 0.32$.

spectral class. Otherwise ($p < p_0$) the pixel was simply left unclassified. Clearly, as the parameter p_0 decreased, the fraction f of pixels in the ROI which belong to the STH-vegetation class increased. This is because pixels which are less and less like the training pixels get included into this class.

It was also investigated how much p_0 needed to be reduced from 1 until the sites which tested positive for *C. immitis* came into the STH-vegetation class. This served to calibrate the method and as a validation step. Clearly, if the *C. immitis*-positive sites get included in the STH-vegetation class for $p_0 \lesssim 1$, then our method is robust. However, if our *C. immitis*-positive sites remain unclassified until $p_0 \gtrsim 0$, then our spectral class is poorly defined.

Lastly, we determined the area (km^2) that was characterized by vegetation that belonged into the STH-vegetation class over the sampling period and until early 2014 using landsat images and the software ENVI 5.1+IDL 8.3.

Surface temperatures

Surface temperature variations across the ROI were of interest as well. In addition to utilizing a vegetation class to assess potential

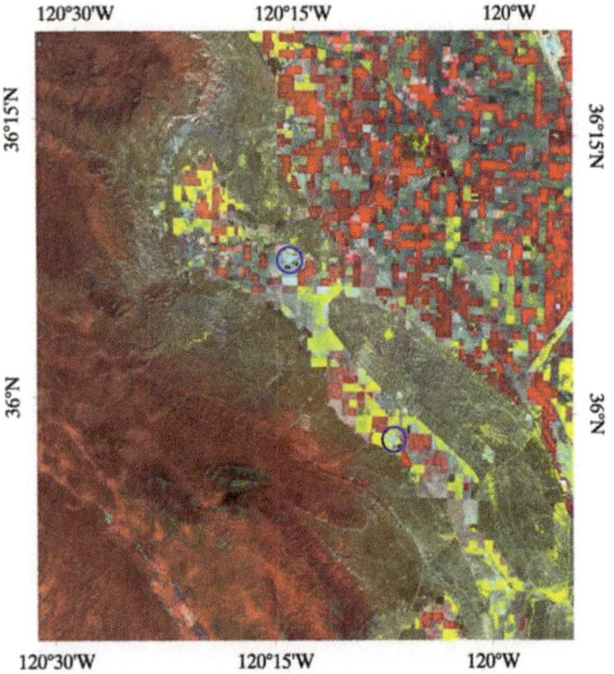

Figure 2. False color image of two San Joaquin Valley prisons. STH-vegetation class pixels are shown in yellow. The circles indicate the location of the prisons. Upper left: Pleasant Valley State Prison in Fresno County, California. Lower right: Avenal State Prison in Kings County, California. Images were taken on April 20, 2008. Maximum Likelihood Classification scheme was used with $p_0 = 0.95$.

sites for *C. immitis* growth, surface temperature variations across the ROI may also help to characterize the niche of this fungus. Landsat-5 TM-6 is an infrared band. This band (from the same April 20, 2008 image) was used as follows to examine the thermal landscape of our ROI. The same area in STH was taken as training pixels, and their average $\bar{x}$ and standard deviation σ were computed. Then, a simple parallelepiped method [49] was used to find other locations in the ROI with similar values. Thus, all pixels whose value was in the ranges$(x-n\sigma,x+n\sigma)$, $n = 1,2,3$ were put into this spectral class, and we referred to this as the "STH-thermal" class. All other pixels were left unclassified. The image's digital numbers were converted to temperatures by applying the procedure described in the National Aeronautics and Space Administration's Landsat 7 Science Data Users Handbook (http://landsathandbook.gsfc.nasa.gov/). See also Chander & Markham [50] for details. As a result, a map was obtained where the surface temperature was close to STH at the time the image was taken.

Weather data

Precipitation data for the Southern San Joaquin Valley was obtained from the California Data Exchange Center (http://cdec.water.ca.gov/snow_rain.html). The cumulative monthly precipitation (inches) over time was assembled from 5 stations (Calaveras Big Trees [CVT], Hetch Hetchy [HTH], Yosemite HQ [YSV], North Fork RS (NFR), and Huntington Lake (HNT]). A more detailed analysis of the weather data for Bakersfield in particular was not the focus of this study.

Physical and chemical soil parameters

To determine physical and chemical soil parameters of all soil samples, the websoilsurvey database of the United States Department of Agriculture (http://websoilsurvey.nrcs.usda.gov/) was used. Furthermore, all sampling sites were characterized by using the soil series extent mapping tool from the website of the Center of Environmental Informatics (CEI) (http://www.cei.psu.edu/cei_wp/). Thus, through agricultural and environmental support tools available from the USDA and CEI websites, our sampling sites were further characterized in regard to land use and vegetation. Additional geological information was obtained as well, such as the distribution of certain soil types and series in California. By using the soil series extent mapping tool, our soil samples were linked to known soil series and soil groups that are characteristic for the Southern San Joaquin Valley and beyond.

Soil sampling sites

Soil physical and chemical parameters that could likely support the growth of the pathogen based on results of the study by Lauer et al. [32] were used to choose 13 new sites that were investigated in winter and spring 2011 (Jan–Apr). Additionally, two sites that were found to be strong growth sites of the pathogen in 2008/2009 were also investigated again in 2011. Six sites were the pathogen was not detected were included in this study as well. Sampling sites included in this study were all non-agricultural silt, clay or sandy loams that differed in regard to physical and chemical parameters. All sites were located within the Central Valley Portion of Kern County. Overall, 23 sites were investigated in this study by satellite imagery and multiplex PCR. Two additional sites from STH were investigated by satellite imagery only (reference sites). Based on information from the USDA websoilsurvey database, the soils belonged to 13 different soil map units. Samples were taken each month in 2008, 2009 and 2011 (some sites were not sampled in 2011) from three different depths (0–2 cm, 5–7 cm, and 18–20 cm), placed on ice during transport to the lab, and frozen at −80°C when not processed immediately. See table 1 for detailed information about all sites, including exact location, soil type, observed rodent activity and indication of the presence or absence of *C. immitis*. Also see the first column of table two for the year they were investigated. Our sampling sites were not chosen based on Landsat imagery. They were chosen mainly based on the percentage of clay in the soil as indicated by the USDA websoilsurvey database. About 30% of clay had been indicative of a potential *C. immitis* positive site based on previous research [32]. After results from the multiplex PCR approach became available, we evaluated if sites where *C. immitis* was detected correlate with sites indicated by Landsat imagery to fall into STH vegetation sites.

DNA extraction and multiplex Polymerase Chain Reaction (PCR)

DNA was extracted from well-mixed soil samples (two replicates) using the MoBio PowerSoil DNA Isolation Kit (MoBio Laboratories, Solana Beach, CA) following the manufacturer's protocol. The multiplex PCR approach developed by Greene et al. [47] and optimized for the detection of *C. immitis* from soil DNA by Lauer et al. [32] was used to determine the presence of fungi in general and specifically *C. immitis* in all soil samples with two primer pairs. Primer pair ITSC1A/ITS C2 (18S ribosomal intertranscribed spacer [ITS] region, 223 bp), which is specific for *C. immitis*, was used in combination with primer pair RDS478/RDS482 (18S ribosomal gene, 650 bp) which amplifies 18S rDNA from all fungi. The ITS region was chosen due to its high

Table 2. Probability that the sites fall in the STH-vegetation class, as predicted by Landsat data.

sampling sites and year sampled	$p_0 = 0.90$				*C. immitis* growth site (multiplex PCR)	$p_0 = 0.95$			
Bakersfield city	4/20/08	4/23/09	4/26/10	4/29/11		4/20/08	4/23/09	4/26/10	4/29/11
1. CSUB Children Center ('08, '09)	N	N	N	N	accumulation site*	N	N	N	N
2. Belle Terrace/P Str. ('11)	Y	N	N	N	growth site*	N	N	N	N
3. Belle Terrace/Gay Str. ('11)	N	N	Y	N	growth site	N	N	Y	N
4. Marella Way ('11)	N	N	N	N	negative site*	N	N	N	N
5. Flood Plain CSUB ('08, '09)	N	N	N	N	accumulation site	N	N	N	N
SW Bakersfield									
6. Bike Path West ('08, '09)	N	N	N	N	negative site	N	N	N	N
7. Lake Webb ('08, '09)	N	N	N	N	accumulation site	N	N	N	N
8. Cole's Levee Rd. I ('08, '09, '11)	Y	Y	Y	Y	growth site	Y	N	N	N
9. Cole's Levee Rd. II ('11)	Y	Y	Y	Y	growth site	Y	Y	Y	N
10. Olen Avenue ('11)	N	N	N	N	growth site	N	N	N	N
11. Valley Street Field ('08, '09)	N	N	N	N	accumulation site	N	N	N	N
NE Bakersfield									
12. Across CALM ('11)	Y	Y	N	N	growth site	Y	Y	N	N
13. Ant Hill Oil Field ('08, '09, '11)	Y	Y	Y	N	growth site	Y	Y	Y	N
14. Round Mt. Rd. I ('08, '09)	Y	Y	Y	N	accumulation site	Y	Y	Y	N
15. Round Mt. Rd. II ('08, '09)	Y	Y	N	N	negative site	Y	Y	N	N
16. Sharktooth hill I	Y	Y	Y	Y	nd*	Y	Y	Y	Y
17. Sharktooth hill 2	Y	Y	Y	Y	nd**	Y	Y	Y	Y
18. Sharktooth hill 3 ('11)	Y	Y	Y	Y	growth site	Y	Y	Y	Y
NW Bakersfield									
19. Acari Rd. ('11)	N	N	N	N	negative site	N	N	N	N
20. Elementary Lne. ('11)	N	N	N	N	growth site	N	N	N	N
21. Beech Str. ('11)	N	Y	N	Y	growth site	N	Y	N	Y
Wasco									
22. Gun Club Rd.('11)	Y	Y	Y	Y	negative site	Y	Y	N	Y
23. McCoy Rd. ('11)	Y	Y	Y	Y	negative site	Y	N	Y	Y
Arvin									
24. Di Giorgio Rd. ('11)	N	Y	N	N	growth site	N	Y	N	N
25. Bear Mt. Rd. ('11)	Y	N	N	N	growth site	Y	N	N	N
Fraction of area covered by class (%)	35	42	24	28		32	39	22	26

(**Y** = in class [indicated in bold], N = not in class).
nd* = not determined in this study.
nd** = not determined in this study, but confirmed as growth site by Swatek (1970).
growth site* = sites were *C. immitis* was detected at least twice in a deeper soil layer during the late winter/spring (February-May).
accumulation site* = the pathogen could only be detected on the surface of the sampling site and never in a deeper soil layer over a several year period.
negative site* = the pathogen could not be detected in any of the soil samples using the multiplex PCR method as described in this study.

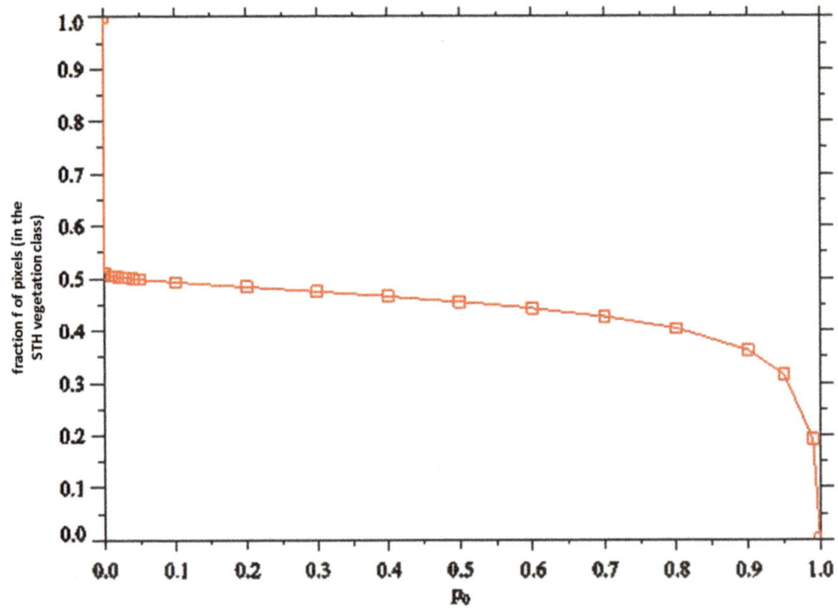

Figure 3. Plot of the fraction *f* of pixels in the STH-vegetation class vs. the threshold value p_0. Pixels whose probability of being in the STH-vegetation class is $p < p_0$ are left unclassified. Pixels with $p \geq p_0$ are put in the class. In this plot, $f = 0.51$ for $p_0 = 10^{-4}$, and $f = 1.0$ for $p_0 = 10^{-5}$.

nucleotide variability. Amplified ITS fragments were extracted from the 2% Agarose Gel, extracted with the Zymo Clean Gel DNA Recovery kit (ZymoResearch, Irvine, CA), and subsequently sequenced to confirm the presence of *C. immitis*. Extracted DNA from a *C. immitis* isolate (M39), obtained from the Laboratory of Medical Mycology at the Universidad Nacional Autónoma de México was used as positive control. Negative controls and positive controls were included in all PCR's to detect contamination and to verify the amplification of a PCR product of the desired size.

Sites were *C. immitis* was detected at least twice in a deeper soil layer during the late winter/spring (February–May) when the soil is moist and the soil temperature increased were referred to as 'growth sites' of the pathogen in this study, assuming that the soil parameters likely supported the growth of the fungus and thus, the pathogen could be detected consecutively over a several year period in the same location, over several growth seasons. Based on the definition provided by Fisher et al. [31], growth sites are sites where physical, chemical, and biological conditions are suitable for completion of the entire growth cycle required by the organism. Thus, it could be assumed that if the pathogen finds supportive environmental conditions, it would likely expand into deeper soil layers, and not just remain on the surface which can be more hostile due to desiccation and increased uv-radiation. In fact, the majority of the soil samples that contained the pathogen in deeper soil layers also contained the pathogen in surface layers. To the contrary, sites were termed 'accumulation sites' in this study when the pathogen could only be detected occasionally on the surface of the sampling site and never in a deeper soil layer over a several year period. This made it likely that arthroconidia had been transported to this location by the wind, but the pathogen was never able to complete its life cycle because of non-supportive environmental conditions. 'Accumulation sites' were also never positive in consecutive years in contrast to 'growth sites'. Fisher et al. [31] defined 'accumulation sites' as sites where arthroconidia of *Coccidioides* may have been deposited on or near the soil surface after being transported from growth sites by wind, water, organisms, or anthropogenic means. We are aware that we did not investigate the activity of the pathogen in the soil or verify its growth, and that finding the pathogen in the surface layer of the soil does not mean that it cannot grow there at all. Therefore, we have to consider that some of our results might have been false negatives.

Results

Remote Sensing Approach

A false color map of our ROI for April 28, 2008 was generated and is presented in figure 1, with indication of all sampling sites. Sites which were similar in vegetation to site STH, a confirmed growth site of *C. immitis*, were indicated in yellow, whereas sites that are characterized by different vegetation types appeared in various shades of green and red (agricultural fields, housing developments with gardens, higher elevated mountain slopes etc.). Results by Landsat imagery indicated large areas west of Bakersfield as potential growth sites of the pathogen, in addition to the STH area east of Bakersfield. The city of Taft southwest of Bakersfield was completely surrounded by vegetation that is similar to the vegetation type that characterizes STH. Landsat imagery furthermore indicated small pockets of potential growth sites of *C. immitis* scattered throughout the Southern San Joaquin Valley and around and within the city of Bakersfield. Overall, the yellow colors indicate that about 15% of the landscape visible in the satellite image was covered with vegetation that has the same reflection pattern as the STH vegetation.

We also applied Landsat imagery to an area northwest of Bakersfield where two prisons are located near the cities of Avenal and Coalinga (Kings County and Fresno County) (Fig. 2). In this area, the incidence of coccidioidomycosis has been observed to be large among prison inmates, so one might hypothesize that *C. immitis* could be present in the neighboring environments. And indeed, yellow areas in the immediate neighborhood of the prisons, as presented by Landsat imagery, indicated the presence of potential growth sites of the pathogen.

Table 3. Probability that the sites fall in the STH-thermal class, as predicted by Landsat data.

sampling sites and year sampled	n = 3				*C. immitis* growth site (multiplex PCR)	n = 2			
Bakersfield city	4/20/08	4/23/09	4/26/10	4/29/11		4/20/08	4/23/09	4/26/10	4/29/11
1. CSUB Children Center ('08, '09)	N	N	N	N	accumulation site	N	N	N	N
2. Belle Terrace/P Str. ('11)	N	N	**Y**	**Y**	growth site	N	N	N	**Y**
3. Belle Terrace/Gay Str. ('11)	N	**Y**	**Y**	N	growth site	N	**Y**	**Y**	N
4. Marella Way ('11)	N	N	N	N	negative site	N	N	N	N
5. Flood Plain CSUB ('08, '09)	N	N	N	N	accumulation site	N	N	N	N
SW Bakersfield									
6. Bike Path West ('08, '09)	N	**Y**	**Y**	**Y**	negative site	N	N	N	N
7. Lake Webb ('08, '09)	N	N	N	N	accumulation site	N	N	N	N
8. Cole's Levee Rd. I ('08, '09, '11)	N	**Y**	**Y**	**Y**	growth site	N	**Y**	N	**Y**
9. Cole's Levee Rd. II ('11)	N	**Y**	**Y**	N	growth site	N	**Y**	**Y**	N
10. Olen Avenue ('11)	N	N	N	N	growth site	N	N	N	N
11. Valley Street Field ('08, '09)	N	N	**Y**	N	accumulation site	N	N	N	N
NE Bakersfield									
12. Across CALM ('11)	**Y**	N	**Y**	N	growth site	**Y**	N	N	N
13. Ant Hill Oil Field ('08, '09, '11)	**Y**	**Y**	**Y**	N	growth site	**Y**	**Y**	N	N
14. Round Mt. Rd. I ('08, '09)	**Y**	**Y**	**Y**	N	accumulation site	**Y**	**Y**	**Y**	N
15. Round Mt. Rd. II ('08, '09)	**Y**	**Y**	**Y**	N	negative site	**Y**	**Y**	N	N
16. Sharktooth hill I	**Y**	**Y**	**Y**	**Y**	nd*	**Y**	**Y**	**Y**	**Y**
17. Sharktooth hill 2	**Y**	**Y**	**Y**	N	nd**	**Y**	**Y**	N	N
18. Sharktooth hill 3 ('11)	**Y**	**Y**	**Y**	N	growth site	**Y**	**Y**	N	N
NW Bakersfield									
19. Acari Rd. ('11)	N	N	N	N	negative site	N	N	N	N
20. Elementary Lne. ('11)	N	N	N	N	growth site	N	N	N	N
21. Beech Str. ('11)	N	**Y**	**Y**	N	growth site	N	**Y**	**Y**	N
Wasco									
22. Gun Club Rd.('11)	N	N	**Y**	**Y**	negative site	N	N	**Y**	**Y**
23. McCoy Rd. ('11)	**Y**	**Y**	**Y**	**Y**	negative site	**Y**	N	**Y**	**Y**
Arvin									
24. Di Giorgio Rd. ('11)	**Y**	**Y**	**Y**	**Y**	growth site	N	**Y**	**Y**	**Y**
25. Bear Mt. Rd. ('11)	N	N	N	N	growth site	N	N	N	N
Fraction of area covered by class (%)	35	42	24	28		32	39	22	26

(**Y** = in class [indicated in bold], N = not in class).
nd* = not determined in this study.
nd** = not determined in this study, but confirmed as growth site by Swatek (1970).

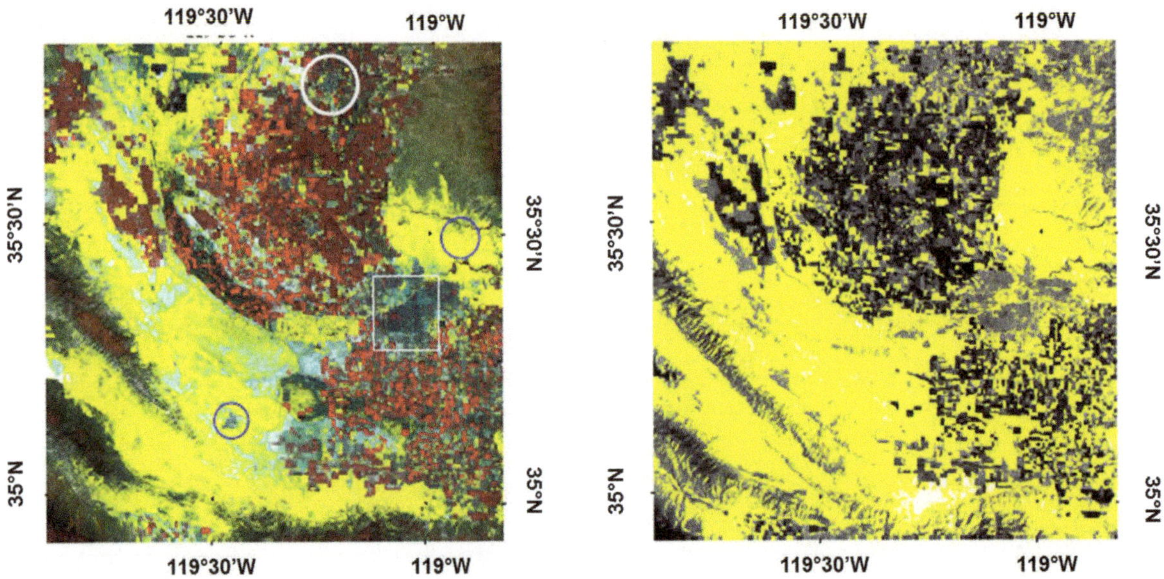

Figure 4. Left: False color image of the ROI on April 20, 2008. Yellow pixels indicate locations in the STH-vegetation class. $p_0 = 0.95$ and $f = 0.32$. The square denotes the location of the city of Bakersfield, the circle on the top indicates the city of Delano, the circle on the right indicates the location of STH, and the circle on the left indicates the location of the city of Taft. **Right**: Spectral class comparison. STH-thermal class is shown in yellow for the same April 20, 2008 image. Parallelepiped scheme was used with thresholds $\bar{x} \mp 20$, with $\bar{x}(=36.4^\circ C)$ the average surface temperature on the STH training pixels, and $\sigma(=1.9^\circ C)$ the corresponding standard deviation.

Overall, 4 false color maps were generated for our ROI showing results for April 2008 to April 2011 (one for each April, see Table 2). For each satellite picture we obtained one STH vegetation profile. Therefore, it did not matter if variation in the vegetation occurred. The April profiles were considered the most definitive for our work, because in early spring, the climate (soil and air temperature and humidity) still support the growth of the vegetation, and grasses and herbs which are characteristic for the STH- vegetation profile have not dried up yet, compared to the summer months. A validation of our approach is shown in figure 3. In this figure, the corresponding plot of f vs. p_0 is presented. This figure also presents the dependence of the fraction f on p_0. It can be pointed out that when p_0 drops from 1, f changes from zero to 0.36 at $p_0 = 0.90$, but then increases slowly with decreasing p_0 until $p \approx 0.04$, where $f \approx 0.50$, and jumps to $f = 1$ for $p_0 = 10^{-5}$. Most likely, this is due to the STH-vegetation class being quite distinct from all other possible spectral classes in the ROI, with a large distance (in pixel space) from those classes. Otherwise, discrete increases in f with decreasing p_0 would be expected as other surface types get merged into the STH-vegetation class.

As a further consistency check, we also wanted to examine the extent to which the STH-vegetation and STH-thermal classes overlapped and if growth sites of the pathogen could be predicted by soil thermal data (tables 2 and 3, figure 4). Thus, we obtained the STH-thermal class by implementing a parallelepiped scheme as described before (see methods). For $n = 1$, only sites 13, 14, 16 and 21 were included in the STH-thermal class (April 2008, data not shown). Sites 8, 9, 12–18, 21, and 22–24 were added when $n = 2$. Site 11 was included when $n = 3$. Site 25 (one of the strongest growth sites of the pathogen) was never included. Figure 4 (right) shows our results for $n = 2$. We present this figure here because site 8, which we identified as a strong growth site of the pathogen, came into the STH-thermal class for this value of n (but site 8 was not in the class for $n = 1$). By evaluating the agreement between satellite imagery (STH-vegetation class and STH-thermal class, data from 4 consecutive years), we found that both data sets disagreed in 12% (sites 6, 11 and 25). Almost all sites that fell into the STH-vegetation class also fell into the STH-thermal class (see tables 2 and 3).

Soil series and soil parameters

By using the soil series extent mapping tool, we found that the soil series and soil groups in which the pathogen was detected around Bakersfield, California, belonged to the Garces (Natragid, sites CLR, Bear Mt. Rd.), Chanac (Haploxerept, site AHOF), and Pleito (Haploxeroll, sites STH1 and 2 and 3) series. These soil series are not restricted to the Southern San Joaquin Valley. See figure 5 for a distribution of these soil series in California. Soils that belonged to the Chanac soil series can also be found in western Arizona and southern Nevada. All soils were of mixed mineralogy, had a superactive cation exchange capacity, and were thermic soils with predominantly fine loamy particles. Soils that belonged to these soil series are among the dominant soils in the Southern San Joaquin Valley, especially Kern County and Kings County, but can also be found in northern and western California. The use of software such as the USDA websoilsurvey database, as well as tools available at the Center of Environmental Informatics (CEI) have been found to be very valuable in obtaining information about physical and chemical parameters of soils that could support the growth of *C. immitis*. Detailed information about soil type, landform, dominant parent material of the soil, as well as soil physical and chemical parameters are listed for all sampling sites in table 4. Using these tools, information about land use, vegetation, mean annual soil temperatures, and geographic setting was accessed as well and is summarized in table S1 in file S1.

Detection of *C. immitis* by multiplex PCR

In addition to the two growth sites of the pathogen that were detected in 2008 and 2009 (Cole's Levee Rd. [CLR], Ant Hill Oil

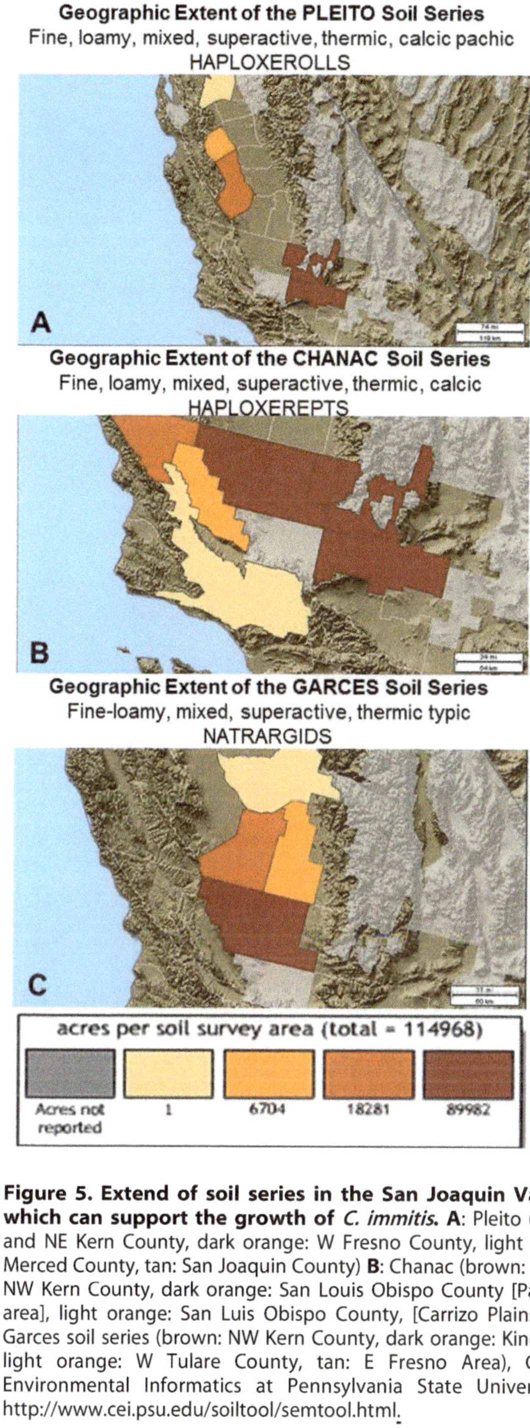

Figure 5. Extend of soil series in the San Joaquin Valley, CA, which can support the growth of *C. immitis.* A: Pleito (brown: SE and NE Kern County, dark orange: W Fresno County, light orange: W Merced County, tan: San Joaquin County) **B**: Chanac (brown: SE, NE and NW Kern County, dark orange: San Louis Obispo County [Paso Robles area], light orange: San Luis Obispo County, [Carrizo Plains]), and **C**: Garces soil series (brown: NW Kern County, dark orange: Kings County, light orange: W Tulare County, tan: E Fresno Area), Center for Environmental Informatics at Pennsylvania State University (CEI), http://www.cei.psu.edu/soiltool/semtool.html.

Field [AHOF], see [32]), we were able to detect additional "hot spots" of the pathogen in 2011. These sites were located within Bakersfield city, in the southwest, northeast and northwest of Bakersfield, and near Arvin, California (see table 1). Sites where the pathogen was detected more than once in a deeper soil layer were considered growth sites, whereas sites where the pathogen was detected occasionally in the surface layer only were considered accumulation sites. Sites were *C. immitis* was never detected were considered negative sites and included areas within Bakersfield city, an area northwest of Bakersfield, and 2 sites near Wasco, California. Fungal DNA could be detected in all soil samples. Of all sites investigated in 2011 (two growth sites investigated in 2008/09 [CLR and AHOF], and 14 new sites, out of 16 sites altogether for 2011), 4 sites (25%) were found negative, and 12 sites (75%) were confirmed as growth sites of the pathogen. No new accumulation sites were discovered in 2011. The site at Bear Mt. Road was the strongest growth site of *C. immitis* for the 2011 sampling set (positive for *C. immitis* from Jan–Apr). For an example of multiplex PCR results, see figure 6.

Correlation between Landsat imagery and multiplex PCR results

By comparing results obtained by satellite imagery (STH-vegetation class) and multiplex PCR, we found that in ~74% (17 out of 23 tested sites) the satellite imagery results and the results obtained by multiplex PCR agreed at least in one year out of four years ($p_0 = 0.90$). When $p_0 = 0.95$, the agreement was ~70% (16 sites). When satellite imagery based on STH-thermal class was compared with multiplex PCR results, we found that both methods agreed only in 61% ($n=3$) or 65% ($n=2$) (table 5). Tables 2 and 3 show the probability that the sites fall in the '*C. immitis* growth area' based on Landsat data in comparison to results obtained by multiplex PCR. We set Red, Green and Blue (RGB) to TM bands 4, 3, and 2 respectively. With this choice, the different depths of red indicated different plant associations. These maps were the result of implementing the MLC method with $p_0 = 0.90$and$p_0 = 0.95$. Two of the three sites at STH listed in table 1 were confirmed growth sites of *C. immitis*. STH site 2 was confirmed as a growth site by Frank Swatek (Fisher FS, personal communication based on [30]), and STH site 3 was confirmed as a growth site by multiplex PCR in this study. Sampling site 8 (Cole's Levee Rd. I), which was determined as a strong growth site of the pathogen by multiplex PCR in every year, became included in the '*C. immitis* growth area' based on Landsat data when p_0 was reduced to 0.95. Sampling site 7 (Lake Webb, accumulation site, located less than 1 mile west of CLR) was added when p_0 was reduced to 0.10. Sampling sites 1, 5, 6, and 11, (near Children Center, Flood Plain, Bike Path West, and Valley Street Field) never got added for $p_0 > 10^{-5}$. This was consistent with results obtained by multiplex which confirmed the absence of the pathogen (site 6, negative site), or which detected the pathogen occasionally in surface samples only (sites 1, 5 and 11). Sites, 1, 5 and 11 were termed accumulation sites, where the arthroconidia had been likely transported to by the wind, and where the presence of the pathogen could not be detected in deeper layers by multiplex PCR. We interpret this to mean that STH was quite representative of the *C. immitis* ecological niche within our ROI. However, in some occasions the prediction made by satellite imagery to indicate soils that could potentially harbor the pathogen could not be confirmed by multiplex PCR. Of all 25 sites included in this study, only sites 8 and 9 (Cole's Levee Rd. I and II) fell in the STH-vegetation class in each year when$p_0 = 0.90$. These sites were confirmed as positive for *C. immitis* by multiplex PCR. Other sites that were confirmed as growth sites of the fungus by our culture independent approach fell in this class at least on one occasion out of four when $p_0 = 0.90$(sites 2, 3 [Belle Terrace/P Str. and Belle Terrace/Gay Str.], 12 [Across CALM], 13 [AHOF], 21 [Beech Str.], 24 [Di Giorgio Rd.], and 25 [Bear Mt. Rd.]). These sites were still included in the STH-vegetation site when p_0 was increased to$p_0 = 0.95$, with the exception of site 25. Of all sites that were found to be *C. immitis* growth sites by multiplex PCR, two sites were never indicated as a potential growth site by the MLC method (site 10 [Olen Ave.] and site 20 [Elementary Lne.]), but the pathogen was present in soil samples from both sites as

Table 4. Detailed physical and chemical information obtained from the USDA websoilsurvey database for all sites included in this study.

soil	sampling sites					
parameters soil parameters	**Elementary Lne.**	**across CALM Ant Hill Oil Field**	**Bear Mt. Rd. Di Georgio Rd. Olen Ave. Cole's Levee Rd.**	**Sharktooth hill**	**Belle Terrace/Gay Str. Belle Terrace/P Str. Marella Way**	**McCoy Rd. Gun Club Rd. Acari Rd. Beech Str.**
soil type	Panoche clay loam	Chanac clay loam	Garces loam	Pleito-Trigo-Chanac complex	Kimberlina-Urban land Cajon complex	Garces silt loam
landform	alluvial fans	fan remnants	Alluvium derived from granitoid	Fan remnants, stream terraces	alluvial fans	rims on basin floors
parent material	alluvium derived from igneous and sedimentary rock	alluvium derived from mixed	alluvium derived from granitoid	Alluvium derived from mixed	alluvium derived from igneous and sedimentary rock	alluvium derived from granite
(map unit symbols)	211	130/131	180	205	180	156
Physical parameters						
Surface texture	clay loam	clay loam	clay loam	clay loam	loamy sand	silt loam
% clay	31	31	25.5	30	12	26.8
% sand	35.4	35.4	38	33.5	71.3	34.2
% silt	33.6	33.6	36.5	36.5	16.7	39.1
Available water capacity (cm/cm)	0.17	0.17	0.21	0.16	0.12	0.11
Available water supply (0–25 cm)	4.25	4.25	5.04	3.69	2.64	2.7
Organic matter	0.25	0.75	0.98	1.5	0.75	0.06
Water content (15 bar)	18.9	18.2	16.7	17.2	8.7	16.2
Water content (1/3 bar)	32	30.1	30.9	27.8	17.7	30.2
Sat. hydraulic conductivity (Ksat) (micrometers/s)	9	9	8.37	2.82	28	0.8362
Chemical parameters						
pH	7.9	7.9	8.5	7.8	7.5	8.9
CaCO3	3	3	3	0	3	3
Cation Exchange Capacity (CEC7)	15	24.4	20.6	24.3	7.5	13.1
Gypsum	0	0	0	0	0	0
Sodium adsorption ratio (SAR)	0	0	2	0	0	14
Electrical conductivity (EC)	1	0	5	0.5	1	10.2

Indicated in cursive are the parameters which seemed to be most important to distinguish *C. immitis* growth sites from negative sites.

confirmed by multiplex PCR (see Discussion). Furthermore, two sites that were indicated as potential growth sites of the pathogen by Landsat imagery at all times when $p_0 = 0.90$, could not be confirmed by multiplex PCR to harbor the pathogen. These sites were located near Wasco, California (NW of Bakersfield), (sites 22 [Gun Club Rd.] and 23 [McCoy Rd.]).

Changes in extend of areas (km^2) that fell into the STH-vegetation class were observed for the sampling period until early 2014 and are displayed in table S2 in file S1. The year with the highest precipitation (2011) had the lowest area of vegetation that belonged into the STH-vegetation class in comparison to the years 2008 and 2009 which were characterized by a significantly reduced amount of precipitation and showed an increased area of vegetation that belonged into the STH-vegetation class (see figure S1).

Discussion

The purpose of our study was to identify soil types in Kern County that could support the growth of *C. immitis* by combining Landsat imagery (based on vegetation and soil temperature), and soil parameter information (from 25 sites) with a culture independent PCR-based method to detect the pathogen. We showed that satellite imagery, combined with soil parameter information, can provide a map of locations within our ROI, where *C. immitis* might reasonably be expected to be found. We were able to verify the presence of the pathogen by a multiplex PCR method in about 74% ($p = 0.90$), when soil samples were investigated over a 4 year period. However, for about a quarter of our sites (26%), results obtained by Landsat imagery and multiplex PCR did not correlate. The reasons for this observation could be multifold. Some main factors to be considered are: 1) The amount of *Coccidioides* DNA extracted from the soil might have been under the detection limit of our PCR based methods (sites 22 and

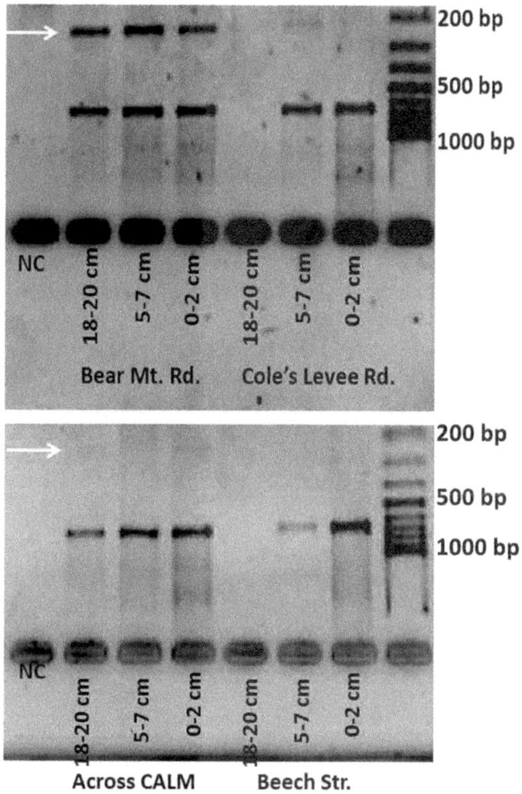

Figure 6. Example of multiplex PCR results. White arrows point on a 223 bp fragment that represents *C. immitis*. Site Bear Mt. Rd. shows the strongest ITS amplicons in all soil layers, whereas sites Cole's Levee Rd. and site Across CALM gave a weaker signal in some soil layers, and site Beech Str. was negative. NC = negative control. Bands that indicate the presence of the pathogen in the 2% Agarose gel were confirmed to origin from *C. immitis* by sequencing.

23), or 2) the resolution of the satellite imagery might not have been detailed enough (site 10, a small site of only 10 m^2), and 3) the distribution of *Coccidioides* in the soil might have been spotty, and the positive site was missed (sites that were not positive for the pathogen in all sampling years). A closer look at sites where satellite imagery and soil parameter data indicated potential growth sites for *C. immitis* also revealed that these sites were not uniform in regard to plant coverage, distribution, and diversity, thus, generating microhabitats for soil microorganisms that most likely would be quiet distinct, especially in and around the rhizosphere [32]. Other factors, such as fluctuation in climate and pollution of the soil might have had an impact on our analyses as well. Furthermore, it has to be considered that *C. immitis* might be able to persist in soils that have been converted to agricultural fields for an unknown amount of time, but its arthroconidia might never germinate and grow into vegetative hyphae. These sites could be termed dormant sites (e.g. site 20, an orchard with young almond trees). To assess these impacts on our results was not the focus of our work, but we are aware of these limitations. In previous research we have investigated the limitation of the multiplex PCR approach to detect *C. immitis*, (see [33] for results of primer efficiency). Briefly, we found that the sensitivity of the diagnostic PCR (ITS primer pair) was reduced compared to the primer pair that amplifies 18S rDNA fragments of all fungi (RDS primer pair).

In previous work [23–25], it was suggested that some environmental fluctuations are a fundamental link missing from coccidioidomycosis incidence statistical modeling schemes. One important aspect to investigate is whether fluctuations in the STH-vegetation class can provide this connection, and be statistically linked to the observed variations in incidence of valley fever. In this regard, one effect to consider is the extent to which this RS approach continues to be valid through the seasons. In the spring, when plants are blooming, the different vegetation types have different spectral signatures. As the weather dries and plants wither, the spectral signatures of the relevant vegetation types may become less distinct. Thus, the vegetation on STH may not be as good a marker for *C. immitis* in the fall, as it is in the spring. The implicit assumption in this study is that the STH environment is the only type of environment which harbors *C. immitis* within the San Joaquin Valley area of Kern County. We presented in this paper arguments to support this assumption; nevertheless, it would be useful to find more similarly suitable test sites to further corroborate our findings, or to find slightly different ecotypes that support *C. immitis*, beside of those detected in this study.

Table 5. Agreement between multiplex PCR and MLC for the STH vegetation class and the STH-thermal class to predict growth sites of *C. immitis* (to agree a prediction by either multiplex PCR or MLC must be confirmed at least once for the four years by the other method).

	STH-vegetation class	
	p_0 = **0.90**	p_0 = **0.95**
multiplex PCR and MLC agree	17 (74%)	16 (70%)
multiplex PCR predicts growth site and MLC disagrees	2 (9%)	3 (13%)
MLC predicts growth site and multiplex PCR disagrees	4 (17%)	4 (17%)
	STH-thermal class	
	n = 3	**n = 2**
multiplex PCR and MLC agree	14 (61%)	15 (65%)
multiplex PCR predicts growth site and MLC disagrees	3 (13%)	4 (17.5%)
MLC predicts growth site and multiplex PCR disagrees	6 (26%)	4 (17.5%)

From altogether 25 sites, only 23 were considered, because no multiplex PCR results were obtained for STH sites I and II.

We also observed changes in the extent of the STH-vegetation class over time. A comparative analysis of precipitation between 2008 and 2011 (up to early 2014) suggest that years with a reduced precipitation (drought) favor plants of the STH-vegetation class, but other factors likely play a role as well, such as development and changes in land use (see figure S1 and table S1 in File S1), which was not assessed in this study.

Compared to the STH-vegetation class data, the STH-thermal data showed considerably more variation for the four different years, as expected. The vegetation on a certain day in each year may be very similar, but soil temperatures might be more variable in different years (data for $n=2$ and $n=3$ can be seen in table 3, no data is shown for$n=1$). Other strong growth sites of the pathogen (sites 8, 9, and 13) were also not consistently included in all years, not even with$n=3$. It should also be noted that sites 6 (negative site) and 11 (accumulation site) were never included into the STH-vegetation class by satellite imagery, but were included in the STH-thermal class when$n=3$. We concluded therefore, that the STH-thermal classes alone might not be sensitive enough to predict growth sites of *C. immits*. Site 25 was never included (thermal class) maybe because of the limited resolution of the satellite imagery, as discussed earlier. It is important to keep in mind that the TM-6 image tells us surface temperatures. It may very well be that what matters is the temperature below the surface [31]. For our ROI, all landcovers were similar on the macroscale; therefore, apparent temperatures were appropriate for comparison purposes.

To improve the value of satellite imagery data, the actual spectral reflectance profiles of various soil components could be included to complement the satellite data (for details see http://www.africasoils.net/data/ldsf-description) in future studies. A time series analysis could also be considered, if feasible. Phenology development throughout the year can make the analysis more specific to a particular vegetation type. Niche modeling rather than automated classification could be considered as well to obtain a richer output that indicates variables of importance. However, a large dataset would be necessary that would include presence and also absence data of the pathogen in a certain type of soil at a certain time with presence or absence of a certain type of vegetation. Furthermore, it should be considered that broad band signatures over larger geographic areas and ecotones might not be precise enough to be useful in predicting growth sites of a pathogen, especially when the pathogen could be adapted to grow in a variety of different ecosystems.

Our results indicated that strong growth sites of the pathogen were likely associated with 3 different USDA soil map units (180, 131, and 205), which were all loamy sands. Several sites around Bakersfield, California, that fell into one of these map units were indeed growth sites of *C. immitis*, as confirmed by multiplex PCR, and were similar in vegetation compared to the STH area. These types of soils are not restricted to the Southern San Joaquin Valley, but can be found in other areas of California as well. One could hypothesize that with a drier and warmer climate, as it is predicted for California in the near future [51,52], *C. immitis* might be able to expand its current range. In our study, we focused on soil samples from only one County, the above mentioned Kern County in the Southern San Joaquin Valley of California, a highly endemic area for *C. immitis*. The soil types investigated here did not comprise all types that can be found in our ROI. Even though Kern County is a hot spot of *C. immitis* with the highest incidence of coccidioidomycosis documented for as long as incidence data is recorded in California, we cannot conclude that soils that predominate in this area are the ones that also predominantly support the growth of the pathogen. A more rigorous sampling framework should be attempted in the future that would include locations beyond Kern County covering as wide a range of habitats as possible to correctly determine growth sites of *C. immitis*, as well as determining sites that are not supporting the growth of the pathogen. Developing such a sampling plan should include stratification, replicate sampling, and determination of important chemical and physical soil parameters, including investigations in other countries where coccidioidomycosis occurs would be of value as well. The ultimate goal would then be to generate a U.S. or America-wide database of occurrence and absence of *Coccidioides* spp. Such a database could be useful for characterizing the ecological niche for both *Coccidioides* species, and could indicate a variety of supporting ecosystems, as well as being an advisory public health tool, to reduce incidence of coccidioidomycosis in Kern County and elsewhere.

In conclusion, the combination of the methods used in our research can be used to generate maps that indicate potential growth sites of *C. immitis*, and thus serve as a tool to further investigate the ecological niche occupied by the pathogen in the Southern San Joaquin Valley and beyond in more detail. Recent advances in computer processing and geographic information system and global positioning system technologies facilitate integration of remote sensing data, such as environmental parameters with disease incidence data, so that models for disease surveillance and control can be developed [53,54].

Supporting Information

Figure S1 Cumulative monthly precipitation (inches) over time for the Southern San Joaquin Valley, assembled from 5 stations (Calaveras Big Trees [CVT], Hetch Hetchy [HTH], Yosemite HQ [YSV], North Fork RS (NFR), and Huntington Lake (HNT]) obtained from the California Data Exchange Center at http://cdec.water.ca.gov/snow_rain.html).

File S1 Supporting tables. Table S1. Detailed soil series descriptions of sites which were found to be growth sites of *C. immitis*. **Table S2.** Extend of STH-vegetation class in our ROI between 2008 and 2011 based on satellite imagery.

Acknowledgments

The authors gratefully acknowledge Tom Mace, retired Senior Science Advisor at NASA Armstrong Flight Research Center (Science Mission Directorate, Airborne Sciences), and to Ned Horning of the Biodiversity Informatics Facility at the American Museum of Natural History's Center for Biodiversity and Conservation (New York, NY) for valuable technical advice on satellite image analysis. The authors also thank Kathleen Szick (CSUB) for helpful discussions and comments to improve this manuscript.

Author Contributions

Conceived and designed the experiments: AL JT. Performed the experiments: AL JT LM JB KC NS. Analyzed the data: AL JT. Contributed reagents/materials/analysis tools: LC KE. Wrote the paper: AL JT LC KE.

References

1. Hector RF, Rutherford GW III, Tsang CA, Erhart LM, McCotter O et al. (2011) The public health impact of coccidioidomycosis in Arizona and California. Int J Environ Res Public Health 8: 1150–1173.
2. Nguyen C, Barker M, Hoover S, Nix DE, Ampel NM et al. (2013) Recent advances in our understanding of the environmental, epidemiological, immunological, and clinical dimensions of coccidioidomycosis, Clin Microbiol Rev 26: 505–525.
3. Lee RV (1944) Coccidioidomycosis: In the western flying training command. Cal West Med 61: 133–134.
4. Pappagianis D (1988) Epidemiology of coccidioidomycosis. Current Topics in Mycology 2: 199–238.
5. Pappagianis D (1994) Marked increase in cases of coccidioidomycosis in California: 1991, 1992, and 1993. Clin Infect Dis 19 (Suppl 1), S14–S18.
6. Negroni R (2008) Evolución de los conocimientos sobre aspectos clinic epidemiológicos de la coccidioidomycosis en las Américas. Rev Argent Microbiol 40: 246–256.
7. Fisher MC, Koenig GL, White TJ, San-Blas G, Negroni R et al. (2001) Biogeographic range expansion into South America by *Coccidioides immitis* mirrors new world patterns of human migration. Proc Nat Acad Sci 98: 4558–4562.
8. Fisher MC, Koenig GL, White TJ, Taylor JW (2002) Molecular and phenotypic description of *Coccidioides posadasii* sp. nov., previously recognized as the non-California population of *Coccidioides immitis*. Mycologia 94: 73–84.
9. Neafsey DE, Barker BM, Sharpton TJ, Stajich JE, Park DJ et al. (2010) Population genomic sequencing of *Coccidioides* fungi reveals recent hybridization and transposon control. Genome Res 20: 938–946.
10. Comrie AC, Glueck MF (2007) Assessment of climate coccidioidomycosis model: Model sensitivity for assessing climatologic effects on the risk of acquiring coccidioidomycosis. Ann NY Acad Sci 1111: 83–95.
11. Kolivras KM, Johnson PS, Comrie AC, Yool SR (2001) Environmental variability and coccidioidomycosis (valley fever). Aerobiologia 17: 31–42.
12. Smith CE, Beard RR, Rosenberger HG, Whiting EG (1946) Effect of season and dust control on coccidioidomycosis. J Am Med Assoc 132: 833–838.
13. Hugenholtz PG (1957) Climate and coccidioidomycosis. In: Proceedings of symposium on coccidioidomycosis, Phoenix, AZ. Public Health Service, Publication no. 575. Atlanta: US Public Health Service, pp. 136–43.
14. Maddy KT (1957) Ecological factors possibly relating to the geographic distribution of *Coccidioides immitis*. In: Proceedings of the Symposium on Coccidioidomycosis (pp. 144–157). Comunicable Disease Center, Atlanta, GA. PHS Pub. No. 575.
15. Pappagianis D, Einstein H (1978) Tempest from Tehachapi takes toll or *Coccidioides* conveyed aloft and afar. West J Med 129: 527–530.
16. Jinadu BA, Welch G, Talbot R, Caldwell J, Johnson R et al. (1994) Update: coccidioidomycosis in California 1991–1993. MMWR 43: 421–423.
17. Mosley D, Komatsu K, Vaz V, Vertz D, England B et al. (1996) Coccidioidomycosis in Arizona 1990–1995. MMWR 45: 1069–1073.
18. Kirkland TN, Fierer J (1996) Coccidioidomycosis: A reemerging infectious disease. Emerg Infect Dis 2: 192–199.
19. Schneider E, Hajjeh RA, Spiegel RA, Jibson RW, Harp EL et al. (1997) A coccidioidomycosis outbreak following the Northridge, Calif. earthquake. J Am Med Assoc 277: 904–908.
20. Komatsu K, Vaz V, McRill C, Colman T, Comrie A et al. (2003) Increase in coccidioidomycosis in Arizona 1998–2001. MMWR 52: 109–112.
21. Kolivras KM, Comrie AC (2003) Modeling valley fever (coccidioidomycosis) incidence on the basis of climate conditions. Int J Biometeor 47: 87–101.
22. Comrie AC (2005) Climate factors influencing coccidioidomycosis seasonality and outbreaks. Environ Health Perspect 113: 688–692.
23. Zender CS, Talamantes J (2006) Climate controls on valley fever incidence in Kern County, California. Int J Biometeorol 50: 174–182.
24. Talamantes J, Behseta S, Zender CS (2007a) Fluctuations in climate and incidence of coccidioidomycosis in Kern County, California: a review. Ann NY Acad Sci 1111: 73–82.
25. Talamantes J, Behseta S, Zender CS (2007b) Statistical modeling of valley fever data in Kern County, California. Int J Biometeorol 51: 301–315.
26. Tamerius JD and Comrie AC (2011) Coccidioidomycosis incidence in Arizona predicted by seasonal precipitation. PLOS One 6: 1–7.
27. Cayan D, Maurer E, Dettinger M, Tyree M, Hayhoe K (2008) Climate change scenarios for the California region. Climate Change 87: 21–42.
28. Greer A, Ng V, Fisman D (2008). Climate change and infectious diseases in North America: the road ahead. CMAJ 11: 178–186.
29. Marsden-Haug N, Goldoft M, Ralston C, Limaye AP, Chua J et al. (2013) BRIEFREPORT - Coccidioidomycosis acquired in Washington State. CID 2013: 56.
30. Swatek FE (1970) Ecology of *Coccidioides immitis*. Mycopath Mycol Applic 40: 3–12.
31. Fisher FS, Bultman MW, Johnson SM, Pappagianis D, Zaborsky E (2007) *Coccidioides* niches and habitat parameters in the southwestern United States: a matter of scale. Ann NYAcad Sci 1111: 47–72.
32. De Boer, Folman LB, Summerbell RC, Boddy L (2005) Living in a fungal world: impact of fungi on soil bacterial niche development. FEMS Microbiol Rev 29: 795–811.
33. Lauer A, Baal JCH, Baal JDH, Verma M, Chen JM (2011). Detection of *Coccidioides immitis* in Kern County, California, by multiplex PCR. Mycologia. doi:10.3852/11-127.
34. Merriam CH (1898) Life zones and crop zones of the United States. Bulletin 10 Division of Biological Survey, United States Department of Agriculture. Washington, DC.
35. Maddy KT (1958) The geographic distribution of *Coccidioides immitis* and possible ecologic implications. Ariz Med 15: 178–188.
36. Maddy KT (1959) Coccidioidomycosis in animals. Vet Med 54: 233–242.
37. Baptista-Rosas RC, Catalan-Dibene J, Romero-Olivares AL, Hinojosa A, Cavazos T et al. (2012) Molecular detection of *Coccidioides* spp. form environmental samples in Baja California: linking Valley Fever to soil and climate conditions. Fung Ecol 5: 177–190.
38. Barker BM, Tabor JA, Shubitz L, Perrill R, Orbach MJ (2012) Detection and phylogenetic analysis of *Coccidioides posadasii* in Arizona soil samples, Fungal Ecol. 5: 163–176.
39. Zak DR, Holmes WE, White DC, Peacock AD, Tilman D (2003) Plant diversity, soil microbial communities, and ecosystem function: Are there any links? Ecology 84: 2042–2050.
40. Nannipieri P, Ascher J, Ceccherini MT, Landi L, Pietramellara G, et al. (2003) Microbial diversity and soil function. Europ J Soil Sci 54: 655–670.
41. Sharpton TJ, Stajich JE, Rounsley SD, Gardner MJ, Wortman JR et al. (2009) Comparative genomic analyses of the human fungal pathogens *Coccidioides* and their relatives. Genome Res. doi:10.1101/gr.087551.108.
42. Egeberg RO, Ely F (1956) *Coccidioides immitis* in the soil of the southern San Joaquin Valley. Am J Med Sci 231: 151–154.
43. Rogers DJ, Randolph SE, Snow RW, Hay SI (2002) Satellite imagery in the study and forecast of malaria. Nature 415: 710–715.
44. Koelle K, Rodó X, Pascual M, Yunus M, Mostafa G (2005) Refractory periods and climate forcing in cholera dynamics. Nature 436: 696–70.
45. Rogers DJ (2000) Satellites, space, time, and the African trypanosomiases. Adv Parasitol 47: 128–171.
46. Goetz S, Prince S, Small J (2000) Advances in satellite remote sensing of environmental variables for epidemiological applications. Adv Parasitol 47: 289–307.
47. Greene DR, Koenig G, Fisher MC, Taylor GW (2000) Soil isolation and molecular identification of *Coccidioides immitis*. Mycologia 92: 406–410.
48. Kemp R (1974) The distribution of *Coccidioides immitis* at Sharktooth hill, Oildale, California [MS Thesis].. Long Beach, California: California State College. 62 p.
49. Richards JA, Jia X (2006) Remote sensing digital image analysis, An Introduction. Springer-Verlag, Germany.
50. Chander G, Markham B (2003) Revised Landsat-5 TM radiometric calibration procedures and postcalibration dynamic ranges. IEEE Trans Geosci Remote Sens 41: 2674–2677.
51. Cayan D, Maurer E, Dettinger M, Tyree M, Hayhoe K (2008) Climate change scenarios for the California region. Climate Change 87: 21–42. Clim Change 87: 21–42.
52. Greer A, Ng V, Fisman D (2008). Climate change and infectious diseases in North America: the road ahead. CMAJ 11: 178–186.
53. Nuckols JR, Ward MH, Jarup L (2004) Using geographic information systems for exposure assessment in environmental epidemiology studies. Environ Health Perspect 11: 1007–1015.
54. Craglia M, Goodchild MF, Annoni A, Camara G, Gould M et al. (2008) Next-Generation Digital Earth. A position paper from the Vespucci Initiative for the Advancement of Geographic Information Science. Int J Spat Data Infrastruct Res 3: 146–167.

Niche-Partitioning of Edaphic Microbial Communities in the Namib Desert Gravel Plain Fairy Circles

Jean-Baptiste Ramond[1], Annelize Pienaar[1], Alacia Armstrong[1], Mary Seely[2], Don A. Cowan[1]*

1 Center for Microbial Ecology and Genomics (CMEG), Genomic Research Institute, University of Pretoria, Pretoria, South Africa, 2 Gobabeb Research and Training Center (GRTC), Walvis Bay, Namibia

Abstract

Endemic to the Namib Desert, Fairy Circles (FCs) are vegetation-free circular patterns surrounded and delineated by grass species. Since first reported the 1970's, many theories have been proposed to explain their appearance, but none provide a fully satisfactory explanation of their origin(s) and/or causative agent(s). In this study, we have evaluated an early hypothesis stating that edaphic microorganisms could be involved in their formation and/or maintenance. Surface soils (0–5cm) from three different zones (FC center, FC margin and external, grass-covered soils) of five independent FCs were collected in April 2013 in the Namib Desert gravel plains. T-RFLP fingerprinting of the bacterial (16S rRNA gene) and fungal (ITS region) communities, in parallel with two-way crossed ANOSIM, showed that FC communities were significantly different to those of external control vegetated soil and that each FC was also characterized by significantly different communities. Intra-FC communities (margin and centre) presented higher variability than the controls. Together, these results provide clear evidence that edaphic microorganisms are involved in the Namib Desert FC phenomenon. However, we are, as yet, unable to confirm whether bacteria and/or fungi communities are responsible for the appearance and development of FCs or are a general consequence of the presence of the grass-free circles.

Editor: Melanie R. Mormile, Missouri University of Science and Technology, United States of America

Funding: This research was funded by the National Research Foundation (http://www.nrf.ac.za/) of South Africa: Grant N °81779, Microbial Ecology of the Namib Desert. The funders had no role in study design, data collection and analysis, decision to publish, or preparation of the manuscript.

Competing Interests: The authors have declared that no competing interests exist.

* Email: don.cowan@up.ac.za

Introduction

The Namib Desert is unique in harboring the enigmatic Fairy Circles (FC) (or Fairy Rings). They occur and have been studied predominantly in the sand dune environment of the eastern Namib but also occur on the gravel plains [1,2]. FCs are circular vegetation-free patterns (generally between 2 to 12 m in diameter), surrounded by grass species of the genus *Stipagrotis* (Figure 1A). Their distribution is restricted to a region in the Pro-Namib zone (60 to 120 km inland), from southern Angola to northern South Africa [2]. The Namib Desert FCs have been described as 'living organisms', as they appear (birth), enlarge (growth) and ultimately disappear (death), with an estimated life span of around 60 years [3].

Since first being reported in the 1970's, many studies and hypotheses have attempted to explain the origins and lifestyles of FCs (most reviewed in [2]): these hypotheses include local radioactivity [4], insect (termite/ant) activities [5–11], the release of volatile chemicals (e.g. allelopathic compounds by dead *Euphorbia damarana* plants [12], semi-volatile products from termite-related activities [8,13] or hydrocarbon-liked compounds of geochemical origins [14]), or plant spatial self-organization [15,16]. The allelopathic compound and radioactivity hypotheses have already been refuted in field studies [2]. The 'abiotic gas leakage' and 'grass harvesting social insect (ant/termite)' hypotheses were contested, based on mathematical modelling (e.g. remote sensing, spatial pattern analysis and vegetation modelling) [16]. The results obtained by Getzin and colleagues [16] suggested that FC distribution was indicative of self-organized models, i.e. in agreement with the 'spatial self-organization' hypothesis [15]. The authors nevertheless noted that more factors should be implemented (e.g. plant root system and soil moisture data) to fully validate the hypothesis. The current trend to include microbiological data when modeling edaphic ecosystems [17,18] may be even more critical in arid edaphic environments where most ecosystem processes are microbially-mediated [19]. It must also be noted that gravel plain FCs are largely unstudied ([1] and [3] only, to our knowledge) compared to the eastern Namib FCs. The patchiness of gravel plain FCs (as shown in Figure 1A) is not consistent with the self-organization pattern hypothesis [15,16].

Despite the fact that the potential role of microorganisms as contributing agents of FC development has not yet been tested experimentally, it has been previously discussed because of circumstantial support for the involvement of microbial processes. Significantly higher alkene emissions have been observed from FC soils than from the vegetated surroundings soils [14]; and alkene evolution has been shown to be indicative of anaerobic microbially-mediated alkane reduction [20]. Using Namib FC soils, bioassay experiments suggested phytotoxicity [2] and germination studies suggested the involvement of a biologically active abiosis factor leading to plant decay [8]. An *'in situ'* pot trial also

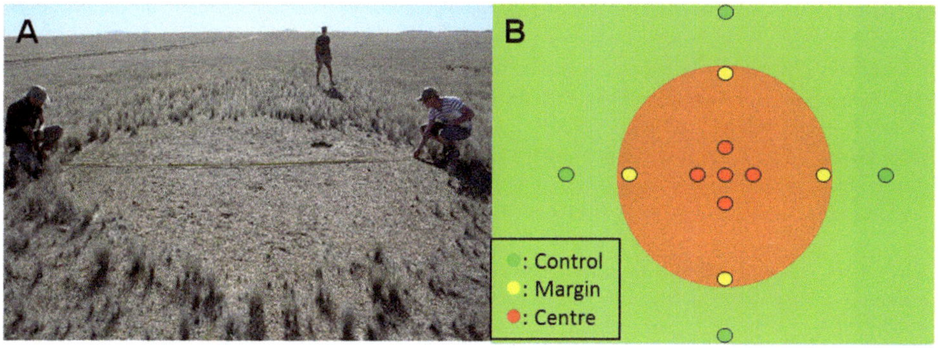

Figure 1. Fairy Circle sampling. Photograph of a Fairy Circle in the gravel plains of the Namib Desert (A) with the schematic of the sampling strategy employed (B).

suggested the possible involvement of either a semi-volatile compound or a (microbial?) toxin in plant-growth inhibition [13].

The current study was initiated with the working hypothesis that edaphic microorganisms could be significant agents in the formation and/or maintenance of Namib Desert FCs. The experimental basis of this study was a comprehensive sampling of surface soils from within FCs (centre and margin) and from external vegetated (controls) sites. Unlike most previous FC studies, which have focused in FC sites situated in the dune sands to the east of the Namib Sand Sea, we selected FCs in the Namib Desert gravel plains. Our analytical approach included T-RFLP fingerprinting of bacterial (16S rRNA gene) and fungal (ITS region) phylogenetic markers and soil chemical analyses (10 different variables), using multivariate statistical analyses to cross-correlate the experimental data.

Materials and Methods

Fairy Circle soil sampling and storage

Namib Desert Fairy Circle soils were sampled under the permits N° P0063476, permit for the importation of controlled goods from the Department of Agriculture, Forestry and Fisheries of the Republic of South Africa, and N° ES 29529, permit for single consignment export of minerals from the Ministry of Mines and Energy of the Republic of Namibia.

Surface soils (0–5 cm; approx. 200g per sample) from five independent Fairy Circles were collected in the gravel plains of Namibia (−23°31′37.09"S/15°11.15"E) near the Gobabeb Research and Training Center (http://www.gobabebtrc.org/) in April 2013, following a strategy shown schematically in Figure 1. A total of thirteen samples per FC were collected: four control soils from outside the FCs and nine within each FCs; 4 at the circle margin and 5 in the center. The samples were stored at 4°C during transport to the CMEG Laboratory (University of Pretoria, South Africa). One gram subsamples were stored at −80°C for subsequent molecular analyses and the residual soil at 4°C for chemical analyses.

Soil Chemistry analyses

Soil chemistry analyses were conducted at the Soil Science Laboratory of the University of Pretoria (South Africa) according to standard quality control procedures [21]. All solutions and reagents used were supplied by Merck Chemicals (South Africa) and, apart for pH measurements, soil samples were sieved (2 mm) prior to analysis.

The slurry technique was used to measure pH recorded with a pH meter (Crison basic 20, Barcelona, Spain) by mixing 4 g of soil with 10 ml of deionized water and allowing it to settle for 1h. The Walkey-Black method [22] was used to determine soil total organic carbon (C) content with minor modifications. 10 mL of 1M potassium dichromate solution was added to 2 g of soil. 10 mL of sulfuric acid (96%) was then added, and the mixture was left to cool at room temperature for 30 min. 150 mL deionized water and 10 mL concentrated (96%) orthophosphoric acid was then added to the mixture which was left to cool at room temperature for 30 min. 1 mL phenylalanine was added and titration was performed by adding iron(II) ammonium sulphate solution until the reaction endpoint was reached; i.e. when the solution changed from purple to green.

Exchangeable ammonium (NH_4^+) and nitrate (NO_3^-) soil content was determined by steam distillation as described [23] with minor modifications. 5 g of soil was mixed with 50 mL of a 2M KCl solution and shaken for 30 min at 220 rpm. Samples were then left to settle for 1 min and the supernatant was filtered through a 110 mm diameter Whatman n°2V filter paper and stored overnight at 4°C. The addition of 0.2 g MgO to the filtrate liberated ammonium, and the residual nitrate was determined by the reduction to nitrite (NO_2) *via* the addition of ~ 2g Devarda alloy powder [24].

Total organic P was determined by the P Bray method described by [25] with minor modifications. 50 mL of P Bray-1 solution was added to 4 g of soil. The mixture was hand-shaken for exactly 1 min prior to filtration using 110 mm diameter Whatman n°2V filter paper. The phosphorous concentration of the filtrate was then determined by Inductively Coupled Plasma-Optical Emission Spectroscopy (ICP-OES) (Spectro Genesis, Germany).

Total ion concentrations were determined by adding 40 mL of a 0.2M ammonium acetate solution to 4 g of soil. The mixture was shaken for 1 hour and the supernatant filtered through 110 mm diameter Whatman n°2V filter paper. 15 mL of the filtrated were used to determine the concentrations of iron (Fe), calcium (Ca), potassium (K), magnesium (Mg) and sodium (Na) by ICP-OES.

Metagenomic DNA extraction

Total soil DNA was extracted from 0.3 g samples using the Powersoil DNA isolation kit according to the manufacturer's instructions (MOBIO laboratories, San Diego, USA). DNA concentrations were estimated with a NanoDrop spectrophotometer (NanoDrop Technologies, Montchanin, DE, USA).

PCR amplification, purification and restriction digestion

All polymerase chain reactions (PCRs) were carried out in a Bio-Rad Thermocycler (T100 TM Thermal Cycler). Bacterial 16S rRNA encoding genes were amplified using the universal bacterial primers 341F (5′-CCTACGGGAGGCAGCAG-3′)/908R (5′-CCGTCAATTCMTTTGAGTTT -3′) [26] and the Fungal ITS regions amplified using the universal primer set ITS1 (5′-CT-TGGTCATTTAGAGGAAGTAA-3′)/ITS4 (5′-TCCTCCGCT-TATTGATATGC-3′) [27]. PCR was carried out in 50 µl reaction volumes, where each reaction contained 1X PCR buffer, 0.2 U DreamTaq polymerase (Fermentas, USA), 200 µM of each dNTP, 0.5 µM of each primer, 4% bovine serum albumin (BSA), 2% to 6% DMSO to increase PCR specificity (the concentration used was sample-dependent) and between 5 and 20 ng of metagenomic DNA.

16S rRNA gene PCR amplifications were carried out as follows: 5 min at 95°C for denaturation; 25 cycles of 30 s at 94°C, 30 s annealing at 54°C and 105 s at 72°C; and a final elongation step of 7 min at 72°C. Fungal ITS region PCR amplifications were performed as follows: 5 min at 95°C for denaturation; 25 cycles of 1 min at 94°C, 50 s annealing at 55°C and 105 s at 72°C; and a final elongation step of 7 min at 72°C.

To perform T-RFLP analyses, the 341F and ITS1 primers were 5′-end FAM-labelled and the PCR products were purified using the GFX TM PCR DNA and gel band purification kit as directed by the supplier (GE Healthcare, UK). 200 ng of purified PCR products for 16S rRNA gene T-RFLP were digested with the restriction enzymes *Msp*I at 37°C overnight, while 400 ng of purified ITS amplicons was digested overnight with *Hae*III.

Terminal-Restriction Fragment Length Polymorphism (T-RFLP) analyses

T-RF size was determined by capillary electrophoresis using a Applied Biosystems DNA Sequencer 3130 (Applied Biosystems, Foster City, California, USA) and according to the molecular weight standard GeneScan-600LIZ V2 (Applied Biosystem), with an error of ±1 bp. Individual T-RFs were considered as Operational Taxonomic Units (OTUs), with recognition that each OTU may comprise more than one distinct bacterial ribotype. Peak height was used to identify each T-RF and characterize their relative abundance in the total T-RFLP profiles, which was used as a proxy for OTU abundance in the microbial populations.

Statistical analyses

Multivariate analyses of T-RFLP and environmental data were performed using the software Primer 6 (Primer-E Ltd, UK). Valid T-RF peaks (between 30 and 567 bp for 16S rRNA gene T-RFLP or 30 and 800 bp for ITS region T-RFLP) from triplicate T-RFLP profiles were identified, compiled and aligned to produce large data matrices using the online software T-REX (http://trex.biohpc.org/) [28]. The community structures obtained were analyzed by ordination using non-metric multidimensional scaling (nMDS) of Bray-Curtis similarity matrices of square-root transformed data. Two-way crossed analysis of similarity (ANOSIM) tests were used to assess significant differences in the structure of assemblages among the Fairy Circles sampled and between the different sampling 'zones' (centre/margin/control vegetated soils) [29]. Multivariate dispersion (MVDISP) was used to measure 'within-zone' assemblage dispersion [30]. For each FC, one-way ANOSIM was used to test for differences in microbial community assemblages in the three FC-zones.

Prior to principal component analysis (PCA), the environmental variables were analyzed using a Draftsman plot [31] to evaluate the need for transformations, i.e. any skewness in the dataset. Total Carbon (% C) and ammonium (NH_4^+) were initially log(x + 1) transformed, and the complete environmental data set was normalized to perform PCA. A resemblance matrix based on Euclidean distances was created using the normalized set prior to two-way crossed ANOSIM.

Results and Discussion

Plant-microbe interactions actively shape plant diversity by various processes (reviewed in [32]), and soil pathogens have been described as active drivers of vegetation-succession [32]. As Namib Desert FCs are circular vegetation-free patterns surrounded by a healthy vegetation-covered matrix, we have hypothesized that a soil-born plant pathogen could be a causative agent of the phenomenon. This would not be unique: a fungus (*Sclerotinia homoeocarpa*) has been shown to be responsible for a plant disease, known as the Dollar Spot, which is morphologically similar to the FC phenomenon. *S. homoeocarpa* infects turfgrass species and, at a much smaller scale (i.e., the size of a dollar coin), creates FC-like circular necrotic patches that grow with time [33]. We therefore investigated the potential involvement of both edaphic bacterial and fungal communities in the origin and/or maintenance of the Namib Desert gravel plain FCs, as members from both phylogenetic groups are known to exhibit phyto-toxicity (e.g., [34-37]).

The measurement of ten edaphic chemical variables (Table 1), analyzed by PCA, provided no clear evidence of an "FC zonation effect" (Figure S1); i.e., the composite soil chemistries of the samples from the three zones tested (FC center/FC margin/external vegetated control) were randomly scattered on the PCA plot. Two-way crossed ANOSIM confirmed this observation with (marginally significant) global R values <0.2, indicating that the different sampling zones and each FC could not clearly be separated [38]. Van Rooyen and colleagues [2] observed a similar trend with no differences between FC microhabitat soil chemistries in each of the sandy FC sites studied. Contrastingly, clear FC-specific patterns of hydrocarbon-like gas emissions were observed in similar FCs [14] which led to the geochemical origin hypothesis of Namib Desert FCs. Taken together, these results could suggest that gas emission rather than different soil chemistries could play a direct, or indirect, role in FC formation or maintenance.

The nMDS plots showing the ordination of the fungal community structures revealed by T-RFLP suggested a similar trend (Figure 2A); i.e., samples from each FC zones were randomly positioned (2D-stress$>$ 0.2). Contrastingly, the nMDS ordination of the bacterial community T-RFLP fingerprints indicated some discrimination between communities from the FC centre and the control soils (virtually separated by the dashed grey line in Figure 2B), and between communities from different FCs (i.e., FC2 and FC5 communities appeared distinct from those of FC3, FC4 and FC5: virtually separated by the dashed black line in Figure 2B). Two-way crossed ANOSIM (Table 2) showed strongly significant differences in the structure of the bacterial and fungal assemblages between the different FC zones (particularly the control and FC-center bacterial communities: $R = 0.702$; $p = 0.01$). The simplest explanation for these results would be habitat filtration due to the absence of rhizosphere and their associated microbial assemblages within the non-vegetated FC soils (e.g., [39,40]). However, highly significant differences between the microbial assemblages of each individual FC were also detected (ANOSIM $R>0.9$ in 6 of the 10 FC pairwise

Table 1. Characterizations of the Fairy Circles sampled and of their edaphic chemical properties.

Fairy Circle	Dimensions (mxm)	Area (m^2)	Zone	% Carbon	NH_4^+ (mg N/kg)	NO_3^- (mg N/kg)	P (mg/kg)	Fe (mg/kg)	Ca (mg/kg)	K (mg/kg)	Mg (mg/kg)	Na (mg/kg)	pH
FC1	4.6×3.7	13.75	Control	0.52 (±0.91)	9.19 (±2.16)	8.47 (±2.54)	13.24 (±1.75)	17.52 (±4.84)	162.95 (±50.98)	17.11 (±0.62)	5.81 (±0.49)	0.95 (±0.39)	8.16 (±0.13)
			Margin	0.88 (±1.54)	13.66 (±4.82)	9.88 (±3.73)	16.79 (±1.26)	14.91 (±2.47)	177.54 (±50.2)	18.42 (±2.53)	6.08 (±0.31)	1.39 (±0.46)	8.07 (±0.08)
			Centre	1.02 (±2.21)	10.3 (±1.71)	10.54 (±4.24)	16.01 (±3.1)	15.72 (±1.32)	138.63 (±11.83)	14.24 (±3.71)	4.85 (±0.46)	2.34 (±1.34)	8.03 (±0.18)
FC2	3.6×3.2	9.05	Control	ND	33.51 (±44.33)	12.86 (±1.38)	13.00 (±1.80)	16.89 (±0.68)	132.54 (±18.56)	14.21 (±2.86)	5.17 (±0.5)	0.83 (±0.57)	8.13 (±0.1)
			Margin	0.64 (±0.79)	18.04 (±10.19)	12.04 (±2.45)	11.90 (±2.97)	13.11 (±6.76)	263.06 (±118.21)	16.72 (±1.78)	6.04 (±0.61)	2.24 (±1.65)	8.15 (±0.09)
			Centre	0.11 (±0.17)	12.29 (±2.19)	12.06 (±2.10)	12.72 (±0.66)	11.35 (±3.97)	193.39 (±42.68)	14.23 (±1.16)	5.68 (±0.27)	1.13 (±0.68)	8.09 (±0.13)
FC3	5×3.7	14.53	Control	0.01 (±0.03)	12.91 (±0.6)	13.76 (±1.80)	13.46 (±1.34)	18.12 (±3.49)	211.71 (±115.24)	15.75 (±1.95)	5.89 (±1.04)	1.43 (±1.70)	8.05 (±0.13)
			Margin	1.84 (±3.67)	9.8 (±2.85)	12.23 (±3.00)	14.62 (±1.56)	17.76 (±3.89)	161.14 (±39.01)	14.13 (±2.26)	5.32 (±0.38)	0.91 (±0.37)	8.19 (±0.05)
			Centre	0.18 (±0.38)	8.91 (±3.04)	9.00 (±1.97)	14.04 (±1.44)	17.53 (±2.05)	158.67 (±20.20)	14.04 (±1.52)	5.38 (±0.14)	0.74 (±0.07)	8.05 (±0.09)
FC4	3.6×4.2	14.7	Control	0.10 (±0.20)	8.50 (±1.30)	6.50 (±1.00)	13.91 (±1.87)	19.59 (±4.38)	131.86 (±29.77)	14.26 (±1.34)	5.69 (±0.25)	0.53 (±0.20)	8.22 (±0.07)
			Margin	0.03 (±0.06)	7.06 (±0.76)	6.63 (±1.22)	14.61 (±1.62)	17.85 (±2.25)	165.94 (±41.3)	13.49 (±1.49)	5.06 (±0.23)	0.63 (±0.18)	8.06 (0.07)
			Centre	ND	8.63 (±2.34)	8.91 (±3.30)	13.01 (±1.1)	15.12 (±3.72)	202.83 (±102.74)	14.02 (±1.46)	5.70 (±0.33)	0.76 (±0.25)	8.09 (±0.09)
FC5	3.8×3.1	9.25	Control	ND	6.16 (±0.87)	6.79 (±0.23)	14.79 (±1.87)	14.2 (±4.62)	191.01 (±80.12)	18.43 (±3.40)	5.43 (±0.17)	1.37 (±0.33)	8.11 (±0.09)
			Margin	ND	11.6 (±6.54)	9.08 (±0.66)	13.65 (±1.33)	17.01 (±6.26)	179.58 (±54.02)	14.34 (±0.70)	5.95 (±0.25)	0.72 (±0.09)	8.13 (±0.14)
			Centre	ND	7.03 (±0.88)	8.58 (±2.30)	12.26 (±2.18)	15.34 (±5.73)	243.54 (±91.70)	14.04 (±0.84)	6.26 (±0.38)	0.88 (±0.15)	8.11 (±0.15)

Values are given as mean ± SD. ND: Not determined as below detection limit.

comparisons, and p consistently <0.05; Table 2). Following this observation, the bacterial and fungal community fingerprint of each FC was analyzed separately by ANOSIM, independently testing for differences in their respective FC zones (Table 3). Excluding the fungal community from FC 2, these analyses showed that both the control and FC-centre communities (pairwise comparisons) and/or the communities from the three zones (global test) were significantly different ($p < 0.05$). These results suggest that each independent FC houses a significantly distinct surface edaphic microbial community which is significantly different from that of the vegetation-covered control soils, and strongly suggest that the absence of rhizosphere/rhizospheric communities cannot satisfactorily explain the structure differences in edaphic microbial communities observed between FCs and external soils.

Recent satellite data have demonstrated that the Namib Desert Fairy Circles could be considered as 'living organisms', as they appeared (birth), grew and disappeared (death) [3]. It was concluded that any study attempting to unravel the origins of FCs should take into account the fact that they are not static environments but vary over time [3,16]. The underlying corollary from this observation is that individual FCs within a site in the Namib Desert (e.g. such as those in our study which were all sampled within one hectare area) might be at different stages in their 'lifespan' and potentially constitute different biotopes (e.g., re-vegetated 'ghost', newly forming or matured/ing FCs; [3]). If we assume that the surface soil microbial communities are not static

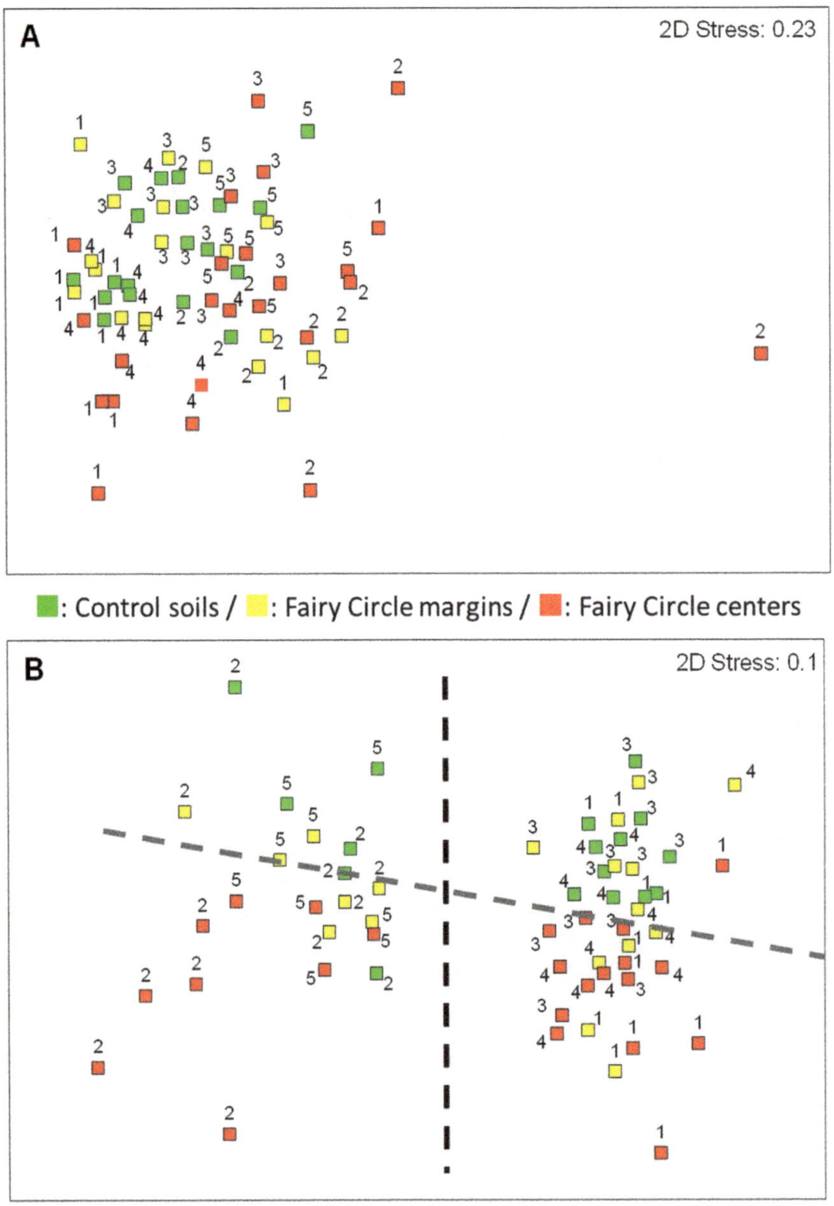

Figure 2. Two dimension nonmetric multidimensiobal scaling (2D-nMDS) plot of Bray–Curtis similarity of fungal (A) and bacterial (B) community structures based on ITS region and 16S rRNA gene square-root transformed T-RFLP profiles respectively. Numbers refer to the respective Fairy Circles. The dashed lines indicate virtual separations between group of samples (Grey: control vs FC centre communities/ Black: FC2 and FC5 vs FC1, FC3 and FC4).

Table 2. Results of two-way crossed ANOSIM tests based on Bray-Curtis similarity matrices from square-root transformed bacterial and fungal T-RFLP profiles.

	Differences among Fairy Circle zones			
	Bacterial communities		**Fungal communities**	
	R	**p**	**R**	**p**
Global Test	0.418	0.001 *	0.26	0.001 *
Control vs Margin	0.103	0.1	0.26	0.001 *
Control vs Centre	0.702	0.001 *	0.329	0.001 *
Margin vs Centre	0.401	0.001 *	0.234	0.007 *
	Differences among Fairy Circles			
	Bacterial communities		**Fungal communities**	
	R	p	R	p
Global Test	0.688	0.001 *	0.477	0.001 *
FC 1 vs FC 2	0.994	0.001 *	0.461	0.002 *
FC 1 vs FC 3	0.466	0.001 *	0.599	0.001 *
FC 1 vs FC 4	0.237	0.004 *	0.234	0.023 *
FC 1 vs FC 5	0.993	0.001 *	0.581	0.001 *
FC 2 vs FC 3	0.969	0.001 *	0.429	0.002 *
FC 2 vs FC 4	0.958	0.001 *	0.522	0.001 *
FC 2 vs FC 5	0.214	0.054	0.274	0.004 *
FC 3 vs FC 4	0.454	0.001 *	0.593	0.001 *
FC 3 vs FC 5	1	0.001 *	0.686	0.001 *
FC 4 vs FC 5	1	0.001 *	0.736	0.001 *

R: ANOSIM statistic; p: probability level. *: Significantly different ($p < 0.05$).

throughout the lifespan of a FC (i.e., change in parallel with the evolution of the FC biotope; [39,41]), we are provided with a possible explanation for the differences observed between individual FC microbial communities (Table 2).

Distinct 'drivers' or environmental disturbances could also lead to alternative FC edaphic communities. For example, fluctuating hydrocarbon-like gas seepages [14], the action of different phytopathogenic microorganisms/pathovars [34] or dissimilar toxin(s) infection stages [37] could all lead to different FC biotopes within a single site. The microbial-mediated hypotheses are supported by the fact that two fungal T-RFs (OTUs 246 and 683; data not shown) were consistently observed within the margin and/or center of all five FCs studied, but never in the external control samples and could therefore constitute molecular signatures of "FC-related" fungi [33].

To test whether FC microbial communities were more heterogeneous than those of the vegetated control soils, multivariate dispersion indices were calculated (MVDISP routine in Primer6; [30]). The MVDISP index was lowest for the control communities and highest for the FC-centre communities, for both the bacterial (MVDISP Control 0.854 <MVDISP FC Margin 0.987 <MVDISP FC Centre 1.08) and fungal (MVDISP Control 0.659 <MVDISP FC Margin 0.92 <MVDISP FC Centre 1.261) communities. As a higher MVDISP index is indicative of higher multivariate dispersal, this clearly demonstrates that the vegetated (control) soil microbial communities exhibited lower intra-variability (or higher homogeneity) than the FC communities; the FC centre communities being the more variable, i.e. the more heterogeneous. These intra-variability differences could be explained by the adaptation of the 'intra-FC' (margin and centre) microbial communities to environmental disturbance leading to their stochastic assembly [41].

Based on the observation that FCs appear to evolve over time [3], we suggest that the control vegetation-covered soils constitute a primary environmental state (or precursor), the FC margins represent a transitional phase, and the centre non-vegetated soils an alternative environment. By applying the three phase community assembly model after disturbance defined by Ferrenberg and colleagues [41] and based on the principles of microbial community dispersal, we propose two models to potentially explain the assembly of Namib Desert gravel plain Fairy Circle microbial communities (Figure 3).

We suggest that an environmental disturbances (e.g., the effect of active microbial phyto-pathogens, (dis)continuous gas leakage or plant spatial self-organization as a response to resource scarcity; [14,33,42]) could lead to plant necrosis which would disrupt the homogenous precursor (control) edaphic communities. Subsequently, FC-margin communities would further develop by stochastic assembly processes (phase 1; Models A and B; Figure 3). Where the disturbance is not continuous in time, the microbial communities of the three FC zones would develop based on niche-partitioning; i.e., vegetation-covered *versus* vegetation-free soils with the FC-margin sites constituting a transitional environment (phase 2, model A). If the disturbance is continuous or multiple sequential disturbances occur over the lifespan of the FC, the FC microbial communities (margin and centre) would initially be determined by stochastic processes. Subsequently, a combination of niche-partitioning (margins constituting a buffered environment separating the well-defined vegetated and not impacted (control) environment and the plant-free FC centers) and stochastic processes (continuous disturbance) would be responsible for

Table 3. One-way ANOSIM statistics comparing the bacterial and fungal community structures the predefined zones of each FC studied.

		Bacterial communities		Fungal communities	
Fairy Circle	**ANOSIM**	**R**	**p**	**R**	**p**
FC1	Global test	0.245	0.057	0.247	0.014 *
	Control vs Centre	0.446	0.036 *	0.281	0.048 *
	Margin vs Centre	0.15	0.198	0.244	0.095
	Control vs Margin	0.185	0.229	0.365	0.029 *
FC2	Global test	0.463	0.001 *	0.115	0.104
	Control vs Centre	0.763	0.008 *	0.163	0.103
	Margin vs Centre	0.663	0.008 *	(-)0.094	0.881
	Control vs Margin	(-)0.083	0.771	0.375	0.086
FC3	Global test	0.511	0.002 *	0.364	0.007 *
	Control vs Centre	0.763	0.008 *	0.400	0.048 *
	Margin vs Centre	0.675	0.008 *	0.488	0.024 *
	Control vs Margin	(-)0.104	0.771	0.188	0.143
FC4	Global test	0.463	0.001 *	0.245	0.04 *
	Control vs Centre	0.788	0.008 *	0.313	0.071
	Margin vs Centre	0.244	0.063	0.206	0.103
	Control vs Margin	0.365	0.029 *	0.25	0.2
FC5	Global test	0.408	0.025 *	0.423	0.015 *
	Control vs Centre	0.714	0.067	0.679	0.133
	Margin vs Centre	0.167	0.2	0.407	0.086
	Control vs Margin	0.333	0.2	0	0.6

R: ANOSIM statistic; p: probability level. *: Significantly different ($p < 0.05$).

microbial community assemblies (phase 2, model B). Finally, as FCs die, i.e. disturbance has ceased, a neutral assembly processes would occur, leading to less variable edaphic microbial communities in the newly vegetation-covered soils (phase 3, models A and B). As the structures of the FC-center communities studied presented higher intra-variability than those of the margins and

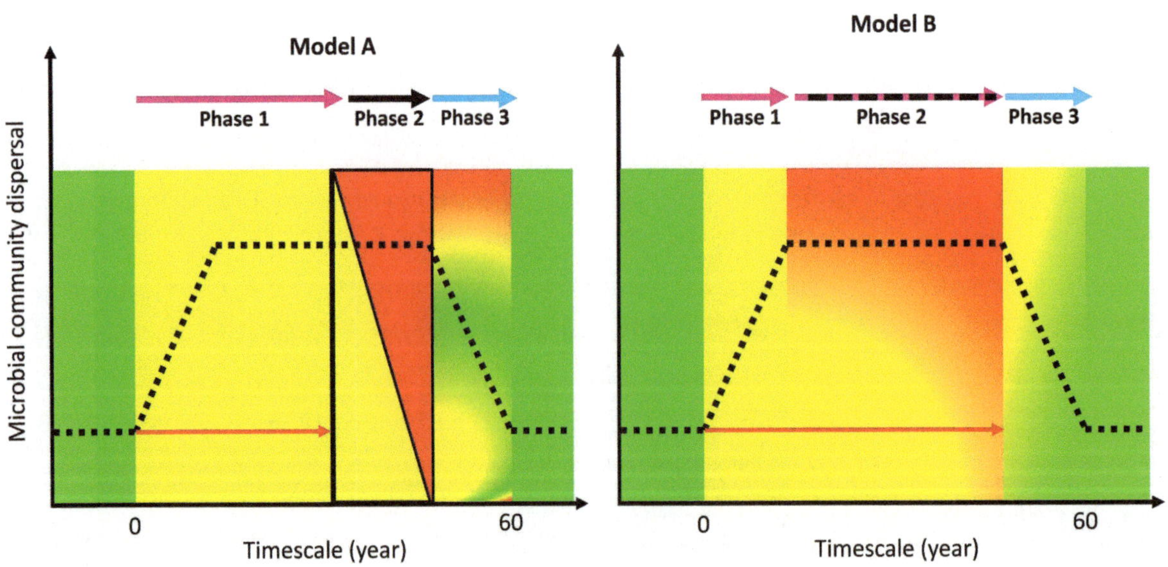

Figure 3. Models hypothesizing microbial community assembly in Namib Desert Fairy Circles. Arrow indicate the assembly processes (purple: stochastic/black: niche-partitioning/blue: neutral). Colors are represent virtually the FC zones (Green: vegetated covered control soils/Yellow: FC margins/Red: FC Centers) where these processes occur. Red arrows indicate the origin in time and length of the environmental disturbance responsible for FC appearance. The x-axis does not reflect proportionally the time scale.

the more homogenous control soils, either the niche-partitioning is still occurring (model A) or FC-centers constitute ever-changing environments (model B).

Conclusion/Perspectives

In ecology, patterns of diversity have been shown to be influenced by four major mechanisms/processes (selection, drift, speciation and dispersal) and by various species interactions (predation, competition and mutualism) [43]. Altogether, our results suggest that FC niche-specific surface edaphic microbial communities are 'selected'. Such selection could originate from (i) habitat/niche-filtration (e.g., rhizospheric communities vs open soil communities; [39]) or (ii) be related to their adaptation to (an) environmental disturbance(s) [41], such as the presence of an active microbial phyto-pathogen [33], a continuous hydrocarbon gas emission [14,20] or vegetation spatial self-organization [15,16,42].

We are yet unable to conclude whether bacteria and/or fungi are actually responsible for the appearance and/or maintenance of Fairy Circles in the Namib Desert. We suggest that more extensive metacommunity/biogeography studies on groups of FCs at different developmental stages [3,44,45] using meta'omic' approaches [46,47] could assist in characterizing the functional basis of FC ecosystems and in potentially identifying the agent(s) responsible for their formation. Such studies should be extended to deep FC soil horizons as it has been shown that (i) sub-surface FC soils are more anoxic than the surrounding vegetation-covered soils [14], (ii) that the abundance of culturable anaerobic bacteria is higher in FC soils [48] and (iii) that sub-surface FC soils accumulate water [11]. Moreover, to fully address the origin and functioning of Namib FC ecosystems, a comparative chemical (soil and gas) and (micro)biological analysis of 'rocky' and 'sandy' FC formations is recommended.

Supporting Information

Figure S1 Principal component analysis (PCA) plot of the normalized environmental variables measured. Numbers refer to the respective Fairy Circles.

Table S1 Results of two-way crossed ANOSIM tests based on Euclidean distance matrices from normalized soil chemistry measurements. R: ANOSIM statistic; p: probability level. *: Significantly different ($p < 0.05$).

Acknowledgments

The authors gratefully acknowledge the National Research Foundation (NRF, South Africa) and the University of Pretoria for supporting this research.

Author Contributions

Conceived and designed the experiments: JBR DC. Performed the experiments: AP AM JBR. Analyzed the data: JBR. Contributed reagents/materials/analysis tools: DC. Wrote the paper: JBR MS DC. Field knowledge: MS.

References

1. Cox GW (1987) The origin of vegetation circles on stony soils of the Namib Desert near Gobabeb, South West Africa/Namibia. J Arid Environ 13: 237–244.
2. Van Rooyen MW, Theron GK, Van Rooyen N, Jankowitz WJ, Matthews WS (2004) Mysterious circles in the Namib Desert: review of hypotheses on their origin. J Arid Environ 57: 467–485.
3. Tschinkel WR (2012) The life cycle and life span of Namibian fairy circles. PloS one. 7: e38056.
4. Fraley L (1987) Response of short grass plains vegetation to gamma radiation. III. Nine years of chronic irradiation. Environ Exp Bot 27: 193–201.
5. Tinley KL (1971) Etosha and the Kaokoveld. African Wild Life. 25: 1–16.
6. Moll EJ (1994) The origin and distribution of fairy rings in Namibia. In: Proceedings of the 13th Plenary Meeting AETFAT, Malawi. 2: 1203–1209.
7. Becker T, Getzin S (2000) The fairy circles of Kaokoland (North-West Namibia) origin, distribution, and characteristics. Basic Appl Ecol 1: 149–159.
8. Albrecht CF, Joubert JJ, de Rycke PH (2001) Origin of the enigmatic, circular, barren patches ('Fairy Rings') of the pro-Namib. S Afr J Sci 97: 23–27.
9. Becker T (2007) The phenomenon of fairy circles in Kaokoland (NW Namibia). Basic Appl Dryland Res 1: 121–137.
10. Picker MD, Ross-Gillespie VERE, Vlieghe K, Moll E (2012) Ants and the enigmatic Namibian fairy circles–cause and effect? Ecol Entomol 37: 33–42.
11. Juergens N (2013) The biological underpinnings of Namib Desert fairy circles. Science 339: 1618–1621.
12. Theron GK (1979) Die verskynsel van kaal kolle in Kaokoland, SuidWes-Afrika. J S Afr Biology Soc 20: 43–53.
13. Jankowitz WJ, Van Rooyen MW, Shaw D, Kaumba JS, Rooyen NV (2008) Mysterious circles in the Namib Desert. S Afr J Bot 74: 332–334.
14. Naudé Y, Van Rooyen MW, Rohwer ER (2011) Evidence for a geochemical origin of the mysterious circles in the Pro-Namib desert. J Arid Environ 75: 446–456.
15. Cramer MD, Barger NN (2013) Are Namibian "Fairy Circles" the Consequence of Self-Organizing Spatial Vegetation Patterning? PloS ONE. 8: e70876.
16. Getzin S, Wiegand K, Wiegand T, Yizhaq H, von Hardenberg J, et al. (2014) Adopting a spatially explicit perspective to study the mysterious fairy circles of Namibia. Ecography 37: 001–011.
17. Wieder WR, Bonan GB, Allison SD (2013) Global soil carbon projections are improved by modelling microbial processes. Nat Clim Change 3: 909–912.
18. Graham EB, Wieder WR, Leff JW, Weintraub SR, Townsend AR, et al. (2014) Do we need to understand microbial communities to predict ecosystem function? A comparison of statistical models of nitrogen cycling processes. Soil Biol Biochem 68:279–282.
19. Pointing SB, Belnap J (2012) Microbial colonization and controls in dryland systems. Nat Rev Microbiol 10: 551–562.
20. Grossi V, Cravo-Laureau C, Guyoneaud R, Ranchou-Peyruse A, Hirschler-Réa A (2008) Metabolism of n-alkanes and n-alkenes by anaerobic bacteria: A summary. Org Geochem 39: 1197–1203.
21. SSSA (1996) Methods of Soil Analysis, Part 3. Soil Science Society of America, Madison.
22. Walkley A (1935) An Examination of Methods for Determining Organic Carbon and Nitrogen in Soils. J Agr Sci 25: 598–609.
23. Bremmer JM, Keeney DR (1966) Determination and isotope-ratio analysis of different forms of nitrogen in soils: 3. Exchangeable ammonium, nitrate and nitrite by extraction-distillation methods. Soil Sci Soc Am J 30: 577–582.
24. Keeney DR, Nelson DW (1982) Nitrogen-inorganic forms. In: Page AL, Miller RH, Keeney DR (eds) Methods of soil analysis, part 2. Madison, Wise, Agronomy 9: 643–698.
25. Bray RH, Kurtz LT (1945) Determination of total, organic and available forms of phosphorous in soils. Soil Sci 59: 39–45.
26. Lane DJ, Pace B, Olsen GJ, Stahl DA, Sogin ML, et al. (1985) Rapid determination of 16S ribosomal RNA sequences for phylogenetic analyses. Proc. Natl. Acad. Sci. 82: 6955–6959.
27. White TJ, Bruns T, Lee S, Taylor J (1990) Amplification and direct sequencing of fungal ribosomal RNA genes for phylogenetics. M. Innis, D. Gelfand, J. Sninsky, T. White (Eds.), PCR Protocols: A Guide to Methods and Applications, San Diego, Academic Press, pp. 315–322
28. Culman SW, Bukowski R, Gauch HG, Cadillo-Quiroz H, Buckley DH (2009) T-REX: software for the processing and analysis of T-RFLP data. BMC Bioinformatics 10: 171.
29. Clarke K (1993) Non-parametric multivariate analysis of changes in community structure. Aust J Ecol 18: 117–143.
30. Clarke KR, Warwick RM (2001) A further biodiversity index applicable to species lists: variation in taxonomic distinctness. Mar Ecol-Prog Ser 216: 265–278.
31. Clarke KR, Warwick RM (1994) Change in Marine Communities: An Approach to Statistical Analysis and Interpretation. Plymouth, UK, Plymouth Marine Laboratory, pp. 144.
32. Reynolds HL, Packer A, Bever JD, Clay K (2003) Grassroots ecology: plant-microbe-soil interactions as drivers of plant community structure and dynamics. Ecology 84: 2281–2291.
33. Walsh B, Ikeda SS, Boland GJ (1999) Biology and management of dollar spot (*Sclerotinia homoeocarpa*); an important disease of turfgrass. HortScience. 34: 13–21.
34. Arrebola E, Cazorla FM, Perez-García A, Vicente AD (2011) Chemical and metabolic aspects of antimetabolite toxins produced by *Pseudomonas syringae* pathovars. Toxins. 3: 1089–1110.

35. Dean R, Van Kan JA, Pretorius ZA, Hammond-Kosack KE, Di Pietro A, et al. (2012) The Top 10 fungal pathogens in molecular plant pathology. Mol Plant Pathol 13: 414–430.
36. Hoagland RE (2001) The genus *Streptomyces*: A rich source of novel phytotoxins. In: Ecology of Desert Environments I Parkash (Ed), 139–169 Scientific Publishers, Jodhpur, India.
37. Horbach R, Navarro-Quesada AR, Knogge W, Deising HB (2011) When and how to kill a plant cell: infection strategies of plant pathogenic fungi. J Plant Physiol 168: 51–62.
38. Clarke KR, Gorley RN (2006) PRIMER v6: User Manual. Plymouth Marine Laboratory: Plymouth, UK.
39. Dumbrell AJ, Nelson M, Helgason T, Dytham C, Fitter AH (2009) Relative roles of niche and neutral processes in structuring a soil microbial community. ISME J 4: 337–345.
40. Ramond J-B, Tshabuse F, Bopda CW, Cowan DA, Tuffin MI (2013) Evidence of variability in the structure and recruitment of rhizospheric and endophytic bacterial communities associated with arable sweet sorghum (*Sorghum bicolor* (L) Moench). Plant Soil. 372: 265–278.
41. Ferrenberg S, O'Neill SP, Knelman JE, Todd B, Duggan S, et al. (2013) Changes in assembly processes in soil bacterial communities following a wildfire disturbance. ISME J 7: 1102–1111.
42. Lejeune O, Tlidi M, Couteron P (2002) Localized vegetation patches: a self-organized response to resource scarcity. Phys Rev E 66: 010901.
43. Vellend M (2010) Conceptual synthesis in community ecology. Q Rev Biol 85: 183–206.
44. Hovatter SR, Dejelo C, Case AL, Blackwood CB (2011) Metacommunity organization of soil microorganisms depends on habitat defined by presence of *Lobelia siphilitica* plants. Ecology 92: 57–65.
45. Martiny JBH, Bohannan BJ, Brown JH, Colwell RK, Fuhrman JA, et al. (2006) Microbial biogeography: putting microorganisms on the map. Nat Rev Microbiol 4: 102–112.
46. Jansson JK, Prosser JI (2013) Microbiology: The life beneath our feet. Nature 494: 40–41.
47. Gunnigle E, Ramond JB, Frossard A, Seeley M, Cowan D (2014) A sequential co-extraction method for DNA, RNA and protein recovery from soil for future system-based approaches. J Microbiol Meth 103: 118–123.
48. Eicker A, Theron GK, Grobbelaar N (1982) 'n Mikrobiologiese studie van "kaal kolle" in die Giribesvlakte van Kaokoland, S.W.A.-Namibia. S Afr J Bot 1: 69–74.

14

Nutrient Limitation in Three Lowland Tropical Forests in Southern China Receiving High Nitrogen Deposition: Insights from Fine Root Responses to Nutrient Additions

Feifei Zhu[1,2,5], Muneoki Yoh[3], Frank S. Gilliam[4], Xiankai Lu[1], Jiangming Mo[1]*

1 Key Laboratory of Vegetation Restoration and Management of Degraded Ecosystems, South China Botanical Garden, Chinese Academy of Sciences, Guangzhou, China, **2** United Graduate School of Agriculture Science, Tokyo University of Agriculture and Technology, Tokyo, Japan, **3** Institute of Agriculture, Tokyo University of Agriculture and Technology, Tokyo, Japan, **4** Department of Biological Sciences, Marshall University, Huntington, West Virginia, United States of America, **5** University of Chinese Academy of Sciences, Beijing, China

Abstract

Elevated nitrogen (N) deposition to tropical forests may accelerate ecosystem phosphorus (P) limitation. This study examined responses of fine root biomass, nutrient concentrations, and acid phosphatase activity (APA) of bulk soil to five years of N and P additions in one old-growth and two younger lowland tropical forests in southern China. The old-growth forest had higher N capital than the two younger forests from long-term N accumulation. From February 2007 to July 2012, four experimental treatments were established at the following levels: Control, N-addition (150 kg N ha^{-1} yr^{-1}), P-addition (150 kg P ha^{-1} yr^{-1}) and N+P-addition (150 kg N ha^{-1} yr^{-1} plus 150 kg P ha^{-1} yr^{-1}). We hypothesized that fine root growth in the N-rich old-growth forest would be limited by P availability, and in the two younger forests would primarily respond to N additions due to large plant N demand. Results showed that five years of N addition significantly decreased live fine root biomass only in the old-growth forest (by 31%), but significantly elevated dead fine root biomass in all the three forests (by 64% to 101%), causing decreased live fine root proportion in the old-growth and the pine forests. P addition significantly increased live fine root biomass in all three forests (by 20% to 76%). The combined N and P treatment significantly increased live fine root biomass in the two younger forests but not in the old-growth forest. These results suggest that fine root growth in all three study forests appeared to be P-limited. This was further confirmed by current status of fine root N:P ratios, APA in bulk soil, and their responses to N and P treatments. Moreover, N addition significantly increased APA only in the old-growth forest, consistent with the conclusion that the old-growth forest was more P-limited than the younger forests.

Editor: Morag Mcdonald, Bangor University, United Kingdom

Funding: This study was funded by National Science Foundation of China(NO.41273143) and National Key Basic Research 973 Program (2010CB833502). The funders had no role in study design, data collection and analysis, decision to publish, or preparation of the manuscript.

Competing Interests: The authors have declared that no competing interests exist.

* E-mail: mojm@scib.ac.cn

Introduction

Many tropical/subtropical regions, eastern Asia in particular, are receiving high levels of atmospheric nitrogen (N) deposition, with such increases projected to continue in the next decades [1,2]. This will add more new N to some lowland tropical forests and may alter soil nutrient availability and carbon (C) balance of these ecosystems. In general, humid tropical forests store approximately 10% of global soil C. Thus, it is crucial to know the current nutrient status of these ecologically important forests and how their structure and function may be altered by increased N deposition [3].

Plant growth in lowland tropical forest is generally considered to be limited by phosphorus (P), rather than N [4,5,6]. Further N inputs from atmospheric deposition are likely to alter soil nutrient conditions negatively, including increases in soil acidification, losses of base cations, and Al mobilization [7,8,9,10], which consequently may lower soil P availability [7,11]. These predictions have been observed in some temperate forests where locally-elevated N deposition contributes to a transition in nutrient limitation to forest growth, i.e., driving forests from N limitation to N saturation and P limitation [12,13,14]. However, such effects of N deposition on ecosystem nutrient limitation have rarely been investigated in tropical forests [15].

Responses of fine root growth to N or P limitation are inherently different, due to the contrasting mobility of nitrate (NO_3) and phosphate (PO_4) in soils [16]. NO_3 is very mobile in the soil and approaches the root via mass flow. Accordingly, new root production may not be required for plants to absorb additional N. Nitrogen limitation ceases when N availability increases beyond N demand. Subsequently, less fine root biomass is needed, according to the allocation theory proposed by Bloom et al. 1985 [17]. Decreases in fine root biomass are frequently observed as N availability increases further [18,19]. In contrast, PO_4 is fairly immobile due to adsorption by organic matter and Al- and Fe-oxides in highly weathered soils. Roots encounter P only by growing through soil rather than having P delivered via diffusion or water flow. As a result, fine roots growing in P-limited soils will theoretically respond to additional P by more growth [20]. Thus, less fine root biomass following N additions is generally expected, whereas more fine root biomass following P additions can imply P limitation.

As indirect proxies for nutrient limitation, N and P concentrations in fine roots have received less attention than those of leaves [21]. It is suggested that leaves have a stronger homeostatic control over N:P ratios than roots [22]. Thus, nutrient concentrations and N:P ratios of fine roots should respond more sensitively to changes in soil nutrient availability. In addition to fine root responses, acid phosphatase, the dominant form of extracellular phosphatase in acid organic-rich soils, has also been suggested to be a useful tool to reflect soil N and P availability. Allocation to enzyme production can increase the availability of nutrient to uptake [23]. In the case of P, its availability in bulk soil depends on acid phosphatase activity (APA), in which plants and microbes produce acid phosphatase to hydrolyze organic P sources. APA has been shown to be tightly coupled with soil P availability, with an increase in APA typically indicating increased P limitation [24,25].

To investigate the effects of N deposition and the potential role of P in mitigating N effects on forest ecosystem processes and functioning, a field experiment employing additions of N and P in a factorial design was initiated in 2007 in three lowland tropical forests in the Dinghushan Biosphere Reserve (DHSBR) in Southern China [26,27,28]. To our knowledge, there have been but four fertilization experiments in lowland tropical forests [29,30,31]. Our sites have lower available soil P (1.7 vs. 5 mg kg^{-1}) and more acidic soil (pH 3.7 vs. 5.2) compared to that of a site in Panama [30], but similar to a Costa Rican forest [31]. The three forests in our study (one old-growth, and two younger) have contrasting soil N status, with the old-growth forest having larger N capital than the two younger forests [32,33], allowing us to determine whether and to what extent soil N status influences ecosystem response to further N deposition and the role of P to mitigate effects of added N.

In the present study, we report responses of fine root biomass, the proportion of live fine roots, fine root nutrient concentrations, and APA in bulk soil to N and/or P treatments under the background of chronically-elevated atmospheric N deposition. Previous studies in the DHSBR from a long-term N addition experiment demonstrated a net loss of 8–16 kg N ha^{-1} yr^{-1} from the soil in the old-growth forest. In total, up to 60 kg N ha^{-1} yr^{-1} was leached from the old-growth forest, indicating that this forest was N saturated from chronic N deposition and long-term N accumulation [34]. Further N addition has been shown to increase soil acidification, base cation losses, and Al mobilization in this forest [9,10]. In contrast, the two younger forests exhibit signs of N limitation, example.g., increased litter decomposition [32], litterfall production [35] and microbial activity [36] following N additions, although N leaching has also been observed [33]. We hypothesize that the old-growth forests have been facing strong P limitation due to long-term N accumulation, whereas the two younger forests are still N limited due to large plant N demand. If so, based on the allocation theory [17] and empirical studies, we expect that (1) N addition would decrease fine root biomass in all the three forests, through N-mediated soil chemical changes in the old-growth forest and through release from N limitation in the two younger forests, (2) P addition would increase fine root biomass in the old-growth forest while having no effects in the two younger forests.

Materials and Methods

Ethics Statement

No specific permits were required for the described field studies. This research station (Dinghushan Biosphere Reserve) belongs to South China Botanical Garden, Chinese Academy of Sciences. This study was also supported by this institute. Data will be made available upon request. We confirm that the location is not privately-owned or protected in any way. We also confirm that the field studies did not involve endangered or protected species.

Study sites

The study was conducted in Dinghushan Biosphere Reserve (DHSBR) (112°33′E and 23°10′N) in the central section of Guangdong Province, southern China. The climate is warm and humid, with annual precipitation of 1927 mm and mean annual temperature of 21.0°C [37]. Soils are oxisols formed from sandstone [38]. In this area, high atmospheric N deposition has been on-going since 1990′s. Nitrogen deposition was 36 kg N ha^{-1} $year^{-1}$ in 1990 and reached to 38 kg N ha^{-1} $year^{-1}$ in 1999 [39,40]. In 2004 and 2005, N deposition in rainfall measured was 34 and 32 kg N ha^{-1} $year^{-1}$, respectively, 60% of which was in the form of NH_4^+-N [34].

Three forest types within the reserve (~2–4 km apart) were used in this study: an old-growth (>400-yr-old) forest, considered the regional climax type, a younger pine forest, and a similarly younger mixed pine/broadleaf forest (~75-yr-old; Table 1). All forest types are of similar elevation range (50–250 m above mean sea level), slope aspect and degree [33]. The notable age of the old-growth forest arises from long-term protection by monks from any form of direct human disturbance. The two younger forests originated from clear-cut harvesting and subsequent planting of *Pinus massoniana* in the 1930s, during which time the sites became badly eroded and degraded [10,32]. Sharply contrasting intensity and frequency of litter and vegetation harvesting during 1930–1998 resulted in contrasting tree species composition. *Pinus massoniana* (*P. massoniana*) dominates the pine forest, whereas several broadleaf species co-dominate with *P. massoniana* the mixed forest. The three forest types differ greatly in N status, with greater N accumulation in the mineral soil and higher N leaching rates in the old-growth forest (Table 1).

Experimental design

A full 2×2 factorial was established in each forest type in 2007, with two levels (with and without addition) of each of two nutrients (N and P). Specifically, four treatments (each with 5 replicates), including control, N-addition (150 kg N ha^{-1} yr^{-1}), P-addition (150 kg P ha^{-1} yr^{-1}) and N+P-addition (150 kg N ha^{-1} yr^{-1} plus 150 kg P ha^{-1} yr^{-1}) were set up in each forest. There were a total of 20 plots of 5 m×5 m in each forest and each plot was surrounded by a 5-m-wide buffer strip. Field plots were laid out randomly and were randomly selected to receive specific treatments. Plot size and fertilizer levels were chosen to resemble Cleveland and Townsend (2006) [41], who studied a tropical forest of Costa Rica. NH_4NO_3 or/and NaH_2PO_4 solutions were sprayed once every other month to the forest floor with a backpack sprayer starting from February 2007 and continued through July 2012. We considered our plots big enough to study soil processes and fine root dynamics, considering that root ingrowth core method is considered suitable to determine nutrient limitation [12,14,20].

Fine root and soil sampling

We sampled fine roots from all plots in the three forests in July 2012 (about five years after the initiation of N and/or P treatments). In each plot, three 5-cm diameter cores of mineral soil were extracted randomly to a 10 cm in depth, then combined to yield one composite sample. From these samples, live and dead

Table 1. General characteristics of the three study forests.

	Old-growth	Pine	Mixed	References
Aboveground				
Age (year)	>400	~75	~75	[57]
Stem density (stems ha^{-1}, DBH>2cm)	1013	767	1933	
Basal area (m^2 ha^{-1})	26	14.0	13.8	
Litter production	7.1	6.0	6.1	[31]
(Mg ha^{-1} $year^{-1}$) in 2009				
Mineral soil (0–10 cm)				
Bulk density (g cm^{-3})	1.0	1.2	1.2	[34]
Organic matter (%)	7.3	5.2	3.7	
Total N concentration (%)	0.2	0.12	0.1	
C:N	21	28	24	
Total P concentration (%)	0.05	0.04	0.04	
Net N mineralization	6.7	8.2	3.9	
(mg kg^{-1} $month^{-1}$ of N)				
Net nitrification (mg kg^{-1} $month^{-1}$ of N)	6.1	7.8	1.1	
Inorganic N leaching in 2006	41.4	20.3	8.9	[35]
(kg ha^{-1} $year^{-1}$ of N, 20 cm below the organic layer)				

fine roots were distinguished by root resilience, brittleness, and color [42]. Roots were sorted, dried at 65°C, and weighed. Live fine root biomass, dead fine root biomass, and live fine root (biomass) proportion were determined. In July 2011, 0–10 cm mineral soil was sampled for analysis of soil chemical properties and APA. The soils were sieved to pass through 2 mm screen prior to chemical analysis.

Chemical analysis

Acid phosphatase activity (APA) in one gram of fresh, sieved soil was analyzed according to the modified procedure by Scheneider et al. (2001) [43]. Concentration of NH_4^+-N was analyzed by the indophenol blue method followed by colorimetry, and NO_3^--N was analyzed after cadmium reduction to NO_2^-, followed by sulfanilamide-NAD reaction [44]. Soil available P was extracted by acid-ammonium fluoride solution (0.025 mol L^{-1} HCL +0.03 mol L^{-1} NH_4F) [45]. Soil pH was measured with soil: water ratio of 1:2.5 [44]. Oven-dried live fine root samples from each plot were ball milled before analyzed for total N concentration by EA-IRMS (EA1112 coupled with Delta-XP, Thermo Fisher Scientific K.K., Yokohama, Japan), and microwave digested with nitric acid before analyzed for total P concentration (inductively-coupled plasma emission spectrophotometry, Optima 2000, Perkin Elmer, USA).

Statistical analysis

We calculated response ratios (RR) of all fine root and soil variables to N, P treatments as the experimental mean divided by the control mean, representing an index of response magnitudes; L_{RR} was calculated as the Ln of RR. Two-way analysis of variance (ANOVA) was used to examine the effects of N and P treatments and their interactions. Differences among forest types were compared using one-way ANOVA. All analyses were conducted using PASW STATISTICS 16.0 for Windows. Level of significance was set at $P<0.05$ unless otherwise stated.

Results

Fine root biomass

In the old-growth forest, N addition significantly reduced live fine root biomass (Fig. 1A; Table 2, 3) and significantly increased dead fine root biomass (Fig. 1B; Table 2, 3), resulting in a significant lower live fine root proportion, compared to respective values in control plots (Fig. 1C; Table 2, 3). In the two younger forests, compared to control plots, dead fine root biomass in N-addition plots was 69% higher in the pine forest (108.0 g m^{-2} vs. 63.8 g m^{-2}, Fig. 1; Table 2, 3), and 66% higher in the mixed forest (240.6 g m^{-2} vs. 144.6 g m^{-2}, Fig. 1). No significant effects of N addition on live fine root biomass or live fine root proportion were observed in the two younger forests, except that live fine root proportion in the pine forest was significantly lower in N-addition plots (Fig. 1A,C; Table 2, 3).

Plots with added P had elevated live fine root biomass in all three forests. Specifically, P addition resulted in increases of 20% in the old-growth forest (212.4 vs. 176.9 g m^{-2}, Fig. 1A; Table 2, 3), of 35% in the pine forest (83 vs. 61.2 g m^{-2}, Fig. 1A; Table 2, 3) and of 76% in the mixed forest (207.8 vs. 117.8 g m^{-2}, Fig. 1A; Table 2, 3), compared to respective controls. Two-way ANOVA showed that P treatments had no significant effects on dead fine root biomass in any forest, but significantly altered live fine root proportion in all three forests (Table 2, 3; Fig. 1B, C).

N and P treatments had no interactive effects on neither live nor dead fine root biomass, or on live fine root proportion in any forest (Table 3). N+P treatment in the two younger forests significantly increased live fine root biomass relative to control (by 74% in the pine forest and by 87% in the mixed forest, Fig. 1A; Table 2), similar to the effects of P addition alone. Higher live fine root proportion in N+P-addition plots comparing to control was also observed in the mixed forest (Fig. 1C; Table 2).

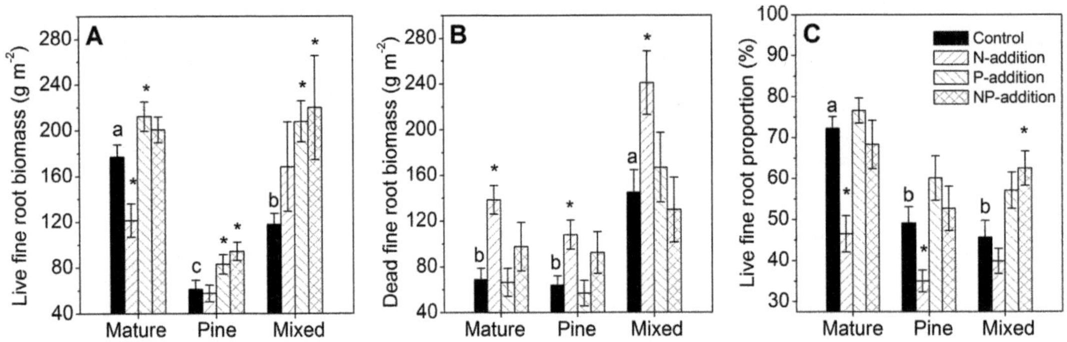

Figure 1. Live, dead fine root biomass and live fine root proportion after five years of nitrogen and phosphorus additions. Data from July 2012. Different letters indicate significant differences among forests ($P<0.05$). * indicates significant differences between each treatment and control ($P<0.05$). Error bars show SE (n = 5).

Fine root nutrient concentrations

The old-growth forest had significantly higher fine root N concentrations and N:P ratios than the two younger forests (Fig. 2A, C). Fine root P concentration was highest in the pine forest (Fig. 2B). N treatments did not alter N or P concentrations in any forest (Fig. 2; Table 2, 3). P treatments had no significant effects on fine root N concentrations (Fig. 2A; Table 3), but significantly elevated fine root P concentrations in all three forests, by 2 to 4 folds (Fig. 2B; Table 2, 3). N:P ratios of fine roots were significantly lower in P-addition and NP-addition plots in all three forests, compared with those in control (Fig. 2C). There were interactions between N and P treatments on fine root P in the old-growth forest (Fig. 2B; Table 3).

Soil chemical properties

Soils in all the three forests were acidic (pH <4.0), with the lowest pH in the old-growth forest (Fig. 3A). The old-growth forest was significantly higher in soil NH_4^+-N and NO_3^--N than the other two forests, whereas soil available P concentrations did not vary among the three forests (Fig. 3B–D).

Soil pH was lower in N-addition plots than in control in the old-growth forest, but not in the two younger forests (Fig. 3A; Table 2, 3). N treatments had no significant effects on soil available P or available N (NH_4^+-N and NO_3^--N) in any forest, except that in the old-growth forest, NO_3^--N in N-addition plot was significantly higher than in control (Table 2, 3; Fig. 3B–D). In contrast, P treatments significantly increased soil pH in the old-growth forest and the mixed forest (Table 2, 3; Fig. 3A), but not in the pine forest. Soil available P increased significantly following P addition by 16-, 31-, and 14-fold, and following NP additions by 13-, 26-, and 12-fold in the old-growth, pine, and mixed forests, respectively, compared with respective controls (Fig. 3B). P treatments decreased NO_3^- in the old-growth forest and decreased NH_4^+-N in the mixed forest (Table 3). Interaction between N and P only existed for soil pH in the mixed forest (Table 3).

APA in bulk soil

APA in control plots was significantly lower in the pine forest than in the other two forests (Fig. 4). Effects of nutrient treatments on APA varied with forest type. Significantly higher APA in N-

Table 2. Response ratios of fine root and soil variables to N and P treatments.

Forests/Treatments	Old-growth			Pine			Mixed		
	N-addition	P-addition	NP-addition	N-addition	P-addition	NP-addition	N-addition	P-addition	NP-addition
Live fine root biomass	−0.38*	0.18*	0.13	−0.06	0.30*	0.43*	0.36	0.57*	0.62*
Dead fine root biomass	0.70*	−0.04	0.35	0.53*	−0.12	0.37	0.51*	0.14	−0.11
Live proportion	−0.44*	0.06	−0.05	−0.34*	0.20	0.07	−0.14	0.22	0.31*
Fine root N	0.05	−0.24	−0.12	0.01	−0.2	0.02	0.02	0.08	0.01
Fine root P	0.29	1.09*	0.85*	0.22	0.92*	1.01*	0.41	1.39*	1.39*
N:P	−0.09	−1.17*	−0.73*	0.12	−0.98*	−0.78*	−0.52	−1.47*	−1.57*
pH	−0.02*	0.03*	0.01	0.01	−0.01	−0.02*	0.01	0.03*	0.02
Available P	0.11	2.85*	2.65*	0.56	3.47*	3.29*	−0.25	2.72*	2.60*
NH_4^+-N	−0.48	0.25	−0.40	−0.44	−0.20	−0.17	−0.01	−0.34	−0.56
NO_3^--N	0.46*	−0.43	−0.43	−0.10	−0.35	−0.27	−0.31	−0.61	−0.41
APA	0.21*	−0.40*	−0.37*	0.01	−0.16	−0.26	−0.05	−0.30*	−0.95*

Values are Ln(treatment mean/control mean). A positive response ratio indicates positive effects of fertilization, whereas a negative one indicates negative effects.
* indicates significant difference at $P<0.05$.

Table 3. Summary of results of two-way ANOVA. Significant effects indicated by *, ** and ***, representing probability at the 5%, 1% and 0.1% levels, respectively; ns, not significant.

	Two-way ANOVA								
Forests/Treatments	**Old-growth**			**Pine**			**Mixed**		
	N	**P**	**N×P**	**N**	**P**	**N×P**	**N**	**P**	**N×P**
Live fine root biomass	*	**	ns	ns	**	ns	ns	*	ns
Dead fine root biomass	**	ns	ns	**	Ns	ns	ns	ns	ns
Live fine root proportion	**	**	ns	**	**	ns	ns	**	ns
Fine root N	ns	ns	ns	ns	Ns	ns	ns	ns	ns
Fine root P	ns	**	**	ns	**	ns	ns	**	ns
N:P	ns	***	ns	ns	***	ns	**	***	ns
pH	***	***	ns	ns	Ns	ns	ns	**	*
Available P	ns	***	ns	ns	***	ns	ns	***	ns
NH_4^+-N	ns	ns	ns	ns	Ns	ns	ns	*	ns
NO_3^--N	ns	**	ns	ns	Ns	ns	ns	ns	ns
APA	*	***	ns	ns	Ns	ns	ns	***	**

addition plots compared with that in control was only observed in the old-growth forest (Fig. 4; Table 2). In contrast, P and N+P treatments resulted in declines in APA in the old-growth and the mixed forests, but not in the pine forest (Fig. 4; Table 2). N+P treatments had interactive effects on APA only in the mixed forest, exhibiting synergistically inhibitive effects (Table 3).

Discussion

Contrasting responses of fine roots to N additions in three tropical forests

The main purpose of this study was to examine the nutrient status of three lowland tropical forests in southern China by quantifying the response of fine roots and APA to five years of N and P additions. Based on the different ages of these forests and their responses to N addition in another N manipulation experiment in DHSBR [8,32], we hypothesized that the old-growth forest was N saturated, whereas productivity in the two younger forests was still N limited, despite chronically elevated N deposition. Thus, we expected N addition to decrease fine root biomass and live fine root proportion (vitality) in the old-growth forest through deleterious changes in soil properties (e.g., increased soil acidification and Al mobilization) [9,10], and in the two younger forests through less carbon allocation to fine roots driven by additional N [18,19].

The results from the old-growth forest support our expectation, in that five years of N addition at 150 kg N ha^{-1} yr^{-1} significantly decreased live fine root biomass and increased dead fine root biomass, resulting in decreased live fine root proportion (vitality). Whereas a decrease in total fine root biomass following two years of N additions had been observed in another N manipulation experiment in DHSBR [8], our results added further evidence of the adverse effects of N on fine root growth in this forest, i.e., different directions of changes in live and dead fine root biomass, and a decreased live fine root proportion, indicative of damage to fine roots [46].

Elevated N input can lead to soil acidification [7,10], which has been associated with significantly lower fine root vitality (live fine root proportion) than that from less acidified soils [47]. In Norway spruce stands in Sweden, acidification caused by addition of ammonium sulfate resulted in faster rates of root death [48,49]. Our results from the old-growth forest supported this observation, with soil pH decreasing significantly following N additions (Fig. 3A). A decline of live fine root biomass and a

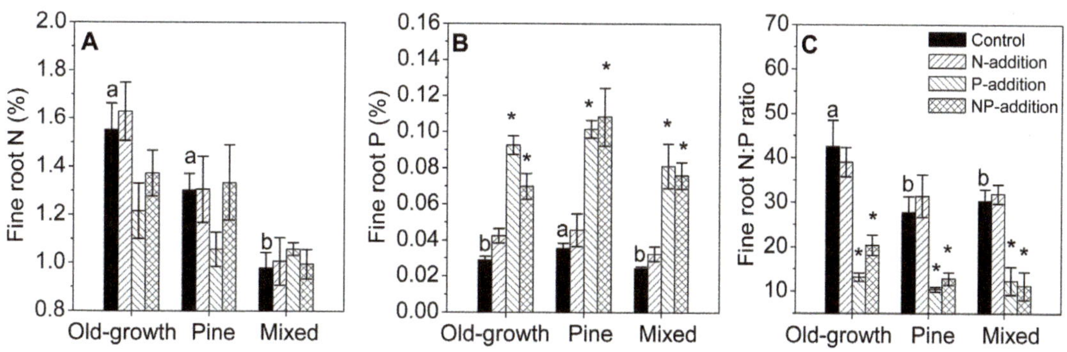

Figure 2. Nutrient concentration in live fine roots after five years of nitrogen and phosphorus additions. Data from July 2012. Different letters indicate significant differences among forests ($P<0.05$). * indicates significant differences between each treatment and control ($P<0.05$). Error bars show SE (n = 5).

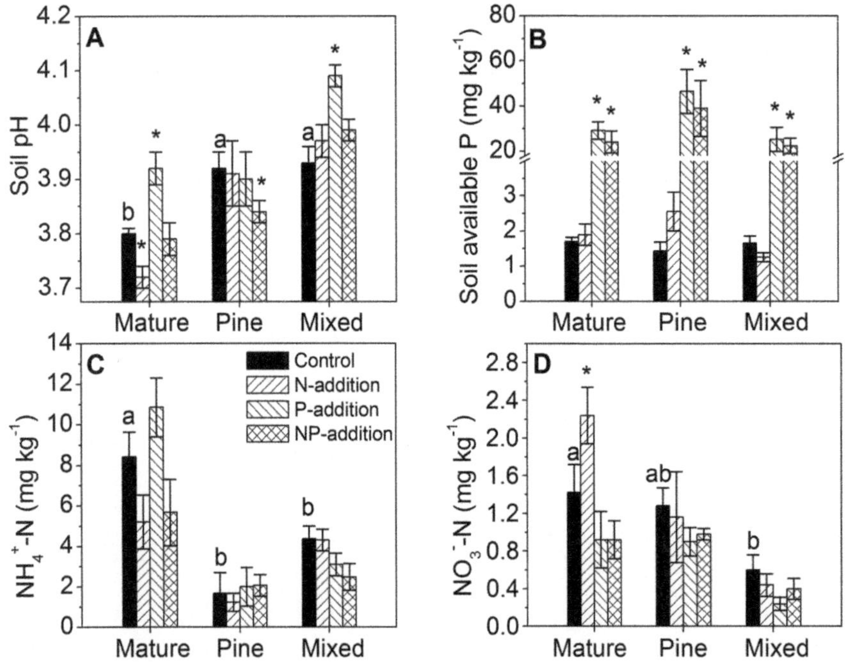

Figure 3. Soil pH, available P and N (NH_4^+-N and NO_3^--N) after four years of nitrogen and phosphorus additions. Data from July 2011. Different letters indicate significant differences among forests (*P*<0.05). * indicates significant differences between each treatment and control (*P*<0.05). Error bars show SE (n = 5).

concurrent increase of dead fine root biomass after N additions have also been observed in other wet tropical forests [50,51], with authors attributing these results to either direct effects of N on belowground root growth/turnover, or to a secondary effect of N fertilization on belowground plant dynamics (e.g. decline in soil pH [50]).

Contrary to our expectations, we failed to observe decreases in fine root biomass following N additions in the two younger forests. Instead, we found increased dead fine root biomass and reduced live fine root proportion in N-addition plots in the two younger forests, which were similar to the observation in the old-growth forest. It appeared that five years of N additions may have moved these forests toward N saturation, when negative secondary effects of N on fine root growth started to emerge. Previous disturbance before the 1990's through understory vegetation and litter harvesting lowered the ecosystem N capital relative to the undisturbed old-growth forest [32]. Furthermore, N additions failed to alter fine root N or P concentrations in any of these three forests, consistent with other studies in tropical forests [16].

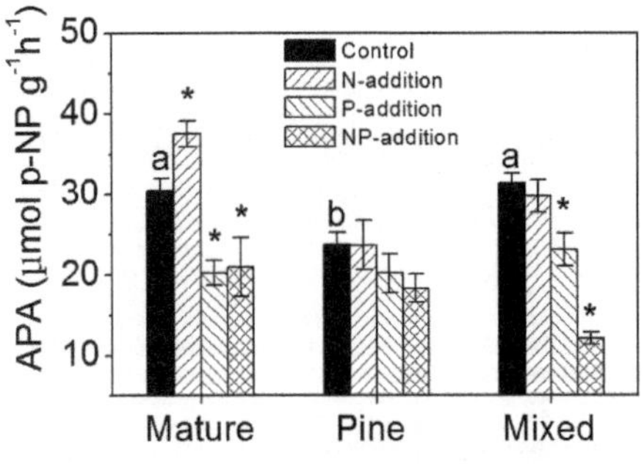

Figure 4. Acid phosphatase activity (APA) in bulk soil after four years of nitrogen and phosphorus additions. Data from July 2011. Different letters indicate significant differences among forests (*P*<0.05). * indicates significant differences between each treatments and control (*P*<0.05). Error bars show SE (n = 5).

The pattern of APA response to N in the three forests paralleled the response of live fine root biomass, with significant increase of APA observed only in the old-growth forest (Fig. 4). Increase of APA suggests further demand for P driven by N addition to the N saturated forest [24,52]. Elevated APA stimulated organic P mineralization and released inorganic P, which in turn likely compensated the low soil P availability. This may explain why available soil P in N-addition plots did not vary significantly from control in the old-growth forest, albeit both were of very low levels (Fig. 3B). In the two younger forests, N addition had no effects on APA or soil pH, suggesting addition of 150 kg N ha^{-1} yr^{-1} did not drive P demand to the extent to induce APA. Thus, they may not be limited by P as the N-saturated old-growth forest. Lower N status may also have constrained phosphatase production, which is highly N-consumptive [25]. Collectively, these results suggest that initial soil N status of the study forests largely influenced fine root responses to increased N inputs.

Variation in P limitation for fine root growth among forests

Following the criteria to assess nutrient limitations stated by Vitousek et al. (2010) [15], we further address P limitation for fine root growth and alleviation of P limitation by P fertilization at the three study forests by combining direct and indirect measurements of (1) fine root growth in response to P fertilization, (2) fine root nutrient concentrations, and (3) nutrient availability and APA in bulk soil.

Contrary to expectation, fine root growth responded positively to P treatments in all three study forests, albeit to a larger extent in old-growth forest and mixed forests, suggesting P limitation to fine root growth in these forests. This pattern corresponded well with notably elevated soil P availability in these forests, indicating that fine roots likely accessed elevated P by increasing growth into the surface soil (0–10 cm) [12]. Significantly higher fine root biomass and fine root N uptake following P additions had been observed in an N-saturated Mediterranean-fir (*Abies pinsapo*) forest in southern Spain which exhibited symptoms of P limitation [53]. A trend toward higher fine root biomass after two years of N+P additions was also observed in the ongoing fertilization experiment in lowland tropical forests in Costa Rica [31]. Fine root biomass represents a balance between fine root production and decomposition. Fine root decomposition, which is microbially mediated [16], may have been stimulated in our study forests, because microbial activities were found to be significantly higher in P-addition plots in the old-growth and the mixed forests [27]. Thus, the observed increases of live fine root biomass and no changes in dead fine root biomass indicate the likelihood of enhanced fine root production following P additions in these forests, which has also been shown in other tropical forests. In an old site (>4 million years) in Hawaii where aboveground net primary production was P-limited, both root net primary production and root turnover (decomposition) increased following P fertilization [16]. In a montane tropical forest, moderate N and P additions enhanced fine root production and turnover at the same time, resulting in a reduced fine root biomass [51]. A meta-analysis research which synthesized data from N and P gradients showed that in lowland tropical forests, fine root production was positively associated with both natural and artificial P gradients [54].

P addition improved fine root nutrient condition. Plant nutrient imbalance can be expected in these forests, with higher fine root N (average 13 vs. 11 g kg^{-1}), but much lower P (average 0.3 vs. 0.9 g kg^{-1}) compared with the global average values compiled by Gordon and Jackson (2000) [55]. Fine root N:P ratios are as high as 43, 28, and 30 in the old-growth, pine and mixed forests, respectively, suggesting high demand of P for tree growth. These ratio values are much higher than the averaged value of 12 from the compiled data set [55], than the values of 26, 17, and 23 of live fine roots in the 300 yr-, 20000 yr- and >4 million yr-old forests along the geologic chronosequence in Hawaii [16], and also higher than the threshold of foliar N:P ratio of 20 beyond which P limitation is suggested [56]. Notably, substantial improvement of fine root nutrient conditions by P addition was observed, i.e., fine root N:P ratios were altered by P addition from high levels of 43, 28, and 30 to 13, 10, and 12 in the old-growth, pine and mixed forests, respectively, which were largely contributed by elevated fine root P concentrations (Fig. 2B, C).

Notably, in spite of high fine root P uptake after P fertilization in all three forests, fine root N concentrations showed no response to P additions. This contrasts with results of Blanes et al. (2012) [53], who found that P addition to an N-saturated Mediterranean-fir (*Abies pinsapo*) forest increased the competitive ability of tree roots for soil N. In our experiment, however, higher microbial N was found following P addition in the old-growth forest [28], implying potentially higher microbial immobilization of P and N. Available soil P in our sites (~2 mg kg^{-1}) was much lower than in the soil of Mediterranean-fir (*Abies pinsapo*, 7.8 mg kg^{-1}), suggesting that the competition between tree roots and free-living microbes for available P may be greatly enhanced in our study sites.

Conclusions

Five years of N and P treatments in relatively high amounts (150 kg ha^{-1} yr^{-1}) substantially altered fine root biomass, nutrient concentrations, APA of bulk soil, and other soil chemical properties in the three study forests. As hypothesized, these forests exhibited diverse responses to N and/or P treatments, largely reflecting contrasting N status. In contrast, fine root growth appeared to be P-limited in all three forests, with the N saturated old-growth forest being more P-limited than the other two forests, based on its APA responses to N addition. These patterns suggest that initial soil nutrient status have large impact on potential responses of belowground structure to further nutrient inputs in these ecosystems. While high N deposition in this region keeps adding N to these tropical forests, P limitation emerges and becomes a key factor to influence ecosystems processes.

Acknowledgments

We wish to thank Xiaoming Fang, Shaofeng Dong, Yuedi Yang for their skillful assistance in field work, Hongying Li,Shaowei Chen for their assistance in laboratory work, and Drs. Keisuke Koba and Xueyan Liu for their invaluable suggestions in earlier version of the manuscript.

Author Contributions

Conceived and designed the experiments: JMM FFZ MY. Performed the experiments: FFZ. Analyzed the data: FFZ. Contributed reagents/materials/analysis tools: FFZ XKL MY FSG. Wrote the paper: FFZ.

References

1. Galloway JN, Dentener FJ, Capone DG, Boyer EW, Howarth RW, et al. (2004) Nitrogen cycles: past, present, and future. Biogeochemistry 70: 153–226.
2. Zheng XH, Fu CB, Xu XK, Yan XD, Huang Y, et al. (2002) The Asian nitrogen cycle case study. Ambio 31: 79–87.
3. Magnani F, Mencuccini M, Borghetti M, Berbigier P, Berninger F, et al. (2007) The human footprint in the carbon cycle of temperate and boreal forests. Nature 447: 848–850.
4. Walker TW, Syers JK (1976) Fate of phosphorus during pedogenesis. Geoderma 15: 1–19.
5. Vitousek PM, Sanford RL Jr (1986) Nutrient cycling in moist tropical forest. Annual Review of Ecology, Evolution, and Systematics 17: 137–167.
6. Tanner EVJ, Vitousek PM, Cuevas E (1998) Experimental investigation of nutrient limitation of forest growth on wet tropical mountains. Ecology 79: 10–22.
7. Matson PA, McDowell WH, Townsend AR, Vitousek PM (1999) The globalization of nitrogen deposition: ecosystem consequences in tropical environments. Biogeochemistry 46: 67–83.
8. Mo JM, Zhang W, Zhu WX, Gundersen P, Fang YT (2008) Nitrogen addition reduces soil respiration in a mature tropical forest in southern China. Glob Chang Biol 14: 403–412.
9. Lu XK, Mo JM, Gundersen P, Zhu WX, Zhou GY, et al. (2009) Effects of simulated N deposition on soil exchangeable cations in three forest types of subtropical China. Pedosphere 19(2): 189–198.
10. Lu XK, Mo JM, Gilliam FS, Zhou GY, Fang YT (2010) Effects of experimental nitrogen additions on plant diversity in an old-growth tropical forest. Glob Chang Biol 16: 2688–2700.
11. Compton J, Cole DW (1998) Phosphorus cycling and soil P fractions in Douglas-fir and red alder stands. For Ecol Manage 110: 101–112.
12. Naples BK, Fisk MC (2010) Belowground insights into nutrient limitation in northern hardwood forests. Biogeochemistry 97: 109–121.
13. Braun S, Thomas VFD, Quiring R, Flickiger W (2010) Does nitrogen deposition increase forest production? The role of phosphorus. Environ Pollut 158: 2043–2052.
14. Gress SE, Nichols TD, Northcraft CC, Peterjohn WT (2007) Nutrient limitation in soils exhibiting differing nitrogen availabilities: what lies beyond nitrogen saturation? Ecology 88: 119–130.
15. Vitousek PM, Porder S, Houlton BZ, Chadwick OA (2010) Terrestrial phosphorus limitation: mechanisms, implications, and nitrogen phosphorus interactions. Ecol Appl 20: 5–15.
16. Ostertag R (2001) Effects of nitrogen and phosphorus availability on fine-root dynamics in Hawaiian montane forests. Ecology 82: 485–499.

17. Bloom AJ, Chapin FS, Mooney HA (1985) Resource limitation in plants – an economic analogy. Annu Rev Ecol Evol Syst 16: 363–392.
18. Hendricks JJ, Hendrick RL, Wilson CA, Mitchell RJ, Pecot SD, et al. (2006) Assessing the patterns and controls of fine root dynamics: an empirical test and methodological review. J Ecology 94: 40–57.
19. Nadelhoffer KJ (2000) The potential effects of nitrogen deposition on fine-root production in forest ecosystems. New Phytol 147: 131–139.
20. Raich JW, Riley RH, Vitousek PM (1994) Use of root ingrowth cores to assess nutrient limitations in forest ecosystems. Can J For Res 24: 2135–2138.
21. Xia JY, Wan SQ (2008) Global response patterns of terrestrial plant species to nitrogen addition. New Phytol 179: 428–439.
22. Garrish V, Cernusak LA, Winter K, Turner BL (2010) Nitrogen to phosphorus ratio of plant biomass versus soil solution in a tropical pioneer tree, Ficus insipid. J Exp Bot 61: 3735–3748.
23. Sinsabaugh RL, Antibus RK, Linkins AE, McClaugherty CA, Rayburn L, et al. (1993) Wood decomposition: nitrogen and phosphorus dynamics in relation to extracellular enzyme activity. Ecology 74: 1586–1593.
24. Olander LP, Vitousek PM (2000) Regulation of soil phosphatase and chitinase activity by N and P availability. Biogeochemistry 49: 175–190.
25. Treseder KK, Vitousek PM (2001) Effects of soil nutrient availability on acquisition of N and P in Hawaiian rain forests. Ecology 82: 946–954.
26. Zhang T, Zhu W, Mo J, Liu L, Dong S (2011) Increased phosphorus availability mitigates the inhibition of nitrogen deposition on CH_4 uptake in an old-growth tropical forest, southern China. Biogeosciences 8: 2805–2813.
27. Liu L, Gundersen P, Zhang T, Mo JM (2012) Effects of phosphorus addition on soil microbial biomass and community composition in three forest types in tropical China. Soil Biol Biochem 44: 31–38.
28. Liu L, Zhang T, Gilliam FS, Gundersen P, Zhang W, et al. (2013) Interactive effects of nitrogen and phosphorus on soil microbial communities in a tropical forest. PloS One 8(4): e61188.
29. Mirmanto E, Proctor J, Green J, Nagy L, Suriantata (1999) Effects of nitrogen and phosphorus fertilization in a lowland evergreen rainforest. Philos Trans R Soc Lond B Biol Sci 354: 1825–1829.
30. Wright SJ, Yavitt JB, Wurzburger N, Turner BL, Tanner EVJ, et al. (2011) Potassium, phosphorus, or nitrogen limit root allocation, tree growth, or litter production in a lowland tropical forest. Ecology 92(8): 1616–1625.
31. Alvarez-Clare S, Mack MC, Brooks M (2013) A direct test of nitrogen and phosphorus limitation to net primary productivity in a lowland tropical wet forest. Ecology 94(7): 1540–1551.
32. Mo JM, Brown S, Xue JH, Fang YT, Li ZA (2006) Response of litter decomposition to simulated N deposition in a disturbed, rehabilitated and mature forest in subtropical China. Plant Soil 282: 135–151.
33. Fang YT, Gundersen P, Mo JM, Zhu WX (2009) Nitrogen leaching in response to increased nitrogen inputs in subtropica monsoon forests in southern China. For Ecol Manage 257: 332–342.
34. Fang YT, Gundersen P, Mo JM, Zhu WX (2008) Input and output of dissolved organic and inorganic nitrogen in subtropical forests of South China under high air pollution. Biogeosciences 5: 339–352.
35. Lu XK, Mo JM, Gilliam FS, Fang H, Zhu FF, et al. (2012) Nitrogen addition shapes soil phosphorus availability in two reforested tropical forests in Southern China. Biotropica 44(3): 302–311.
36. Wang H, Mo JM, Lu XK, Xue JH, Li J, et al. (2008) Effects of elevated nitrogen deposition on soil microbial biomass carbon in the main subtropical forests of Southern China. Acta Ecol Sin 28: 470–478 (in Chinese with English abstract).
37. Huang ZF, Fan ZG (1982) The climate of Dinghushan. Tropical and Subtropical Forest Ecosystems 1: 11–23 (in Chinese with English abstract).
38. He JH, Chen ZQ, Liang YA (1982) The soil of Dinghushan. Tropical and Subtropical Forest Ecosystems 1: 25–38 (in Chinese with English abstract).
39. Huang ZL, Ding MM, Zhang ZP, Yi WM (1994) The hydrological processes and nitrogen dynamics in a monsoon evergreen broad-leafed forest of Dinghushan. Acta Phytoecol Sin 18: 194–199 (in Chinese with English abstract).
40. Zhou GY, Yan JH (2001) The influences of regional atmospheric precipitation characteristics and its element inputs on the existence and development of Dinghushan Forest ecosystems. Acta Ecol Sin 21: 2002–2012 (in Chinese with English abstract).
41. Cleveland CC, Townsend AR (2006) Nutrient additions to a tropical rain forest drive substantial soil carbon dioxide losses to the atmosphere. Proc Natl Acad Sci U S A 103(27): 10316–10321.
42. Yavitt JB, Harms KE, Garcia MN, Mirabello AJ, Wright SJ (2011) Soil fertility and fine root dynamics in response to 4 years of nutrient (N, P, K) fertilization in a lowland tropical moist forest, Panama. Austral Ecol 36: 433–445.
43. Scheneider K, Turrion M-B, Grierson PF, Gallardo JF (2001) Phosphorus activity, microbial phosphorus, and fine root growth in forest soils in the Sierra de Gata, western central Spain. Biol Fertil Soils 34: 151–155.
44. Liu GS, Jiang NH, Zhang LD (1996) Soil physical and chemical analysis and description of soil profiles. Standards Press of China, Beijing.
45. Bray RH, Kurtz LT (1945) Determination of total, organic, and available forms of phosphorus in soils. Soil Sci 59: 39–45.
46. Majdi H (1994) Effects of nutrient applications on fine root dynamics and root/rhizosphere chemistry in a Norway spruce stand. Ph.D. thesis, Department of Ecology and Environmental Research, Swedish University of Agricultural Science.
47. Godbold DL, Heinz-Werner F, Jentschke G, Meesenburg H, Rademacher P (2003) Root turnover and root necromass accumulation of Norway spruce (Picea abies) are affected by soil acidity. Tree Physiol 23: 915–921.
48. Clemensson-Lindell A, Asp H (1995) Fine-root morphology and uptake of ^{32}P and ^{35}S in a Norway spruce (Picea abies (L.) Karst.) stand subjected to various nutrient and water supplies. Plant Soil 173: 147–155.
49. Majdi H, Kangas P (1997) Demography of fine roots in response to nutrient applications in a Norway spruce stand in southwestern Sweden. Ecoscience 4: 199–205.
50. Cusack DF, Silver WL, Torn MS, Mcdowell WH (2011) Effects of nitrogen additions on above- and belowground carbon dynamics in two tropical forests. Biogeochemistry 104: 203–225.
51. Homeier J, Hertel D, Camenzind T, Cumbicus NL, Maraun M, et al. (2012) Tropical Andean forests are highly susceptible to nutrient inputs-Rapid effects of experimental N and P addition to an Ecuadorian montane forest. PloS One 7(10): e47128.
52. Marklein AR, Houlton BZ (2012) Nitrogen inputs accelerate phosphorus cycling rates across a wide variety of terrestrial ecosystems. New Phytol 193: 696–704.
53. Blanes MC, Emmet BA, Viñegla B, Carreira JA (2012) Alleviation of P limitation makes tree roots competitive for N against microbes in a N-saturated conifer forest: A test through P fertilization and N^{15} labelling. Soil Biol Biochem 48: 51–59.
54. Yuan Z Y, Chen HYH (2012) A global analysis of fine root production as affected by soil nitrogen and phosphorus. Proc R Soc B 279: 3796–3802.
55. Gordon WS, Jackson RB (2000) Nutrient concentrations in fine roots. Ecology 81(1): 275–282.
56. Güsewell S (2004) N:P ratios in terrestrial plants: variation and functional significance. New Phytol 164: 243–266.
57. Fang YT, Zhu WX, Mo JM, Zhou GY, Gundersen P (2006) Dynamics of soil inorganic nitrogen and their responses to nitrogen additions in three subtropical forests, south China. J Environ Sci 18: 752–759.

15

Community Functional Responses to Soil and Climate at Multiple Spatial Scales: When Does Intraspecific Variation Matter?

Andrew Siefert*, Jason D. Fridley, Mark E. Ritchie

Department of Biology, Syracuse University, Syracuse, New York, United States of America

Abstract

Despite increasing evidence of the importance of intraspecific trait variation in plant communities, its role in community trait responses to environmental variation, particularly along broad-scale climatic gradients, is poorly understood. We analyzed functional trait variation among early-successional herbaceous plant communities (old fields) across a 1200-km latitudinal extent in eastern North America, focusing on four traits: vegetative height, leaf area, specific leaf area (SLA), and leaf dry matter content (LDMC). We determined the contributions of species turnover and intraspecific variation to between-site functional dissimilarity at multiple spatial scales and community trait responses to edaphic and climatic factors. Among-site variation in community mean trait values and community trait responses to the environment were generated by a combination of species turnover and intraspecific variation, with species turnover making a greater contribution for all traits. The relative importance of intraspecific variation decreased with increasing geographic and environmental distance between sites for SLA and leaf area. Intraspecific variation was most important for responses of vegetative height and responses to edaphic compared to climatic factors. Individual species displayed strong trait responses to environmental factors in many cases, but these responses were highly variable among species and did not usually scale up to the community level. These findings provide new insights into the role of intraspecific trait variation in plant communities and the factors controlling its relative importance. The contribution of intraspecific variation to community trait responses was greatest at fine spatial scales and along edaphic gradients, while species turnover dominated at broad spatial scales and along climatic gradients.

Editor: Suzannah Rutherford, Fred Hutchinson Cancer Research Center, United States of America

Funding: AS was supported by National Science Foundation grant DEB-1210609. JDF was supported by National Science Foundation grant DEB-1119743. The funders had no role in study design, data collection and analysis, decision to publish, or preparation of the manuscript.

Competing Interests: The authors have declared that no competing interests exist.

* Email: asiefert@syr.edu

Introduction

Understanding and predicting how communities respond to environmental variation is a central goal of ecology, and ecologists are increasingly adopting trait-based approaches to study these responses [1–3]. Because an organism's functional traits directly influence its responses to and effects on the environment [4], information about the traits of individuals in a community (i.e. community trait distributions) offers insights into community assembly mechanisms and can be used to predict community composition and ecosystem functioning [3,5,6]. Understanding how community trait distributions, particularly community-weighted mean trait values (CWMs), respond to environmental variation has therefore become a major focus in community ecology [5,7,8].

Community trait distributions may change along environmental gradients through a combination of species turnover (changes in species presence and relative abundance) and intraspecific trait responses, including genetic adaptation and phenotypic plasticity [8,9]. Most studies examining trait-environment relationships in plant communities have assigned a single, fixed trait value to each species, thereby accounting only for trait variation due to species turnover and ignoring intraspecific variation [10]. However, recent studies have shown that intraspecific variation may also play an important role in community trait responses to environmental variation. For example, Jung et al. [11] found that intraspecific variation accounted for up to 44% of the change in CWMs of several key functional traits across an elevation gradient in flood meadow communities. Similarly, Lepš et al. [10] found that community-level responses of multiple traits to fertilization and mowing in grassland communities were primarily driven by intraspecific responses. Results of these and other recent studies [12–14] demonstrate that accounting for intraspecific trait variation may be crucial for quantifying community trait responses to the environment, but the relative magnitude of intraspecific variation has varied strongly among and within studies. The next step beyond simply quantifying intraspecific trait variation is to understand the factors controlling its relative role and importance in communities. Determining when and where intraspecific variation matters at the community level is a major concern for plant ecology, with important implications for predicting community and ecosystem responses to global change [15].

One factor that may influence the relative importance of intraspecific trait variation is the spatial scale of the studied communities. Previous studies have shown that interspecific trait variation increases with increasing spatial scale due to species turnover driven by dispersal limitation and environmental filtering along gradients of increasing breadth [16,17]. Intraspecific variation is also expected to increase with increasing spatial scale, as more genetic and plastic variability within species is included, but it is expected to saturate at large scales once the entire range and thus potential trait variation of individual species is reached [15,18]. The relative magnitude of interspecific vs. intraspecific variation is therefore expected to increase at broad spatial scales encompassing strong environmental gradients [15]. This hypothesis has not been tested to our knowledge, in part due to the lack of studies measuring intraspecific variation at broad scales.

Another factor that may influence the relative importance of intraspecific trait variation is the type of environmental gradient considered. Previous studies have shown that intraspecific variation may be important for community responses to local-scale edaphic variation [8,11], but its role in responses to broad-scale climatic variation has not been examined. Determining whether the relative importance of intraspecific variation differs between edaphic and climatic gradients is a useful step towards a more general understanding of when intraspecific variation matters at the community level. Knowledge of the role of intraspecific variation is also relevant for predicting responses of communities to climate change [19]. If community trait responses to climate are driven by species turnover, climate change will result in large changes in community composition and species distributions. On the other hand, if species are able to cope with climatic variation through genetic adaptation or phenotypic plasticity, community composition may remain stable [20,21]. Assessing the degree to which intraspecific variation contributes to community trait responses to strong spatial climatic gradients will provide insights into which of these scenarios is most likely in the face of future climate change.

In this study, we examined community functional responses to environmental variation in old-field plant communities across eastern North America. The study was conducted on a broad spatial extent (1200 km in latitude), allowing us to test the relative importance of species turnover vs. intraspecific variation to community trait patterns along strong edaphic and climatic gradients at multiple spatial scales. Specifically, we addressed the following questions: 1) What is the relative importance of species turnover vs. intraspecific variation to among-site trait variation, and how is this influenced by spatial scale? We hypothesized that the relative importance of intraspecific trait variation would decrease with increasing spatial scale and breadth of environmental gradients. 2) How do community mean trait values respond to edaphic and climatic variation, and what are the relative contributions of species turnover and intraspecific variation to these responses? We hypothesized that intraspecific variation would be more important for community responses to edaphic compared to climatic factors.

Materials and Methods

Study site

We surveyed vegetation and functional traits in 22 old fields across the eastern United States in June-August, 2012. The study area extended from central South Carolina (30°40″N) to central New York (43°10″N), spanning approximately 1200 km of latitude (see Fig. S1 for map of study sites). The study sites were located on private and public land. A list of authorities who granted permission to conduct field work and should be contacted for future permissions is found in the Supporting Information (Table S5). The study area encompassed strong variation in both climatic and edaphic factors (Table 1), making it a useful system for comparing the influence of these factors on community functional composition. Moving from south to north, there was a strong decrease in mean annual temperature (17.9 to 6.9°C) and growing season length (263 to 156 annual frost-free days) and a weaker decrease in mean annual precipitation (1330 to 976 mm). In addition, with increasing latitude there was a strong increase in soil fertility and shift from coarse to fine-textured soils driven primarily by recent glaciation history [22].

The fields sampled had different histories of agricultural land use, but all had been abandoned for at least 5 years prior to sampling. Fields were maintained in early stages of succession by mowing in late summer or fall once every 1–2 years. Time since mowing was not significantly related to any environmental variable or response variable measured and was not included in the analyses. No burning, livestock grazing, or herbicide application had occurred in any of the fields within the past 5 years. Vegetation in the fields was almost entirely herbaceous, including a mix of grasses and forbs. Dominant species included goldenrods (e.g., *Solidago altissima*, *S. rugosa*) and grasses (e.g., *Andropogon virginicus*, *Schedonorus pratensis*, *Poa pratensis*). While there was considerable turnover in species composition across the study area, many of the dominant species were widely distributed, creating the potential for intraspecific variation to play an important role in community trait patterns.

Vegetation and environmental data

In each field, we recorded the percent cover of vascular plant species in 20 1-m^2 quadrats arrayed along transects. The number and arrangement of transects and spacing between quadrats varied depending on the size and shape of the field, with the goal of obtaining a representative sample of the overall vegetation. Cover values in the 20 quadrats were pooled to obtain the relative cover of each species in each field. We recorded a total of 227 species in the 22 sampled sites, with a mean richness of 21.4 species per site. Most species occurred in three or fewer fields, and there were large differences in species composition among sites (mean Bray-Curtis dissimilarity based on relative cover = 0.76; based on presence/absence = 0.75).

We collected soil samples in four randomly selected quadrats in each field and pooled samples for physical and chemical analysis. Percent sand, silt, and clay were determined using the hydrometer method [23]. Percent organic matter was measured as loss on ignition at 360°C. Soil samples were analyzed for cation exchange capacity (CEC), pH, available nitrogen (nitrate plus ammonium; KCl extraction/cadmium reduction method), available (Bray II) phosphorus [24], and available (Mehlich 3 extractant) sulfur, calcium, magnesium, potassium, iron, manganese, and aluminum [25]. Soil analyses were performed by Brookside Laboratories, Inc., New Bremen, OH, USA. We accessed daily precipitation and temperature data (1980–2010) for each site from Daymet (http://www.daymet.org). Using these data, we derived mean annual temperature, mean temperature of the coldest month, mean temperature of the warmest month, temperature seasonality (standard deviation of monthly mean temperature), annual frost-free days, annual growing-degree days (base of 5°C and cap of 30°C), annual precipitation, precipitation in the driest month, precipitation in the wettest month, and precipitation seasonality

Table 1. Mean, standard deviation, and range of environmental variables included in regression analyses.

Variable	Abbreviation	Unit	Mean	Standard deviation	Range (max-min)
Mean annual temperature	MAT	°C	11.95	3.0	11.0
Mean annual precipitation	MAP	mm	1132	99.2	354
Cation exchange capacity	CEC	mEq kg^{-1}	79.4	30.0	11.23
Soil pH	pH		5.6	0.37	1.5
Soil available phosphorus	P	mg kg^{-1}	49.2	53.8	200.0
Soil available nitrogen	N	ppm	6.31	2.87	10.4
Soil organic matter	OM	%	4.98	1.73	7.38
Sand	Sa	%	41.75	18.82	71.91

(CV of monthly precipitation). Environmental data were log transformed as necessary to improve normality.

Trait data

We focused on four traits that relate to different aspects of plant functional strategy: vegetative height, leaf area, specific leaf area (SLA), and leaf dry matter content (LDMC). Vegetative height is related to light acquisition and competitive ability [26,27]. Leaf area relates to energy and water balance and tolerance to environmental stress [28]. SLA is a central component of the leaf economics spectrum, which captures the tradeoff between rapid growth and resource conservation [29]. LDMC is also associated with the leaf economics spectrum as well as leaf water balance and resistance to physical stress [30,31].

In each field, we measured functional traits of species that collectively accounted for 80–100% of the total vegetation cover. This sampling threshold has been shown to provide robust estimates of community mean trait values [32] but excludes some rare species. In each field, we selected five mature- and healthy-looking individuals of each species from different areas of the field for trait measurements. Vegetative height was measured as the distance (cm) from the base to the highest part of the general canopy of the plant. We selected one young, fully-expanded, upper canopy leaf per individual for leaf trait measurements. We measured the one-sided surface area and fresh mass of each leaf following full rehydration [33] and dry mass after oven drying at 80°C for 48 hours [31]. SLA was calculated as leaf area divided by dry mass (mm^2/mg), and LDMC was calculated as leaf dry mass divided by fresh mass. We measured traits on 61 species total (mean 7.5 per field), representing on average 87% of the total cover of each field.

Data analysis

Our first analysis partitioned the contributions of species turnover and intraspecific variation to among-site variation in functional traits following the approach of Lepš et al. [10]. For each field, we calculated three types of community-weighted mean trait values. 1) "Total CWM", calculated as the abundance-weighted average of site-specific species mean trait values:

$$\text{Total CWM}_j = \sum_{i=1}^{S} p_{ij} x_{ij}$$

where p_{ij} is the relative cover of species i in site j, x_{ij} is the mean trait value of species i measured in site j, and S is the number of species sampled in the site. Among-site variation in total CWMs may be generated by a combination of changes in species composition (i.e. species turnover) and changes in species mean trait values (i.e. intraspecific variation) among sites. 2) "Interspecific CWM", calculated as the abundance-weighted average of overall species mean trait values:

$$\text{Interspecific CWM}_j = \sum_{i=1}^{S} p_{ij} x_i$$

where x_i is the mean trait value of species i across all sites where it is measured. Variation in interspecific CWMs is generated by species turnover only. 3) "Intraspecific CWM", calculated as the difference between the total CWM and interspecific CWM for a given site:

$$\textit{Intraspecific CWM}_j = \textit{Total CWM}_j - \textit{Interspecific CWM}_j$$

Variation in intraspecific CWMs is generated by intraspecific trait variation only. Using the total, interspecific, and intraspecific CWMs, we partitioned trait variation among sites into species turnover, intraspecific variation, and covariation components using the sum of squares decomposition approach of Lepš et al. [10]. Positive covariation indicates that species turnover and intraspecific variation reinforce each other (i.e., sites dominated by species with high trait values also have individuals with high trait values for their species), while negative covariation indicates that species turnover and intraspecific variation oppose each other (i.e., sites dominated by species with high trait values have individuals with low trait values for their species).

Our second analysis assessed community trait responses to edaphic and climatic variation and partitioned the contributions of species turnover and intraspecific variation to these responses. Due to strong correlations among variables, we attempted to reduce the dimensionality of the environmental data using principal components analysis, but the primary axes identified left much unexplained variation and were poorly related to community trait values. We therefore selected subsets of climatic (mean annual temperature and annual precipitation) and edaphic variables (sand, pH, CEC, organic matter, available nitrogen, and available phosphorus) that were expected to be important drivers of community functional structure and were not strongly correlated with each other (r<0.5). We modeled relationships between

Table 2. Percentages of total variation in community-weighted mean trait values due to species turnover, intraspecific variation, and their covariation.

Trait	Species turnover	Intraspecific variation	Covariation
Height	52%	22%	26%
Leaf area	89%	7%	4%
SLA	70%	28%	2%
LDMC	52%	27%	21%

Positive covariation indicates that sites dominated by species with high trait values also have individuals with higher than average trait values for their species.

community-weighted mean trait values and edaphic and climatic factors using multiple linear regressions with the full subset of edaphic or climatic variables as predictors. We then performed stepwise model selection by AIC_c to select the best edaphic and climatic model for each trait. To quantify the contributions of species turnover, intraspecific variation, and their covariation to overall community trait responses, we partitioned the variance explained by the edaphic and climatic models using the sum of squares decomposition described above [10]. To examine community trait responses to specific environmental factors, we conducted a similar variance partitioning analysis using single environmental variables as predictors. We also used regression analyses to quantify and compare trait responses of the five most abundant and widely distributed species in the study area: forbs *Solidago altissima* and *Solidago rugosa*, and grasses *Schedonorus pratensis*, *Poa pratensis*, and *Andropogon virginicus*.

Our third analysis tested whether the between-site trait variation and the relative contribution of species turnover vs. intraspecific variation increased with increasing spatial and environmental distance. First, we calculated the geographic distance and environmental distance between each pair of sites in the study area (22 sites, resulting in 231 pairs). Geographic distance between sites was calculated as great circle distance and ranged from 6.5 to 1,151 km (see Fig. S2 for distribution of between-plot distances). Environmental distance was calculated as Euclidean distance using scaled environmental variables (Fig. S2). Next, for each pair of plots we calculated the total dissimilarity in CWMs, dissimilarity due to species turnover, and dissimilarity due to intraspecific variation by applying the sum of squares decomposition described above [10] to each plot pair. We quantified the relative importance of species turnover vs. intraspecific variation by taking the log of the ratio of the species turnover and intraspecific variation components. This created a symmetric measure of the relative contribution of species turnover vs. intraspecific variation to between-site trait dissimilarity, with positive values indicating a greater contribution of species turnover and negative values a greater contribution of intraspecific variation. We tested whether total between-site trait dissimilarity, dissimilarity due to species turnover, dissimilarity due to intraspecific variation, and the relative importance of turnover vs. intraspecific variation varied as a function of geographic and environmental distance using Mantel tests. Because geographic and environmental distance were strongly correlated (Mantel $r = 0.70$), we also used partial Mantel tests to test for the effect of either geographic or environmental distance while controlling for the other.

All analyses were conducted in R [34] using the fields [35], vegan [36], and ecodist [37] packages.

Results

Partitioning among-site trait variation

Among-site variation in community mean trait values was generated by a combination of species turnover and intraspecific variation, with species turnover making the greater contribution for each trait (Table 2). For height, SLA, and LDMC the contribution of species turnover was 2–2.5 times greater than that of intraspecific variation. For leaf area, almost all variation was due to species turnover (Table 2). There was positive covariation between species turnover and intraspecific variation for height and LDMC, indicating that the effects of species turnover and intraspecific variation reinforced each other (i.e., sites dominated by species with high values of those traits also tended to have individuals with high trait values for their species). There was little covariation between species turnover and intraspecific variation for leaf area or SLA (Table 2).

Community trait responses to soil and climate and relative contributions of species turnover vs. intraspecific variation

The best linear models relating environmental variables to community mean trait values varied among traits, but for all traits edaphic factors explained more variation (16–50%) than climatic factors (<20%; Table 3). For example, CWM height was best explained by a combination of soil pH, CEC, organic matter, and available phosphorus, and the best model for CWM leaf area included soil organic matter and available phosphorus (Table 3). CWM SLA was the trait most strongly influenced by climate, showing a significant decrease with increasing mean annual temperature (Table 3).

Community trait responses to the environment were generated primarily by species turnover (Fig. 1), but the importance of intraspecific variation depended on the trait and environmental factor. Intraspecific variation contributed most strongly to community responses of height. For all traits, intraspecific variation was more important, both in terms of total magnitude and relative to species turnover, for community responses to edaphic compared to climatic factors (Fig. 1). Similar results were obtained when using single environmental variables as predictors (Table S2). For example, the only cases in which intraspecific variation contributed more than species turnover were the responses of CWM height to soil CEC and available P (Table S2). Positive covariation between species turnover and intraspecific variation contributed strongly to community responses of height and to a lesser extent SLA, indicating that changes due to species turnover and intraspecific variation reinforced each other (Fig. 1). In contrast, there was weak or even negative covariation between

Table 3. Relationships between community-weighted mean trait values and environmental variables measured in the study.

Trait	Model	Predictors	R^2
Height	Edaphic	CEC (+), pH (−), OM (−), P (+)	0.50
	Climatic	MAP (−)	0.17
Leaf area	Edaphic	OM (−), P (−)	0.25
	Climatic	MAT (+)	0.16
SLA	Edaphic	pH (−), OM (+)	0.26
	Climatic	MAT (−)	0.20
LDMC	Edaphic	N (−), sand (+)	0.16
	Climatic	ns	-

Results are shown for the best edaphic and climatic models for each trait as determined by stepwise selection, including the predictor variables retained in each model and the direction of their effects (negative or positive) on the community trait value. Abbreviations: SLA, specific leaf area; LDMC, leaf dry matter content; predictor variables abbreviated as in Table 1.

species turnover and intraspecific variation effects for leaf area and LDMC.

Individual species had strong trait responses to environmental variables in many cases, but these responses tended to be highly idiosyncratic, differing in strength and direction among species (Fig. 2; Table S3). There were no obvious patterns in the strength or consistency of intraspecific trait responses across species, traits, or environmental variables.

Effects of geographic and environmental distance on between-site trait dissimilarity and species turnover vs. intraspecific variation effects

The influence of geographic and environmental distance on community trait dissimilarity varied among traits. Between-site dissimilarity in CWM SLA increased significantly with increasing geographic ($r = 0.23$; $P = 0.02$) and environmental distance ($r = 0.20$; $P = 0.04$; Fig. 3), indicating that nearby and environmentally similar sites had similar mean SLA. There was also a marginally significant increase in between-site dissimilarity in CWM leaf area with increasing environmental distance ($r = 0.17$; $P = 0.07$; Fig. 3). For both traits, increases in functional dissimilarity were driven by increases in species turnover effects, whereas intraspecific variation was insensitive to both geographic and environmental distance (Table S4). As a result, the relative importance of species turnover vs. intraspecific variation increased with increasing geographic ($r = 0.33$; $P < 0.01$) and environmental distance ($r = 0.24$; $P < 0.01$) for SLA and environmental distance for leaf area ($r = 0.12$; $P = 0.08$; Fig. 3). In contrast, between-site dissimilarity in CWMs and the relative importance of turnover vs. intraspecific variation were not related to geographic or environmental distance for height or LDMC (Fig. 3).

Due to the strong correlation between geographic and environmental distance, it was generally not possible to separate the effects of space vs. environment on community trait dissimilarity (partial Mantel tests not statistically significant; Table S4). The only exception was a positive relationship between geographic distance and the relative importance of species turnover for SLA even after removing the effect of environmental distance (partial $r = 0.24$; $P = 0.01$; Table S4).

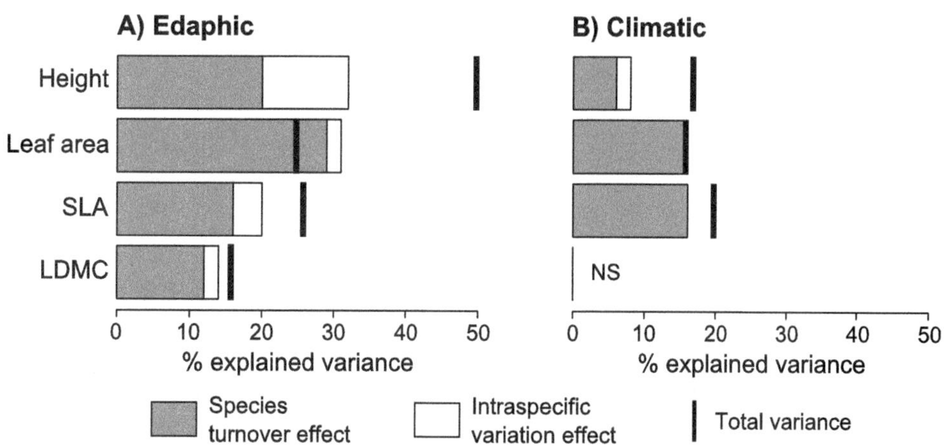

Figure 1. Decomposition of community trait responses to environment. Among-site variance in community-weighted mean trait values explained by (A) edaphic and (B) climatic variables is partitioned into species turnover, intraspecific variation, and covariation effects. Covariation is represented by the difference between the total variance and the sum of the species turnover and intraspecific variation effects. Total variance greater than the sum of species turnover and intraspecific variation effects indicates positive covariance. Total variance less than the sum of species turnover and intraspecific variation effects indicates negative covariance. Abbreviations: SLA, specific leaf area; LDMC, leaf dry matter content.

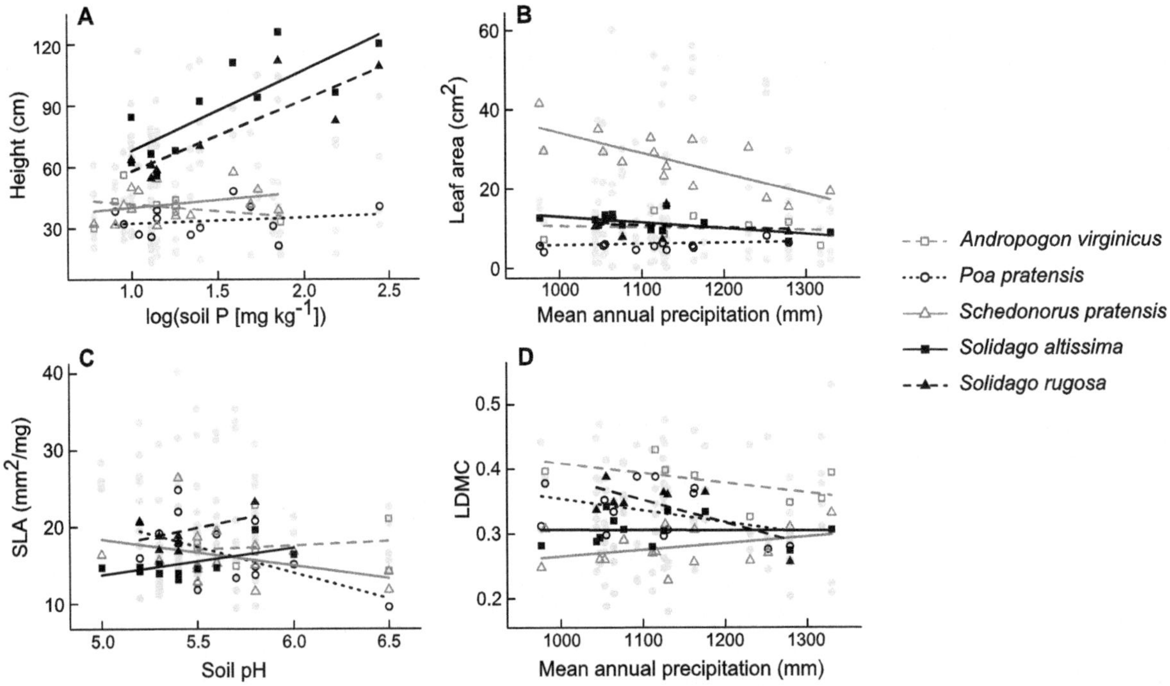

Figure 2. Intraspecific trait responses to environment. Examples of relationships between trait values of individual species and important environmental predictors are shown for each trait measured in the study: (A) height vs. soil P; (B) leaf area vs. mean annual precipitation; (C) specific leaf area vs. soil pH; (D) leaf dry matter content vs. mean annual precipitation. Points represent site-specific mean trait values of all species sampled in each site. Mean trait values and best fit lines from linear regressions are indicated for the five most abundant and widespread species in the study area: *Andropogon virginicus*, *Poa pratensis*, *Schedonorus pratensis*, *Solidago altissima*, and *Solidago rugosa*.

Discussion

Recent studies have shown that intraspecific trait variation may play a fundamental role in responses of plant communities to environmental variation [10–12,19], but many questions remain about when intraspecific variation matters at the community level and the factors controlling its relative importance. By examining intraspecific trait variation in plant communities at an unprecedented spatial extent (1200 km), we were able to address unresolved questions about the relative role of intraspecific variation in community responses to the environment. We found that community-level trait variation was driven primarily by species turnover, but the relative importance of intraspecific variation depended strongly on the trait, environmental factor, and spatial scale considered. In particular, intraspecific variation was more important for responses to edaphic compared to climatic factors, and its relative importance decreased with increasing spatial scale and the related increase in breadth of environmental gradients for two of the four traits examined. These findings contribute to a more general understanding of the role intraspecific trait variation in plant communities, with implications for understanding community assembly and predicting community responses to global change.

Role of intraspecific variation in community trait responses to environment

Community trait responses to environmental variation were primarily driven by species turnover, but intraspecific variation tended to play a larger role in community trait responses to edaphic compared to climatic variation. This finding may reflect differences in the spatial scale on which edaphic and climatic factors vary. Edaphic factors tend to vary on finer spatial scales than climatic factors [38], so the geographic ranges of individual species are likely to include a relatively larger proportion of the total edaphic variation than the total climatic variation found in a region. As a result of species being distributed widely along edaphic gradients but narrowly along climatic ones, there is greater potential for strong intraspecific trait responses to drive community-level trait shifts along edaphic gradients. In addition, fine-scale environmental heterogeneity is known to promote the evolution of adaptive phenotypic plasticity [39,40], raising the possibility that plants have stronger plastic trait responses to fine-scale edaphic compared to broad-scale climatic factors. Assessing the strength of intraspecific trait responses to different environmental factors and how this relates to the spatial scale at which those factors vary is an important aim for future research.

The relative importance of intraspecific variation also varied among traits. The contribution of intraspecific variation to total among-site variation and responses to environmental factors was greatest for vegetative height. Intraspecific variation contributed to community-level variation in height directly and through strong positive covariation with species turnover effects, consistent with the expectation that interspecific and intraspecific trait responses should be in the same direction to increase individual plant fitness [8]. Intraspecific variation played a particularly large role in responses of height to edaphic factors such as available phosphorus. This finding is in line with previous studies showing that plant height is highly responsive to soil resource availability within species and that intraspecific variation contributes strongly to community-level variation in height [10,41,42]. In contrast, the relative contribution of intraspecific variation was smallest for leaf area and LDMC, traits that are known to be less plastic and for which intraspecific variation tends to be much smaller than interspecific variation [30,43].

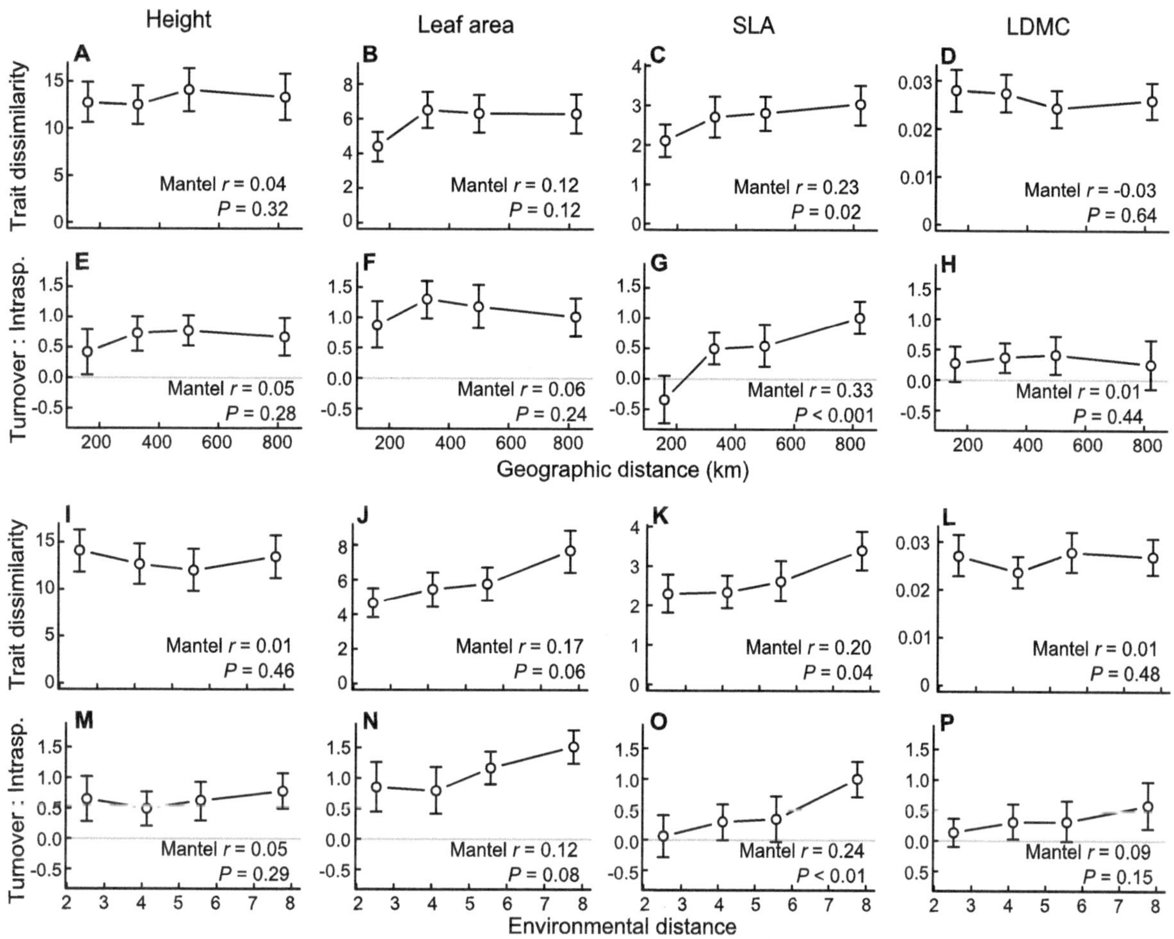

Figure 3. Effects of geographic and environmental distance on between-site variation in plant functional traits. Panels A-C and I-L show the trait dissimilarity (difference in community-weighted mean trait values) between sites as a function of geographic or environmental distance. Panels E-H and M-P show the relative contribution of species turnover vs. intraspecific variation (measured as log of species turnover effect divided by intraspecific variation effect) to the total trait dissimilarity between sites. Values were calculated for each pair of sites in the study area and binned by geographic or environmental distance (n = 57–58 pairs per bin) for ease of interpretation. Points represent mean values for each distance bin; error bars represent 95% bootstrap confidence intervals. Abbreviations: SLA, specific leaf area; LDMC, leaf dry matter content.

What explains the relatively weak contribution of intraspecific variation to most community-level trait-environment relationships observed in this study? The effect of intraspecific trait responses seen at the community level is the aggregate of responses of many individual species. Although traits of individual species responded strongly to the environment in many cases, these responses were highly idiosyncratic, differing in both strength and direction. As a result, they often cancelled out and generally failed to scale up to the community level. Previous studies examining trait variation in multiple species along an environmental gradient have also found intraspecific trait responses to be highly variable [12,44,45]. This variability may have several causes. First, species may respond to environmental variation by altering their multivariate functional strategies rather than single traits in isolation [19]. Functional tradeoffs can result in multiple trait combinations that are equally adaptive in a given environment [46], so responses of single traits may be variable. Second, traits respond not only to the abiotic environment, but also to biotic interactions [47]. The trait response of a particular species to an environmental change will therefore depend not only on the direct influence of the environment, but also on changes in biotic interactions mediated by traits of the target species and its neighbors [12]. Third, some traits may have unimodal responses to environmental gradients, such that trait values are maximized at a species' environmental optimum. The observed direction of the intraspecific trait responses may therefore vary depending on the part of the gradient examined in relation to the environmental ranges and optima of the sampled species [44].

It should be noted that community trait-environment relationships observed in this study were generally weak, with much unexplained variation. This is probably due in part to a strong influence of site history on community assembly [48]. Sites with similar edaphic and climatic conditions were often dominated by functionally dissimilar species, suggesting that unknown past management regimes, along with dispersal history and stochastic effects may have played a large role in determining the functional composition of these early-successional communities. In addition, it is likely that key environmental variables, such as nitrogen supply rate and midsummer water potential, were not measured.

Influence of spatial scale on the relative importance of species turnover vs. intraspecific variation

The broad spatial extent of our study allowed us to examine the influence of spatial scale on the relative importance of species turnover vs. intraspecific variation. Albert et al. [15] hypothesized that the relative importance of intraspecific variation should decrease with increasing spatial scale and the accompanying increase in environmental heterogeneity. We found limited support for this hypothesis for two of the four traits examined in the study, leaf area and SLA. Trait variation due to species turnover increased with increasing geographic and environmental distance for both traits, reflecting the effects of environmental filtering, dispersal limitation, or both [17]. In contrast, intraspecific trait variation was insensitive to both geographic and environmental distance, indicating that intraspecific variation in the studied communities was for the most part spatially random or driven by unmeasured fine-scale environmental variation. This agrees with the findings of Albert et al. [45] that most intraspecific variance in plant traits was found at fine spatial scales, i.e. within rather than between populations along an environmental gradient.

This study is the first to our knowledge to assess the contribution of intraspecific variation to community-level trait variation across a broad geographic extent encompassing strong climatic variation. The relative importance of intraspecific variation observed in this study was generally less than that observed for the same traits in previous local-scale studies of grassland communities. For example, Li et al. (unpublished data) showed that intraspecific variation was the primary source of community-level changes in height, SLA, and LDMC along a soil moisture gradient in subalpine meadow communities. Similarly, Kichenin et al. [12] found that intraspecific variation drove community shifts in SLA along an elevation gradient. At relatively small distances (<200 km), we also found that intraspecific variation contributed more than species turnover to among-site variation in SLA. For height, leaf area, and LDMC, species turnover was the primary source of between-community variation even at the smallest distances resolvable in this study (6.5 km), but it is possible that a similar transition from intraspecific variation to species turnover as the main source of among-community trait variation occurred at finer spatial scales.

Implications

Our results have several potentially interesting implications for plant community ecology. First, with the growing recognition that intraspecific variation may play an important role in plant communities [49], and the great effort required to measure it [50], there is a need for information to help researchers decide when it should be considered in plant ecology studies. To this end, Albert et al. [15] proposed that intraspecific variation might be negligible and therefore safely ignored at very broad spatial scales encompassing strong environmental gradients. Our results provide limited support for this recommendation by showing for the first time that the relative importance of intraspecific variation in some traits decreases with increasing geographic and environmental distance between sites, so that species turnover dominates at very broad scales. In such cases, it may be acceptable to use species mean trait values to characterize community-level trait patterns, whereas measuring each species in each site where it occurs may be necessary for studies conducted on finer scales and shorter gradients [50]. Second, community trait-environment relationships are commonly used to infer community assembly mechanisms, particularly trait-mediated environmental filtering [8], but most studies have only considered interspecific trait variation. Our findings suggest that different environmental filters or selection pressures operate at different levels of organization within communities. Specifically, climatic factors may filter species based on their mean trait values, resulting in species turnover along broad-scale climatic gradients, whereas intraspecific variation may be more important for responses to edaphic filters. Third, the dominance of species turnover in driving community trait responses to spatial climatic gradients suggests that species turnover will also play a large role in responses of the studied communities to future predicted climate change [51].

To conclude, intraspecific trait variation may play an important role in community trait responses to the environment in some situations, and there is a need for empirical data to generalize when and to what extent it matters at the community level. We found that functional responses of old-field plant communities to environmental variation at broad spatial scales were primarily driven by species turnover, but several factors influenced the relative importance of intraspecific variation. Specifically, intraspecific variation was more important for responses of vegetative height compared to leaf traits, responses to edaphic compared to climatic gradients, and at fine compared to broad spatial scales. Future research should examine whether our findings extend to other plant communities and types of environmental gradients.

Supporting Information

Figure S1 Map of study sites and location of study area within North America.

Figure S2 Distributions of geographic and environmental distances between study sites.

Table S1 Mean trait values of sampled species.

Table S2 Decomposition of variation in community-weighted mean trait values explained by single environmental variables.

Table S3 Slopes of intraspecific trait-environment relationships for the five most frequently occurring study species.

Table S4 Results of Mantel and partial Mantel tests for effects of geographic and environmental distance on between-site trait dissimilarity.

Table S5 List of authorities issuing permits for field work at study sites.

Acknowledgments

We are grateful to Josh Lynn and Luka Negoita for assistance with field work and data collection. We thank the following agencies and organizations for providing access to field sites: National Park Service, South Carolina State Park Service, North Carolina Division of Parks and Recreation, Duke Forest, VCU Rice Center, Virginia Department of Conservation and Recreation, Pennsylvania Department of Conservation and Natural Resources, Hawk Mountain Sanctuary, SUNY Cortland, Cary Institute of Ecosystem Studies, Mohonk Preserve and Cornell Biological Field Station. We thank Francesco de Bello for providing helpful comments on the manuscript.

Author Contributions

Conceived and designed the experiments: AS JDF MER. Performed the experiments: AS. Analyzed the data: AS. Contributed to the writing of the manuscript: AS JDF MER.

References

1. Lavorel S, Garnier E (2002) Predicting changes in community composition and ecosystem functioning from plant traits: revisiting the Holy Grail. Funct Ecol 16: 545–556.
2. McGill BJ, Enquist BJ, Weiher E, Westoby M (2006) Rebuilding community ecology from functional traits. Trends Ecol Evol 21: 178–185.
3. Suding KN, Lavorel S, Chapin FS III, Cornelissen JHC, et al. (2008) Scaling environmental change through the community-level: a trait-based response-and-effect framework for plants. Glob Chang Biol 14: 1125–1140.
4. Diaz S, Cabido M (2001) Vive la différence: plant functional diversity matters to ecosystem processes. Trends Ecol Evol 16: 646–655.
5. Shipley B, Vile D, Garnier E (2006) From plant traits to plant communities: a statistical mechanistic approach to biodiversity. Science 314: 812–814.
6. Chapin FS III, Zavaleta ES, Eviner VT, Naylor RL, Vitousek PM, et al. (2000) Consequences of changing biodiversity. Nature 405: 234–242.
7. Ricotta C, Moretti M (2011) CWM and Rao's quadratic diversity: a unified framework for functional ecology. Oecologia 167: 181–188.
8. Cornwell WK, Ackerly DD (2009) Community assembly and shifts in plant trait distributions across an environmental gradient in coastal California. Ecol Monogr 79: 109–126.
9. Ackerly DD (2003) Community assembly, niche conservatism, and adaptive evolution in changing environments. Int J Plant Sci 164: S165–S184.
10. Lepš J, de Bello F, Šmilauer P, Doležal J (2011) Community trait response to environment: disentangling species turnover vs intraspecific trait variability effects. Ecography 34: 856–863.
11. Jung V, Violle C, Mondy C, Hoffmann L, Muller S (2010) Intraspecific variability and trait-based community assembly. J Ecol 98: 1134–1140.
12. Kichenin E, Wardle DA, Peltzer DA, Morse CW, Freschet GT (2013) Contrasting effects of plant inter- and intraspecific variation on community-level trait measures along an environmental gradient. Funct Ecol 27: 1254–1261.
13. Auger S, Shipley B (2012) Inter-specific and intra-specific trait variation along short environmental gradients in an old-growth temperate forest. J Veg Sci 24: 419–428.
14. Pérez-Ramos I, Roumet C, Cruz P, Blanchard A, Autran P, et al. (2012) Evidence for a "plant community economics spectrum" driven by nutrient and water limitations in a Mediterranean rangeland of southern France. J Ecol 100: 1315–1327.
15. Albert CH, Grassein F, Schurr FM, Vieilledent G, Violle C (2011) When and how should intraspecific variability be considered in trait-based plant ecology? Perspect Plant Ecol Evol Syst 13: 217–225.
16. Swenson NG, Anglada-Cordero P, Barone JA (2011) Deterministic tropical tree community turnover: evidence from patterns of functional beta diversity along an elevational gradient. Proc R Soc B 278: 877–884.
17. Siefert A, Ravenscroft C, Weiser MD, Swenson NG (2013) Functional beta-diversity patterns reveal deterministic community assembly processes in eastern North American trees. Glob Ecol Biogeogr 22: 682–691.
18. Violle C, Jiang L (2009) Towards a trait-based quantification of species niche. J Plant Ecol 2: 87–93.
19. Jung V, Albert CH, Violle C, Kunstler G, Loucougaray G, et al. (2014) Intraspecific trait variability mediates the response of subalpine grassland communities to extreme drought events. J Ecol 102: 45–53.
20. Lloret F, Escudero A, Iriondo JM, Martínez-Vilalta J, Valladares F (2012) Extreme climatic events and vegetation: the role of stabilizing processes. Glob Chang Biol 18: 797–805.
21. Grime JP, Fridley JD, Askew AP, Thompson K, Hodgson JG, et al. (2008) Long-term resistance to simulated climate change in an infertile grassland. Proc Natl Acad Sci USA 105: 10028–10032.
22. Wright JP, Fridley JD (2010) Biogeographic synthesis of secondary succession rates in eastern North America. J Biogeogr 37: 1585–1596.
23. Bouyoucos GJ (1962) Hydrometer method improved for making particle size analyses of soils. Agron J 54: 464–465.
24. Bray RH, Kurtz LT (1945) Determination of total, organic, and available forms of phosphorus in soils. Soil Sci 59: 39–46.
25. Mehlich A (1984) Mehlich 3 soil test extractant: A modification of Mehlich 2 extractant. Commun Soil Sci Plant Anal 15: 1409–1416.
26. Gaudet CL, Keddy PA (1988) A comparative approach to predicting competitive ability from plant traits. Nature 334: 242–243.
27. Wright IJ, Reich PB, Cornelissen JHC, Falster DS, Groom PK, et al. (2005) Modulation of leaf economic traits and trait relationships by climate. Glob Ecol Biogeogr 14: 411–421.
28. Ackerly D, Reich P (1999) Convergence and correlations among leaf size and function in seed plants: a comparative test using independent contrasts. Am J Bot 86: 1272–1281.
29. Wright IJ, Reich PB, Westoby M, Ackerly DD, Baruch Z, et al. (2004) The worldwide leaf economics spectrum. Nature 428: 821–827.
30. Wilson P, Thompson K, Hodgson J (1999) Specific leaf area and leaf dry matter content as alternative predictors of plant strategies. New Phytol 143: 155–162.
31. Perez-Harguindeguy N, Diaz S, Garnier E, Lavorel S, Poorter H, et al. (2013) New handbook for standardised measurement of plant functional traits worldwide. Aust J Bot 61: 167–234.
32. Pakeman RJ, Lepš J, Kleyer M, Lavorel S, Garnier E (2009) Relative climatic, edaphic and management controls of plant functional trait signatures. J Veg Sci 20: 148–159.
33. Garnier E, Shipley B, Roumet C, Laurent G (2001) A standardized protocol for the determination of specific leaf area and leaf dry matter content. Funct Ecol 15: 688–695.
34. R Core Development Team (2012) R: A Language and Environment for Statistical Computing. Available: http://www.r-project.org.
35. Furrer R, Nychka D, Sain S (2012) fields: Tools for spatial data. Available: http://cran.r-project.org/package=fields.
36. Oksanen J, Blanchet FG, Kindt R, Legendre P, Minchin P, et al. (2012) vegan: Community ecology package. Available: http://cran.r-project.org/package=vegan.
37. Goslee S, Urban D (2007) The ecodist package for dissimilarity-based analysis of ecological data. J Stat Softw 22: 1–19.
38. Lechowicz MJ, Bell G (1991) The ecology and genetics of fitness in forest plants II. Microspatial heterogeneity of the edaphic environment. J Ecol 79: 687–696.
39. Via S, Lande R (1985) Genotype-environment interaction and the evolution of phenotypic plasticity. Evolution 39: 505–522.
40. Baythavong BS (2011) Linking the spatial scale of environmental variation and the evolution of phenotypic plasticity: selection favors adaptive plasticity in fine-grained environments. Am Nat 178: 75–87.
41. Dantas VDL, Pausas JG, Batalha MA, Loiola PDP, Cianciaruso MV (2013) The role of fire in structuring trait variability in Neotropical savannas. Oecologia 171: 487–494.
42. Gross N, Börger L (2013) Uncovering multiscale effects of aridity and biotic interactions on the functional structure of Mediterranean shrublands. J Ecol: 637–649.
43. Roche P, Díaz-Burlinson N, Gachet S (2004) Congruency analysis of species ranking based on leaf traits: which traits are the more reliable? Plant Ecol 174: 37–48.
44. Albert CH, Thuiller W, Yoccoz NG, Soudant A, Boucher F, et al. (2010) Intraspecific functional variability: extent, structure and sources of variation. J Ecol 98: 604–613.
45. De Frenne P, Graae B (2013) Latitudinal gradients as natural laboratories to infer species' responses to temperature. J Ecol 101: 784–795.
46. Marks CO, Lechowicz MJ (2006) Alternative designs and the evolution of functional diversity. Am Nat 167: 55–66.
47. Callaway RRM, Pennings SSC, Richards CLC (2003) Phenotypic plasticity and interactions among plants. Ecology 84: 1115–1128.
48. Chase JM (2003) Community assembly: when should history matter? Oecologia 136: 489–498.
49. Violle C, Enquist BJ, McGill BJ, Jiang L, Albert CH, et al. (2012) The return of the variance: intraspecific variability in community ecology. Trends Ecol Evol 27: 244–252.
50. Baraloto C, Paine CET, Patiño S, Bonal D, Hérault B, et al. (2010) Functional trait variation and sampling strategies in species-rich plant communities. Funct Ecol 24: 208–216.
51. Northeast Climate Impacts Assessment (2006) Climate change in the US Northeast. Union of Concerned Scientists Publications, Cambridge, MA.

16

Biotic and Abiotic Soil Properties Influence Survival of *Listeria monocytogenes* in Soil

Aude Locatelli[1], Aymé Spor[1], Claudy Jolivet[2], Pascal Piveteau[1,2,3], Alain Hartmann[1]*

1 INRA, UMR1347 Agroécologie, Dijon, France, **2** INRA, US-1106 InfoSol, BP20619, Orléans, France, **3** Université de Bourgogne, UMR1347 Agroécologie, Dijon, France

Abstract

Listeria monocytogenes is a food-borne pathogen responsible for the potentially fatal disease listeriosis and terrestrial ecosystems have been hypothesized to be its natural reservoir. Therefore, identifying the key edaphic factors that influence its survival in soil is critical. We measured the survival of *L. monocytogenes* in a set of 100 soil samples belonging to the French Soil Quality Monitoring Network. This soil collection is meant to be representative of the pedology and land use of the whole French territory. The population of *L. monocytogenes* in inoculated microcosms was enumerated by plate count after 7, 14 and 84 days of incubation. Analysis of survival profiles showed that *L. monocytogenes* was able to survive up to 84 days in 71% of the soils tested, in the other soils (29%) only a short-term survival (up to 7 to 14 days) was observed. Using variance partitioning techniques, we showed that about 65% of the short-term survival ratio of *L. monocytogenes* in soils was explained by the soil chemical properties, amongst which the basic cation saturation ratio seems to be the main driver. On the other hand, while explaining a lower amount of survival ratio variance (11%), soil texture and especially clay content was the main driver of long-term survival of *L. monocytogenes* in soils. In order to assess the effect of the endogenous soils microbiota on *L. monocytogenes* survival, sterilized versus non-sterilized soils microcosms were compared in a subset of 9 soils. We found that the endogenous soil microbiota could limit *L. monocytogenes* survival especially when soil pH was greater than 7, whereas in acidic soils, survival ratios in sterilized and unsterilized microcosms were not statistically different. These results point out the critical role played by both the endogenous microbiota and the soil physic-chemical properties in determining the survival of *L. monocytogenes* in soils.

Editor: Nancy E. Freitag, University of Illinois at Chicago College of Medicine, United States of America

Funding: Ph.D. grant of Aude Locatelli was funded by the French Agency for Environment and Energy Management (ADEME) and INRA (Environnement Agronomie Department). Part of this work was funded by the Conseil Régional de Bourgogne. This work was also funded by the 'Agence Nationale de la Recherche' (ANR) (programs 07 SEST project 018-01, 05 SEST project 009-01). The funders had no role in study design, data collection and analysis, decision to publish, or preparation of the manuscript.

Competing Interests: The authors have declared that no competing interests exist.

* E-mail: alain.hartmann@dijon.inra.fr

Introduction

Listeria monocytogenes is a food-borne pathogen responsible for listeriosis a potentially fatal disease that results in meningitis, septicemia or abortion [1,2]. This disease can affect humans and a large range of wild and domestic animals [3]. Outbreaks of human listeriosis have been reported worldwide and are mainly associated with consumption of various contaminated food such as meat, dairy products, vegetables and fish [4–7]. Ready-to-eat food products, which are consumed without further cooking, are most likely at the origin of listeriosis outbreaks [8–11]. Although listeriosis infections are uncommon, mortality rates can reach 30% in at-risk people [12–14]. As a consequence, *L. monocytogenes* is recognized as one of the most important food-borne pathogen.

L. monocytogenes is widely distributed in nature including vegetation [15,16], water [17], sediment [18,19] and soil [15,20]. Although *L. monocytogenes* is ubiquitous in the environment, human and animals are likely to be an important reservoir [3,21,22]. *L. monocytogenes* has been isolated from livestock, domestic and wild animals in both infections and latent states [3,23], in animal feces and in the close environment of animals [24]. The incidence of *L. monocytogenes* is generally higher in fecal sample of healthy cattle (33%) than in sheep (8%) or pig (5.9%) [3]. This finding is consistent with other studies reporting a significantly higher prevalence of *L. monocytogenes* positive samples in bovine farms than in small ruminant farms without listeriosis cases [24]. In British-fresh livestock manure, prevalence of *Listeria* spp. (including *L. monocytogenes* and *L. ivanovii*) is globally higher in cattle (29.8%) and sheep (29.2%) than in pig (19.8%) and poultry (19.4%) wastes with levels ranging from 2×10^2 to 1×10^3 *Listeria* spp. per gram of manure [25].

Farm environments are potential sources of *L. monocytogenes* and may contribute to the contamination of vegetables at the pre-harvest stage. *L. monocytogenes* is frequently isolated from a large variety of vegetables collected in farms [26,27]. One of the first potential sources of vegetable contamination at the preharvest stage (in the field) is soil when seeds are sown. In addition, some agricultural practices such as recycling animal feces as crop fertilizers or irrigation with contaminated water may increase the risk of soil and vegetable contamination. Soil fertilized with sludge cake can contaminate parsley seeds with *L. monocytogenes* which can be detected until plant harvesting [28]. Finally, direct transfer of *L. monocytogenes* from amended soil to seeds of carrots, lettuce, radish, spinach and tomato has been described [29]. Recent field experiment has shown that the transfer of the pathogen surrogate

Listeria innocua from contaminated soil to parsley leaves can occur by splashing due to rain and irrigation [30].

Public health hazard linked with transmission of pathogens from soil to plants and vegetables is relevant only if the pathogen is able to survive long enough in soil. Previous studies have investigated which soil properties might impact *L. monocytogenes* survival. Survival studies were performed either by direct inoculation of *L. monocytogenes* in soil or by adding contaminated fertilizer in soil. First, survival of *L. monocytogenes* is not significantly affected by the type of livestock manure added to soil [31,32]. Soil type had a strong effect on *L. monocytogenes* survival. *L. monocytogenes* survived better in a fertile soil (up to 295 days) than in a clay soil (at 24–26°C) [33]. *L. monocytogenes* population was stable in clay soils, significantly decrease in sandy soils, while displaying an intermediary survival in sandy loam soils up to 30 days [31]. Regardless of the type of manure spread on soil, *L. monocytogenes* persisted over 32 days in a clay loam grassland soil while survival was lower in a sandy arable soil [32]. Soil pH seems to be determinant for *L. monocytogenes* persistence which can survive more than 32 days in 2 soils harboring pH of 6.5 and 6.9 [32]. On the contrary, *L. monocytogenes* EGDe did not persist more than 6 days at 25°C in a forest soil characterized by a low pH (5.22) [34]. The rapid decline observed in this study, can be explained by the low pH of the soil. Higher survival of *L. monocytogenes* was observed at low temperature [34]. Soil microflora appears to have an impact on *L. monocytogenes* survival. Generally, suppression of microflora via soil sterilization allowed a better growth of *L. monocytogenes* than in the presence of a competitive microflora [31,34,35]. Biotic and abiotic soil parameters also affect the persistence in soils of pathogenic bacteria belonging to the Enterobacteriaceae family, for example, *Escherichia coli* and *Salmonella enterica* [32,36,37].

Generally, studies reporting the survival of pathogenic bacteria in soil, including *L. monocytogenes*, were focused on a limited number of poorly characterized soils [31,32,35–37]. Identification of the soil abiotic and biotic parameters influencing the survival of pathogenic bacteria in terrestrial ecosystems will help understanding their cycle of contamination in the environment. The objective of this study was to assess the survival of *L. monocytogenes* in a large collection (n = 100) of well characterized soils collected throughout France and representative of the pedology and land use of the whole territory. ANOVA and variance partitioning were used to correlate 40 soil parameters with *L. monocytogenes* survival ratios in order to identify parameters that determine the fate of *L. monocytogenes* in soil.

Materials and Methods

Soil Samples

One hundred soil samples were randomly chosen among the soil library of the French Soil Quality Monitoring Network (RMQS, Réseau de Mesure de la Qualité des Sols) [38,39]. Soils were collected from 2001 to 2010 following a single sampling procedure. Soil sample preparation and storage were achieved according to ISO and AFNOR standards : NF ISO 10381-1 [40] and ISO 10381-6 [41]. A large range of physical (particle-size) and chemical parameters (pH, organic C, N, exchangeable cations and cation exchange capacity (CEC)...) were measured for each soil by the Soil Analysis Laboratory of INRA (Arras, France, http://www.lille.inra.fr/las). In addition, 2 variables were calculated from measured parameters: (i) C/N ratio and (ii) Base Cation Saturation Ratio (%) determined by the sum of Ca^{2+}, Mg^{2+}, Na^{+}, K^{+} divided by CEC and multiplied by 100. Supplementary information such as climatic data (monthly rain, evapotranspiration (ETP) and temperature) and detailed land cover are available at the DONESOL database (http://www.gissol.fr/programme/rmqs/RMQS_manuel_31032006.pdf [42,43]). All parameters characterizing the 100 soils used in this study, including soil texture, soil chemistry, land-use, climatic data and spatial localization are detailed as supporting information (Table S1). Moreover, the textural classification of the 100 soils is represented in supplementary figure (Figure S1).

Bacterial Strain and Culture Conditions

The rifampicin-resistant (Rif^R) mutant L9 of wild-type strain *L. monocytogenes* EGD-e was used in this study [44]. Stock cultures were prepared by growing *L. monocytogenes* Rif^R in Tryptone Soy Broth (TSB, AES Chemunex, Bruz, France) at 37°C. After washing in sterile water, the cell pellet was suspended in a Brain Heart Infusion (BHI, AES Chemunex, Bruz, France) with 25% of glycerol, aliquoted (200 µl in microtubes) and frozen −80°C until further utilization. Pre-cultures were prepared by inoculating 20 ml of TSB inoculated with 200 µl of the stock culture. After 48H of incubation at 20°C, pre-cultures were centrifuged (10,000 g, 5 min), pellets were washed and re-suspended in 20 ml of physiological saline solution (8.5 g/L NaCl). Cell density was estimated spectrophotometrically (a cell suspension with an OD_{600nm} of 1 was considered to contain 1.9×10^9 *L. monocytogenes* cells per ml), and pre-cultures were diluted with physiological saline solution to a final inoculum concentration of 1.6×10^7 *L. monocytogenes* per ml. The exact concentration of the final inoculum was determined by plate count on TSB medium.

Microcosms Preparation and Inoculation

Microcosms were prepared in sterile flasks (40 ml) using 2 grams of carefully mixed and homogenized soil taken from the initial stock (100 grams).

To test the effect of biotic factors, a subset of nine soils were sterilized by gamma-radiation. Two grams of soil were conditioned in a 40 ml-flask. For each soil, 9 individual microcosms were prepared (3 sampling days and 3 replicates). All 81 (9×9) soil microcosms were packed in a box and send to Ionisos for gamma radiation sterilization (Dagneux, France). The entire box was sterilized without being open by receiving an external minimum dose of 45 KGray and an external maximum dose of 60 KGray.

Soil microcosms were adjusted to 80% of the water field capacity one week before inoculation. Water field capacity was determined by granulometric method taking into account clay, fine silt and organic matter content of each soil. For all tested soils, one individual microcosm was prepared for each sampling time. One hundred and twenty µl of inoculum was added to each microcosm giving concentrations of 1×10^6 *L. monocytogenes* per gram of dry soil. Due to the large number of soils tested only one repeat was realized, however the reproducibility of the method was evaluated on a subset of 9 soils for which three repeats were realized. On the subset of nine soils, three independent flasks were inoculated for each soil condition (sterilized or non-sterilized). For each sterilized soil, uninoculated microcosms remained sterile for the whole duration of the experiment (84 days), thus proving that gamma irradiation eradicated all microorganisms and that no recontamination of microcosms occurred during the experiment. Soil microcosms were incubated at 20°C and were sampled 1 h (*i.e.* survival ratio at t_0), 7, 14 and 84 days after inoculation. All viable counts were expressed per gram of dry soil.

Total cultivable soil bacteria were enumerated on one tenth-strength Nutrient Agar medium supplemented with 100 mg/L of cycloheximide for non-sterile soil microcosms at each sampling time (from day 0 to day 84 after inoculation).

Monitoring *L. monocytogenes* Survival by Viable Plate Count

Eighteen ml of Tryptone Salt (TS, 1 g/L tryptone, 8.5 g/L NaCl) was added to each microcosm and bacteria were resuspended by shaking 10 min, 150 rpm, 20°C. This suspension was serially diluted and plated on Polymyxin-Acriavin-Lithium-Chloride-Ceftazidime-Aesculin-Mannitol agar (PALCAM, AES Chemunex, Bruz, France) supplemented with rifampicin and cycloheximide each at 100 mg/L and incubated at 37°C. Appropriate dilutions were plated on PALCAM medium. When high levels of *L. monocytogenes* are expected in soil, 10×10 µL were spotted for each dilution level (Detection limit = 100 bacteria per gram of dry soil), while when less than 100 *L. monocytogenes* per gram of soil was expected, 1 ml of the soil suspension was poured into the medium (Detection limit = 10 bacteria per gram of dry soil). Uninoculated control experiments were processed accordingly to verify that *L. monocytogenes* was not initially detected in uninoculated soils.

Statistical Analyses of the Survival Data

Survival ratios were calculated as follows prior to the analyses to make them closely conform to a Gaussian distribution:

$$SR_{t_x} = \log_{10}\left(\frac{CFU_{t_x}}{CFU_{t_0}}\right)$$

Where SR_{t_x} is the survival ratio at time t_x ($x = 7$ days, 14 days or 84 days), CFU_{t_x} is the number of Colony Forming Units at time t_x and CFU_{t_0} is the number of Colony Forming Units at time t_0.

Replicated Experiment : pH, Land Use and Sampling Day Accounting for Variation of *L. monocytogenes* Survival Ratio in Sterilized and Non-sterilized Soils

A repeated-measures ANOVA was performed on the survival ratios. Factors accounting for variation of the survival ratio were the pH of the soils (treated as an ordinal variable: pH <5.5; $5.5<$pH <7; pH >7), the presence or absence of soil microflora (non-sterilized versus sterilized soil), the land use (culture, forest and grassland) and the sampling time, the latter being the repeated factor. Pair-wise differences between levels of factors were assessed using a *t*-test. Bonferroni corrections were used to take into account multiple comparisons.

Ecological Survey: Partitioning of the Biological Variation for Survival Ratio in the 100 Soils

Forty different soil properties and environmental parameters were used in this analysis (Table S1). To identify the soil edaphic and the environmental factors driving *L. monocytogenes* survival in soils, all the explaining variables measured were first grouped into five categories: soil texture, soil chemistry, land use, climate and spatial coordinates.

Spatial vectors were constructed using the Principal Coordinates of a Neighbor Matrix approach (PCNM, [45]). This method was applied to the geographic coordinates of the different sampled sites and yielded 59 spatial vectors. All quantitative explanatory variables were transformed using Box-Cox transformation prior to analyses (the corresponding lambda parameters were estimated by maximum likelihood [46]).

For each SR_{t_x}, significant explanatory variables, as well as PCNM vectors, were chosen by model selection (forward and backward) and by minimizing the Akaike Information Criterion (AIC). Statistical significance was assessed by 1999 permutations of the reduced model. The respective contribution of contextual variables (explanatory variables or combinations of explanatory variables) was assessed using redundancy analysis ordination [47]. All these analyses were performed with R using the vegan package (functions *pcnm*, *varpart* and *rda*).

Results

Identification of Soil Abiotic (Physico-chemical) Properties Impacting *L. monocytogenes* Survival in a Panel of 100 Soils

Total cultivable bacteria were enumerated for each soil at day 0, day 7, day 14 and day 84. At day 0, total number of bacteria range from 4.78×10^6 to 1.07×10^8 CFU per gram of soil. We found that the total bacterial community counts remain stable (within the same order of magnitude) over the 2-month experiment period for all investigated soils (Student's t test, $p = 0.265$).

As soils were stored for varying periods, a regression analysis between *L. monocytogenes* survival rates and soil sampling date (year) was done. This analysis revealed no correlation between these two variables (data not shown). The age of soil samples doesn't not seem to influence *L. monocytogenes* survival.

Survival of *L. monocytogenes* was determined in a collection of 100 soils. *L. monocytogenes* population globally declined with time in the 100 soils. In 71% of the soils, *L. monocytogenes* was detected until the end of the experiment (i.e. 84 days after inoculation) and final populations ranged from 10 to 1.27×10^4 CFU per gram of soil. In 21% of the soils, survival was observed only at short term (up to day 7 or up to day 14). Finally, in 8% of the soils, *L. monocytogenes* was not detected 7 days after inoculation.

To identify soil edaphic and environmental factors driving *L. monocytogenes* survival in soils, partial regression models were calculated for the 5 categories of explaining variables (Table 1). Using this approach, we were able to explain from 46.6% to 79.5% of the survival observations. Most of the variance of the survival ratios was explained by 3 groups of variables, *i.e.* soil chemistry, soil texture and spatial localization. Climate and land use do not appear in the model (Table 1) because they do not explain any variance of the survival ratios of *L. monocytogenes*. Soil chemistry was relevant to explain short-term survival ratios at days 7 and 14 (64.5% and 65.4%) of *L. monocytogenes* in soils (Table 1). When studying the effect of each variable independently, Basic Cation Saturation Ratio (BCSR) was identified as the major soil chemical characteristic determining short-term survival profiles (day 7 and day 14, Table 2) and differences in soil BCSR explained up to 55.4% of the variability of *L. monocytogenes* survival. Cationic exchange capacity (10.3%) in one hand and exchangeable Ca (11%) in the other hand further explained survival ratio of *L. monocytogenes* at day 7 and day 14, respectively. Soil texture was the contextual variable with the highest weight for explaining long-term survival profiles (up to day 84), as it explained 11% of the observed variance (Table 1). When studying the effect of each variable independently, clay content explaining 30.9% of the observed variance seems to be the major contributor followed by exchangeable Al (Table 2). Interestingly, pH was not found to be a significant contributing factor to *L. monocytogenes* survival but this is explained by the fact that BCSR and pH are strongly but non-linearly correlated as shown in supporting figure (Figure S2). Hence, variation in survival ratio captured by the BCSR cannot be attributed to the pH in the partial regression analysis. In other terms, if the BCSR was removed from the analysis, pH would be selected as a significant explaining variable.

Table 1. Partitioning of the variation of survival ratios of *Listeria monocytogenes* as a function of contextual variables.

		Overall model			% explained variance of the contextual variables		
		N[a]	F-ratio	Explained variance[b] (%)	Soil Chemistry	Soil Texture	Spatial Distance
Survival at	Day 7	13	30.64***	79.5	64.5***	2**	0.5°
	Day 14	5	43.02***	67.1	65.4***	-	1.6°
	Day 84	8	27.64***	46.6	1.3[NS]	11.1***	4.2*

[a]N is the number of explanatory variables retained after selecting the most parsimonious explanatory variables (by minimizing the Akaike Information Criterion).
[b]the % explained variance corresponds to the adjusted R^2 values of the overall model using partial redundancy analysis.
[NS]Non Significant, °p<0.1, *p<0.05, **p<0.01 and ***p<0.001.
Note that the covariation between the contextual variables is not reported in this table, therefore summing over the different contextual variables does not give the % explained variance of the overall model.
- Soil texture do not explain any variance of the survival ratio of *L. monocytogenes* at day 14.

Evaluating the Impact of the Interactions between Biotic and Abiotic Soil Parameters on *L. monocytogenes* Survival

Focusing on a subset of 9 soils chosen for their contrasted physico-chemical properties and land use characteristics (Table 3), we evaluated the impact of the soil endogenous microbiota on the survival ratio of *L. monocytogenes* in a replicated complete block design. Statistical analysis by ANOVA allowed first to validate the reliability of inoculation and counting method and the repeatability between experiments. *L. monocytogenes* population globally declined with time in non-sterile soils. However, significant differences in the survival profiles were observed between the nine different soils (Figure 1, panel A). Long-term survival was observed after inoculation in 5 soils (n° 1492, 921, 755, 2191 and 1133) and the population of *L. monocytogenes* range from about 10^2 to 4×10^3 bacteria per gram of soil at the end of the experiment (84 days). In three other soils (soils n°1500, 765, and 1709) survival did not exceed 7 to 14 days after inoculation, while in soil n° 881, *L. monocytogenes* was no longer detected as soon as 7 days after inoculation. Results showed that the survival of *L. monocytogenes* was different between sterile soils (Figure 1, panel B). Indeed, 1 to 3 log growth was observed in soil n° 1492, 921 and 755. For the others soils, no growth of *L. monocytogenes* was observed and *L. monocytogenes* abundance decreased. In these soils, both short and long-term survival profiles were observed as already noticed with-non sterile soils.

Using repeated-measures ANOVA, we investigated whether pH or land use could be factors interacting with the soil microbiological status and accounting for the different survival profiles observed over the experiment time course (Table 4). We found a strong and significant interaction between pH and the microbiological status of the soil ($F_{2,132}=55.03$, $p<0.001$) that explains a large part of the observed variance in *L. monocytogenes* survival ratio. The 3-way interaction integrating time is also significant ($F_{4,132}=4.47$, $p=0.002$) indicating that differences in survival ratio between pH classes for the sterilized and the unsterilized soils vary over time. Globally, *L. monocytogenes* survival is higher in the highest pH class of soils, than in lower pH classes of soils. Moreover, the suppressive role of the endogenous microbiota on its survival is clearly evidenced in the high pH class and growth of *L. monocytogenes* populations was observed in sterilized soils grouped in the highest pH class (Figure 2).

We also found a significant 3-way interaction between land use, microbiological status of the soil and time of sampling ($F_{4,132}=4.14$, $p=0.003$) on *L. monocytogenes* survival. The highest survival was observed in soils collected from culture fields while it is lowest in grassland (sterilized or unsterilized). However, land use classes overlapped to some extent with pH classes. For example, two out of the three culture soils belong to the high pH class and two out of the three grassland soils belong to the low pH class, the forest soils being the only well-spread out in the different pH classes. For these reasons, pH appeared to be more likely (than land use) an explaining factor determining *L. monocytogenes* survival in soils. This study based on a small subset of soils allowed us to identify the pH and the endogenous microbial communities as important factors influencing *L. monocytogenes* survival in soils.

Discussion

Circulation of pathogens in the farm environment may generate health hazards [26,27,48]. Regarding food-borne pathogens, pre-harvest contamination may be a source of food contamination, thus increasing the risk of outbreaks. The presence of the food-borne pathogen *L. monocytogenes* in soil has been connected with pre-harvest food contamination [29,30,49] and its survival in soil

Table 2. Contribution of the five most important explanatory variables to the variation in survival ratios of *Listeria monocytogenes*.

		% Variance explained by:				
Survival at	*Day 7*	BCSR (55.4%)	CEC (10.3%)	Coarse Silt (4.6%)	Sp. $Dist_{V6}$ (2.6%)	Al_{exch} (1.8%)
	Day 14	BCSR (47.7%)	Ca_{exch} (11%)	Mn_{exch} (4.9%)	Al_{exch} (4.6%)	Sp. $Dist_{V13}$ (1.1%)
	Day 84	Clay (30.9%)	Al_{exch} (5.5%)	$CaCO_{3tot}$ (3.5%)	$Sp.Dist_{V3}$ (2.9%)	$Temp_{month}$ (2.7%)

The respective contributions of each variable were calculated by taking into account all the significant variables in the model using partial redundancy analyses (significance assessed with 1999 data permutations).
BCSR : Basic Cation Saturation Ratio, CEC : Cation-Exchange Capacity, Sp. $Dist_V$: Spatial Distance correspond to the spatial vector *x* from the PCNM analysis, Al_{exch} : exchangeable aluminum, Ca_{exch} : exchangeable calcium, Mn_{exch} : exchangeable manganese, $CaCO_{3tot}$: Total calcareous content, $Temp_{month}$: mean temperature per month.

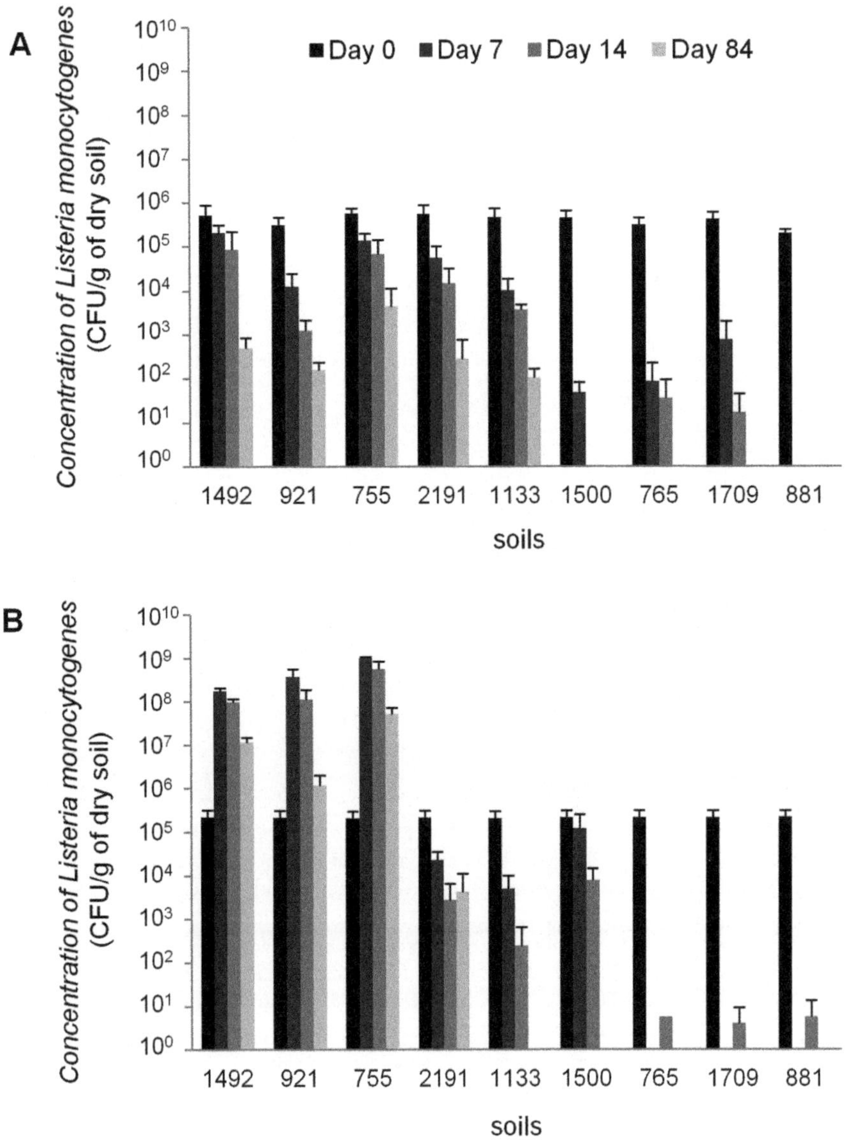

Figure 1. Survival of *Listeria monocytogenes* in nine soils: non-sterile (panel A) and sterile (panel B). Black, dark grey, medium grey and light grey bars represent *L. monocytogenes* population density (CFU per gram of dry soil) at Days 0, 7, 14 and 84 after inoculation, respectively. Error Bars indicate the mean ± standard deviation over three replicated measurements.

Table 3. Land use and main edaphic factors of the subset of nine soils.

Soil n°	Land use	pH	Clay content (g/kg soil)	Silt content (g/kg soil)	Sand content (g/kg soil)
765	Grassland	4.7	121	164	715
1709	Grassland	4.7	185	288	527
2191	Grassland	5.9	374	440	186
921	Culture	7	454	508	38
1492	Culture	7.9	403	223	374
1500	Culture	5.3	150	210	640
755	Forest	7	650	334	16
881	Forest	4.7	153	446	401
1133	Forest	5.6	819	92	89

Table 4. Analysis of variance for *Listeria monocytogenes* survival ratios in a subset of nine soils.

	Survival Ratio		
Sources of Variation	*df*	*Sum of Squares*	*F*
pH	2	608.15	471.88***
Microbiological status[a]	1	99.20	153.94***
Land use	2	21.16	16.42***
Day of sampling	2	90.77	70.44***
*Land use*Microbiological status*	2	13.31	10.33***
*Land use*Day of sampling*	4	1.85	0.72
*pH*Microbiological status*	2	70.92	55.03***
*pH*Day of sampling*	4	25.77	9.99***
*Microbiological status*Day of sampling*	2	0.27	0.21
*Land use*Microbiological status*Day of sampling*	4	10.68	4.14**
*pH*Microbiological status*Day of sampling*	4	11.53	4.47**

df: degrees of freedom, F: Fisher's F, .
[a]correspond to sterilized versus non-sterilized soils.
***p<0.001,
**p<0.01.

has been documented [31–34] but limited information is available on the soils used in these studies and edaphic factors that may affect its survival are poorly understood. Previous studies focused on a limited number of soils including from 1 to 3 soils [32,33]. Moreover, in these studies, soils were mainly characterized by their textural classes and only a few chemical characteristics were available and were used in statistical analysis [31,32]. In this study, we investigated the survival of *L. monocytogenes* in a large set of 100 well-characterized soils (chemical and textural characteristics) representative of the French territory and occurring worldwide. Indeed, the variation of the age of the soil samples is inherent to a study based on a very large set of soils. In our study, the age of soil samples did not explain any variation of the survival rates of *L. monocytogenes*. This result suggests that soil storage did not modify significantly soil composition and thus did not introduce a major bias in the experiment. Moreover, there is no significant quantitative variation of total soil bacterial populations over the 2 month's experiment for all soil tested.

We observed a decline of *L. monocytogenes* populations with time in all 100 soils that is consistent with *L. monocytogenes* behavior reported in previous studies [31,33–35]. Moreover, in the present study, we demonstrated that the survival ratio depended on the soil under scrutiny and overall three trends were evidenced: long-term, short-term and lack of survival. Indeed, in most soils (71%), *L. monocytogenes* was still detected at the end of the experiment (84 days); this long-term survival is in agreement with previous studies relating *L. monocytogenes* survival until 200 days in a clay soil and 295 days in a "fertile" (as stated by authors) soil [31,33–35]. In 21% of soils, the pathogen was no longer detected after two weeks. Finally, in 8% of soils, a dramatic decline to undetectable levels occurred within the first week of incubation. A similar observation was reported in a forest soil by McLaughlin *et al.* [34]. Identification of edaphic factors that may explain these trends is difficult as biotic and abiotic parameters are intertwined, however our study demonstrated clearly and for the first time that *L. monocytogenes* is able to survive in a majority of soils.

We used intentionally large numbers of bacteria for the inoculation of soil samples to facilitate short term monitoring of declining populations. This might introduce a bias since the survival of larger bacterial population requires higher quantities of nutrients. However, Dowe *et al.* (1997), using two levels of inoculum (10^2 and 10^6 bacteria per gram of soil) have demonstrated that after a short period, *L. monocytogenes* counts

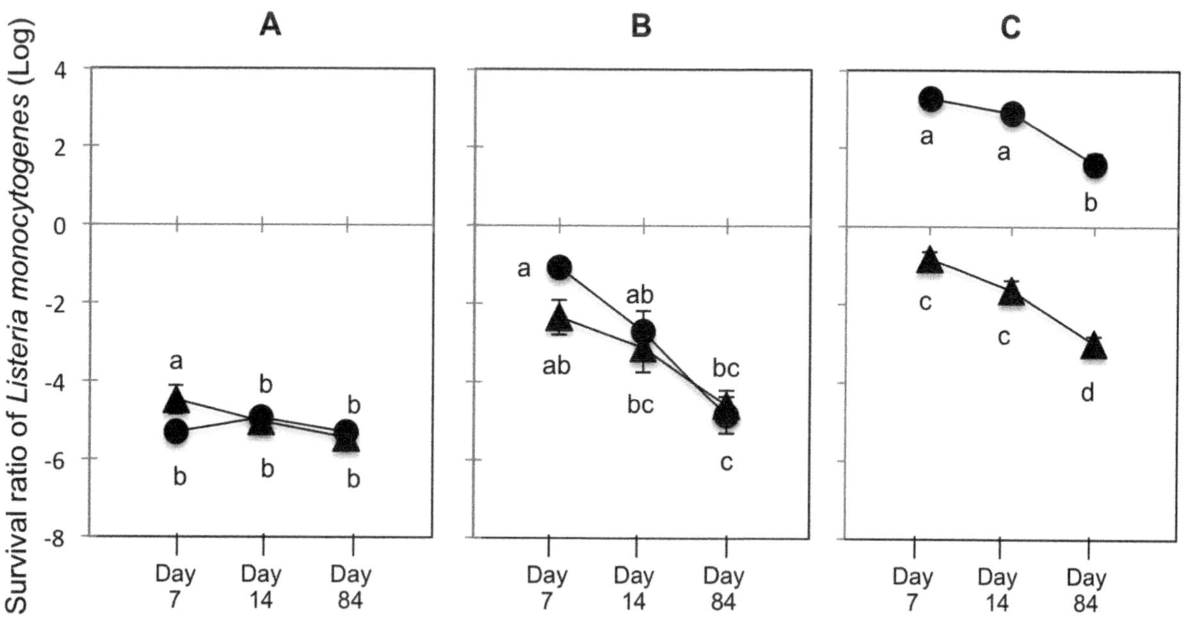

Figure 2. Mean survival ratios of *L. monocytogenes* in non-sterile and sterile soils depends on soil pH. The nine soils were distributed in three equal classes function of their pH with (A) 3 soils with pH <5.5, (B) 3 soils with 5.5<pH <7 and (C) 3 soils with pH >7. Mean survival ratios were calculated between soils belonging to the same pH classes for each sampling time. Triangles represents mean survival ratio in non-sterile soils and circles represented mean survival ratio in sterile soils. Error bars represent the mean ± standard deviation of triplicate experiments.

reached similar levels with both inocula [31]. This result indicates that *L. monocytogenes* survival in soil is not durably affected by the size of the inoculum.

The statistical model we developed could explain most of the variance of *L. monocytogenes* short-term survival (67% at day 14 and 80% at day 7) suggesting that the comprehensive characterization of our soil samples is adequate to explain short-term survival ratio. The soil chemistry was the most significant contextual variable and, among these variables, the Base Cation Saturation Ratio (BCSR) was the most impacting on short-term survival. BCSR was calculated as a ratio between the sum of exchangeable base cations (Ca^{2+}, Mg^{2+}, K^{+} and Na^{+}) and CEC. The CEC of a soil is the total quantity of exchangeable cations that a soil can fix and release at a specific pH.

Contrarily to short-term survival, the model was less efficient in explaining variations of long-term survival (47% at day 84) this suggests that the soil abiotic characteristics that fed the model were not sufficient to explain long-term survival. We can hypothesize that soil microflora is one of the variables missing and that biotic factors are critical for long term survival. However, soil texture and more precisely clay content appear to be the principal variables explaining long-term survival variance among the abiotic factors analyzed. (Figure S1).

Until now, only a few studies have shown that soil texture had an impact on *L. monocytogenes* survival [31–33]. The survival of other pathogens such as *E. coli* or *Salmonella* spp. was also shown to depend on the soil texture [32,36,37,50]. In general, these studies demonstrated that finer-textured (clayey) soils result in prolonged survival of introduced bacterial pathogens compared to coarser-textured (sandy) soil.

The experiment conducted on a subset of 9 soils (comparison of sterile versus non sterile soil) evidenced the major role of soil microflora in controlling *L. monocytogenes* survival. Our results are in agreement and reinforce previous studies [31,34,35] showing that suppression of the microflora allowed the growth of *L. monocytogenes* in soils. Unexpectedly, no *L. monocytogenes* growth was observed in sterilized microcosms for 6 out of the 9 soils tested in our study. This result strongly suggests that abiotic characteristics of these soils were non-permissive for the growth (and survival) of *L. monocytogenes*. The repeated-measures ANOVA pointed out to pH as the major factor explaining this lack of growth. Indeed, we found a strong and significant interaction between pH and soil microflora on *L. monocytogenes* survival. More precisely, the suppressive effect of soil microflora on *L. monocytogenes* survival was the strongest in soils with high pHs. Soil pH has already been identified as a structuring and even predictive parameter for the composition and structure of global bacterial communities' [51,52]. Bacterial richness and diversity appears to be higher in neutral soils and lower in acidic soils [51,53]. Similarly, pH seems to be critical for the fate of *L. monocytogenes* in soil, since the survival ratio is higher in neutral soils.

In our study, we demonstrate that chemical properties of soils (in particular BCSR) explain most of the variability of short-term survival (soil texture explaining mostly long-term survival). However, it is established that soil chemistry is tightly linked to soil texture. The BCSR reflects the amount of base cations present in soil as well as the number of negative sites supplied by the soil matrix. Clay particles and organic matter are both constituents of soil that harbor negative charges, which can fix and release positively charged nutrients including cations. So, clay and organic matter content largely determine the CEC and the BCSR of a soil. Most of the studies investigating the effect of soil texture on bacterial pathogens survival highlight the fact that finer-textured soils with high clay content are more favorable to bacterial pathogens survival than coarser-textured soils. This trend can be explained by a higher availability of pore spaces protecting bacteria from protozoan predation. Our result further complete this explanation proving that the finer soil texture with high clay content will also help to maintain a sufficient base cations pool essential for bacterial life. This hypothesis can be confirmed by the result of the long-term survival, which is correlated with clay content.

Only plate counts were used to monitor *L. monocytogenes* survival. This might lead to underestimate *L. monocytogenes* concentration if Viable But Not Culturable (VBNC) cells of *L. monocytogenes* are present [54,55]. Indeed, VBNC state can be induced for *L. monocytogenes* in response to low pH and nutrient limitation [55–57]. These conditions (low pH and starvation) may occur in some soil microcosms. Occurrence of VBNC *L. monocytogenes* may induce inconsistencies of the model to explain variations of *L. monocytogenes* survival (especially for survival at 84 days). Counting VBNC bacteria in non sterile soil samples is extremely difficult, however, this might be achieved in the future.

Overall, this is the first extensive study of the survival of *L. monocytogenes* in a large collection of well-characterized soils. We found that the pool of cations that soil can exchange is an indicator of *L. monocytogenes* short-term survival, that population decline is faster in acidic soils and finally that the presence of the microflora participates to the barrier effect of soil towards invasion by *Listeria monocytogenes*. Further work should aim at deciphering which members of the soil microflora are critical to explain *L. monocytogenes* survival. This might be realized on a limited subset of soils chosen for contrasted characteristics.

Supporting Information

Figure S1 Distribution of the 100 soils in the textural triangle. Each dot corresponds to one soil. Survival ratio of *L. monocytogenes* in soils at day 84 are expressed as the grey level of each dot (light grey corresponding to low survival ratios and dark grey corresponding to high survival ratios).

Figure S2 Relationship between pH and corresponding BCSR of the 100 non-sterile soils. Each dot corresponds to one of the 100 soils tested.

Table S1 Soil parameters, including soil texture, soil chemistry, land-use, and climatic data characterizing the 100 soils used in this study.

Acknowledgments

The RMQS network is supported by a French Scientific Group of Interest on soils: the "GIS Sol", involving the French Ministries in charge of Ecology and Agriculture, ADEME, INRA, the Institute for Research and Development (IRD) and the National Institute for Geographic and Forest Information (IGN). We thank all the soil surveyors and technical assistants involved in sampling the RMQS sites. Special thanks are addressed to the technical assistants from the National French Soil Bank for sample handling and preparation.

Author Contributions

Conceived and designed the experiments: AL AS CJ PP AH. Performed the experiments: AL AS CJ PP AH. Analyzed the data: AL AS CJ PP AH. Contributed reagents/materials/analysis tools: AL AS CJ PP AH. Wrote the paper: AL AS CJ PP AH.

References

1. Farber JM, Peterkin PI (1991) *Listeria monocytogenes*, a food-borne pathogen. Microbiol Rev 55: 476–511.
2. Low JC, Donachie W (1997) A review of *Listeria monocytogenes* and listeriosis. Vet J 153: 9–29.
3. Wesley I (1999) Listeriosis in animals. In: Ryser E, Marth E, editors. *Listeria*, Listeriosis and Food Safety. New York: Dekker Marcel. 39–73.
4. Beuchat LR (1996) *Listeria monocytogenes*: Incidence on vegetables. Food Control 7: 223–228.
5. Jacquet C, Catimel B, Brosch R, Buchrieser C, Dehaumont P, et al. (1995) Investigations related to the epidemic strain involved in the French listeriosis outbreak in 1992. Applied and Environmental Microbiology 61: 2242–2246.
6. Mead PS, Dunne EF, Graves L, Wiedmann M, Patrick M, et al. (2006) Nationwide outbreak of listeriosis due to contaminated meat. Epidemiology and Infection 134: 744–751.
7. Rocourt J, Jacquet C, Reilly A (2000) Epidemiology of human listeriosis and seafoods. International Journal of Food Microbiology 62: 197–209.
8. Sim J, Hood D, Finnie L, Wilson M, Graham C, et al. (2002) Series of incidents of *Listeria monocytogenes* non-invasive febrile gastroenteritis involving ready-to-eat meats. Letters in applied microbiology 35: 409–413.
9. Guenther S, Huwyler D, Richard S, Loessner MJ (2009) Virulent bacteriophage for efficient biocontrol of *Listeria monocytogenes* in ready-to-eat foods. Applied and environmental microbiology 75: 93–100.
10. Gaul LK, Farag NH, Shim T, Kingsley MA, Silk BJ, et al. (2013) Hospital-acquired Listeriosis outbreak caused by contaminated diced celery - Texas, 2010. Clinical Infectious Diseases 56: 20–26.
11. Aureli P, Fiorucci GC, Caroli D, Marchiaro G, Novara O, et al. (2000) An outbreak of febrile gastroenteritis associated with corn contaminated by *Listeria monocytogenes*. New England Journal of Medicine 342: 1236–1241.
12. Siegman-Igra Y, Levin R, Weinberger M, Golan Y, Schwartz D, et al. (2002) *Listeria monocytogenes* infection in Israel and review of cases worldwide. Emerging infectious diseases 8: 305.
13. Wing EJ, Gregory SH (2002) *Listeria monocytogenes*: clinical and experimental update. Journal of Infectious Diseases 185: S18–S24.
14. Slutsker l, Schuchat A (1999) Listeriosis in Humans. In: Ryser E, Marth E, editors. *Listeria*, Listeriosis and Food Safety. New York: Dekker Marcel. 75–95.
15. Weis J, Seeliger HPR (1975) Incidence of *Listeria monocytogenes* in nature. Applied Microbiology 30: 29–32.
16. Welshimer HJ (1968) Isolation of *Listeria monocytogenes* from Vegetation. Journal of Bacteriology 95: 300–303.
17. Lyautey E, Lapen DR, Wilkes G, McCleary K, Pagotto F, et al. (2007) Distribution and Characteristics of *Listeria monocytogenes* Isolates from Surface Waters of the South Nation River Watershed, Ontario, Canada. Applied and Environmental Microbiology 73: 5401–5410.
18. Colburn KG, Kaysner CA, Abeyta C, Wekell MM (1990) *Listeria* species in a California coast estuarine environment. Applied and Environmental Microbiology 56: 2007–2011.
19. El Marrakchi A, Boum'handi N, Hamama A (2005) Performance of a new chromogenic plating medium for the isolation of *Listeria monocytogenes* from marine environments. Letters in Applied Microbiology 40: 87–91.
20. Welshimer HJ (1975) *Listeria* in nature. Problems of listeriosis: 59–60.
21. Grif K, Patscheider G, Dierich MP, Allerberger F (2003) Incidence of fecal carriage of *Listeria monocytogenes* in three healthy volunteers: A one-year prospective stool survey. European Journal of Clinical Microbiology & Infectious Diseases 22: 16–20.
22. Borucki MK, Gay CC, Reynolds J, McElwain KL, Kim SH, et al. (2005) Genetic diversity of *Listeria monocytogenes* strains from a high-prevalence dairy farm. Applied and Environmental Microbiology 71: 5893–5899.
23. Gray ML, A. H Killinger (1966) *Listeria monocytogenes* and listeric infections. BacteriologyRevews 30: 309–382.
24. Nightingale KK, Schukken YH, Nightingale CR, Fortes ED, Ho AJ, et al. (2004) Ecology and transmission of *Listeria monocytogenes* infecting ruminants and in the farm environment. Applied and Environmental Microbiology 70: 4458–4467.
25. Hutchison ML, Walters LD, Avery SM, Munro F, Moore A (2005) Analyses of livestock production, waste storage, and pathogen levels and prevalences in farm manures. Applied and Environmental Microbiology 71: 1231–1236.
26. Aytac SA, Ben U, Cengiz C, Taban BM (2010) Evaluation of *Salmonella* and *Listeria monocytogenes* contamination on leafy green vegetables. J Food Agric Environ 8: 275–279.
27. Jeyaletchumi P, Tunung R, Selina PM, Chai LC, Radu S, et al. (2011) Evaluation of *Listeria* spp. and *Listeria monocytogenes* in selected vegetable farms. J Trop Agric and Fd Sc 39: 255–266.
28. Al-Ghazali MR, Al-Azawi SK (1990) *Listeria monocytogenes* contamination of crops grown on soil treated with sewage sludge cake. Journal of Applied Bacteriology 69: 642–647.
29. Jablasone J, Warriner K, Griffiths M (2005) Interactions of *Escherichia coli* O157:H7, *Salmonella typhimurium* and *Listeria monocytogenes* plants cultivated in a gnotobiotic system. International Journal of Food Microbiology 99: 7–18.
30. Girardin H, Morris CE, Albagnac C, Dreux N, Glaux C, et al. (2005) Behaviour of the pathogen surrogates *Listeria innocua* and *Clostridium sporogenes* during production of parsley in fields fertilized with contaminated amendments. FEMS Microbiology Ecology 54: 287–295.
31. Dowe MJ, Jackson ED, Mori JG, Bell CR (1997) *Listeria monocytogenes* survival in soil and incidence in agricultural soils. Journal of Food Protection 60: 1201–1207.
32. Nicholson FA, Groves SJ, Chambers BJ (2005) Pathogen survival during livestock manure storage and following land application. Bioresource Technology 96: 135–143.
33. Welshimer HJ (1960) Survival of *Listeria monocytogenes* in soil. Journal of Bacteriology 80: 316–320.
34. McLaughlin H, Casey P, Cotter J, Gahan C, Hill C (2011) Factors affecting survival of *Listeria monocytogenes* and *Listeria innocua* in soil samples. Archives of Microbiology 193: 775–785.
35. Botzler RG, Cowan AB, Wetzler TF (1974) Survival of *Listeria monocytogenes* in soil and water. Journal of Wildlife Diseases 10: 204–212.
36. Franz E, Semenov AV, Termorshuizen AJ, de Vos OJ, Bokhorst JG, et al. (2008) Manure-amended soil characteristics affecting the survival of *E. coli* O157 : H7 in 36 Dutch soils. Environmental Microbiology 10: 313–327.
37. Mubiru DN, Coyne MS, Grove JH (2000) Mortality of *Escherichia coli* O157 : H7 in two soils with different physical and chemical properties. Journal of Environmental Quality 29: 1821–1825.
38. Jolivet C, D Arrouays, L Boulonne, C Ratié and S Saby (2006) Le Réseau de mesures de la Qualité des sols de France (RMQS) Etat d'avancement et premiers résultats. étude et Gestion des Sols 13: 149–164.
39. Ranjard L, Dequiedt S, Jolivet C, Saby NPA, Thioulouse J, et al. (2009) Biogeography of soil microbial communities: a review and a description of the ongoing french national initiative. Agronomy for Sustainable Development 30: 359–365.
40. ISO10381–1 (2002) Soil quality - Sampling - Part 1: Guidance on the design of sampling programmes.
41. NFISO10381–6 (2009) Soil quality - Sampling - Part 6: Guidance on the collection, handling and storage of soil under aerobic conditions for the assessment of microbiological processes, biomass and diversity in the laboratory.
42. Grolleau E, L Bargeot, A Chafchafi, R Hardy, J Doux, A Beaudou, H. Le Martret, J. C Lacassin, J. L Fort, P Falipou and D Arrouays (2004) Le système d'information national sur les sols : DONESOL et les outils associés. Étude et Gestion des Sols 11: 255–269.
43. Ranjard L, Dequiedt S, Prévost-Bouré NC, Thioulouse J, Saby NPA, et al. (2013) Turnover of soil bacterial diversity driven by wide-scale environmental heterogeneity. Nature communications 4: 1434.
44. Lemunier M, Francou C, Rousseaux S, Houot S, Dantigny C, et al. (2005) Long-term survival of pathogenic and sanitation indicator bacteria in experimental biowaste composts. Applied and Environmental Microbiology 71: 5779–5786.
45. Borcard D, Legendre P (2002) All-scale spatial analysis of ecological data by means of principal coordinates of neighbour matrices. Ecological Modelling 153: 51–68.
46. Cook R, Weisberg S (1999) Applied Regression, Including Computing and Graphics; John Wiley & Sons I, editor: Wiley-Interscience.
47. Peres-Neto PR, Legendre P, Dray Sp, Borcard D (2006) Variation partitioning of species data matrices: estimation and comparison of fractions. Ecology 87: 2614–2625.
48. Jones DL (1999) Potential health risks associated with the persistence of *Escherichia coli* O157 in agricultural environments. Soil Use and Management 15: 76–83.
49. Vanrenterghem B, Huysman F, Rygole R, Verstraete W (1991) Detection and prevalence of *Listeria monocytogenes* in the agricultural ecosystem. Journal of Applied Bacteriology 71: 211–217.
50. Fenlon DR, Ogden ID, Vinten A, Svoboda I (2000) The fate of *Escherichia coli* and *E. coli* O157 in cattle slurry after application to land. Journal of Applied Microbiology 88: 149S–156S.
51. Hartman WH, Richardson CJ, Vilgalys R, Bruland GL (2008) Environmental and anthropogenic controls over bacterial communities in wetland soils. Proceedings of the National Academy of Sciences 105: 17842–17847.
52. Lauber CL, Hamady M, Knight R, Fierer N (2009) Pyrosequencing-based assessment of soil pH as a predictor of soil bacterial community structure at the continental scale. Applied and Environmental Microbiology 75: 5111–5120.
53. Fierer N, Jackson RB (2006) The diversity and biogeography of soil bacterial communities. Proceedings of the National Academy of Sciences of the United States of America 103: 626–631.
54. Besnard V, Federighi M, Cappelier JM (2000) Development of a direct viable count procedure for the investigation of VNBC state in *Listeria monocytogenes*. Letters in Applied Microbiology 31: 77–81.
55. Lindbäck T, Rottenberg ME, Roche SM, Rorvik LM (2010) The ability to enter into an avirulent viable but non-culturable (VBNC) form is widespread among *Listeria monocytogenes* isolates from salmon, patients and environment. Veterinary research 41: 1–10.
56. Besnard V, Federighi M, Declerq E, Jugiau F, Cappelier J-M (2002) Environmental and physico-chemical factors induce VBNC state in *Listeria monocytogenes*. Veterinary research 33: 359–370.
57. Cunningham E, O'Byrne C, Oliver JD (2009) Effect of weak acids on *Listeria monocytogenes* survival: Evidence for a viable but nonculturable state in response to low pH. Food control 20: 1141–1144.

17

Cross-Species Extrapolation of Prediction Models for Cadmium Transfer from Soil to Corn Grain

Hua Yang[1,2], Zhaojun Li[1]*, Lu Lu[3], Jian Long[2], Yongchao Liang[1]

1 Institute of Agricultural Resources and Regional Planning, Chinese Academy of Agricultural Sciences, Key Laboratory of Plant Nutrition and Fertilizer, Ministry of Agriculture, Beijing, China, 2 Guizhou Key Laboratory of Mountain Environment, Guizhou Normal University, Guiyang, China, 3 Graduate School of the Chinese Academy of Agricultural Sciences, Beijing, China

Abstract

Cadmium (Cd) is a highly toxic heavy metal for both plants and animals. The presence of Cd in agricultural soils is of great concern regarding its transfer in the soil-plant system. This study investigated the transfer of Cd (exogenous salts) from a wide range of Chinese soils to corn grain (Zhengdan 958). Through multiple stepwise regressions, prediction models were developed, with the combination of Cd bioconcentration factor (BCF) of Zhengdan 958 and soil pH, organic matter (OM) content, and cation exchange capacity (CEC). Moreover, these prediction models from Zhengdan 958 were applied to other non-model corn species through cross-species extrapolation approach. The results showed that the pH of the soil was the most important factor that controlled Cd uptake and lower pH was more favorable for Cd bioaccumulation in corn grain. There was no significant difference among three prediction models in the different Cd levels. When the prediction models were applied to other non-model corn species, the ratio ranges between the predicted BCF values and the measured BCF values were within an interval of 2 folds and close to the solid line of 1:1 relationship. Furthermore, these prediction models also reduced the measured BCF intra-species variability for all non-model corn species. Therefore, the prediction models established in this study can be applied to other non-model corn species and be useful for predicting the Cd bioconcentration in corn grain and assessing the ecological risk of Cd in different soils.

Editor: Xianglin Shi, University of Kentucky, United States of America

Funding: This work was jointly supported by the Special Fund of Public Industry in China (Agriculture, project No. 200903015; website: http://www.most.gov.cn/) and the Natural Science Foundation of Institute of Agricultural Resources and Regional Planning, Chinese Academy of Agricultural Sciences (Project No.2013-9; website:http://www.iarrp.cn). The funders had no role in study design, data collection and analysis, decision to publish, or preparation of the manuscript.

Competing Interests: The authors have declared that no competing interests exist.

* E-mail: lizhaojun@caas.cn

Introduction

Trace metal elements accumulation in crop of agricultural soils is of extensive concern due to its potential risks to human health and detrimental effects on soil ecosystems [1–4]. Cadmium (Cd) is one of the most dangerous trace elements because excessive dietary intake of Cd and its accumulation in human organs over a lifetime can lead to kidney malfunction [5–8].

Numerous studies have demonstrated that a number of factors affect Cd bioavailability in soils, including soil pH, organic matter (OM), cation exchange capacity (CEC), cultivars of crop plants, plant age, and so on [9–13]. Among these soil properties, the pH plays the most important role in determining metal speciation, due to its strong effects on solubility and speciation of metals in the soil and solution [14–15]. Dramatic increases in Cd desorption from soil constituents and its dissolution ability in solution have been observed with the decreasing soil pH [16–17]. The bioavailability of Cd in soil also increases when soil pH decreases [18–23]. Apart from soil pH, OM in soil is another important factor affecting heavy metal availability. It was reported that heavy metal adsorption onto soil constituents declined when organic matter content in soils decreased [24–25], and that the ability of plants to Cd uptake decreased with an increase in soil organic matter [26–27].

The influence of soil factors derived from the soil types on plant uptake of heavy metals and the difficulties in assessing the ecological risk of heavy metals in complex soil types have urged the research on prediction models that can predict trace element transfer to plant [28–30]. These models provide great opportunities to carry out ecological risk assessments and establish soil quality criteria for heavy metals [31–32]. However, the influences of soil factors may result in variability of ecological risk assessments in different soils and of the soil quality criteria for heavy metals [33]. Therefore, when the soil environmental quality standard is established based on the species sensitivity distributions (SSD), which are routinely used in ecological risk assessment procedures [34–35], the toxicology data from the soils with different properties need to be normalized using toxicity prediction models to eliminate the influence of soil factors [36]. The normalization of toxicology data can also improve the accuracy of the sensitivity distribution of species and environmental quality standard values [37]. Ideally, separate species-specific models should be developed for each single species. However, this is unrealistic due to the expenses and efforts involved. Thus, for practical reason, the cross-species extrapolation of biological toxicity model was applied in some cases of pollutants risk assessment. For example, Deleebeeck et al. [38–39] have applied biotic bigand model (BLM) from O. *mykiss* to other fish species and from D. *magna* to other cladocera species. Schlekat et al. [33] applied chronic nickel BLMs

Table 1. Selected soil properties of the 17 soils used in the cadmium phytotoxicity test.

Soil NO.	Location*	pH	OM ($g.kg^{-1}$)	CEC ($cmol. kg^{-1}$)	TN ($g.kg^{-1}$)	TP ($mg.kg^{-1}$)	TK ($g.kg^{-1}$)	Background Cd ($mg.kg^{-1}$)
S1	Hunan	4.90	15.52	10.85	1.14	468.5	15.26	0.18
S2	Chongqing	5.74	17.48	21.34	1.00	547.8	22.61	0.18
S3	Liaoning	5.74	25.84	12.19	1.00	726.6	23.94	0.16
S4	Yunnan	5.92	34.26	11.10	2.01	811.2	4.77	0.26
S5	Jiangxi	6.01	11.69	8.70	0.51	518.0	9.96	0.14
S6	Anhui	6.25	20.04	19.08	0.99	347.1	15.41	0.17
S7	Heilongjiang	6.27	35.69	28.59	1.74	481.1	24.7	0.18
S8	Jilin	6.82	32.85	31.11	1.75	349.2	24.58	0.17
S9	Jiangsu	6.93	47.69	26.20	2.44	689.2	21.03	0.15
S10	Shaanxi	7.90	16.49	22.37	1.36	984.0	24.37	0.24
S11	Hebei	7.98	8.57	8.12	0.68	529.9	24.22	0.19
S12	Henan	8.07	17.79	16.01	1.07	746.1	19.86	0.19
S13	Xinjiang	8.12	19.43	25.25	1.32	780.0	25.49	0.16
S14	Shanxi	8.24	23.17	16.80	1.13	950.7	23.70	0.20
S15	Tianjin	8.29	22.02	24.67	1.42	916.3	24.63	0.19
S16	Shandong	8.65	11.84	13.09	0.93	965.6	21.37	0.18
S17	Neimeng	8.80	16.30	11.61	0.96	376.3	26.40	0.21

*Soil sample locations are listed in the order of increasing pH.

developed for the cladocera such as *Daphnia magna* and *Ceriodaphnia dubia* to predict chronic toxicity of nickel to three other invertebrates including snail (*Lymnaea stagnalis*), insect (*Chironomus tentans*), and rotifer (*Brachionus calyciflorus*). However, little effort has been devoted to establishing models to describe the relationship between Cd uptake of corn grain and soil properties. In addition, the cross-species extrapolation approach has rarely been applied to terrestrial ecosystem. The objectives of this study are: 1) to develop the models for predicting Cd accumulation in corn grain on seventeen soils with greatly different properties, 2) to assess the feasibility of applying these models to other non-model corn species, and 3) to investigate their accuracy in predicting accumulation of Cd in non-model corn species.

Materials and Methods

None of these 17 soil samples were collected from national parks or other protected areas. It is confirmed that no specific permissions were required for the soils sampling activities in the 17 locations in China. It is also can be confirmed that the field studies did not involve endangered or protected species. No tested corn species are under first- or second-class state protection, and they are not listed in the Inventory of Rare and Endangered Plants of China (http://zrbhq.forestry.gov.cn/portal/zrbh/s/3053/content-457748.html), or the Key Protected Inventory of Wild Plants of China (http://zrbhq.forestry.gov.cn/uploadfile/zrbh/2010-10/file/2010-10-14-bb296addeaa047798d6b6c476aaa1da9.doc). These corn species were used for only scientific research as permitted by the Ministry of Agriculture of China.

Table 2. Soil Environmental Quality Standards of China (GB15618–1995) and the content of exogenous Cd ($mg.kg^{-1}$).

pH	<6.5	6.5~7.5	>7.5
Grade Two Standard	0.3	0.3	0.6
Low Cd	0.3	0.3	0.6
High Cd	0.6	0.6	1.0

Soil samples

A set of 17 soils covering a wide range of soil properties was collected from typical locations in China. In each sample location, the soils were collected from the top 20 cm of the soil profile. For analysis of their phys-chemical characteristics, the soils were air dried and sorted to pass a 2-mm sieve. Soil pH was measured in deionized water (soil:solution ratio, 1:5) [40]. Cation exchange capacity (CEC) was determined by the unbuffered silver-thiourea method [41]. OM was measured by dry combustion [42]. The total phosphorus in soils (TP) was measured by colorimetric method [43]. The total nitrogen in soils was determined by the Kjeldahl's method [44]. The background Cd content in the soils was determined by aqua regia (1:3 fresh mixture of concentrated HNO_3 and HCl) digestion [45]. The selected properties of the 17 soil samples were shown in Table 1.

Experimental design

Bioaccumulation factors for Cd by corn with one species from seventeen soils. According to Grade Two Standard for Cd in the Soil Environmental Quality Standards of China (GB15618–1995), the tested levels of Cd in soils were shown in Table 2. To obtain the tested levels, a certain amount of exogenous Cd ($3[CdSO_4].8H_2O$) was added to the 8 kg air-dried soils. All the soils were thoroughly mixed and placed into pots. Then the soils were moistened with deionized water to the 60% of field moisture capacity. Each tested level of Cd was carried out in triplicate. The soils containing Cd at different levels were covered by plastic film with several pores. For Cd aging, all the pots were

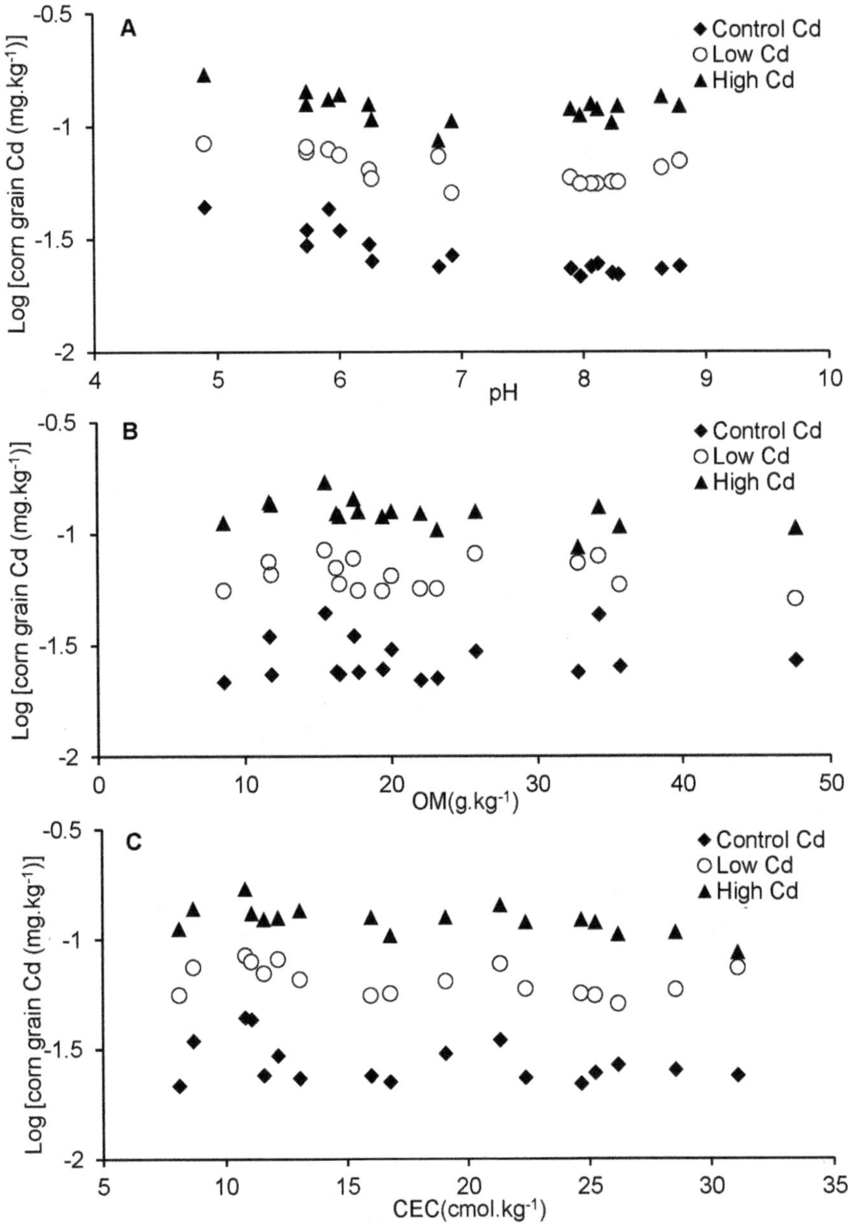

Figure 1. Relationships between Cd in corn grain and (A) soil pH, (B) organic matter (OM) and (C) cation exchange capacity (CEC).

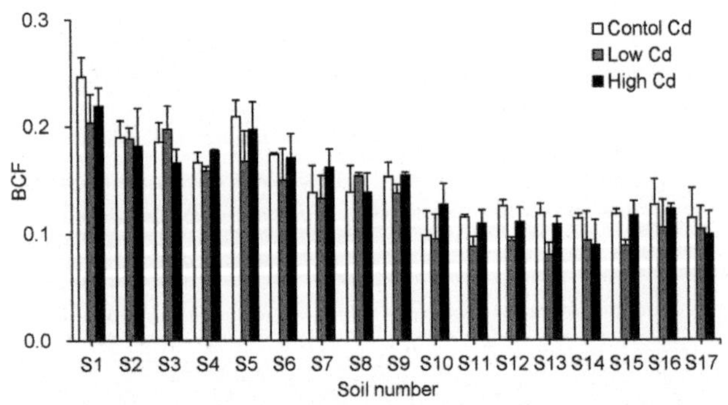

Figure 2. Effects of soil types on the transfer factor based on different Cd treatments in soils.

kept for 3 months at a greenhouse at the temperature of 25±3°C during the daytime and 20±3°C at night with a natural light photoperiod. During the period of aging, water contents of each soil sample were retained by adding deionized water every 3 days.

Table 3. Prediction models for the different Cd levels.

Model No.	Cadmium levels	Prediction models	R^2	*P*
Model 1	Control Cd	Log[BCF] = −0.081pH-0.254	0.728	<0.001
Model 2	Low Cd	Log[BCF] = −0.104pH-0.170	0.811	<0.001
Model 3	High Cd	Log[BCF] = −0.079pH-0.280	0.713	<0.001

Table 4. Intrinsic sensitivity (*k*) for non-model species fitted by models from cultivar Zhengdan 958.

	Intrinsic sensitivities (*k*)					
Prediction Model	**Jingketian 183**	**Chuandan 30**	**Liaodan 565**	**Tunyu 88**	**Zhongdan 808**	**Nongda 84**
Model 1	−0.284	−0.284	−0.339	−0.418	−0.324	−0.317
Model 2	−0.128	−0.123	−0.175	−0.254	−0.164	−0.158
Model 3	−0.302	−0.299	−0.353	−0.432	−0.338	−0.331

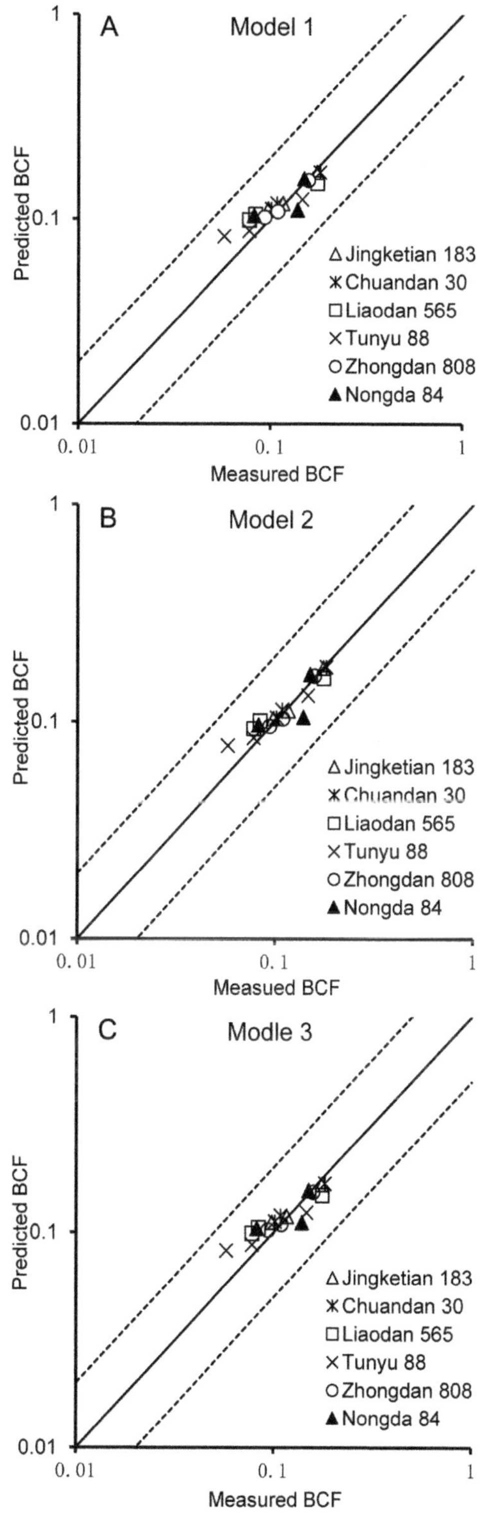

Figure 3. Relativity between measured and predicted BCF values for Cd in non-model corn grains. The predicted BCF values in Figure 3 were estimated by (A) model 1, (B) model 2, and (C) model 3 in Table 3. The solid line represents a 1:1 relationship; the dashed lines indicate a 2-fold prediction interval between the predicted and measured values.

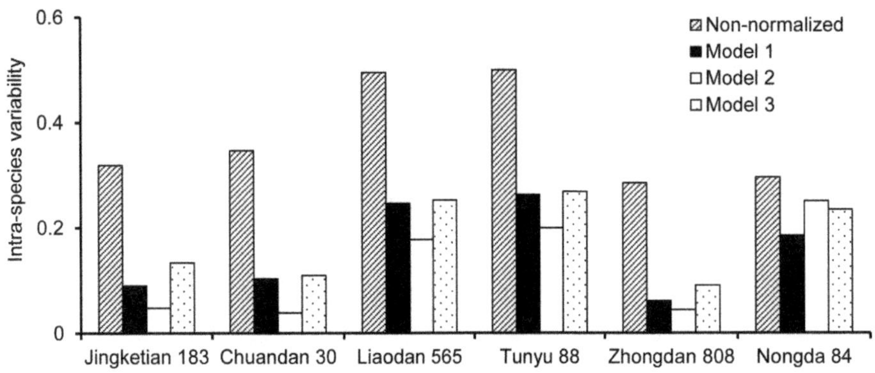

Figure 4. Intra-species variability of Cd BCF The data from all non-model corn species were normalized with the models listed in Table 3.

After 3 months' aging, uniformed seeds of corn (*Zea may* L. cv. Zhengdan 958) were sown in each soil containing Cd.

Bioaccumulation factors for Cd of eight corn species from three soils. Three soils with typical properties sampled from Jiangxi, Shaanxi, and Shanxi provinces were selected for this experiment. The tested levels of Cd in soils and exogenous Cd addition were the same as those in the experiment of bioaccumulation factors for Cd by corn from different soils. Triplicate was conducted on each tested level of Cd. After three months' aging of soils, the seeds of corn (cv. Jingketian 183, Chuandan 30, Liaodan 565, Tunyu 88, Zhongdan 808 and Nongda 84) were surface sterilized with 1% (v/v) NaOCl, rinsed, and soaked in distilled water for 24 h at 33°C in the dark. Then these seeds were sown in each soil containing Cd. All these pots were placed in the greenhouse at the temperature of 25±3°C during the daytime and 20±3°C at night with a natural light photoperiod.

Experimental soil sampling and analysis

After three months' aging, soils in the pots of the above two experiments were sampled, air-dried, passed through a <0.20 mm sieve, and then thoroughly mixed. 0.5 g of the soil sample was weighed and added to digestion tubes containing 9 ml HNO_3 and 3 ml HF. The corn grains were harvested through being oven-dried at 105°C for 30 minutes and kept at 70°C until the weight of the grains was stable. The samples (0.5 g) were digested in 6 ml concentrated HNO_3 (70% w/v) and 3 ml H_2O_2 in a CEM Mars X microwave oven (CEM Mars) at a pressure of 3.1 MPa. The Cd concentrations in the digestion solution from soil and corn grain samples were measured by inductively coupled plasma-mass spectrometry (ICP-MS, Agilent 7500a, Agilent Technologies Co. Ltd., USA).

Data analysis

The bioconcentration factor (BCF) was calculated as the ratio of the content of Cd in the corn grain to that in the soil (Equation 1) [46–47].

$$BCF = C_{grain}/C_{soil} \quad (1)$$

Where C_{grain} is the Cd concentration in the corn grain and C_{soil} is Cd concentration in the soil.

SPSS 16.0 for Windows® 107 (SPSS Inc, Chicago, IL, USA) was used for the regression analysis and statistical analysis of the significant differences. Origin 8.0 (OriginLab Co., Northampton, MA, USA) was employed for figure rendering.

Model

The prediction models for Cd transfer were established through multiple stepwise regression of the Cd BCF from Zhengdan 958 in a wide range of soil properties. These models are based on equation 2.

$$Log[BCF] = a \times pH + b \times \log[OM] + c \times \log[CEC] + k \quad (2)$$

Where Log[BCF], log[OM], and log[CEC] were the logarithm base 10 of the BCF values, the content of organic matter ($g.kg^{-1}$) in soils, and the cation exchange capacity ($cmol.kg^{-1}$) of soils, respectively. The slope of soil property parameters such as a, b and c indicate the impacts of soil properties on heavy metal. The intercept *k* is the intrinsic sensitivity that characterizes the species absorbing Cd.

Cross-species extrapolation

In the process of cross-species extrapolation, an interim alternative was to assume that interactions among Cd and pH, OM, CEC were the same among related species. In other words, the models stability constants including a, b, c were assumed the same among related species, and the only difference between related species was assumed to be their intrinsic sensitivity (*k*) [33]. The variation of intrinsic sensitivities within a species among

Table 5. Soil properties and BCF values for Cd in non-model wheat grain [61].

				BCF			
Cultivation soil	**pH**	**OM(%)**	**CEC(mequiv./100g)**	**Anomol**	**Tj-83**	**Abadgar**	**Mehran**
Soil	7.53	75.2	15.0	0.241	0.332	0.266	0.417
Sewage	7.70	87.6	20.8	0.179	0.311	0.287	0.375

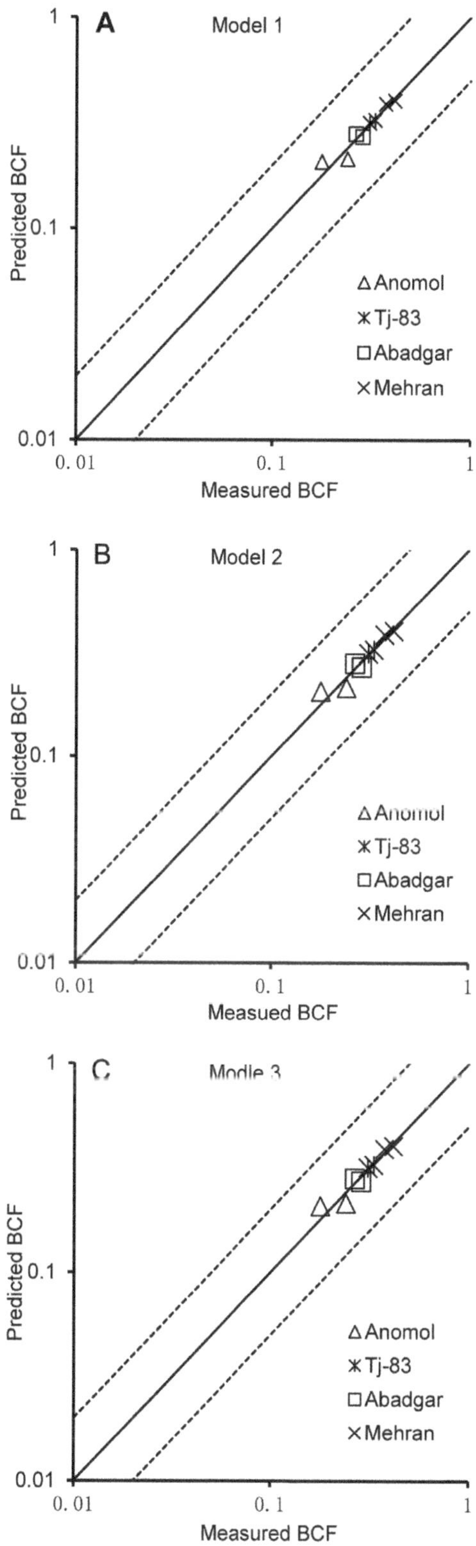

Figure 5. Relativity between measured and predicted BCF values for Cd in non-model wheat grains. The predicted BCF values in Figure 5 were estimated by (A) model 1, (B) model 2, and (C) model 3 in Table 3. The solid line represents a 1:1 relationship; the dashed lines indicate a 2-fold prediction interval between the predicted and measured values.

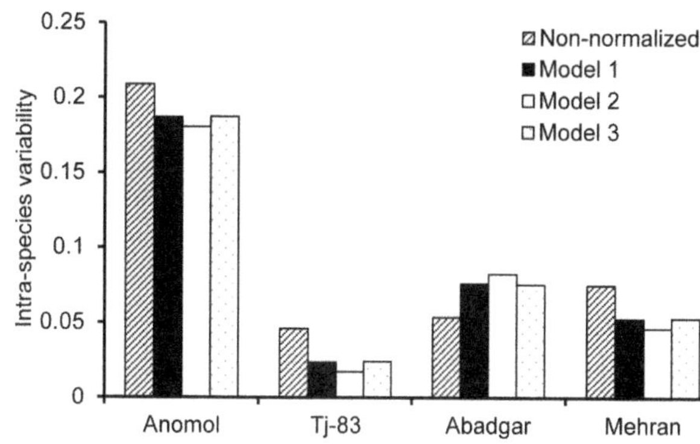

Figure 6. Intra-species variability of BCF for Cd. The data from all non-model wheat species were normalized with the models listed in Table 3.

plants reflects residual variation [36]. According to the minimum squared error between predicted BCF value and the measured BCF value $\sum_{i=1}^{n}(measured\,BCF_i - predicited\,BCF_i)^2$, the intercept ($k$) for different models corresponding to various species were obtained through Excel Solver for linear optimization [32].

The accuracy of the model predictions was evaluated by comparing the measured BCF of the other non-model corn species with the predicted BCF from the model. The predictions for other non-model corn species were calculated with the prediction model developed for Zhengdan 958 [48].

Analysis of the reduction of intra-species variability

The Cd BCF values of non-model corn species were normalized to the specific soil conditions by means of the obtained model for Zhengdan 958 and equation 3.

$$\sqrt{\frac{\sum_{i=1}^{n}(BCFsi-\overline{BCFs})^2}{(n-1)\times(\overline{BCFs})^2}} \tag{3}$$

Where the BCF of i-th conditions for specific corn specie was normalized to specific soil conditions. *BCFsi* is the BCF of i-th conditions for specific corn species. $\overline{BCFs}$ is the mean of n BCFs. *n* is the number of different conditions for specific corn species.

The BCFs that Cd for different corn species was normalized by the prediction models to a set soil condition should be equal. Therefore, the decrease of intra-species variability indicates that normalization processing eliminated soil properties to a certain extent.

Results

Major factors affecting accumulation of Cd in corn grain in different soils

Simple relationships between log [corn grain Cd] at different Cd levels and soil pH, OM and CEC are shown in Figure 1. At the treatment of low concentration Cd, the Cd content in maize grain decreased with an increase of pH. Similar trends were found in the treatments of control and high concentration Cd. The significantly negative correlations were observed between Cd accumulation in corn grain and soil pH in the treatments with control Cd, low concentration Cd, and high concentration Cd ($P<0.001$). There was no significant correlation between the log [corn grain Cd] and soil OM and CEC. These results indicate that soil pH is the main factor that controls the accumulation of Cd in corn grain.

Effects of different soil types on Cd transfer in soil-corn system

As shown in Figure 2, the BCF values from cultivar Zhengdan 958 with different treatments were significantly different in 17 soils with a wide range of soil properties in China. In general, the variation tendency of BCF values from cultivar Zhengdan 958 was alike over all soil samples from S1 to S17 at different treatments. These BCF values decreased with the increase of soil pH. The BCF values detected in S1 to S9 (pH from 4.90 to 6.93) were significantly higher than those in S10 to S17 (pH from 7.90 to 8.80). In the control soil, the maximum value for BCF (0.24) was achieved in S1 (pH 4.90), and the minimum value for BCF (0.09) in S10. The maximum value was 2.67 folds as high as the minimum value. At the treatment of low Cd, the maximum BCF (0.20) was observed in S1, and the minimum value for BCF (0.08) was found in S13. The maximum value was 2.50 folds as high as the minimum. At the treatment of high Cd, the highest BCF (0.22) values among the 17 soils were once again obtained in S1, and the minimum value for BCF (0.10) was in S17. The maximum was 2.22 folds as high as the minimum. These results indicate that Cd under acidic soil conditions is more highly bioavailable, leading to increased absorption by plants.

Prediction models

The prediction models established by stepwise regression are shown in Table 3. A significantly negative correlation existed between the Log[BCF] and soil pH in the Cd treatments including control, low Cd, and high Cd with the R^2 values ranging from 0.713–0.811 ($P<0.001$). All models displayed the same trend in which Log[BCF] of Cd in corn grain was negatively related to soil pH. There was no significant difference among three prediction models at the different Cd levels.

Cross-species extrapolation

The intercept (k) for different corn species deduced by three different models are shown in Table 4. The intercept (k) means the sensitivity of corn species to Cd accumulation. No significant differences were found to exist either among the k values of all tested non-model corn species deduced by the same model (model 1, 2, or 3) or among those in each of the tested non-model corn species deduced by different models (model 1, 2, and 3).

On the basis of the k values, the BCF values of non-model corn species were predicted by different models developed from cultivar Zhengdan958. The relationship between the predicted BCF values and the measured BCF values of non-model corn species was shown in Figure 3. The ratio between the predicted BCF values and the measured BCF values was within 2 folds interval and close to the solid line of 1:1 relationship, indicating that these models from cultivar Zhengdan 958 can be applied to predict the Cd BCF of non-model corn species.

Reduction of intra-species variability

Figure 4 shows the intra-species variability of Cd BCF of non-model species which were normalized with the models listed in Table 3. The prediction models can reduce the measured BCF intra-species variability for all non-model corn species. The intra-species variability of Cd BCF fitted by model 2 was found much lower than that by model 1 and model 3 except for cultivar Nongda 84. Interestingly, the intra-species variability of Cd BCF for cultivar Nongda 84 normalized with model 1 was much lower than that with model 2 and model 3. No significant differences existed in the intra-species variability of cultivars Chuandan 30, Liaodan 565 and Tunyu 88 which were normalized with model 1 and model 3. Since all models are effective in reducing the uncertainty caused by the soil property differences, they all can be used to establish the soil Cd SSD curves and provide an ecological benchmark in China.

Discussion

The bioavailable concentration of Cd in soil is more important than its total concentration in terms of its uptake and accumulation in plants [49]. In this study, we found that Cd uptake into corn grain increased with an increase of Cd concentration in soil. However, the BCF values tended to be similar when the concentration of Cd in soil was changed responding to the same soil. These findings were in agreement with previous studies [28–30,50–51].

We also found that the soil pH was the major factor influencing the bioavailability of Cd in soil. This may be attributed to Cd uptake by plants from soils through diverse reactions such as absorption, ionic exchange, redox reactions, and precipitation-dissolution etc. [52]. These reactions are mainly affected by soil pH. The pH effects also include its influence on the solution activity of Cd and the distribution of Cd between the soil phase and the solution phase [53–54]. The increased sorption of Cd at high pH values could reduce the solution concentration, which tends to decrease the bioavailable concentration of Cd in soil. In this study, the BCF values observed in acidic soils (S1 to S9) were higher than those in alkaline soils (S10 to S17). These results suggest that Cd is much more bioavailable and leads to increased absorption by plants under acidic conditions [55–57]. No significant differences were observed among three prediction models at the different Cd levels. There were no significant differences among Cd BCF values of corn grain at different Cd levels, either. These results were consistent with the findings of Rezvani et al. [50].

More importantly, we confirmed that the Cd BCF prediction models from cultivar Zhengdan 958 could be used to predict Cd BCF of other non-model corn species according to the soil pH values through cross-species extrapolation approach. Although many prediction models were developed to describe the relationship between concentration of heavy metals in plants and soil properties in terrestrial ecosystems, these models were only suitable for single species [31–32,58–60]. Only a few studies were carried out to develop the prediction models for one species in aquatic ecosystems and subsequently used to normalize ecotoxicity data for other non-model species [33,38,39].

In order to further validate the feasibility of the prediction modes developed in the present study for other non-model plant species, the measured Cd BCF values of wheat grain from soils were selected from the paper of Jamali et al. [61], and shown in Table 5. Models presented in Table 3 were applied to these non-model wheat species. The relativity between measured and predicted BCF values for Cd in non-model wheat grains is shown in Figure 5. The ratios between all of the predicted BCF values and the measured BCF values were also within 2 folds interval and close to 1:1 relationship. In addition, the intra-species variability of Cd BCF of all non-model wheat species was decreased except for Abadgar (Figure 6). This indicates that the models from cultivar Zhendan 958 can be applied not only to non-model corn species but also to non-model wheat species.

Conclusions

The soil pH was the most important factor that can control Cd uptake and lower pH was more favorable for Cd bioaccumulation in corn grain. The BCF of Cd in corn grain can be predicted by the prediction models such as equation 4. There was no significant difference among three prediction models in the different Cd sources. These prediction models could be applied to other non-model corn species,

$$Log[BCF] = a \times pH + k \quad (4)$$

Acknowledgments

We are indebted to Prof. Yibing Ma and Dr. Xiaoqing Wang for their help in the data processing and model derivation.

Author Contributions

Conceived and designed the experiments: ZJL. Performed the experiments: HY LL. Analyzed the data: JL. Contributed reagents/materials/analysis tools: YCL. Wrote the paper: HY ZJL.

References

1. McLaughlin MJ, Singh BR (1999) Cadmium in soil and plants: a global perspective. Kluwer Academic Publishers, Dordrecht, 13–21.
2. Li ZJ, Xu JM, Tang CX, Wu JJ, Muhammad A, et al. (2006) Application of 16S rDNA-PCR amplification and DGGE fingerprinting for detection of shift in microbial community diversity in Cu⁻, Zn⁻, and Cd⁻ contaminated paddy soils. Chemosphere 26: 1374–1380.
3. Zhang CB, Huang LN, Luan TG, Jin J, Lan CY (2006) Structure and function of microbial community during the early stages of revegetation near Shaoguan Pb/Zn Smelter, Guangdong, P.R. China. Geoderma 136: 555–565.
4. Chien CC, Kuo YM, Chen CC, Hung CW, Yeh CH, et al. (2008) Microbial diversity of soil bacteria in agricultural field contaminated with heavy metals. Journal of environmental sciences 20: 359–363.
5. Tudoreanu L, Phillips CJC (2004) Empirical models of cadmium accumulation in maize, rye grass and soya bean plants. J Sci Food Agric 84: 845–852.
6. Mushtakova VM, Fomina VA, Rogovin VV (2005) Toxic effect of heavy metals on human blood neutrophils. Biol Bull 32: 276–278.
7. Huang ML, Zhou SL, Sun B, Zhao QG (2008) Heavy metals in wheat grain: Assessment of potential health risk for inhabitants in Kunshan, China. Science of the Total Environment 405: 54–61.
8. Hu JL, Wu FY, Wu SC, Sun XL, Lin XG, et al. (2013) Phytoavailability and phytovariety codetermine the bioaccumulation risk of heavy metal from soils, focusing on Cd-contaminated vegetable farms around the Pearl River Delta, China. Ecotoxicology and Environmental Safety 91: 18–24.
9. Harter RD, Naidu R (2001) Anassessment of environmental and solution parameter impact on trace-metal sorption by soil. Soil Sci Soc Am J 65: 597–612.
10. Jung MC (2008) Heavy metal concentration in soils and factors affecting metal uptake by plants in the vicinity of a Korean Cu-W mine. Sensors 8: 2413–2423.
11. Wang J, Zhang CB, Jin ZX (2009) The distribution and phytoavailability of heavy metal fractions in rhizosphere soils of paulowniu fortune (seem) Hems near a Pb/Zn smelter in Guangdong, PR China. Geoderma 148: 299–306.
12. Chapman EEV, Dave G, Murimboh JD (2013) A review of metal (Pb and Zn) sensitive and pH tolerant bioassay organisms for risk screening of metal-contaminated acidic soils. Environmental Pollution 179: 326–342.
13. Barančíková G, Madaras M, Rybár O (2004) Crop contamination by selected trace elements. Journal of Soils and Sediments 4: 37–42.
14. Mühlbachová G, Simon T, Pechová M (2005) The availability of Cd, Pb and Zn and their relationships with soil pH and microbial biomass in soils amended by natural clinoptilolite. Plant, Soil and Environment 51: 26–33.
15. Zhao KL, Liu XM, Xu JM, Selim HM (2010) Heavy metal contaminations in a soil-rice system: identification of spatial dependence in relation to soil properties of paddy fields. Journal of Hazardous Materials 181: 778–787.
16. Sukreeyapongse O, Holm PE, Strobel BW, Magid SJ, Panichsakpatana S, et al. (2002) pH-dependent release of cadmium, copper, and lead from natural and sludge-amended soils. Journal of Environmental Quality 31: 1901–1909.
17. Bang J, Hesterberg D (2004) Dissolution of trace element contaminants from two coastal plain soils as affected by pH. Journal of Environmental Quality 33: 891–901.
18. Badawy SH, Helal MID, Chaudri AM, Lawlor K, McGrath SP (2002) Soil solidphase controls lead activity in soil solution. Journal of Environmental Quality 31: 162–167.
19. Wang AS, Angle JS, Chaney RL, Delorme TA, Reeves RD (2006) Soil pH effects on uptake of Cd and Zn by Thlaspi caerulescens. Plant and Soil 281: 325–337.
20. Du Laing G, Rinklebe J, Vandecasteele B, Meers E, Tack FMG (2009) Heavy metal mobility and availability in estuarine and riverine floodplain soils and sediments: a review. Science of the Total Environment 407: 3972–3985.
21. Kirkham MB (2006) Cadmiumin plantson polluted soils: Effect of soil factors, hyperaccumulation and amendments. Geoderma 137: 19–32.
22. Kim KR, Owens G, Naidu R (2009) Heavy metal distribution, bioaccessibility and phytoavailability in long-term contaminated soils from Lake Macquarie, Australia. Aust J Soil Res 47: 166–176.
23. Zeng F, Ali S, Zhang H, Quyang Y, Qiu B, et al. (2001) The influence of pH and organic matter content in paddy soil on heavy metal availability and their uptake by rice plants. Environmental pollution 159: 84–91.
24. Hettiarachchi GM, Ryan JA, Chaney RL, La Fleur CM (2003) Sorption and desorption of cadmium by different fractions of biosolids-amended soils. Journal of Environmental Quality 32: 1684–1693.
25. Antoniadis V, Robinson JS, Alloway BJ (2008) Effects of short-term pH fluctuations on cadmium, nickel, lead, and zinc availability to ryegrass in a sewage sludge-amended field. Chemosphere 71: 759–764.
26. Eriksson JE (1988) The effects of clay, organic matter and time on adsorption and plant uptake of cadmium added to the soil. Water Air Soil Pollut 40: 359–373.
27. Pinto AP, Mota AM, De Varennes A, Pinto FC (2004) Influence of organic matter on the uptake of cadmium, zinc, copper and iron by sorghum plants. Science of the total environment.326: 239–247.
28. Brown SL, Chaney RL, Angle JS, Ryan JA (1998) The phytoavailability of cadmium to lettuce in long-term biosolids-amended soils. J Environ Qual 27: 1071–1078.
29. Kim SJ, Chang AC, Page AL, Warneke JE (1988) Relative concentrations of cadmium and zinc in tissue of selected food plants grown on sludge-treated soils. Journal of Environmental Quality 17: 568–573.
30. Ge Y, Murray P, Hendershot WH (2000) Trace metal speciation and bioavailability in urban soils. Environ Pollut 107: 137–144.
31. Li Q, Xu XH, Zuo YB (2009) Effects of biosolid on the accumulation and transfer of trace elements in soil-wheat/maize system. Journal of agro-environment science 28: 2042–2049.
32. Wang XQ (2012) Ecological Thresholds for Copper and Nickel in Chinese Agricultural Soils, Beijing, China Mining University.
33. Schlekat CE, Genderen EV, UGent KDS, Antunes PMC, Rogevich EC, et al. (2010) Cross-species extrapolation of chronic nickel Biotic Ligand Models. Science of the Total Environment 408: 6148–6157.
34. Solomon KR, Baker DB, Richards RP, Dixon KR, Klaine SJ, et al. (1996) Ecological risk assessment of Atrazine in North American surface waters. Environmental Toxicology and Chemistry 15: 31–76.
35. Steen RJCA, Leonards PEG, Brinkman UAT, Barcelo D, Tronczynski J, et al. (1999) Ecological risk assessment of agrochemicals in European estuaries. Environmental Toxicology and Chemistry 18: 1574–1581.
36. Van Sprang PA, Verdonck FAM, Van Assche F, Regoli L, De Schamphelaere KAC (2009) Environmental risk assessment of zinc in European freshwaters: a critical appraisal. Sci Tot Environ 407: 5373–5391.
37. Allard P, Fairbrother A, Hope BK, Hull RN, Johnson MS, et al. (2010) Recommendations for the development and application of wildlife toxicity reference values. Intergrated environmental assessment and management 6: 28–37.
38. Deleebeeck NME, De Schamphelaere KAC, Janssen CR (2007a) A bioavailability model predicting the toxicity of nickel to rainbow trout (*Oncorhynchus mykiss*) and fathead minnow (*Pimephales promelas*) in synthetic and natural waters. Ecotoxicol Environ Saf 67: 1–13.
39. Deleebeeck NME, Muyssen BTA, De Laender F, Janssen CR, De Schamphelaere KAC (2007b) Comparison of nickel toxicity to cladocerans in soft versus hard surface waters. Aquat Toxicol 84: 223–35.
40. Sparks DL, Page AL, Helmke PA, Loeppert RH, Soltanpour PN, et al. (1996) Methods of Soil Anlysis. Part 3. Chemical Methods. Soil Science Society of America, Inc, Madison. WI.
41. Dohrmann R (2006) Cation exchange capacity methodology II: A modified silver–thiourea method. Applied clay science 34: 38–46.
42. Ball DF (1964) Loss-on-ignition as an estimate of organic matter and organic carbon in noncalcareous soils. Journal of Soil Science 15: 84–92.
43. Bray RH, Kurtz LT (1945) Determination of total, organic, and available forms of phosphorus in soils. Soil science 59: 39–46.
44. Bremner JM (1996) Nitrogen total. Methods of soil analysis. Part 3: Chemical methods, American Society of Agronomy-Soil Science Society of America, 1996, 1085–1121.

45. Zarcinas BA, McLaughlin MJ, Smart MK (1996) The effect of acid digestion technique on the performance of nebulisation systems used in inductively coupled plasma spectrometry. Commun Soil Sci Plant Anal 27: 1331–1354.
46. Baker AJM, Reeves RD, Hajar ASM (1994) Heavy metal accumulation and tolerance in British populations of metallophyte Thlaspi caerulescens. New Phytol 127: 61–68.
47. Raskin I, Kumar PBAN, Dushenkov S, Salt DE (1994) Bioconcentration of heavy metals by plants. Curr Opin Biotechnol 5: 285–290.
48. De Schamphelaere KAC, Heijerick D, Janssen C (2006) Cross-phylum comparison of a chronic biotic ligand model to predict chronic toxicity of copper to a freshwater rotifer, Brachionus calyciflorus (Pallas). Ecotoxicol Environ Saf 63: 189–195.
49. Sarwar N, Saifullah, Malhi SS, Zia MH, Naeem A, et al. (2010) Role of mineral nutrition in minimizing cadmium accumulation by plants. J Sci Food Agric 90: 925–937.
50. Rezvani M, Zaefarian F (2011). Bioaccumulation and translocation factors of cadmium and lead in Aeluropus littoralis. Australian Journal Agricutural Engineering 2: 114–119.
51. Mertens J, Luyssaert S, Verheyen K (2005) Use and abuse of trace metal concentrations in plants tissue for biomonitoring and phytoextraction. Environ Pollut 138: 1–4.
52. Smical AI, Hotea V, Oros V, Juhasz J, Pop I (2008) Studies on transfer and bioaccumulation of heavy metals from soil into lettuce. Environmental Engineering and Management Journal 7: 609–615.
53. Christensen TH (1984) Cadmium soil sorption at low concentrations: I. Effect of time, cadmium load, pH, and calcium. Water. Air. Soil Pollut 21: 105–114.
54. Naidu R, Bolan NS, Kookana RS, Tiller KG (1994) Ionic-strength and pH effects on the sorption of cadmium and the surface charge of soils. Euroean Journal of Soil Science 45: 419–429.
55. Hinesly TD, Redborg KE, Pietz RI, Ziegler EL (1984) Cadmium and zinc uptake by corn (Zea mays L.) with repeated applications of sewage sludge. J Agric Food Chem 32: 155–163.
56. Castilho PD, Chardon WJ (1995) Uptake of soil cadmium by three field crops and its prediction by a pH-dependent Freundlich sorption model. Plant Soil 171: 263–266.
57. Adams ML, Zhao FJ, McGrath SP, Nicholsonb FA, Chambersb BJ (2004) Predicting Cadmium Concentrations in Wheat and Barley Grain Using Soil Properties. Journal of Environmental Quality 33: 532–541.
58. Hao XZ, Zhou DM, Li DD (2012) Growth, cadmium and zinc accumulation of ornamental sunflower (*Helianthus annuus* L.) in contaminated soil with different amendments. Pedosphere 22: 631–639.
59. McBride MB (2002) Cadmium uptake by crops estimated from soil total Cd and pH. Soil Sci. 167: 62–67.
60. Wang XP, Shan XQ, Zhang SZ, Wen B (2004) A model for evaluation of the phytoavailability of trace elements to vegetables under the field conditions. Chemosphere. 55: 811–822.
61. Jamali MK, Kazi TG, Arain MB, Afridi HI, Jalbani N, et al. (2009) Heavy metal accumulation in different varieties of wheat (*Triticum aestivum* L.) grown in soil amended with domestic sewage sludge. Journal of Hazardous Materials 164: 1386–1391.

18

Long-Term Effects of Liming on Health and Growth of a Masson Pine Stand Damaged by Soil Acidification in Chongqing, China

Zhiyong Li[1,2]*, Yanhui Wang[2], Yuan Liu[3], Hao Guo[4], Tao Li[5], Zhen-Hua Li[2], Guoan Shi[1]

1 College of Agriculture, Henan University of Science and Technology, Luoyang, China, **2** Institute of Forest Ecology, Environment and Protection, Chinese Academy of Forestry, Beijing, China, **3** University of Chinese Academy of Sciences, Beijing, China, **4** Institute of Desertification Studies, Chinese Academy of Forestry, Beijing, China, **5** Hebei University, Baoding, China

Abstract

In the last decades, the Masson pine (*Pinus massoniana*) forests in Chongqing, southwest China, have increasingly declined. Soil acidification was believed to be an important cause. Liming is widely used as a measure to alleviate soil acidification and its damage to trees, but little is known about long-term effects of liming on the health and growth of declining Masson pine forests. Soil chemical properties, health condition (defoliation and discoloration), and growth were evaluated following application of limestone powder (0 (unlimed control), 1, 2, 3, and 4 t ha^{-1}) in an acidified and declining Masson pine stand at Tieshanping (TSP) of Chongqing. Eight years after liming, in the 0–20 cm and 20–40 cm mineral soil layers, soil pH values, exchangeable calcium (Ca) contents, and Ca/Al molar ratios increased, but exchangeable aluminum (Al) levels decreased, and as a result, length densities of living fine roots of Masson pine increased, with increasing dose. Mean crown defoliation of Masson pines (dominant, codominant and subdominant pines, according to Kraft classes 1–3) decreased with increasing dose, and it linearly decreased with length densities of living fine roots. However, Masson pines (Kraft classes 1–3) in all treatments showed no symptoms of discoloration. Mean current-year twig length, twig dry weight, needle number per twig, needle length per twig, and needle dry weight per twig increased with increasing dose. Over 8 years, mean height increment of Masson pines (Kraft classes 1–3) increased from 5.5 m in the control to 5.8, 6.9, 8.3, and 9.5 m in the 1, 2, 3, and 4 t ha^{-1} lime treatments, and their mean DBH (diameter at breast height) increment increased from 3.1 to 3.2, 3.8, 4.9, and 6.2 cm, respectively. The values of all aboveground growth parameters linearly increased with length densities of living fine roots. Our results show that liming improved tree health and growth, and these effects increased with increasing dose.

Editor: Raffaella Balestrini, Institute for Plant Protection (IPP), CNR, Italy

Funding: This study was sponsored by the State Forestry Administration of China (SFA) with the project (201304301-05), the Chinese-Norwegian cooperation project "Forest in south China: an important sink for reactive nitrogen and a regional hotspot for N2O?" (209696/E10), and the Key Laboratory of Forest Ecology and Environment of SFA. The funders had no role in study design, data collection and analysis, decision to publish, or preparation of the manuscript.

Competing Interests: The authors have declared that no competing interests exist.

* E-mail: pphdll@126.com

Introduction

Atmospheric emissions of acidifying compounds resulting from anthropogenic activities since the beginning of the industrial revolution have led to extensive soil acidification in terrestrial ecosystems [1–3]. Soil acidification has been blamed for causing indirect forest damage in Europe, North America, and Asia in the past few decades [4–10]. Some experiments have been implemented with liming in forests in Europe and the northeastern United States. Investigations on forest liming have shown that it enhanced cation exchange capacity (CEC), base saturation (BS), and contents of exchangeable Ca and Mg in the forest floor and the mineral soil, while Al levels and acidity decreased [11–15]. Effects of liming on the growth of tree fine roots were reported to be positive [16–18], although some researchers almost found the contrary, depending on tree species, tree age, site, soil conditions, form and dose of liming, and the time lapse since liming [19–21]. In contrast to expectations, however, liming has not made an overall improvement in aboveground growth of trees; instead, it has often produced either no effect or a detrimental effect on stand growth [22–26]. But, positive effects of liming on stand growth have also been observed [27–29]. The effect of liming on tree growth is dependent on the soil chemical properties, particularly the deficiency of Ca and Mg [29]. In addition, the reaction of forest stands to liming can be affected by such factors as tree species and age [30].

Masson pine (*Pinus massoniana*) is one of the most widespread forest species across subtropical China and it is highly sensitive to acid deposition [31]. Chongqing, located in the upper reaches of the Yangtze River, is an industrial and commercial municipality in southwest China. Total areas of pure Masson pine forests are in excess of 1.2 million hectares and estimated to account for more than 50% of the total forests in the region [32]. Obviously, this conifer species plays a very important role in protecting the environment of the middle and lower reaches of the Yangtze River. On the other hand, Chongqing ranks amongst the regions with the highest levels of acid deposition in China [33–35]. Due to extensive use of coal with high sulphur (S) contents, basin topography conducive to the accumulation of air pollutants, and low wind velocity, Chongqing has undergone severe deposition of

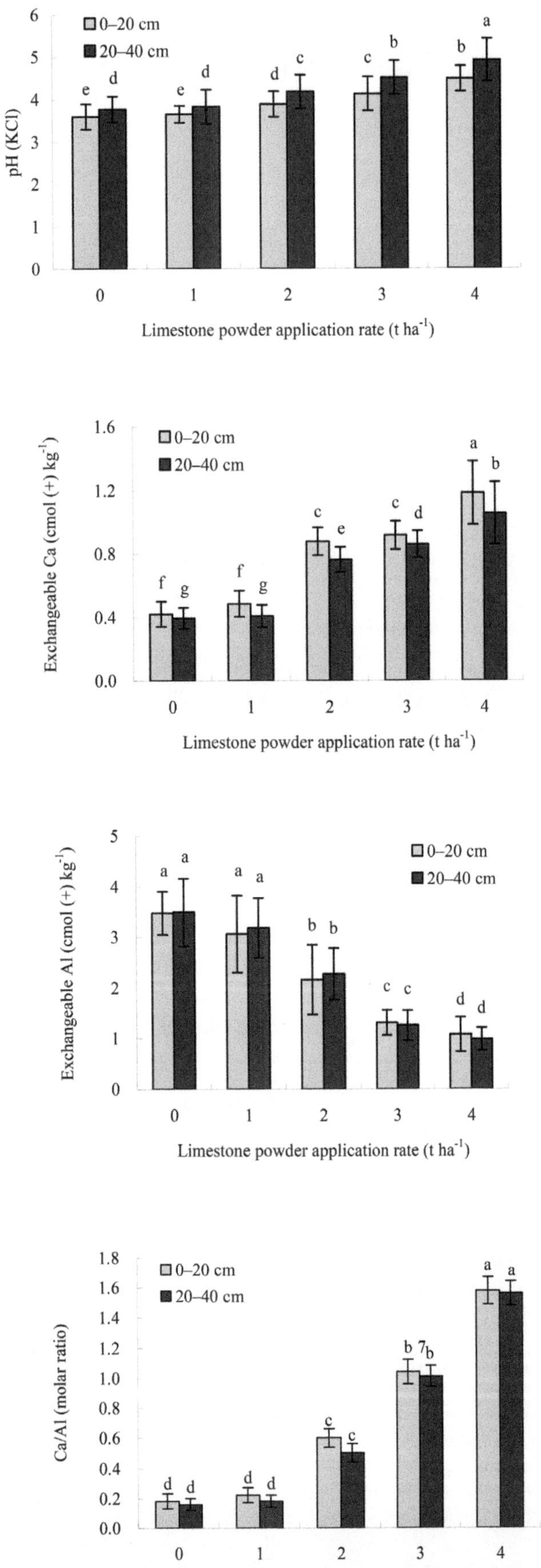

Figure 1. Soil pH values, exchangeable Ca contents, exchangeable Al levels, and Ca/Al molar ratios in the 0–20 cm and 20–40 cm of the Masson pine stand 8 years after liming. Error bars represent the standard deviation (SD) of the mean. Different letters above the error bars indicate significant differences at the 0.05 level (ANOVA and Duncan's multiple range test), n = 3.

acidifying S compounds since the late 1970s, when the Chinese Economic Reform Policy started [1,33,36–38]. More than half of 40 districts and counties within the region was designated as being part of so-called acid rain control areas, i.e. parts of China where pH value of rain water is lower than 4.5 and sulfur deposition exceeds the critical load, and the acid rain problem has been addressed by the national authorities [39]. The soil of Masson pine forests in Chongqing is generally acidic, with low amounts of exchangeable Ca and high amounts of exchangeable Al [33,35,40,41]. Masson pine is highly sensitive to Al [42]. Since the 1980s, large patches of Masson pine forest in some areas have exhibited serious decline symptoms, mainly including fine root death, tip necrosis of needles, thinned crown, reduced needle length, premature needle abscission, dieback of twigs and branches, and reduced growth, which was believed to be caused by soil acidification [4,33,35,43,44].

A common measure to counteract soil acidification is liming of soils [45–47]. In the last decades, however, only a few studies have been conducted to investigate the effects of liming on Masson pine forests growing on acidified soil, mainly with a single lime dose over relatively short time spans. For example, Du and Tian [44] found increased pH values in the 0–25 cm mineral soil on a limed young Masson pine forest 2 years after liming. Huang et al. [48] observed increased Ca contents in current-year needles at limed plots established in a 40-year-old pure Masson pine forest about 1 year after liming with 3 t ha^{-1} limestone powder. Up to now, little is known about the long-term effects of different lime doses on Masson pine. Thus, fifteen permanent plots treated with five doses of limestone powder were set up in a Masson pine stand damaged by soil acidification in Chongqing in 2004, and an investigation was carried out 8 years after liming. The aim of this work was to assess the effects of the five doses on the health and growth of Masson pine.

Study Sites and Methods

Site description

This study was conducted at Tieshanping (TSP) (29°38′N, 106°41′E, 512–579 m a.s.l.). TSP, covering about 1200 ha, is an almost pure Masson pine forest located on a sandstone ridge approximately 25 km northeast from the urban centre of Chongqing (28°10′–32°13′N, 105°17′–110°11′E) in southeast Sichuan Basin, southwest China. Masson pines have been planted since the early 1960s. The study site has a subtropical humid climate, with a mean annual air temperature of 18°C, a mean annual precipitation of 1100 mm that mainly occurs from April to October, and a mean annual relative humidity of 80%. TSP has been experiencing a long-term severe acid deposition, with low mean annual pH (4.0–4.2) and high frequency of acid rain (rainfall of pH<5.6 being about 90.0%) [35,38,49]. High deposition of S has been frequently reported. For example, during 2002 and 2003, throughfall S input was reported to be 16.0 g S m^{-2} a^{-1}, and throughfall N input was measured at 4.0 g N m^{-2} a^{-1} [50]. The forest soil is yellow mountain soil (Haplic Acrisol, WRB) developed on sandstone and prone to acidification. The mean mineral soil depth is between 40 and 80 cm, and the soil was rather homogeneous [33,34]. This acidic soil is representative for

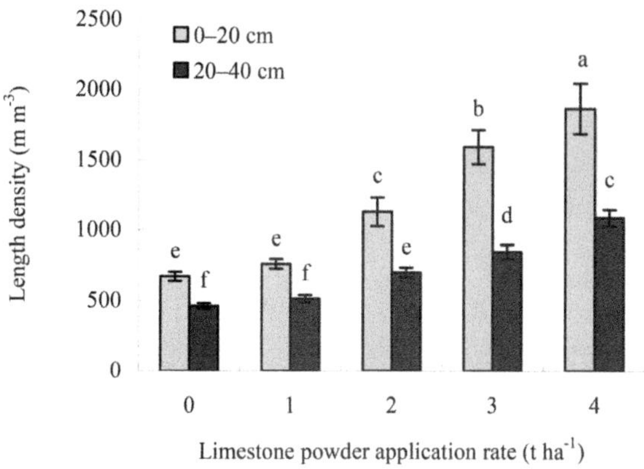

Figure 2. Length densities of living fine roots of Masson pine in the 0–20 cm and 20–40 cm of the Masson pine stand 8 years after liming. Error bars represent the SD of the mean. Different letters above the error bars indicate significant differences at the 0.05 level (ANOVA and Duncan's multiple range test), n = 3.

Chongqing and characterized by low levels of exchangeable Ca and high levels of exchangeable Al [33,35,41]. In the last decades the Masson pine forest ecosystem has declined due to severe soil acidification [33,35].

Plot establishment and liming treatments

In early May 2004, we selected a 26-year-old Masson pine stand, damaged (defoliation >25%) by soil acidification [35] for this limestone powder dose experiment. This stand grew on almost level ground and was composed of evenly-spaced and similarly-sized Masson pines, which were similar in defoliation. Therein, only a few scattered individuals of undergrowth species, e.g. *Schima superba*, *Cunninghamia lanceolata*, *Schoepfia jasminodora*, and *Lithocarpus glaber* grew. The density of Masson pines for this stand was 1192 trees ha^{-1}. In early June 2004, the mean defoliation, discoloration, height, and DBH of Masson pines (dominant, codominant and subdominant pines, Kraft classes 1–3) for this stand were 34%, 1%, 13.3 m, and 15.5 cm, respectively.

The trial was established in this stand as a randomized complete block experiment with five limestone powder (49.5% CaO, particle size <0.25 mm) treatments (0 (unlimed control), 1, 2, 3, and 4 t ha^{-1}) and three replicates. There were total fifteen plots. Each plot measured 10 m×10 m, with about 1 m buffer strips around each plot. The overall size of the stand within which all plots were installed was about 0.3 ha. Limestone powder was broadcast one time and evenly on the forest floor by hand in early June 2004.

Soil and living fine root sampling and measurements

In mid June 2012, two dominant Masson pines (Kraft class 1) were selected near the centre of each plot for sampling. Humus and soil were collected from the east, south, west, and north directions 1.0 m away from the main stem of each selected tree. Samples of the humus layer and the 0–20 cm, 20–40 cm and 40–60 cm mineral soil layers were taken using a cylindrical auger with an inner diameter of 10 cm and a height of 25 cm. The four samples collected around each selected tree in each plot were pooled to produce one composite sample per layer. In all, there were 120 composite samples representing five treatments and three field replicates. All composite samples were placed in plastic bags and transported to the laboratory, where they were bulked and crumbled manually. Living roots of different diameters of Masson pine were picked out according to the characteristics of appearance, color, elasticity, and odor [35] and placed into 0.2-mm sieves using tweezers. Meanwhile, roots of other plant species and gravel were removed. The living roots of Masson pine were carefully cleaned and washed in slow running water to remove soil from roots, spread out on clean sheets of paper for 5–10 minutes in the shade, kept in sealed bags, and stored at −20°C until later measurements. After being air-dried at room temperature (25°C), all soil samples were ground, passed through a 2-mm sieve, and stored at 4°C in the dark before chemical analysis.

Soil pH was determined in a 1:2.5 slurry of soil:1 M KCl solution using a combination glass electrode, after shaking for 0.5 h and then equilibrating for 0.5 h. The accuracy was 0.01 pH. Exchangeable Ca and Al were extracted with 1 M KCl solution and determined using an atomic absorption spectrometer (AAnalyst 300, Perkin-Elmer, CT, USA). The Ca/Al molar ratios were calculated from exchangeable Ca and Al contents of all humus and soil samples. Lengths of living fine roots (diameter ≤2 mm) were measured using an automated root scanning and analysis system (WinRHIZO, Regent Instruments Inc., Canada). Length densities (m m^{-3}) of living fine roots were calculated from all samples according to the volumes of the humus and soil samples. During sampling and measurements, polypropylene gloves were used to avoid contamination. In this study, we chose to analyze the soil chemical properties and fine root growth of Masson pine in the 0–20 cm and 20–40 cm mineral soil layers where the roots were mainly distributed [33,35].

Table 1. Relationships between the length densities (*y*, m m^{-3}) of living fine roots and the soil chemical properties (*x*) in the 0–20 cm and 20–40 cm.

Soil horizon (cm)	*x*	Unit of *x*	Regression equation *y* = a*x*+b	Correlation coefficient R^2	Degree of freedom *df*	*F*-value	*P*-value
0–20	pH (KCl)	-	*y* = 1408.5*x*-4356.6	0.9741	4	150.44	<0.05
	Exchangeable Ca	cmol (+) kg^{-1}	*y* = 1562.4*x*-7.207	0.9212	4	46.76	<0.05
	Exchangeable Al	cmol (+) kg^{-1}	*y* = −485.98*x*+2284.2	0.9707	4	132.52	<0.05
	Ca/Al (molar ratio)	-	*y* = 866.51*x*+579.7	0.9757	4	160.61	<0.05
20–40	pH (KCl)	-	*y* = 528.91*x*-1519.6	0.9970	4	1329.33	<0.05
	Exchangeable Ca	cmol (+) kg^{-1}	*y* = 864.05*x*+122.69	0.9564	4	87.74	<0.05
	Exchangeable Al	cmol (+) kg^{-1}	*y* = −220.62*x*+1217.6	0.9342	4	56.79	<0.05
	Ca/Al (molar ratio)	-	*y* = 86.627*x*+456.97	0.9913	4	455.77	<0.05

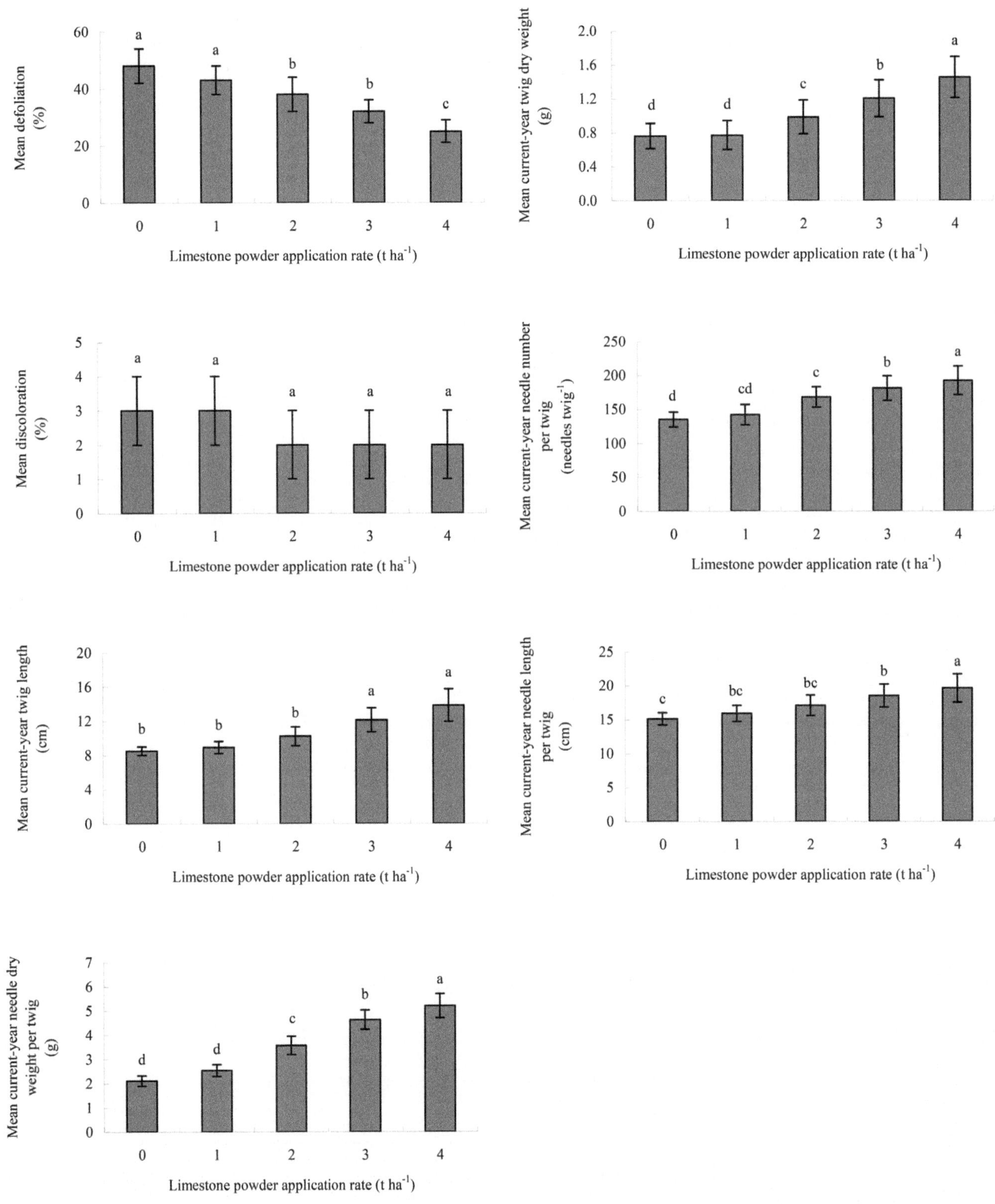

Figure 3. Crown condition of pines (Kraft classes 1–3) in the Masson pine stand 8 years after liming. Error bars represent the SD of the mean. Different letters above the error bars indicate significant differences at the 0.05 level (ANOVA and Duncan's multiple range test), n = 3.

Health and growth monitoring

In early June 2004 and in early June 2012, the assessments of crown health condition (defoliation and discoloration), height, and DBH were performed twice on all dominant, codominant and subdominant Masson pines (Kraft classes 1–3) per plot according to the ICP–Forests manual [51]. Defoliation and discoloration in the assessable crown were estimated based on 5% classes as compared with a local reference tree, which is defined as the healthiest tree that could grow at a particular site, taking into

Table 2. Relationships between the crown condition (*y*) and the length densities (*x*, m m^{-3}) of living fine roots in the 0–20 cm and 20–40 cm.

Soil horizon (cm)	*y*	Unit of *y*	Regression equation $y=ax+b$	Correlation coefficient R^2	Degree of freedom *df*	*F*-value	*P*-value
0–20	Mean defoliation	%	$y=-0.0172x+57.915$	0.9720	4	138.86	<0.05
	Mean current-year twig length	cm	$y=0.0043x+5.5405$	0.9903	4	408.37	<0.05
	Mean current-year twig dry weight	g	$y=0.0006x+0.354$	0.9845	4	254.06	<0.05
	Mean current-year needle number per twig	Needles $twig^{-1}$	$y=0.0465x+107.49$	0.9658	4	112.96	<0.05
	Mean current-year needle length per twig	cm	$y=0.0035x+12.984$	0.9892	4	366.37	<0.05
	Mean current-year needle dry weight per twig	g	$y=0.0025x+0.5518$	0.9922	4	508.82	<0.05
20–40	Mean defoliation	%	$y=-0.0351x+62.617$	0.9816	4	213.39	<0.05
	Mean current-year twig length	cm	$y=0.0087x+4.4043$	0.9890	4	359.64	<0.05
	Mean current-year twig dry weight	g	$y=0.0012x+0.2003$	0.9920	4	496.00	<0.05
	Mean current-year needle number per twig	Needles $twig^{-1}$	$y=0.0936x+95.809$	0.9456	4	69.53	<0.05
	Mean current-year needle length per twig	cm	$y=0.0071x+12.075$	0.9776	4	174.57	<0.05
	Mean current-year needle dry weight per twig	g	$y=0.0051x-0.068$	0.9631	4	104.40	<0.05

Table 3. Relationships between the mean height and DBH increments (y) and the length densities (x, m m^{-3}) of living fine roots in the 0–20 cm and 20–40 cm.

Soil horizon (cm)	y	Unit of y	Regression equation $y=ax+b$	Correlation coefficient R^2	Degree of freedom df	F-value	P-value
0–20	Mean height increment	m	$y=306.23x-998.03$	0.9951	4	812.33	<0.05
	Mean DBH increment	cm	$y=387.93x-438.03$	0.9565	4	87.95	<0.05
20–40	Mean height increment	m	$y=150.05x-356.26$	0.9893	4	369.83	<0.05
	Mean DBH increment	cm	$y=192.68x-92.876$	0.9771	4	170.67	<0.05

account factors such as altitude, latitude, tree age, site conditions and social status. The local reference tree has 0% defoliation and 0% discoloration. Pines with defoliation of 0–10%, >10–25%, >25–60%, >60–<100%, and 100% are classified as no defoliation, slight defoliation, moderate defoliation, severe defoliation, and death, respectively. Pine trees with defoliation >25% were classified as "damaged" [52]. Pine trees with discoloration of 0–10%, >10–25%, >25–60%, and >60% are categorized as no discoloration, slight discoloration, moderate discoloration, and severe discoloration, respectively. Tree height was measured with a Haga hypsometer, and DBH using a diameter tape.

At TSP, current-year needles make up the majority of tree foliage. In early June 2012, five current-year twigs with needles were sampled from the upper third of the crown on the south side of each of the two selected trees near the centre of each plot. The five current-year twigs with needles collected from each selected tree per plot were pooled to form a sample. Thirty samples were transported in clean plastic bags to the laboratory, where for each sample we separated needles from twigs, and measured twig length, twig dry weight, needle number per twig, needle length per twig, and needle dry weight per twig. Twig and needle lengths were measured to the nearest millimeter. The samples were ovendried at 60°C for about one week to a constant weight and weighed. Mean twig lengths, mean twig dry weights, mean needle numbers per twig, mean needle lengths per twig, and mean needle dry weights per twig were calculated from all samples.

Statistical analysis

The data were expressed as mean ± standard deviation (SD). They were assessed using one-way analysis of variance (ANOVA) with the SPSS 12.0 software package. The fixed model was used in ANOVA. Multiple comparisons of means among treatments were performed using Duncan test. $P<0.05$ was considered significant.

Results

Soil chemistry

Liming increased the pH, exchangeable Ca, and Ca/Al molar ratio, but decreased the exchangeable Al in the 0–20 cm and 20–40 cm mineral soil layers (Fig. 1). The pH, exchangeable Ca, and Ca/Al molar ratio in the same layers were found to be in the following order of lime treatments: 4 t ha^{-1}>3 t ha^{-1}>2 t ha^{-1}>1 t ha^{-1}>0 t ha^{-1}, while the exchangeable Al exhibited the reverse pattern. Overall, there were no statistical differences in the soil chemical parameters in the same layers between the 0 and 1 t ha^{-1} lime treatments, whereas significant differences were observed among the 1, 2, 3, and 4 t ha^{-1} lime treatments ($P<0.05$).

Fine root growth

Liming improved the growth of fine roots of Masson pine in the 0–20 cm and 20–40 cm mineral soil layers (Fig. 2). The length densities of living fine roots in the same layers were observed to be in the following order of lime treatments: 4 t ha^{-1}>3 t ha^{-1}>2 t ha^{-1}>1 t ha^{-1}>0 t ha^{-1}. There were no statistical differences in the length density in the same layers between the 0 and 1 t ha^{-1} lime treatments, whereas significant differences were found among the 1, 2, 3, and 4 t ha^{-1} lime treatments ($P<0.05$). The length densities of living fine roots were significantly positively correlated with the pH values, exchangeable Ca contents, and Ca/Al molar ratios, but significantly negatively with the exchangeable Al levels in both soil layers (Table 1).

Crown condition

Liming decreased the defoliation of Masson pines (Kraft classes 1–3), but increased their current-year twig and needle growth (Fig. 3). The mean defoliation varied with lime treatments, following an order of 0 t ha^{-1}>1 t ha^{-1}>2 t ha^{-1}>3 t ha^{-1}>4 t ha^{-1}. In contrast, the mean current-year twig length, twig dry weight, needle number per twig, needle length per twig, and needle dry weight per twig showed the reverse behavior. The mean defoliation of Masson pines (Kraft classes 1–3) in the 4 t ha^{-1} lime treatment was 25%, which is the lower limit of the moderate defoliation range (>25%–60%). However, the mean discoloration of Masson pines (Kraft classes 1–3) in all treatments lay within no discoloration range (0–10%). The mean defoliation was significantly negatively correlated with the length densities of living fine roots in both soil layers, but the values of current-year twig and needle growth parameters were significantly positively correlated with them (Table 2).

Height and DBH growth

Liming increased the height and DBH of Masson pines (Kraft classes 1–3) from early June 2004 to early June 2012. Among all treatments, the 4 t ha^{-1} lime treatment was the greatest in terms of both the mean height increment and the mean DBH increment of Masson pines (Kraft classes 1–3), it was followed by, in decreasing order, the 3, 2, 1, and 0 t ha^{-1} lime treatments. The mean height increments of Masson pines (Kraft classes 1–3) were 5.5±0.4, 5.8±0.5, 6.9±0.5, 8.3±0.6, and 9.5±0.6 m in the 0, 1, 2, 3, and 4 t ha^{-1} lime treatments, and their mean DBH increments 3.1±0.2, 3.2±0.2, 3.8±0.3, 4.9±0.4, and 6.2±0.4 cm, respectively. There were no statistical differences in both parameters between the 0 and 1 t ha^{-1} lime treatments, but significant differences were detected among the 1, 2, 3, and 4 t ha^{-1} lime treatments ($P<0.05$). The mean height increment and mean DBH increment were significantly positively correlated with the length densities of living fine roots in the 0–20 cm and 20–40 cm mineral soil layers (Table 3).

Discussion

Liming in early June 2004 increased the soil pH and exchangeable Ca and decreased the exchangeable Al to a 40 cm depth on the Masson pine stand; this effect increased with increasing dose. It is worth noting that in 0–20 cm and 20–40 cm in the 4 t ha^{-1} lime treatment the soil pH values increased to 4.48 and 4.92, respectively (Fig. 1), which are very near to or in the optimum pH (KCl) range (4.50–6.00) for the growth of Masson pine [53]. As a result, amelioration of the decline of the Masson pine stand has been observed 8 years after liming; this effect also increased with increasing dose. From the viewpoint of forest health restoration, the dose of 4 t ha^{-1} limestone powder is suggested, since this dose induced the best health and growth of Masson pine. More long-term observations are needed to fully assess the effects of liming on the health and growth of the Masson pine stand.

Fine roots as uptake organs to sustain aboveground health and growth of forests are susceptible to changes in the soil environment [54]. Due to soil acidification, toxic Al is released, which adversely impacts the growth of fine roots and inhibits the uptake of water and cations [55–57]. Al stress on the growth of fine roots is governed by the Ca/Al molar ratio [58]. According to Cronan and Grigal [56], the Ca/Al molar ratio in soil solution provided an indicator for identification of thresholds beyond which the risk of fine root damage from Al stress and nutrient imbalances increased. They estimated that there was a 50% risk of Al toxicity when the Ca/Al molar ratio was as low as 1.0, a 75% risk when the ratio was as low as 0.5, and nearly a 100% risk when the ratio was as low as 0.2.

This study showed that in 0–20 cm and 20–40 cm the Ca/Al molar ratios increased with increasing dose (Fig. 1) and the length densities of fine roots of Masson pine linearly increased with the ratios (Table 1). Al stress on the growth of fine roots of Masson pine was probably excluded in both layers in the 3 and 4 t ha^{-1} lime treatments, where the Ca/Al molar ratios were >1.0, and the best growth of fine roots was found in the 4 t ha^{-1} lime treatment. However, risks seemed to exist in both layers in the 0, 1, and 2 t ha^{-1} lime treatments, where the Ca/Al molar ratios were <1.0. The stress indication was very noticeable in both layers in the 0 and 1 t ha^{-1} lime treatments, where the Ca/Al molar ratios were ≤0.22, and the worst growth of fine roots was observed in the 0 t ha^{-1} lime treatment (Figs. 1 and 2). The application of limestone powder ameliorated Al toxicity on fine roots of Masson pine; this effect increased with increasing dose.

Liming ameliorates upper soil acidity in a relatively short term, but is generally slow in reducing deeper soil acidity [29,47]. The movement of lime to greater depths depends on site, soil type, form of lime, timing and rate of liming, and weather conditions [20,59–62]. Matzner et al. [63] showed a clear increase in exchangeable Ca and Mg in the 0–50 cm mineral soil depth in an experiment in *Fagus sylvatica* and *Picea abies* stands about 10 years after fertilization of NH_4NO_3 and KCl followed by Ca and Ma fertilization by 2 liming applications. Moore et al. [64] reported that in an uneven-aged *Acer saccharum* forest 10 years after liming, exchangeable Ca and Mg contents down to 20 cm depth increased obviously with the amount of added lime (powdered and pelletized dolomitic lime) while Ca contents in 20–40 cm depth were unaffected. Guckland et al. [47] found a significant increase in exchangeable Mg in the forest floor and exchangeable Ca in the 0–20 cm mineral soil depth in stands of *P. abies*, *F. sylvatica*, *Quercus petraea*/*Q. robur*, and *Pinus sylvestris* 2–9 years after one-time liming with 3–5 t ha^{-1} lime (mainly calcite and dolomite), and they suggested that the element input, CEC, C stock of the forest floor, and groundwater recharge were the most important parameters connected with the change of Ca and Mg stocks after liming in 0–10 cm, but in 10–20 and 20–40 cm, the C concentration of this depth and the year after first liming gained more influence. In our study conducted at TSP in the humid subtropical zone, in the 0–40 cm mineral soil depth of the limed Masson pine stand 8 years after one-time liming with 2–4 t ha^{-1} limestone powder, significant effects of liming were observed on the chemical properties and the growth of fine roots (Figs. 1 and 2). The fast downward movement of dissolved Ca cations into deeper mineral soil depth (the 20–40 cm) might be mainly related to the amount of finely ground limestone powder applied, rather thin forest floor layer (approximately 1-cm-thick humus layer), and relatively high annual and seasonal rainfall in Chongqing [33,34]. Undoubtedly, the promotion of the growth of fine roots of Masson pine in the 20–40 cm in the 2–4 t ha^{-1} lime treatments, especially the application of high lime doses, is beneficial to enhancing tree resistance to drought and frost.

The growth of aboveground parts (stem, branches, twigs, and leaves) of plants is coordinated with the growth of belowground part (roots) [20,65–67]. This lime-induced stimulation of the growth of fine roots of Masson pine favored Ca uptake by trees [48], resulting in the improvement of aboveground health and growth of trees (Fig. 3). The results in this study showed that the mean tree crown defoliation linearly decreased with the length densities of living fine roots in both soil layers, whereas the values of all aboveground growth parameters linearly increased with them (Tables 2 and 3). Pine trees with more fine roots grew more healthily and faster than those with less fine roots.

In summary, in the acidified and declining Masson pine stand of Chongqing, southwest China, application of limestone powder proved to be effective for ameliorating the soil acidity, improving the crown health condition, and stimulating the tree growth; these effects increased as the amount of applied limestone powder increased.

Acknowledgments

We are grateful to Forestry Administration of Chongqing Municipality and Tieshanping Forest Farm of Chongqing for their help in field works. We acknowledge two anonymous referees for their valuable comments on our manuscript. We used the Masson pine (*Pinus massoniana*) stand in this study with the permission of Forestry Administration of Chongqing Municipality and Tieshanping Forest Farm of Chongqing and the field studies did not involve endangered or protected species.

Author Contributions

Conceived and designed the experiments: ZL YW. Performed the experiments: ZL YL HG TL. Analyzed the data: ZL YW ZHL GS. Wrote the paper: ZL YW.

References

1. Zhao DW, Seip HM (1991) Assessing effects of acid deposition in southwestern China using the magic model. Water, Air and Soil Pollution 60: 83–97.
2. Hruška J, Oulehle F, Šamonil P, Šebesta J, Tahovská K, et al. (2012) Long-term forest soil acidification, nutrient leaching and vegetation development: linking modelling and surveys of a primeval spruce forest in the Ukrainian Transcarpathian Mts. Ecological Modelling 244: 28–37.
3. Liang GH, Liu XZ, Chen XM, Qiu QG, Zhang DQ, et al. (2013) Response of soil respiration to acid rain in forests of different maturity in southern China. PLoS One 8: e62207.
4. Liu HT, Zhang WP, Shen YW, Du XM, Zou XY, et al. (1988) Relationship between acid rain and the decline of a Masson pine forest in Nanshan,

Chongqing. Acta Scientiae Circumstantiae 8: 331–339. (In Chinese with English abstract).
5. Ulrich B (1990) Waldsterben: forest decline in West Germany. Environmental Science & Technology 24: 436–441.
6. Matzner E, Murach D (1995) Soil changes induced by air pollutant deposition and their implication for forests in central Europe. Water, Air and Soil Pollution 85: 63–76.
7. Harrison AF, Carreira J, Poskitt JM, Robertson SMC, Smith R, et al. (1999) Impacts of pollutant inputs on forest canopy condition in the UK: possible role of P limitations. Forestry 72: 367–377.
8. Tkacz B, Moody B, Castillo JV, Fenn ME (2008) Forest health conditions in North America. Environmental Pollution 155: 409–425.
9. Elias PE, Burger JA, Adams MB (2009) Acid deposition effects on forest composition and growth on the Monongahela National Forest, West Virginia. Forest Ecology and Management 258: 2175–2182.
10. Ito K, Uchiyama Y, Kurokami N, Sugano K, Nakanishi Y (2011) Soil acidification and decline of trees in forests within the precincts of shrines in Kyoto (Japan). Water, Air and Soil Pollution 214: 197–204.
11. Durka W, Schulze E-D, Gebauer G, Voerkeliust S (1994) Effects of forest decline on uptake and leaching of deposited nitrate determined from ^{15}N and ^{18}O measurements. Nature 372: 765–767.
12. Burke MK, Raynal DJ (1998) Liming influences growth and nutrient balances in sugar maple (*Acer saccharum*) seedlings on an acidic forest soil. Environmental and Experimental Botany 39: 105–116.
13. Meiwes KJ, Mindrup M, Khanna PK (2002) Retention of Ca and Mg in the forest floor of a Spruce stand after application of various liming materials. Forest Ecology and Management 159: 27–36.
14. Prietzel J, Rehfuess KE, Stetter U, Pretzsch H (2008) Changes of soil chemistry, stand nutrition, and stand growth at two Scots pine (*Pinus sylvestris* L.) sites in Central Europe during 40 years after fertilization, liming, and lupine introduction. European Journal of Forest Research 127: 43–61.
15. Rizvi SH, Gauquelin T, Gers C, Guérold F, Pagnout C, et al. (2012) Calcium–magnesium liming of acidified forested catchments: effects on humus morphology and functioning. Applied Soil Ecology 62: 81–87.
16. Matzner E, Murach D, Fortmann H (1986) Soil acidity and its relationship to root growth in declining forest stands in Germany. Water, Air and Soil Pollution 31: 273–282.
17. Hahn G, Marschner H (1998) Effect of acid irrigation and liming on root growth of Norway spruce. Plant and Soil 199: 11–22.
18. Bakker MR, Garbaye J, Nys C (2000) Effect of liming on the ectomycorrhizal status of oak. Forest Ecology and Management 126: 121–131.
19. Persson H, Ahlström K (1990/91) The effects of forest liming and fertilization on fine-root growth. Water, Air and Soil Pollution 54: 365–375.
20. Bakker MR, Kerisit R, Verbist K, Nys C (1999) Effects of liming on rhizosphere chemistry and growth of fine roots and of shoots of sessile oak (*Quercus petraea*). Plant and Soil 217: 243–255.
21. Helmisaari HS, Hallbäcken L (1999) Fine-root biomass and necromass in limed and fertilized Norway spruce (*Picea abies* (L.) Karst.) stands. Forest Ecology and Management 119: 99–110.
22. Huettl RF, Zoettl HW (1993) Liming as a mitigation tool in Germany's declining forests—reviewing results from former and recent trials. Forest Ecology and Management 61: 325–338.
23. Ljungström M, Nihlgård B (1995) Effects of lime and phosphate additions on nutrient status and growth of beech (*Fagus sylvatica* L.) seedlings. Forest Ecology and Management 74: 133–148.
24. Sikström U (1997) Effects of low-dose liming and nitrogen fertilization on stemwood growth and needle properties of *Picea abies* and *Pinus sylvestris*. Forest Ecology and Management 95: 261–274.
25. Lundström US, Bain DC, Taylor AFS, van Hees PAW (2003) Effects of acidification and its mitigation with lime and wood ash on forest soil processes: a review. Water, Air and Soil Pollution: Focus 3: 5–28.
26. Børja I, Nilsen P (2009) Long term effect of liming and fertilization on ectomycorrhizal colonization and tree growth in old Scots pine (*Pinus sylvestris* L.) stands. Plant and Soil 314: 109–119.
27. Tveite B, Abrahamsen G, Stuanes AO (1990/91) Liming and wet acid deposition effects on tree growth and nutrition: experimental results. Water, Air and Soil Pollution 54: 409–422.
28. Kreutzer K, Weiss T (1998) The Höglwald field experiments—aims, concept and basic data. Plant and Soil 199: 1–10.
29. Jonard M, André F, Giot P, Weissen F, Van der Perre R, et al. (2010) Thirteen-year monitoring of liming and PK fertilization effects on tree vitality in Norway spruce and European beech stands. European Journal of Forest Research 129: 1203–1211.
30. Kakei M, Clifford PE (2002) Short-term of lime application on soil properties and fine-root characteristics for a 9-year-old Sitka spruce plantation growing on a deep peat soil. Forestry 75: 37–50.
31. Kuang YW, Wen DZ, Li J, Sun FF, Hou EQ, et al. (2010) Homogeneity of $\delta^{15}N$ in needles of Masson pine (*Pinus massoniana* L.) was altered by air pollution. Environmental Pollution 158: 1963–1967.
32. Teng XR (2005) Status quo of forest resources and management strategy in Chongqing city. Forest Inventory and Planning 30: 73–76. (In Chinese with English abstract).
33. Li ZY, Wang YH, Yu PT, Zhang ZJ (2007) A comparative study of resistance to soil acidification and growth of fine roots between pure stands of *Pinus massoniana* and *Cinnamomum camphora*. Acta Ecologica Sinica 27: 5245–5253. (In Chinese with English abstract).
34. Wang YH, Solberg S, Yu PT, Myking T, Vogt RD, et al. (2007) Assessments of tree crown condition of two Masson pine forests in the acid rain region in south China. Forest Ecology and Management 242: 530–540.
35. Li ZY, Wang YH (2009) Forest Health under Acidification Stress—A Case Study of Tieshanping in Chongqing. China Agriculture Press, Beijing. (In Chinese).
36. Rodhe H, Herrera R (1988) Acidification in Tropical Countries. John Wiley & Sons Ltd., New York.
37. Zhao DW, Xiong JL, Xu Y, Chan WH (1988) Acid rain in southwestern China. Atmospheric Environment 22: 349–358.
38. Jin L, Shao M, Zeng LM, Zhao DW, Tang DG (2006) Estimation of dry deposition fluxes of major inorganic species by canopy throughfall approach. Chinese Science Bulletin 51: 1818–1823.
39. Hao JM, Wang SX, Liu BJ, He KB (2000) Control strategy for sulfur dioxide and acid rain pollution in China. Journal of Environmental Sciences 12: 385–393.
40. Jiang WH, Zhang S, Chen GC, Xiong HQ, Ding Y, et al. (2002) Effect of acid deposition on soil and vegetation of forest ecosystem in Nanshan of Chongqing. Research of Environmental Sciences 15: 8–11. (In Chinese with English abstract).
41. Guo JH, Zhang XS, Vogt RD, Xiao JS, Zhao DW, et al. (2007) Evaluating main factors controlling aluminum solubility in acid forest soils, southern and southwestern China. Applied Geochemistry 22: 388–396.
42. Larssen T, Carmichael GR (2000) Acid rain and acidification in China: the importance of base cation deposition. Environmental Pollution 110: 89–102.
43. Chinese Society of Forestry (1989) Acid Rain and Agriculture. China Forestry Press, Beijing. (In Chinese).
44. Du XM, Tian RS (1996) The relation between aluminium poisoning and decline of Masson pine forest in Nanshan Mountain, Chongqing. Research of Environmental Sciences 9: 21–25. (In Chinese with English abstract).
45. Bäckman JSK, Hermansson A, Tebbe CC, Lindgren PE (2003) Liming induces growth of a diverse flora of ammonia-oxidising bacteria in acid spruce forest soil as determined by SSCP and DGGE. Soil Biology & Biochemistry 35: 1337–1347.
46. Rosenberg W, Nierop KGJ, Knicker H, de Jager PA, Kreutzer K, et al. (2003) Liming effects on the chemical composition of the organic surface layer of a mature Norway spruce stand (*Picea abies* [L.] Karst.). Soil Biology & Biochemistry 35: 155–165.
47. Guckland A, Ahrends B, Paar U, Dammann I, Evers J, et al. (2012) Predicting depth translocation of base cations after forest liming: results from long-term experiments. European Journal of Forest Research 131: 1869–1887.
48. Huang YM, Duan L, Jin T, Yang YS, Hao JM (2006). Effect of limestone and magnesite applications on Masson pine (*Pinus massoniana*) forest growing on acidified soil. Acta Ecologica Sinica 26: 786–792. (In Chinese with English abstract).
49. Zhao DW, Larssen T, Zhang DB, Gao SD, Vogt RD, et al. (2001) Acid deposition and acidification of soil and water in the Tie Shan Ping Area, Chongqing, China. Water, Air and Soil Pollution 130: 1733–1738.
50. Vogt RD, Guo JH, Luo JH, Peng XY, Xiang RJ, et al. (2007) Water chemistry in forested acid sensitive sites in sub-tropical Asia receiving acid rain and alkaline dust. Applied Geochemistry 22: 1140–1148.
51. ICP Forests website. Available: http://www.icp-forests.org/Manual.htm. Accessed 2013 July 31.
52. Fischer R, Mues V, Ulrich E, Becher G, Lorenz M (2007) Monitoring of atmospheric deposition in European forests and an overview on its implication on forest condition. Applied Geochemistry 22: 1129–1139.
53. Wu ZY (1980) Plants of China. Science Press, Beijing. (In Chinese).
54. Seftigen K, Moldan F, Linderholm HW (2013) Radial growth of Norway spruce and Scots pine: effects of nitrogen deposition experiments. European Journal of Forest Research 132: 83–92.
55. Hutchinson TC, Bozic L, Munoz-Vega G (1986) Response of five species of conifer seedlings to aluminium stress. Water, Air and Soil Pollution 31: 283–294.
56. Cronan CS, Grigal DF (1995) Use of calcium/aluminum ratios as indicators of stress in forest ecosystems. Journal of Environmental Quality 24: 209–226.
57. Staszewski T, Kubiesa P, Łukasik W (2012) Response of spruce stands in national parks of southern Poland to air pollution in 1998–2005. European Journal of Forest Research 131: 1163–1173.
58. Gao JX, Cao HF (1991) Effects of ionic strength, pH and Ca/Al ratio on aluminum toxicity of Masson pine seedlings. Acta Scientiae Circumstantiae 11: 194–198. (In Chinese with English abstract).
59. Ponette Q, Dufey JE, Weissen F (1997) Downward movement of dolomite, kieserite or a mixture of $CaCO_3$ and kieserite through the upper layers of an acid forest soil. Water, Air and Soil Pollution 95: 353–379.
60. Frank J, Stuanes AO (2003) Short-term effects of liming and vitality fertilization on forest soil and nutrient leaching in a Scots pine ecosystem in Norway. Forest Ecology and Management 176: 371–386.
61. Caires EF, Garbuio FJ, Churka S, Barth G, Corrêa JCL (2008) Effects of soil acidity amelioration by surface liming on no-till corn, soybean, and wheat root growth and yield. European Journal of Agronomy 28: 57–64.
62. Löfgren S, Cory N, Zetterberg T, Larsson PE, Kronnäs V (2009) The long-term effects of catchment liming and reduced sulphur deposition on forest soils and

runoff chemistry in southwest Sweden. Forest Ecology and Management 258: 567–578.
63. Matzner E, Khanna PK, Meiwes KJ, Ulrich B (1985) Effects of fertilization and liming on the chemical soil conditions and element distribution in forest soils. Plant and Soil 87: 405–415.
64. Moore JD, Duchesne L, Ouimet R (2008) Soil properties and maple–beech regeneration a decade after liming in a northern hardwood stand. Forest Ecology and Management 255: 3460–3468.
65. Misra RK, Turnbull CRA, Cromer RN, Gibbons AK, LaSala AV (1998) Below- and above-ground growth of *Eucalyptus nitens* in a young plantation I. Biomass. Forest Ecology and Management 106: 283–293.
66. Štofko P (2010) Relationships between the parameters of aboveground parts and the parameters of root plates in Norway spruce with respect to soil drainage. Journal of Forest Science 56: 353–360.
67. Dhief A, Abdellaoui R, Tarhouni M, Belgacem AO, Smiti SA, et al. (2011) Root and aboveground growth of rhizotron-grown seedlings of three Tunisian desert *Calligonum* species under water deficit. Canadian Journal of Soil Science 91: 15–27.

19

Soil Bacterial Community Response to Differences in Agricultural Management along with Seasonal Changes in a Mediterranean Region

Annamaria Bevivino[1]*, Patrizia Paganin[1], Giovanni Bacci[2,3], Alessandro Florio[2], Maite Sampedro Pellicer[1], Maria Cristiana Papaleo[3], Alessio Mengoni[3], Luigi Ledda[4], Renato Fani[3], Anna Benedetti[2], Claudia Dalmastri[1]

1 ENEA (Italian National Agency for New Technologies, Energy and Sustainable Economic Development) Casaccia Research Center, Technical Unit for Sustainable Development and Innovation of Agro-Industrial System, Rome, Italy, **2** Consiglio per la Ricerca e la Sperimentazione in Agricoltura - Research Centre for the Soil-Plant System, Rome, Italy, **3** Laboratory of Microbial and Molecular Evolution, Department of Biology, University of Florence, Florence, Italy, **4** Dipartimento di Agraria, University of Sassari, Sassari, Italy

Abstract

Land-use change is considered likely to be one of main drivers of biodiversity changes in grassland ecosystems. To gain insight into the impact of land use on the underlying soil bacterial communities, we aimed at determining the effects of agricultural management, along with seasonal variations, on soil bacterial community in a Mediterranean ecosystem where different land-use and plant cover types led to the creation of a soil and vegetation gradient. A set of soils subjected to different anthropogenic impact in a typical Mediterranean landscape, dominated by *Quercus suber* L., was examined in spring and autumn: a natural cork-oak forest, a pasture, a managed meadow, and two vineyards (ploughed and grass covered). Land uses affected the chemical and structural composition of the most stabilised fractions of soil organic matter and reduced soil C stocks and labile organic matter at both sampling season. A significant effect of land uses on bacterial community structure as well as an interaction effect between land uses and season was revealed by the EP index. Cluster analysis of culture-dependent DGGE patterns showed a different seasonal distribution of soil bacterial populations with subgroups associated to different land uses, in agreement with culture-independent T-RFLP results. Soils subjected to low human inputs (cork-oak forest and pasture) showed a more stable bacterial community than those with high human input (vineyards and managed meadow). Phylogenetic analysis revealed the predominance of *Proteobacteria*, *Actinobacteria*, *Bacteroidetes*, and *Firmicutes* phyla with differences in class composition across the site, suggesting that the microbial composition changes in response to land uses. Taken altogether, our data suggest that soil bacterial communities were seasonally distinct and exhibited compositional shifts that tracked with changes in land use and soil management. These findings may contribute to future searches for bacterial bio-indicators of soil health and sustainable productivity.

Editor: Jack Anthony Gilbert, Argonne National Laboratory, United States of America

Funding: This research was funded by MIUR (Integrated Special Fund for Research - FISR) in the frame of the Italian National Project SOILSINK "Climate change and agro-forestry systems, impacts on soil carbon sink and microbial diversity", and partially supported by MIUR (Research Department of Italian Government) in the framework of the Agreement Program ENEA-CNR (Articolo 2, comma 44, Legge 23.12.2009 n. 191 - Legge Finanziaria 2010). The funders had no role in study design, data collection and analysis, decision to publish, or preparation of the manuscript.

Competing Interests: The authors have declared that no competing interests exist.

* Email: annamaria.bevivino@enea.it

Introduction

Soil microorganisms play an important role as regulators of major biogeochemical cycles and can significantly affect the ecosystem functioning [1], being involved in organic matter dynamics, nutrient cycling and decomposition processes [2]. The anthropogenic activities affect the diversity of natural habitats modifying the number of species occurring in the environment at the landscape scale. Soil management strongly influences soil biodiversity in agricultural ecosystems. Different practices can alter the below-ground ecosystem, often leading to depletion of soil carbon and loss of biodiversity, and thus affecting the structure of the resident microbial communities [3]. Therefore, characterizing genetic and functional diversity of soil bacterial communities in response to agricultural practices and/or climate is fundamental to better understand and manage the ecosystem processes.

The Mediterranean area is one of the most important biodiversity hotspots in the world and is increasingly threatened by intensive land use [4]. The high environmental diversity that characterizes the Mediterranean region is related to the integration of natural ecosystems and traditional human activities such as the agroforestry practices [5]. The collapse of the traditional agro-silvo-pastoral system that occurred during the past century has led to major changes in the extension of woodlands dominated by typical Mediterranean species, i.e. cork oak (*Quercus suber*) and/ or holm oak (*Quercus ilex*) woodlands [6], [7]. Research on the influence of management practices on the biodiversity of these agro-silvo-pastoral systems is increasing but it has focused mostly on plants [8] and vertebrates [5], [9]. In the frame of the Italian Project SOILSINK (Climatic changes and agricultural and forest systems: impact on C reservoirs and on soil microbial diversity), a hilly basin in Gallura (Berchidda site, Sardinia, Italy) was selected as a reference Mediterranean site for studying the influence of land-use changes on diversity, function and seasonal variations of soil microbial communities [10], [11]. The site is within an area of about 1,450 ha and is characterised by extensive agro-silvo-pastoral systems, typical of north-eastern Sardinia (Italy) and similar areas of the Mediterranean basin [12]. The chosen site represents a sustainable balance between human activities and natural resources that have created a landscape of high heterogeneity and cultural value, whose importance has been recognized at the European level [13], [14]. Indeed, it is considered climatically (Mediterranean zone) and pedologically homogeneous with vegetation patterns similar to those called *dehesas* or *montados* of the south-western Iberian Peninsula [15], [16], [17]. In the past, this area was covered by cork-oak forests, which gradually were subjected to increasing under-storey grazing and usage for the extraction of cork. Today, there are different land-use and plant cover types that lead to a soil and vegetation gradient with an ecological progression: from a cork-oak forest undergoing minimum disturbance to managed vineyards with an intensive agricultural practice (grass covered and ploughed), passing through areas with temporary grassland, and pasture.

The different land uses altered soil potential, making possible to discriminate the role of human management on soil functioning. When forests are converted to grasslands, and grasslands turned into agricultural lands, a sharp switch from one type of soil microbial community to another one occurs. Since the ability of an ecosystem to withstand serious disturbances may partly depend on its microbial component(s), characterizing bacterial community composition and/or structure might help to better understand and manipulate ecosystem processes. The aim of the present study was to investigate the effects of soil characteristics and different agricultural managements on soil bacterial community in two seasons. Sampling was carried out in spring and autumn 2007, when the plant cover-growing season usually starts and ends. A combination of culture-based and molecular techniques along with statistical analysis of data obtained was applied to interrogate the diversity, function, and ecology of soil bacterial communities. The results obtained in the present study, along with the other ones obtained within the SOILSINK Project [11], [18], [19], [20], provide useful data on the impact of soil type, cover vegetation, and human activities on the distribution of the bacterial genetic resources in soil communities for this Mediterranean region. Our results confirm that the environments with low inputs (cork-oak forest and pasture) show a more stable soil microbial community than those subjected to increasing human input (vineyards and managed meadow) and suggest that soil bacterial communities are seasonally distinct with compositional shifts that track with changes in land use and soil management.

Materials and Methods

Ethics Statement

We carried out the study on the hilly basin in Gallura (Olbia-Tempio municipalities, Sardinia, Italy). Five soils uses were identified in private farms within Berchidda site (40°49′ 15″N, 9°17′ 32″ E) and were obtained. The soil sampling was carried out in the frame of a national research project (SOILSINK Project) and soils used in this study were collected under consent of the landowners. The responsible of the study site was Prof. P.P. Roggero (University of Sassari, Italy). We confirm that our study did not harm the environment and did not involve endangered or protected species. Specific geographic coordinates (referred to World Geodetic System, 1984) of our study area are reported in Table S1.

Sampling site

The study area (Olbia-Tempio) is representative of the climate, vegetation type and management of some of the most common agro-forestry systems in the Mediterranean basin [16].

The Berchidda site is made up of hydromorphic and granitic soil with a loamy sand texture. The altitude ranges from 275 m to 300 msl. This area is referred to as a meso-thermo Mediterranean, subhumid phytoclimatic belt with a mean annual rainfall of 862 mm and the mean annual temperature of 13.8°C [10]. In the

Figure 1. Long-term effects of different land-use with increasing level of intensification in spring and autumn. Both pasture and managed meadow included spotted cork oak trees, which are key components of the Dehesatype landscape typical of this area of Sardinia. The cork-oak formation, pasture, and managed meadow have been converted to the current use and maintained unchanged for more than 30 years, whereas the non-tilled cover cropped vineyard and the tilled one were planted in 1985 and 1994, respectively. From the left to right: cork-oak forest (CO), hayland pasture rotation (PA), managed meadow (MM), grass covered vineyard (CV), tilled vineyard (TV).

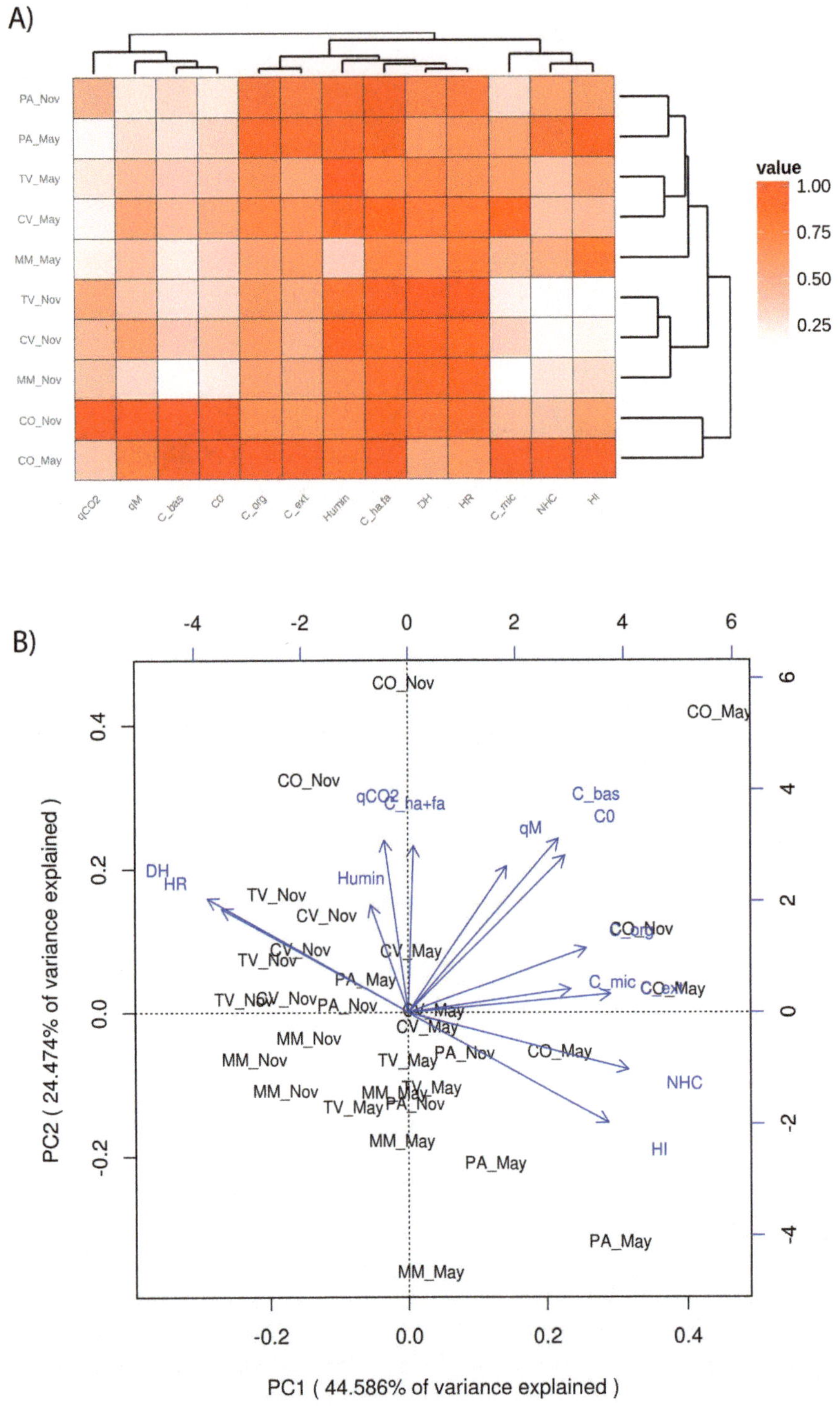

Figure 2. Effect of land-use and season on soil physical-chemical and biological parameters. A) Heat map with hierarchal clustering of physical-chemical and biological parameters across the five Sardinia soils with different land uses at the two different sampling time points (May and November). The heat map was constructed using a maximum-minimum normalization of the data in order to represent each value in a range between 0 and 1. Higher values are represented by darker colors whereas lower ones are represented by lighter colors. CO = cork-oak forest; PA = hayland-pasture rotation; MM = managed meadow; TV = tilled vineyard; CV = grass covered vineyard. B) PCA ordination of data (axes 1 and 2) generated from physical-chemical and biological properties of the different types of land use in May and November.

past, the Berchidda area was covered by cork-oak forests (dominated by *Quercus Suber* L.), which were subjected to intense usage for the extraction of cork. Today, there are five main different land-use units close together: cork-oak forest (CO), hayland-pasture rotation (PA), managed meadow (MM), and tilled (TV) and grass covered vineyard (CV). The cork-oak formation, pasture, and managed meadow have been converted to the current use and maintained unchanged for more than 30 years,

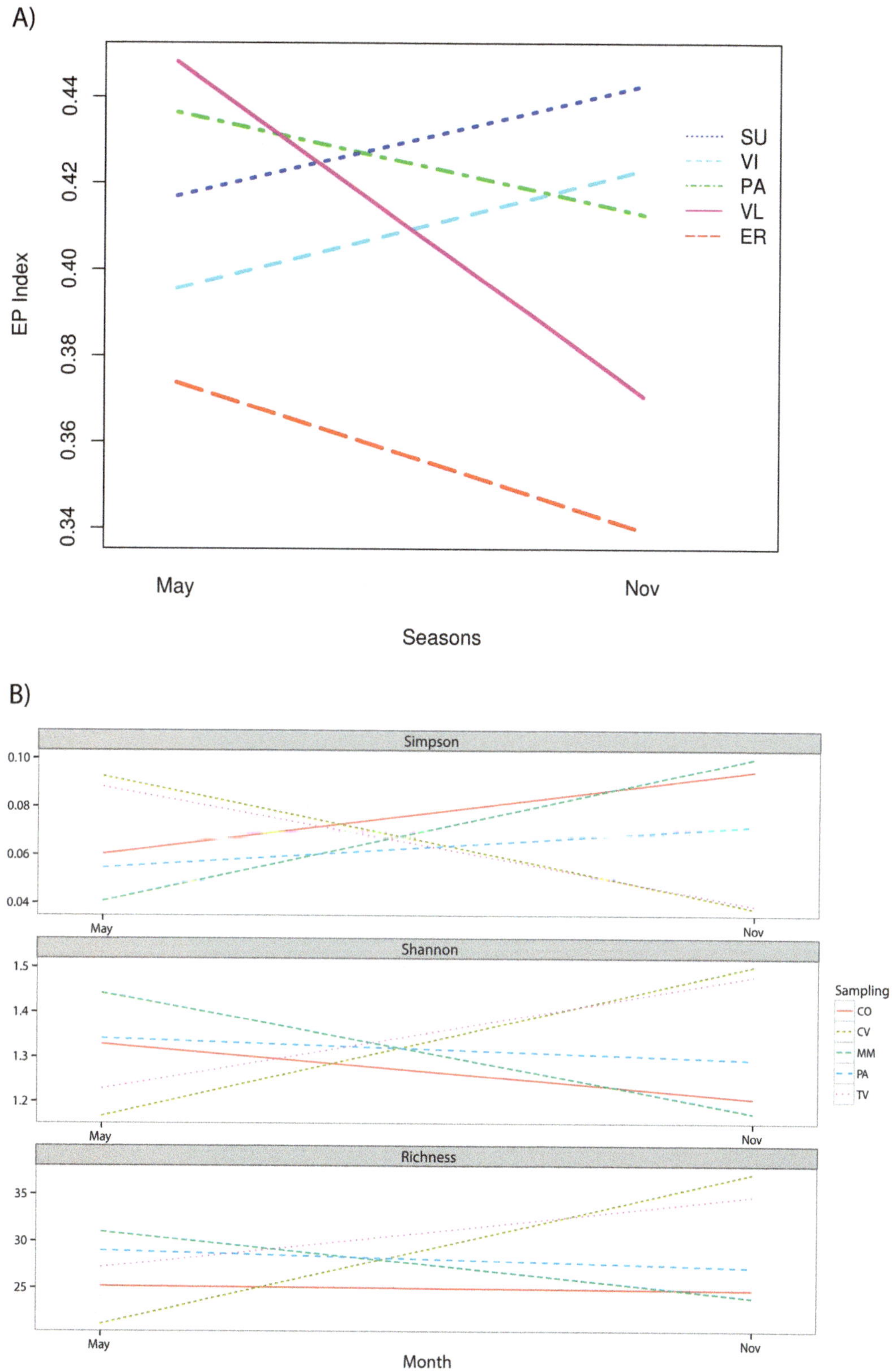

Figure 3. Effect of land-use and season on Eco-Physiological (EP) index of culturable bacteria (A) and diversity indices from CD-DGGE profiles (B).

whereas the non-tilled cover cropped vineyard and the tilled one were planted in 1985 and 1994, respectively. The five soils are located inside an area of 161.5 km^2. Detailed characteristic and management of the five soils have been previously described [12], [20], [21], [22], [23].

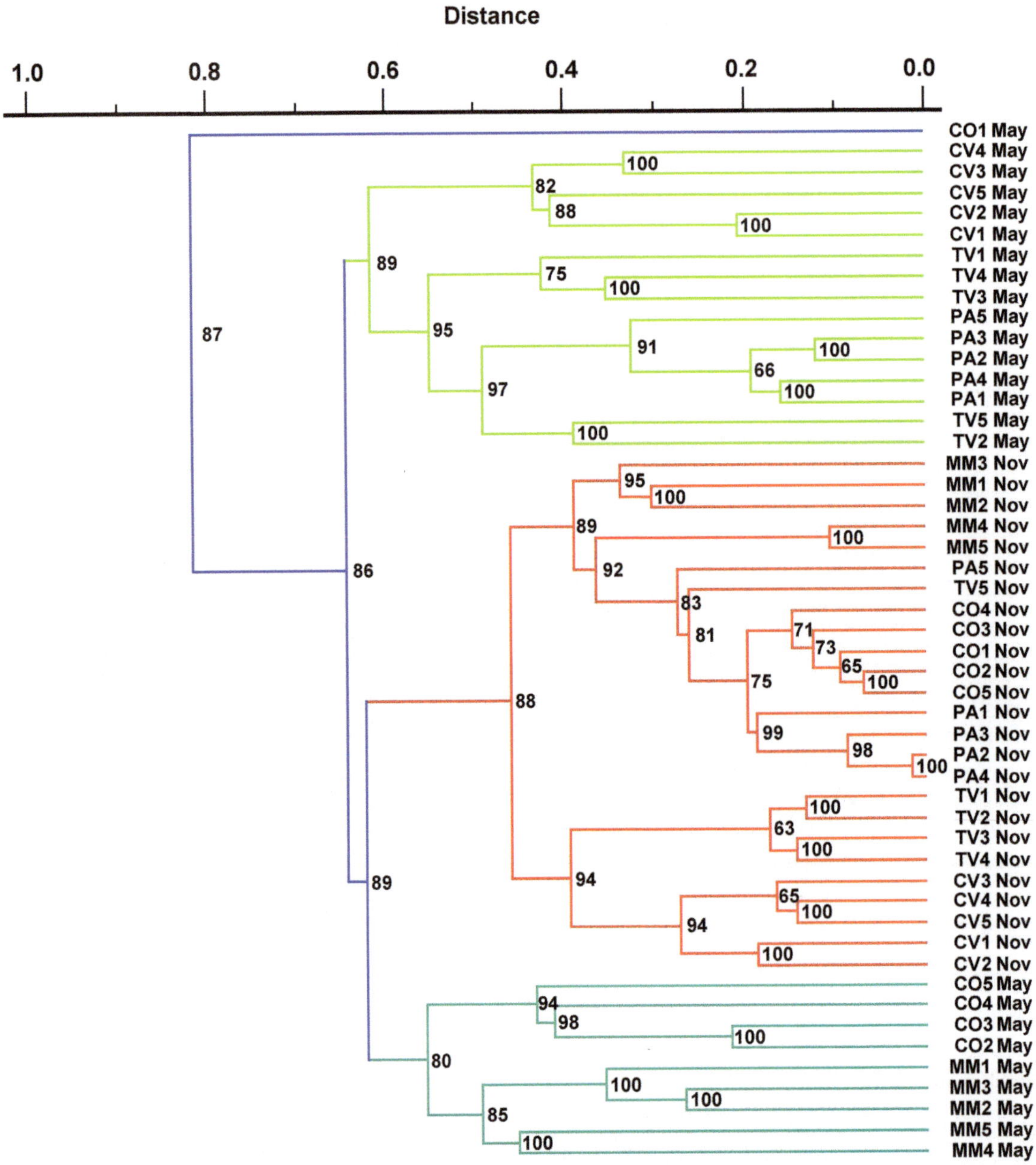

Figure 4. UPGMA dendrogram of DGGE profiles of amplified 16S rDNA of bacterial communities recovered in CO, PA, MM, CV and TV, in spring and autumn, generated using Phoretix ID advanced analysis package. Designation of samples is the same as in Fig. 1. The scale bar represents dissimilarity among samples. Consistency of each cluster was measured by the Cophenetic correlation coefficient shown at each node.

Pedological characterization of the study area

The pedogenic substrate of the study area consists of medium-grained granite, affected by localized presence of veins of quartz and porphyry. The morphology of the Berchidda area varies from flat to undulating. The processes of soil erosion by water channeled are evident only in the short-term forage crops made on soil with more than 15% of slope. All profiles have a horizons sequence of the type A-Bw-C or A-Bw-BC-C or more rarely, A-Bw-C-R (Table S1). The power of the profiles, limited to horizons A and Bw, varies from a minimum of 38 cm to a maximum of 100 cm. In three soil profiles in TV, the sequence between the horizon A and Bw and the substrate is gradually altered by the presence of a horizon BC, characterised by coarse texture, the power of which varies from about 30 cm to 90 cm. Direct contact with the unaltered rock, R horizon, was observed only in one profile in CV. The prevailing textural classes are sandy and sandy loam. The content of organic substance in the horizons A is never very high, with average value around to about 3%. The maximum values of 11.8% and 8.6% were observed on the A horizons in two soil profiles in CO. The exchange complex, in agreement with the reduced content in clays, is never high. The degree of base saturation is predominantly less than 60% (Dystric conditions of the USDA Soil Taxonomy). The profiles with the exchange complex with a degree of base saturation below 60% in all horizons between 25 and 75 cm (TV, CV, MM, and PA) were

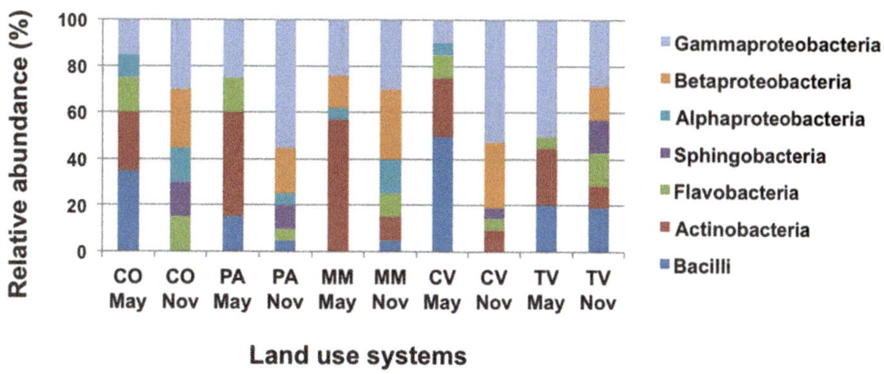

Figure 5. Relative abundances of major taxonomic groups across land use systems in spring and autumn. Detailed data of each class are listed in Additional file 1: Table S6. CO = cork-oak forest; PA = hayland-pasture rotation; MM = managed meadow; TV = tilled vineyard; CV = grass covered vineyard. Values presented are the mean percent.

classified as Typic Dystroxerepts [24]. The other one (CO) was classified as Lithic Xerorthents.

Soil sampling

Five soil replicates were collected from bulk soil of the five different managements (CO, PA, MM, CV, and TV) (Fig. 1). Soils were collected in May and November 2007. After removal of litter layer, soil core samples (50 to 100 g; diameter, 5 cm) were taken from each of the five locations, using a 5-on-dice sampling pattern with ca. 70 m distance between each sampling point. Sampling was performed at 20 cm depth, where most microbial activity is known to occur [25], [26]. In the vineyard soils, samples from along the rows and between the rows were pooled together to form a field replicate. The other soil samples (CO, PA, MM) were collected out of trees influence. At each season and for each soil type, five randomly field replicates were collected for a total of 25 soil samples (5 replicates × 5 land uses), each one being a composite sample of five soil cores. A total of 50 composite samples were taken for the two seasons. Soil samples were immediately sieved (<2 mm) to remove fine roots and large organic debris, air dried, and transported to the labs for microbiological analysis. The moisture content was adjusted to 60% of their water holding capacity (WHC) and soil samples were then left to equilibrate at room temperature in the dark for one day prior to analyses, in order to restore, within limits [27], the microbial activity of air- dried soils to that of soils in the field.

Chemical and biochemical analyses of soil samples

The chemical and biochemical analyses were performed on three replicates for each land use for a total of 15 soil samples (3 replicates × 5 land uses), each one being a composite sample of five soil cores. Total organic carbon (C_{org}) was estimated after oxidation with $K_2Cr_2O_7$ and subsequent titration of unreduced $Cr_2O_7^{2-}$ with $Fe(NH_4)_2(SO_4)_2$, as reported by Springer and Klee [28]. Soil Organic Matter (SOM) was determined by C_{org} multiplied by 1.724 van Belem coefficient. The C_{org} fractionation was set up as reported by Ciavatta and co-workers [29]. In particular, solid samples were extracted at 65°C for 24 h using 0.1 mol l^{-1} NaOH plus 0.1 mol l^{-1} $Na_4P_2O_7$ solution (1:50, solid:liquid ratio). The samples were then centrifuged at 5000×*g* and the supernatants were filtered through a 0.20-µm Millipore filter (Millipore, Billerica, MA) (total extractable C, C_{ext}). The humic-like acid (HA) fraction was separated from the fulvic-like acid (FA) and the non-humified carbon fractions (NHC) by precipitation after acidification of the alkaline solution (supernatant) to pH<2. Chromatography on a column of polyvinylpyrrolidone (PVP, Aldrich, Germany) was used to separate the NHC from the FA. The FA was then combined with the HA to obtain total humified fraction (HA+FA). Total extractable C (C_{ext} %), humic and fluvic acid C (C_{HA+FA} %) were determined by the dichromate oxidation method. The non-humified carbon (C_{NH}) was determined as the difference between C_{ext} and C_{HA+FA}. Humification indexes HI, DH, and HR were determined according to previous works [29], [30].

Microbial biomass C (C_{mic}) was determined by the fumigation-extraction method of Vance and co-workers [31] with some slight modifications. The measurements were performed on air-dried soils, pre-conditioned by a 10-d incubation in open glass jars, at −33 kPa water tension, and 30°C. The incubation was employed for restoring, within limits [27], the microbial activity of air-dried soils to that of soils in the field. Four replicates of each soil sample were used. Average values are given in mg C kg^{-1} of soil. For measuring microbial respiration, 20 g (oven-dry basis) of moist sample were placed in 1 L stoppered glass jars. The CO_2 evolved was trapped, after 1, 2, 4, 7, 10, 14, 21, 28 days of incubation, in 2 ml 1 M NaOH and determined by titration of the excess NaOH with 0.1 M HCl [32].

Non-linear least square regression analysis was used to calculate parameters affecting C mineralization from daily CO_2 evolution data (Stat Win 4.0 for Windows). The best fit was obtained with the exponential model of CO_2-C accumulation according to the negative exponential decay model:

$$C_m = C_0\left(1 - e^{-kt}\right),$$

where C_m is the cumulative value of mineralized C during *t* days, *k* is the rate constant, and C_o is the potentially mineralizable C [33]. The CO_2 emitted in 28 days of incubation was used as cumulative respiration (C_{cum}). The CO_2 evolved during the 28th day of incubation was used as the basal respiration value (C_{bas}). Microbial indices were calculated as follows [34], [35], [36]:

$$qCO_2 = \left[\left(\mu g\, CO_2 - C_{bas} \times \mu g^{-1} C_{mic}\right) h^{-1}\right] 10^3,$$
$$qM_{cum} = \left(\mu g\, CO_2 - C_0 \times \mu g^{-1} C_{org}\right),$$
$$qM_{bas} = \left(\mu g\, CO_2 - C_{bas} \times \mu g^{-1} C_{org}\right)$$

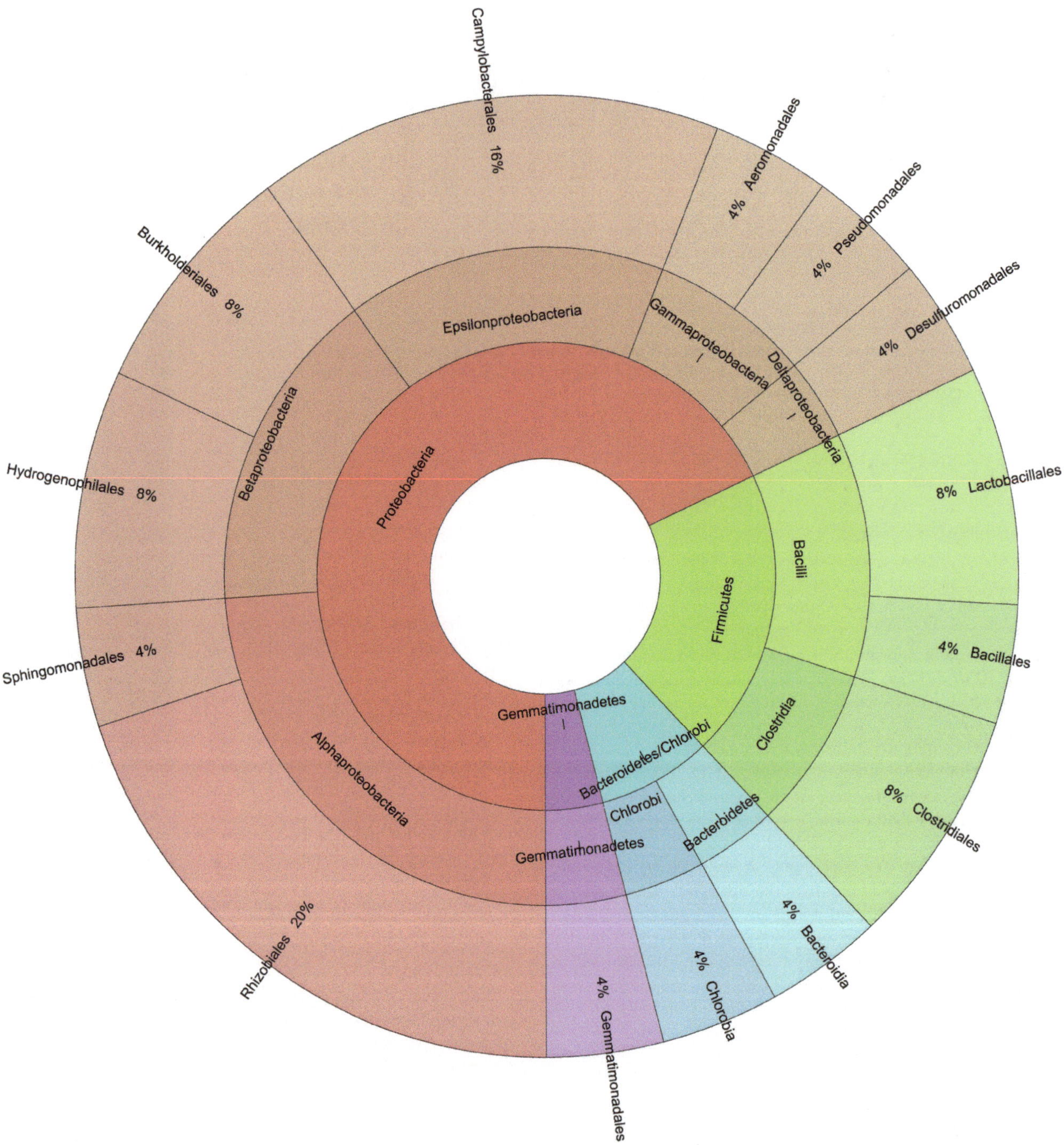

Figure 6. Plot of the taxonomic composition of total bacterial community as inferred from taxonomic interpretation of T-RFLP profiles.

Recovery of cultured bacterial cells from soil samples

Soil bacterial community analysis was performed on five soil replicates for each land use, as described above. Bacterial cell extraction was performed according to the recommendations of Smalla et al. [37] with minor modifications. Briefly, 1 g of soil was placed in a sterile 15 ml plastic tube containing 10 ml of phosphate buffered saline (PBS, pH 7.0). This mixture was homogenized for 30″ at low speed by using the Ultra-Turrax Thyristor Regle 50 (Janke & Kunkel IKA-Labortechnik). After homogenization, suspension was placed into a Erlenmeyer flask (100 ml) containing 10 g of glass beads (0,2 mm) and shaken for 1 h at 180 r.p.m. and 28°C to disperse bacteria. The flasks and glass beads were autoclaved for 20 min at 121°C before use. The soil suspension was transferred into a sterile 15 ml plastic tube and serially diluted with sterile saline solution (9 g l^{-1}NaCl) from 10^{-1} up to 10^{-7}. The, 100 μl aliquots of serially diluted soil suspensions were plated in triplicate on 0.1 Triptic Soy Broth (TSB; Difco) amended with 15 g l^{-1} agar (0.1 TSA) and 100 g μl^{-1} of cycloheximide (Sigma) to inhibit fungal growth. Plates were incubated at 28°C for 6 days.

Growth strategy and total bacterial populations

To determine the changes in the structure of culturable fraction of soil bacteria, the r/K-strategy concept proposed by De Leij and co-workers [38] was used. Bacterial colonies appearing within 48 h were designated as r-strategists, and the remaining as K-strategists. Colonies were enumerated at 1, 2 and 6 days of growth on 0.1 TSA; in this way, three counts (or classes) were generated per sample. Plates containing between 30 and 300 colonies were then selected for enumeration. Total bacterial counts obtained were expressed as colony forming units (CFU) per gram of soil. Distribution of bacteria in each class as a percentage of the total counts gave insight into the distribution of r- and K-strategists in each sample.

To evaluate the changes in the biodiversity of bacterial populations in soils, the eco-physiological (EP) index [38] was used. The EP index of each soils tested was calculated using the equation:

$$H' = -\sum(P_i)(log_e P_i),$$

where P_i represents the CFU on each day (1, 2 and 6 days of incubation) as a proportion of the total CFU in that sample after 6 days incubation, i.e., the proportion of colonies appearing on counting day i ($i = 1, 2, 6$) with $EP_{min} = 0$. Higher values of EP index imply a more even distribution of proportions of bacteria developing on different days (i.e., different classes of bacteria).

Terminal-Restriction Fragment Length Polymorphism (T-RFLP)

DNA was extracted from soil samples by using the FastDNA SPIN Kit for Soil (QBiogene). Terminal-Restriction Fragment Length Polymorphism (T-RFLP) was performed on 16SrRNA genes amplified from extracted DNA with primer pairs P0 and P6 as previously reported [39], [40]. Purified amplification products were digested separately with restriction enzymes *Rsa*I and *Msp*I and digestions were resolved by capillary electrophoresis on an ABI310 Genetic Analyzer (Applied Biosystems, Foster City, CA, USA) using LIZ 500 (Applied Biosystems) as size standard. T-RFLP analysis was performed as previously reported [41]. Diversity indices were calculated with PAST software [42] as previously reported [19], taking into account peak intensities of T-RFLP fragments. Taxonomic interpretation of T-RFLP profiles was performed by querying the Ribosomal Database Project Database by using MiCA3 web tool (http://mica.ibest.uidaho.edu/), as previously described [40].

Culture-dependent DGGE (CD DGGE) analysis

Culture-dependent DGGE (CD DGGE) fingerprinting of 16S rRNA gene was used to characterize mixed bacterial communities recovered on agar plates.

Collection of cultured bacterial communities for DGGE analysis. Cultured bacterial communities were collected following procedure proposed by Duineveld and co-workers [43] with minor modifications. Briefly, after one week of incubation at 28°C, colonies were removed from plates containing between 100 and 1000 colonies by adding 3.0 ml sterile physiological solution (0.9% NaCl) on each plate and scraping off all grown colonies with a sterile Drigalski spatula. The cell suspensions thus obtained were aliquoted into 1.5 ml Eppendorf tubes and centrifuged at 8,000 r.p.m. for 10 minutes. The pellets were stored at −80°C for subsequent DNA extraction and PCR-DGGE analysis.

DNA extraction and PCR amplification of 16S rRNA genes. Genomic DNA was extracted with sodium dodecyl sulfate-proteinase K lysis buffer, followed by a treatment with cetyl-trimethylammonium bromide (CTAB) as described in *Current Protocols for Molecular Biology* [44]. Briefly, the pellet was resuspended in 567 µl of TE [10 mM Tris-HCl - 1 mM EDTA (pH 8.0)] buffer, and glass beads 0.3 mm in diameter (250 mg) were added, followed by bead beating for 20 s. Then, 30 µl of 10% sodium dodecyl sulfate and 3 µl proteinase K 20 mg/ml (Sigma) were added and samples were incubated at 37°C for 1 h. Glass beads were removed by centrifugation 2 min at 2,800 rpm and samples were then incubated at 65°C for 10′ with100 µl of 5 M NaCl prepared with sterile water and 80 µl of CTAB/NaCl (10% CTAB in 0.7 M NaCl). Following incubation, extracts were purified by using phenol/phenol-chloroform/isoamyl alcohol (49.5:49.5:1) extraction and DNA was recovered by isopropanol precipitation at 4°C o/n. Pelleted DNA was washed twice with cold 70% ethanol, allowed to air dry, and re-suspended in 50 µl of sterile water. Quantity and purity of DNA were checked by NanoDrop (NanoDrop Technologies, USA) and gel electrophoresis. The DNA samples were stored at –20°C until required for use.

The 16S rRNA gene was amplified using 20 ng of lysate suspension and the universal bacterial primers P0 and P6 [45]. Dilutions 1:100 (2 µl) of the1450 bp PCR products were then used as template for the second PCR amplification with the forward primer 63F (5′- AGGCCTAACACATGCAAGTC -3′), with a GC clamp (5′-CGCCCGCCGCGCGCGGCGGGCGGGGCG-GGGGCACGGGGGG -3′) incorporated at the 5′ end, and the reverse primer 518R (5′-ATTACCGCGGCTGCTGG-3′), to produce 495 bp fragments suitable for DGGE analysis [46]. Both PCR reactions were performed in Qiagen Taq buffer (10X) containing 1.5 mM $MgCl_2$, with 150 ng of each primer, 250 µM (each) deoxynucleoside triphosphates, and 0.5 U of *Taq* DNA polymerase (Qiagen, Hilden, Germany) in a 25 µl reaction volume. Cycle parameters for PCR with the primer pairs P0–P6 and 63F-GC and 518R were previously described by Di Cello and co-workers [45] and El Fantroussi and co-workers [46], respectively.

Denaturing Gradient Gel Electrophoresis. 16S rRNA gene amplicons were separated by double gradient denaturing gradient gel electrophoresis (DG-DGGE) as described by Cremonesi et al. [47], in a DCode universal mutation detection system (Bio-Rad, CA, USA). Separation of purified PCR products (700 ng) was achieved in6%–12% polyacrylamide (acrylamide: N,N-methylenebisacrylamide, 37.5:1) gels containing an increasing linear gradient of denaturants ranging from 30% to 60% (100% denaturant corresponds to 7 M urea and40% deionized formamide). Each gel also included marker lanes represented by DGGE profiles containing a large number of discrete bands spanning the entire gradient, suitable for within- and between-gel alignment. Electrophoresis were carried out for 16 h at 75 mV in 1X TAE buffer at 60°C, stained with 50 µg/ml ethidium bromide for 30 min, destained in water and photographed with the UVIpro Platinum Gel Documentation System (GAS7500/7510; Eppendorf, Cambridge,UK).

Cluster analysis and diversity indices

Quantity One software package (Bio-Rad) and Phoretix 1D PRO software (Phoretix International, Newcastle upon Tyne, United Kingdom) were used for CD-DGGE profile analysis. The cluster analysis and dendrogram generation were carried out by using the Phoretix 1D Pro software according to the manufacturer's instructions (Phoretix International, Newcastle upon Tyne,

United Kingdom). Bands of CD-DGGE patterns were aligned and normalized using reference lanes. Background noise was subtracted by rolling ball algorithm with a radius of 50 pixels; the automatic band detection was performed with a minimum slope of 200 and a noise reduction of 10, and peaks smaller than 2% of the maximum peak were discarded. Bands were manually corrected and matched to create an absent/present binary matrix. The similarity between the band patterns was calculated using the Dice coefficient and the clustering analysis was performed with the unweighted pair group method with arithmetic averages (UPGMA) to generate a dendrogram by using mathematic averages algorithm programs integral to the Phoretix 1D Pro software. Coefficient of cophenetic correlation was used to measure the consistency of clusters.

DGGE banding data were used to estimate three diversity indices by treating each band as an individual operational taxonomic unit (OTU). The number of DGGE bands present in each sample was used to measure the Richness index (*R*). The Shannon-Weaver index of general diversity (H') [48] and the Simpson index of dominance (*D*) [49] were calculated from the number of bands present and the relative intensities of each band (P_i) in each lane. Relative signal intensities of detected bands, in each gel track, were determined by using the Quantity One software package (Bio-Rad) and calculated from the peak area of the densitometric curves.

The Shannon-Weaver diversity (H') was calculated using the following equation:

$$H' = -\sum (P_i)(log_e P_i),$$

where P_i (the proportion of abundances of the ith band) is measured as:

$$P_i = n_i / N,$$

where n_i is the peak height of a band, and N is the sum of all peak heights in the densitometry profiles.

The Simpson index (D) was calculated with the formula:

$$D = \sum (P_i x P_i);$$

it measures the strength of dominance because it weights towards the abundance of the OTUs and varies inversely with species diversity [50].

Bacterial isolation and identification

A total of 100 bacterial colonies were randomly picked up for each soil sample from 0.1 TSA plates (containing approximately 50 to 500 colonies), previously used for the determination of the CFU counts and EPI-index, and repeatedly streaked onto 0.1 TSA fresh plates to obtain pure cultures. Isolated colonies were then grown overnight (o/n) in TSB medium at 28°C and 200 r.p.m., and stored at −80°C in 30% glycerol until further analysis. From all five soil samples, 500 colonies were isolated in each season, for a total of 1000 bacterial colonies. A total of 203 colonies with different morphologies (about 20 colonies per each sample) were taken up to investigate their taxonomic affiliation.

Genomic DNA and PCR amplification of the 16S rRNA gene were performed as described above. Sequencing reactions were prepared from PCR products using an Applied Biosystem Big Dye Terminator sequencing kit version 3.1, according to the manufacturer's instructions and analysed using a 3730 DNA Analyzer Applied Biosystem apparatus. The sequences were compared with those in the GenBank databases by using the BLAST program and Seqmatch tool of the RDP (http://www.ncbi.nlm.nih.gov/BLAST/and http://rdp.cme.msu.edu/, respectively) and aligned with the closest relatives with the Clustal W function of the BioEdit package [51]. Bacterial identification by 16S rRNA gene sequences assignment was performed using the RDP Classification Algorithm (http://rdp.cme.msu.edu/classifier/classifier.jsp).

Statistical analysis

Bacterial population data (CFU/g of soil) were log transformed and subsequently analysed by one-way ANOVA (STATISTICA, Release 3.0b, Copyright StatSoft Inc., CA, USA). Percentage data of EP index value were *logit*-transformed, as follows:

$$Logit\,(p) = log[p/(1-p)]$$

for the proportion *p*, and compared using one-way ANOVA (STATISTICA, Release3.0b, Copyright StatSoft Inc., CA, USA).

Analyses of variance (ANOVA) on biodiversity indexes (Shannon-Weaver, Richness and Simpson), principal component analysis (PCA) and clustering analysis on biochemical data were performed using R packages "stats" and "vegan" (http://cran.r-project.org/and http://cran.r-project.org/web/packages/vegan/index.html). All data clustering were performed using the "UPGMA" algorithm implemented in the "hclust" function of the R "stats" package. Distances among samples were calculated using "Bray-Curtis" distance implemented in "vegan" package as:

$$i = sample\ "i";\ j = sample\ "j",\ BC_{ij} = 2C_{ij}/\left(S_i + S_j\right),$$

where C_{ij} = sum of the smaller value for species in common between samples i and j, and S_i and S_j = total number of species in samples i and j, respectively [52]. Variation of biodiversity indexes of cultured bacteria was inspected using ANOVA analysis. Biodiversity indexes were first divided into groups depending on sampling season and different managements of soils and then the analysis was performed. Clustering analysis (UPGMA) on biochemical parameters was performed. Each parameter was first divided in groups, in the same way of previous ANOVA analysis, and averaged. Then, each result obtained was normalized using the maximum-minimum normalization technique in order to make the data comparable. PCA analysis using each biochemical data was performed.

Results and Discussion

Effect of land use on soil chemical and biochemical properties

Soil organic matter (SOM) represents a dynamic system influenced by several factors, including climate, clay content, mineralogy and soil management, which all affect the processes of organic matter transformation and evolution in soil [53], [54]. Both soil fertility and stability are related to the organic matter content of soil. Many functions of SOM are due to its more stabilised fractions, humified materials and balance between the labile and the stabilised fractions [55]. Changes in SOM content are related to changes in microbial biomass turnover, because they reflect the balance between rates of microbial organic matter accumulation and rates of organic matter degradation. The extent of organic matter's organization not only impacts the amount of carbon mineralized but also the type of carbon that is consumed by the microorganisms.

In this study two categories of soil quality indicators were used: organic matter quality indicators and microbial biomass activity indicators (Tables S2 and S3). Our results (Fig. S1) revealed that both land use and sampling season affect the chemical and structural composition of the most stabilised fractions of SOM. A higher content of C_{org} occurred in CO soil in May, and C_{ext} and C_{HA+FA} showed the same pattern, whereas the average values of these parameters were slightly higher in PA soil in November. These results were reflected in the humification parameters, where the humification rate (HR%) can provide quantitative information about the humic substances content normalised with respect to total SOM, while the degree of humification (DH%) provides the amount of the humified carbon in the extracted organic fraction and the humification index (HI) can be considered as an index of soil humification activity as well as of availability of non humified labile fractions [29]. Overall, land use change reduced soil C stocks and labile organic matter at both sampling times except for PA in November. Pasture has a great potential soil organic C stock and, in the long term, grass management systems have nearly equivalent potential to store soil organic C as forest [56]. Potentially mineralizable C (C_0), which indicates the amount of C in the labile fraction of soil organic matter, decreased over sampling time in all soils, and C_{mic} similarly declined. Cultivated soils are characterised by low microbial activity, mainly due to the disappearance of easily decomposable organic compounds through tillage and soil disturbance. As previously found for chemical properties, the metabolic activity responds to the different land uses; in both May and November samplings, qCO_2 was higher in CO soil when compared to the others, resulting in increases stress. Hence, unfavourable conditions result in a decrease in the size of the microbial biomass and the efficiency of C substrates degradation, conducting to an increase in respiration rate per unit of microbial biomass [57].

As beneficial and negative effects of soil management practices are strongly linked to microbial activities and regulate soil quality and functioning [58] the relationship existing between soil management practices and variation in chemical and biochemical parameters was evaluated, by performing a clustering analysis (UPGMA) as reported in Tables S2 and S3. The dendrogram obtained was linked to a heat-map representing all the biochemical parameters variation (Fig. 2A). Data analysis revealed that chemical and biochemical parameters clustered in two different groups each of which corresponding to one of the two parameters analyzed (chemical and biochemical). In both sampling times, TV and CV clustered together, suggesting that seasonal change rather than management regime was a major driving force contributing to vineyard soil fertility. Furthermore, according to the UPGMA clustering, CO formed a separate cluster from that of the other soil uses (Fig. 2A). Interestingly, the chemical and biochemical parameters clustered together in both CO and PA, regardless of the season, in contrast to MM, CV and TV, where they clustered separately in relation to the sampling season. This finding suggests that soils subjected to low human inputs (pasture and cork-oak forest) showed a more stable chemical and biochemical soil composition than those with high human input (managed meadow and vineyard).

The different positions of the variables in the plane of the first two principal components, as revealed by PCA analysis (Fig. 2B), indicated that chemical and biochemical parameters were differentially affected by the various land use types. The first and the second principal components (PC1 and PC2) accounted for 44.59% and 24.47% of the total variance in the data, respectively (Fig. 2B). Both chemical and biochemical parameters were positively affected by PC1 except for HR and DH, whereas PC2 was able to discriminate between the two variable sets, with the most of chemical and biochemical parameters being positively and negatively affected, respectively. The above analyses revealed an important effect of land uses on both chemical and biochemical; on the contrary, only biochemical parameters were affected by seasons.

Influence of land use on soil bacterial community

Community structure analysis by EP-index and r-K strategy. Environmental conditions select organisms that either grow rapidly in uncrowded, nutrient-rich conditions (r-strategists), or can efficiently exploit resources in crowded conditions (K-strategists). In this work, we applied the method developed by De Leij et al. [38], who used the r/K-strategy concept for the characterization of soil bacterial communities. First, we compared the microbial community structures found in the different land uses during each of the two seasons. Since sampling was performed at 20 cm depth, where most microbial activity is known to occur [26], the main changes are expected through conversion from one soil management system to another one [59]. Microbial community structure found in the different land uses during each of the two seasons was investigated by means of EP index that is a measure of both richness (i.e. total number of species in the community) and evenness (i.e. how evenly individuals in the community are distributed over the different species) of groups of microorganisms with similar developmental characteristics.

The cultured bacteria belonging to fast-growing organisms, especially the r-strategists detected on day 2, dominated in all land uses and in both seasons (Table S4). The MM soil exhibited a lower EP index and a lower percentage of r-strategists in both seasons, possibly due to amensalism from a dominant bacterial group in the community. Variation of EP index of cultured bacteria was inspected using an ANOVA analysis of variance (Fig. 3A). Interestingly, statistical analysis revealed a significant effect of land uses on bacterial community structure ($P<0.001$) as well as an interaction effect between land uses and season change ($P<0.001$) on EP index. Variation of EP index due to seasonal changes was significant in the soil with higher human impact (TV) when compared with the other soils. Total bacterial concentrations varied significantly in respect of season ($P<0.001$) being higher in spring than in autumn (Table S2) in all but CO soils, with significant differences between CO and PA in spring, and between CO and MM, and MM and PA in autumn.

Overall results suggest that land uses affected cultured bacterial communities. Soils with low human impact (cork-oak forest) have a more stable bacterial density and show a less variation of bacterial community structure across seasons than soils subjected to high human impact such as tilled vineyard.

Bacterial community profiling by CD-DGGE and T-RFLP analyses. Culture dependent DGGE (CD-DGGE) fingerprinting of the 16S rRNA was used to characterize mixed bacterial communities recovered on agar plates. CD-DGGE represents a useful technique to follow the dynamics of distinct culturable fractions of the soil bacterial community in relation to physical, chemical and biological changes in the soil environment [60]. Since culture-dependent and culture-independent methods likely profile distinct fractions of the soil bacterial community with unique ecological roles [61], culturable bacteria may provide an ecologically relevant complement to culture-independent community characterization [62].

By pooling the bacterial cells growing on individual agar plates, we obtained a culture-dependent bacterial community. The analysis of DGGE profiles revealed clear banding patterns for each land-use and plant-cover types of sufficient complexity to investigate differences in soil microbial communities and identified stable communities with highly reproducible profiles (Fig. S2). A different community composition among land-use types was found as evidenced by the presence of different dominant signals; this finding indicated a compositional shift among soils examined in spring and autumn and subjected to different anthropogenic impact. Bacterial diversity was investigated through richness (*R*), Shannon-Weaver (*H'*), and Simpson (*D*) indices. Analysis of variance (ANOVA) confirmed the differences in the distribution of bacterial species due to land uses and sampling seasons ($P<0.001$). Both *R* and *H'* indices decreased from May to November in MM samples and increased in TV and CV samples, and the complementary opposite trend was observed for *D* index (Fig. 3B). Otherwise, cultured bacterial community in CO and PA soils did not vary significantly over seasons. Most likely, these results reflect the impact of both land-use and vegetation type and coverage on soil microbial communities. In particular, the low shift of biodiversity indices observed in CO and PA seems to be correlated to a higher stability of bacterial populations in natural habitats with low human impact, whereas populations inhabiting more anthropogenic areas tend to be more variable. Land-use type and, in particular, differences in vegetation dynamics may have a large role in modulating the temporal variability in soil bacterial communities. As observed by Lauber and co-workers [3], soils from the different land-use types did not exhibit identical temporal dynamics even though all the soils were located in close proximity and exposed to the same climatic conditions. Diversity indices obtained from T-RFLP analysis performed on total bacterial DNA (Fig. S3) partially confirmed the trends of diversity shown by CD-DGGE, though observed differences were not statistically significant (Table S5).

The unweighted pair-group method using arithmetic averages (UPGMA) of bacterial community profiling by CD-DGGE revealed a high diversity of the bacterial communities in each land-use and plant-cover soils. The similarity between the DGGE patterns of the soil bacterial communities revealed three distinct clusters (Fig. 4). Samples collected in autumn grouped into a separate cluster with about 52% similarity, whereas samples collected in spring grouped into two separate clusters, each of which composed by samples sharing about 40% similarity. The highest similarity values were shown by clusters based on soil land uses, suggesting a low bacterial variation within each soil type. This finding is consistent with previous reports showing that land-use type was the most important factor in determining the composition of soil microbial community [3].

The clustering results based on CD-DGGE were in agreement with culture-independent T-RFLP analysis that confirmed a different seasonal distribution of soil bacterial populations with subgroups associated to different land uses. Most of samples retrieved from the same soil (CO, PA, and MM) clustered together, whereas samples retrieved from TV and CV were often intermixed in the UPGMA dendrogram (Fig. S4). As culture-dependent and culture-independent profiles can separately resolve unique, diverse, and equally complex fractions of the soil bacterial community [60], our combined results from both CD-DGGE and T-RFLP methods, applied on culturable and total fractions, respectively, revealed an interaction effect between land uses and season change in affecting soil bacterial communities.

Taxa responses to land use determined by phylogenetic affiliation of soil bacterial isolates. Taxonomic affiliation was investigated on a total of 203 bacterial isolates. In detail, 20–21 colonies recovered from each soil sample at each season and showing different morphologies (for a total of 101 bacterial isolates in spring and 102 in autumn) were subjected to DNA extraction and PCR amplification of 16S rRNA gene. An amplicon of the expected size (about 1500 bp) was obtained from each isolate, and its nucleotide sequence was determined and submitted to GenBank (Table S6). Although it is generally accepted that not all bacteria, including types of soil bacteria, are culturable, the isolation of bacteria by agar plate cultivation and subsequent phylogenetic analysis permit to isolate and identify previously uncultured representatives or even new members of certain bacterial species for further analysis of their metabolic function. Therefore, even if the metagenome sequencing is becoming the most powerful tool to investigate microbial communities, the ability to isolate indigenous strains actually remains the unique way to further characterize and select soil bacteria showing interesting properties. Additionally, standard cultivation techniques have been shown to be able to capture members of the soil rare biosphere which could not be detected by metagenome sequencing [63].

The 203 sequences obtained were compared with those present in the GenBank databases by using the BLAST [64] program. Results showed that most of 16S rRNA gene sequences matched NCBI database sequences at 99–100% of similarity at the genus level, with *Arthrobacter*, *Bacillus*, *Stenotrophomonas*, *Pseudomonas* and *Burkholderia* as the most representative genera. So, identification at the genus level was achieved in 103 isolates and at the species level in 72 isolates, while 12 isolates were only affiliated to taxa level higher than genus and 16 remained unidentified (Table S6). Further comparison with GenBank databases by using the Seqmatch tool of the RDP indicated that 16S rRNA gene sequences were affiliated to four phyla: *Proteobacteria* (classes α, β and γ), *Bacteroidetes* (classes *Flavobacteria* and *Sphingobacteria*), *Actinobacteria* and *Firmicutes*. Even though the RDP analysis does not permit to affiliate a bacterial isolate to a given species, it allowed to assess the genus of almost all isolates, including those not identified through the BLAST search. In fact, a boostrap value of at least 80% is satisfactory for RDP requirement. Almost all our sequences gave rise to 100% boostrap, 12 ranged from 89% to 99%, and only two of them only resulted in low values (24 and 54% respectively, referred to two spring isolates).

Data obtained from both approaches allowed to assess that Berchidda soil is colonized by bacteria included in the classes of γ-*Proteobacteria* (with a prevalence of *Pseudomonas* and *Stenotrophomonas* genera), *Actinobacteria* (in particular, the genus *Arthrobacter*), β-*Proteobacteria* (with a prevalence of *Burkholderia* genus), *Bacilli* (with the genus *Bacillus* as the dominant one), Flavobacteria and Sphingobacteria. The relative abundances of the different classes identified by RDP across the different samples related to soil uses and season are represented in Figure 5. A putative taxonomic description of total bacterial community, derived from the interpretation of T-RFLP data (Fig. S3) is reported in Figure 6, in which the largest fraction of the community is represented by members of *Proteobacteria*, with α-*Proteobacteria* as the most abundant class.

Differences in class composition across the site were observed suggesting that the microbial composition changes in response to land uses. In fact, all the seven classes (α, β and γ *Proteobacteria*, *Sphingobacteria Flavobacteria*, *Actinobacteria* and *Bacilli*) were present in CV, PA, and CO soils, while all classes but α-*Proteobacteria* in TV and all classes but *Sphingobacteria* in MM were found. The observed large diffusion of *Proteobacteria* in all soil-uses is in agreement with previously reported data concerning soil bacterial communities in different land-use systems [65]. Within *Proteobacteria*, the majority of the isolates fell into the gamma subgroup, which showed higher relative abundance compared to that of any other taxa at the class level while a few were alfa proteobacteria, especially in vineyards and pasture. In all soil-uses, were also found *Actinobacteria*, in agreement with previous work [65], and β-*Proteobacteria*, like *Burkholderia* sp., previously detected by Pastorelli and co-workers [19] who analysed the denitrifying bacterial communities present in the same Berchidda soil samples. Among Bacteroidetes, it has to be noted the relative abundance of the genera *Chryseobacterium* sp. (in all soil uses) and *Flavobacterium* sp. (in all soil uses, but MM), which include isolates already observed in Korea soils [66], and bacteria with plant growth promoting properties recovered in Iran soil [67], respectively. As already pointed out by Fierer and co-workers [68], the β-*Proteobacteria* and *Bacteroidetes* follow copiotrophic lifestyles and their relative abundance were highest in soils with high C availability. In general, copiotrophic bacteria should have higher growth rates and traditional culturing methods are likely to select for microorganisms that can grow rapidly in high resource environments.

Strong differences in class composition were observed in each soil when sampled in the two seasons: for instance, *Bacilli* and *Actinobacteria* dominate in spring (particularly, *Bacilli* in CV and CO, while *Actinobacteria* in MM and PA), whilst β-*Proteobacteria* tend to dominate in autumn. Six out of the seven identified classes (β and γ-*Proteobacteria*, *Flavobacteria*, *Actinobacteria* and *Bacilli*) were recovered in both seasons whilst *Sphingobacteria* were recovered only in autumn. In all soils, *Actinobacteria* were prevalent in spring, β *Proteobacteria* predominated in autumn, while β and γ-*Proteobacteria* in both spring and autumn. When genera composition was used to infer the diversity indices of cultured bacteria (Table S7) an increase of diversity (estimated as Shannon H and Evenness) from May to November was present for all soils, but CO, where Shannon H was slightly higher in May than in November. In particular, MM soil showed the highest increases for most indices, especially for alpha diversity (*i.e.* the species richness and evenness within a sample) that has often been correlated with ecosystem stability and functionality [69]. It must be noted that cultivated bacterial populations did not fluctuate with seasonal changes in soils with low human input (cork-oak forest). The bacterial communities of the cork-oak forest soil were similar in richness and composition, furthering the point that in a community with moderate disturbance, new individuals and groups could be introduced in a manner that promotes competition and diversity of the community, thus establishing a more stable community [65].

These data suggest that shifts from forest to managed meadow and vineyard result in changes of bacterial communities composition. Previous studies have also shown that shifts from forest to grassland soil [70], from cultivated system to pasture [71] and from grazed pine forest to cultivated crop and grazed pasture [65] resulted in significant changes in bacterial community composition. Our results suggest that the use of culture-dependent 16S rRNA gene sequencing along with traditional analysis of soil physiochemical properties may provide insight into the ecological relevance of soil bacterial taxa.

Conclusion

Overall, data obtained in this work revealed an important effect of land uses on both chemical and biochemical soil parameters. Soil bacterial communities were seasonally distinct and exhibited compositional shifts that tracked with changes in land use and soil management. This study, combining the pedological and biochemical data with microbiological and molecular analysis, furnishes a good methodological approach to describe the influence of different soil managements on soil microbial community structure. In fact, the results demonstrate that, in the same pedological conditions, long-term soil management influence the community structure; i.e., soils subjected to low human inputs (cork-oak forest and pasture) showed a more stable chemical and biochemical soil composition as well as bacterial community than those with high human input (vineyards and managed meadow). Further research is required to determine whether the observed shifts in bacterial community composition produce parallel changes in the functional attributes of these communities across soil types under different long-term management regimes. The use of culture-independent approaches, like metabarcoding and metagenome sequencing, will make it possible to identify the specific drivers of land-use dynamics exhibited by soil bacterial communities and to give a complete picture of the bacterial communities in a typical Mediterranean agro-silvo-pastoral system.

Supporting Information

Figure S1 Box-plot analysis showing the frequency distribution of physical-chemical and biological properties of the five Sardinia soils.

Figure S2 Examples of CD-DGGE profiles of the soil bacterial communities associated to the different land uses in May and November. A) From the left to right: hayland pasture rotation (PA), tilled vineyard (TV), grass covered vineyard (CV) in May; B) managed meadow (MM), cork-oak forest (CO), hayland pasture rotation (PA) in May; C) grass covered vineyard (CV) in May, grass covered vineyard (CV) in November, tilled vineyard (TV) in May; D) cork-oak forest (CO), hayland pasture rotation (PA), managed meadow (MM) in November.

Figure S3 Examples of T-RFLP profiles obtained after digestion with *Msp*I (A) and *Rsa*I (B) restriction enzymes of amplified of 16S rRNA gene sequences.

Figure S4 Cluster analysis of T-RFLP patterns generated by *Msp*I and *Rsa*I digestion of 16S rRNA gene sequences. The UPGMA cluster analysis based on Jaccard similarity matrix was calculated for each set of samples using the "hclust" function of the R "stats" package. The scale bar represents the percent of dissimilarity.

Table S1 Pedological profiles and classification of the soils investigated.

Table S2 Determination of total organic carbon soil (C_{org}), extractable carbon (C_{ext}), humified carbon (C_{HA+FA}), non humified carbon (C_{NH}) and humification parameters of the five Sardinian soils.

Table S3 Biochemical parameters measured in the five Sardinian soils.

Table S4 r/k bacterial strategists, total culturable bacteria and EPI index measured in soils under different long-term management practices.

Table S5 Diversity indices of total bacterial communities as inferred from T-RFLP profiles.

Table S6 Phylogenetic affiliations of 203 randomly selected soil bacterial isolates based on comparative analysis of their 16S rRNA gene sequences.

Table S7 The ratio of diversity indices related to the abundance of the different genera detected in cultured isolates between November and May.

Acknowledgments

We are grateful to P.P. Roggero (University of Sassari) and his team (G. Seddaiu, G. Urracci and L. Doro) for their contribution to collect the Sardinian soils and for providing data on experimental area. The experimental site was chosen on the basis of vegetation and soil surveys made in collaboration with S. Madrau, S. Bagella, R. Filigheddu, M.C. Caria, and E. Farris (University of Sassari). We thank R. Pastorelli for her help in soil sampling, S. Tabacchioni and L. Chiarini for bacterial isolation, C. Cantale and M. Sperandei for figure preparation, M.T. Rubino for chemical-biochemical analysis, and S. Cesarini and L. Pirone for their valuable suggestions. The research was carried out in the context of the FISR SOILSINK research project coordinated by R. Francaviglia (CRA-RPS, Rome) (?http://soilsink.entecra.it).

Author Contributions

Conceived and designed the experiments: A. Bevivino RF CD A. Benedetti. Performed the experiments: A. Bevivino PP MSP MCP CD LL. Analyzed the data: A. Bevivino PP GB AF AM CD LL. Contributed reagents/materials/analysis tools: A. Bevivino PP GB AF MSP AM LL MCP RF CD A. Benedetti. Contributed to the writing of the manuscript: A. Bevivino RF GB AF CD LL A. Benedetti.

References

1. Tiedje JM, Asuming-Brempong S, Nusslein K, Marsh TL, Flynn SJ (1999) Opening the black box of soil microbial diversity. Appl Soil Ecol 13: 109–122. doi:10.1016/S0929-1393(99)00026-8.
2. Nannipieri P, Ascher J, Ceccherini MT, Landi L, Pietramellara G, et al. (2003) Microbial diversity and soil functions. Eur J Soil Sci 54: 655–670. doi:10.1046/j.1365-2389.2003.00556.x.
3. Lauber CL, Ramirez KS, Aanderud Z, Lennon J, Fierer N (2013) Temporal variability in soil microbial communities across land-use types. ISME J 7: 1641–1650. doi:10.1038/ismej.2013.50.
4. Myers N, Mittermeier RA, Fonseca GAB, Fonseca GAB, Kent J (2000) Biodiversity hotspots for conservation priorities. Nature 403: 853–858. doi:10.1038/35002501.
5. Puddu G, Falcucci A, Maiorano L (2011) Forest changes over a century in Sardinia: implications for conservation in a Mediterranean hotspot. Agrofor Syst 85: 319–330. doi:10.1007/s10457-011-9443-y.
6. Médail F, Quézel P (1999) Biodiversity Hotspots in the Mediterranean Basin: Setting Global Conservation Priorities. Conserv Biol 13: 1510–1513. doi:10.1046/j.1523-1739.1999.98467.x.
7. Blondel J, Aronson J (1999) Biology and wildlife of the Mediterranean region. Oxford University Press, Oxford.
8. Salis L, Marrosu M, Bagella S, Sitzia M RP (2010) Grassland management, forage production and plant biodiversity in a Mediterranean grazing system. In: Porqueddu C, Ríos S, editors. The contributions of grasslands to the conservation of Mediterranean biodiversity. Zaragoza: CIHEAM/CIBIO/FAO/SEEP, Vol. 185. 181–185.
9. Gonçalves P, Alcobia S, Simões L, Santos-Reis M (2012) Effects of management options on mammal richness in a Mediterranean agro-silvo-pastoral system. Agrofor Syst 85: 383–395. doi:10.1007/s10457-011-9439-7.
10. Bacchetta G, Bagella S, Biondi E, Farris E, Filigheddu R, et al. (2004) A contribution to the knowledge of the order Quercetalia ilicis Br. -Bl. ex Molinier 1934 of Sardinia. Fitosociologia 41: 29–51.
11. Orgiazzi A, Lumini E, Nilsson RH, Girlanda M, Vizzini A, et al. (2012) Unravelling soil fungal communities from different Mediterranean land-use backgrounds. PLoS One 7: e34847. doi:10.1371/journal.pone.0034847.
12. Francaviglia R, Benedetti A, Doro L, Madrau S, Ledda L (2014) Influence of land use on soil quality and stratification ratios under agro-silvo-pastoral Mediterranean management systems. Agric Ecosyst Environ 183: 86–92.
13. Council of the European Communities (1992) Council Directive 92/43/EEC of 21 May 1992 on the conservation of natural habitats and of wild fauna and flora. Off J Eur Communities 35: 7–50.
14. Council of the European Communities (2001) Commission Regulation (EC) No 1808/2001 of 30 August 2001 laying down detailed rules concerning the implementation of Council Regulation (EC) No 338/97 on the protection of species of wild fauna and flora by regulating trade therein. Off J Eur Communities L250: 1–43.
15. Aru A, Baldaccini P, Delogu G, Dessena M, Madrau S, et al. (1990) Carta dei Suoli della Sardegna, scala 1/250.000. Assessorato alla programmazione e all'assestamento del territorio, Centro Regionale Programmazione, Dip. Sc. della Terra, Università di Cagliari, Italy.
16. Bacchetta G, Bagella S, Biondi E, Farris E, Filigheddu R, et al. (2009) Forest vegetation and serial vegetation of Sardinia (with map at the scale 1:350,000). Fitosociologia 46: 3–82.
17. Bagella S, Caria MC (2011) Vegetation series: a tool for the assessment of grassland ecosystem services in Mediterranean large-scale grazing systems. Fitosociologia 48: 47–54.
18. Lumini E, Orgiazzi A, Borriello R, Bonfante P, Bianciotto V (2010) Disclosing arbuscular mycorrhizal fungal biodiversity in soil through a land-use gradient using a pyrosequencing approach. Environ Microbiol 12: 2165–2179. doi:10.1111/j.1462-2920.2009.02099.x.
19. Pastorelli R, Landi S, Trabelsi D, Piccolo R, Mengoni A, et al. (2011) Effects of soil management on structure and activity of denitrifying bacterial communities. Appl Soil Ecol 49: 46–58. doi:10.1016/j.apsoil.2011.07.002.
20. Lagomarsino A, Benedetti A, Marinari S, Pompili L, Moscatelli MC, et al. (2011) Soil organic C variability and microbial functions in a Mediterranean agro-forest ecosystem. Biol Fertil Soils 47: 283–291. doi:10.1007/s00374-010-0530-4.
21. Francaviglia R, Coleman K, Whitmore AP, Doro L, Urracci G, et al. (2012) Changes in soil organic carbon and climate change - Application of the RothC model in agro-silvo-pastoral Mediterranean systems. Agric Syst 112: 48–54.
22. Seddaiu G, Porcu G, Ledda L, Roggero PP, Agnelli A, et al. (2013) Soil organic matter content and composition as influenced by soil management in a semi-arid Mediterranean agro-silvo-pastoral system. Agric Ecosyst Environ 167: 1–11. doi:http://dx.doi.org/10.1016/j.agee.2013.01.002.
23. Lai R, Lagomarsino A, Ledda L, Roggero PP (2014) Variation in soil C and microbial functions across tree canopy projection and open grassland microenvironments. Turkish J Agric For 38: 62–69. doi:10.3906/tar-1303-82.
24. Soil Survey Staff (2006) Keys to soil taxonomy. In: United States Department of Agriculture NRCS, editor. Natural Resoiurces Concervation Service.
25. Doran JW, Elliott ET, Paustian K (1998) Soil microbial activity, nitrogen cycling, and long-term changes in organic carbon pools as related to fallow tillage management. Soil Tillage Res 49: 3–18.
26. O'Brien HE, Parrent JL, Jackson JA, Moncalvo J-M, Vilgalys R (2005) Fungal community analysis by large-scale sequencing of environmental samples. Appl Environ Microbiol 71: 5544–5550.
27. Stotzky G, Goos RD, Timonin MI (1962) Microbial changes occurring in soil as a result of storage. Plant Soil 16: 1–18. doi:10.1007/BF01378154.
28. Springer U KJ (n.d.) Prüfung der Leistungsfähigkeit von einigen wichtigeren Verfahren zur Bestimmung des Kohlenstoffs mittels Chromschwefelsäure sowie Vorschlag einer neuen Schnellmethode. J Plant Nutr Soil Sci 64: 1–26 (in German).
29. Ciavatta C, Govi M, Vittori Antisari L, Sequi P (1990) Characterization of humified compounds by extraction and fractionation on solid polyvinylpyrrolidone. J Chromatogr 509: 141–146.
30. Sequi P, De Nobili M, Leita L, Cercignani G (1986) A new index of humification. Agrochimica 30: 175–179.
31. Vance ED, Brookes PC, Jenkinson DS (1987) An extraction method for measuring soil microbial biomass C. Soil Biol Biochem 19: 703–707. doi:10.1016/0038-0717(87)90052-6.
32. Badalucco L, Grego S, Dell'Orco S, Nannipieri P (1992) Effect of liming on some chemical, biochemical, and microbiological properties of acid soils under spruce (Picea abies L.). Biol Fertil Soils 14: 76–83. doi:10.1007/BF00336254.
33. Riffaldi R, Saviozzi A, Levi-Minzi R (1996) Carbon mineralization kinetics as influenced by soil properties. Biol Fertil Soils 22: 293–298. doi:10.1007/BF00334572.
34. Dilly O, Munch J-C (1998) Ratios between estimates of microbial biomass content and microbial activity in soils. Biol Fertil Soils 27: 374–379. doi:10.1007/s003740050446.
35. Anderson T-H, Domsch KH (1989) Ratios of microbial biomass carbon to total organic carbon in arable soils. Soil Biol Biochem 21: 471–479. doi:10.1016/0038-0717(89)90117-X.
36. Pinzari F, Trinchera A, Benedetti A, Sequi P (1999) Use of biochemical indices in the mediterranean environment: Comparison among soils under different forest vegetation. J Microbiol Methods 36: 21–28.
37. Smalla K, Wieland G, Buchner A, Zock A, Parzy J, et al. (2001) Bulk and Rhizosphere Soil Bacterial Communities Studied by Denaturing Gradient Gel Electrophoresis: Plant-Dependent Enrichment and Seasonal Shifts Revealed. 67: 4742–4751. doi:10.1128/AEM.67.10.4742.

38. De Leij FAAM, Whipps JM, Lynch JM (1994) The use of colony development for the characterization of bacterial communities in soil and on roots. Microb Ecol 27: 81–97.
39. Grifoni A, Bazzicalupo M, Di Serio C, Fancelli S, Fani R (1995) Identification of Azospirillum strains by restriction fragment length polymorphism of the 16S rDNA and of the histidine operon. FEMS Microbiol Lett 127: 85–91. doi:10.1016/0378-1097(95)00042-4.
40. Trabelsi D, Mengoni A, Elarbi Aouani M, Mhamdi R, Bazzicalupo M (2009) Genetic diversity and salt tolerance of bacterial communities from two Tunisian soils. Ann Microbiol 59: 25–32. doi:10.1007/BF03175594.
41. Mengoni A, Tatti E, Decorosi F, Viti C, Bazzicalupo M, et al. (2005) Comparison of 16S rRNA and 16S rDNA T-RFLP approaches to study bacterial communities in soil microcosms treated with chromate as perturbing agent. Microb Ecol 50: 375–384. doi:10.1007/s00248-004-0222-4.
42. Hammer Ø, Harper DAT, Ryan PD (2001) Past: Paleontological Statistics Software Package for Education and Data Analysi. Palaeontol Electron 4: 1–9. doi:10.1016/j.bcp.2008.05.025.
43. Duineveld B, Kowalchuk G, Keijzer A, van Elsas J, van Veen J (2001) Analysis of bacterial communities in the rhizosphere of chrysanthemum via denaturing gradient gel electrophoresis of PCR-amplified 16S rRNA as well as DNA fragments coding for 16S rRNA. Appl Environ Microbiol 67: 172–178.
44. Ausubel FM, Brent R, Kingston RE, Moore DD, Seidman JG, et al. (1987) Unit 2.4 Preparation of genomic DNA from bacteria. Curr Protoc Mol Biol: 2.4.1–2.4.2.
45. Di Cello F, Pepi M, Baldi F, Fani R (1997) Molecular characterization of an n-alkane-degrading bacterial community and identification of a new species, Acinetobacter venetianus. Res Microbiol 148: 237–249.
46. El Fantroussi S, Verschuere L, Verstraete W, Top EM (1999) Effect of phenylurea herbicides on soil microbial communities estimated by analysis of 16S rRNA gene fingerprints and community-level physiological profiles. Appl Environ Microbiol 65: 982–988.
47. Cremonesi L, Firpo S, Ferrari M, Righetti PG, Gelfi C (1997) Double-gradient DGGE for optimized detection of DNA point mutations. Biotechniques 22: 326–330.
48. Shannon C, Weaver W (1963) The mathematical theory of communication. University of Illinois Press, Urbana. doi:10.1145/584091.584093.
49. Simpson E (1949) Measurement of diversity. Nature 163: 688–688. doi:10.1038/163688a0.
50. Whittaker RH (1972) Evolution and measurement of species diversity. Taxon 21: 213. doi:10.2307/1218190.
51. Hall TA (1999) BioEdit: a user-friendly biological sequence alignment editor and analysis program for Windows 95/98/NT. Nucleic Acids Symp 41: 95–98. doi:citeulike-article-id:691774.
52. Bray J, Curtis J (1957) An ordination of the upland forest communities of southern Wisconsin. Ecol Monogr 27: 325–349.
53. Haider K (1992) Problems related to the humification processes in soils of temperate climates. In: Stotzky G, Bollag J-M, editors. Soil Biochemistry. Marcel Dekker, New York. 55–94.
54. Oades JM (1995) An overview of processes affecting the cycling of organic carbon in soils. In: Zepp R, Sonntag C, editors. Role of non living organic matter in the earth's carbon cycle. John Wiley, New York. 293–303.
55. Stevenson FJ, Cole MA (1999) Cycles of soil: carbon, nitrogen, phosphorus, sulfur, micronutrients. 2nd ed. John Wiley, New York.
56. Franzluebbers AJ, Stuedemann JA, Schomberg HH, Wilkinson SR (2000) Soil organic C and N pools under long-term pasture management in the Southern Piedmont USA. Soil Biol Biochem 32: 469–478. doi:10.1016/S0038-0717(99)00176-5.
57. Anderson T, Domsch KH (1993) The metabolic quotient for CO2 (qCO2) as a specific activity parameter to assess the effects of environmental conditions, such as pH, on the microbial biomass of forest soils. Soil Biol Biochem 25: 393–395.
58. Benedetti A, Brookes P, Lynch J (2005) Conclusive remarks. Part I: Approaches to defining, monitoring, evaluating and managing soil quality. In: Bloem J, Hopkins D, Benedetti A, editors. Soil Quality. CABI; Wallingford. 63–70.
59. Conant RT, Paustian K, Elliott ET (2001) Grassland management and conversion into grassland: effects on soil carbon. Ecol Appl 11: 343–355. doi:10.1890/1051-0761(2001)011[0343:GMACIG]2.0.CO;2.
60. Edenborn SL, Sexstone AJ (2007) DGGE fingerprinting of culturable soil bacterial communities complements culture-independent analyses. Soil Biol Biochem 39: 1570–1579.
61. Garland JL (1999) Potential and limitations of BIOLOG for microbial community analysis. In: Bell C, Brylinsky M, Johnson-Green P, editors. Microbial Biosystems: new frontiers. Proceedings of the 8th international symposium on microbial ecology. Atlantic Canada Society for Microbial Ecology. Halifax, Canada. 1–7.
62. Ellis RJ, Morgan P, Weightman AJ, Fry JC (2003) Cultivation-dependent and -independent approaches for determining bacterial diversity in heavy-metal-contaminated soil. Appl Environ Microbiol 69: 3223–3230. doi:10.1128/AEM.69.6.3223.
63. Shade A, Hogan CS, Klimowicz AK, Linske M, McManus PS, et al. (2012) Culturing captures members of the soil rare biosphere. Environ Microbiol 14: 2247–2252. doi:10.1111/j.1462-2920.2012.02817.x.
64. Altschul SF, Madden TL, Schäffer AA, Zhang J, Zhang Z, et al. (1997) Gapped BLAST and PSI-BLAST: a new generation of protein database search programs. Nucleic Acids Res 25: 3389–3402.
65. Shange RS, Ankumah RO, Ibekwe AM, Zabawa R, Dowd SE (2012) Distinct soil bacterial communities revealed under a diversely managed agroecosystem. PLoS One 7: e40338. doi:10.1371/journal.pone.0040338.
66. Weon H-Y, Kim B-Y, Yoo S-H, Kwon S-W, Stackebrandt E, et al. (2008) Chryseobacterium soli sp. nov. and Chryseobacterium jejuense sp. nov., isolated from soil samples from Jeju, Korea. Int J Syst Evol Microbiol 58: 470–473. doi:10.1099/ijs.0.65295-0.
67. Soltani A, Khavazi K, Asadi Rahmani H, Omidvari M, Abaszadeh Dahaji P (2010) Plant growth promoting characteristics in some Flavobacterium spp. isolated from soils of Iran. J Agri Sci 2: 106–115. Available: http://www.ccsenet.org/journal/index.php/jas/article/view/8404/6221.
68. Fierer N, Bradford MA, Jackson RB (2007) Toward an ecological classification of soil bacteria. Ecology 88: 1354–1364. doi:10.1890/05-1839.
69. Girvan MS, Campbell CD, Killham K, Prosser JI, Glover LA (2005) Bacterial diversity promotes community stability and functional resilience after perturbation. Environ Microbiol 7: 301–313. doi:10.1111/j.1462-2920.2005.00695.x.
70. Nacke H, Thurmer A, Wollherr A, Will C, Hodac L, et al. (2011) Pyrosequencing-based assessment of bacterial community structure along different management types in German forest and grassland soils. PLoS One 6: e17000. doi:10.1371/journal.pone.0017000.
71. Acosta-Martínez V, Dowd S, Sun Y, Allen V (2008) Tag-encoded pyrosequencing analysis of bacterial diversity in a single soil type as affected by management and land use. Soil Biol Biochem 40: 2762–2770. doi:10.1016/j.soilbio.2008.07.022.

20

Variation in Yield Gap Induced by Nitrogen, Phosphorus and Potassium Fertilizer in North China Plain

Xiaoqin Dai*, Zhu Ouyang, Yunsheng Li*, Huimin Wang

Key Laboratory of Ecosystem Network Observation and Modeling, Institute of Geographic Sciences and Natural Resources Research, Chinese Academy of Sciences, Beijing, China

Abstract

A field experiment was conducted under a wheat-maize rotation system from 1990 to 2006 in North China Plain (NCP) to determine the effects of N, P and K on yield and yield gap. There were five treatments: NPK, PK, NK, NP and a control. Average wheat and maize yields were the highest in the NPK treatment, followed by those in the NP plots among all treatments. For wheat and maize yield, a significant increasing trend over time was found in the NPK-treated plots and a decreasing trend in the NK-treated plots. In the absence of N or P, wheat and maize yields were significantly lower than those in the NPK treatment. For both crops, the increasing rate of the yield gap was the highest in the P omission plots, i.e., 189.1 kg ha^{-1} yr^{-1} for wheat and 560.6 kg ha^{-1} yr^{-1} for maize. The cumulative omission of P fertilizer induced a deficit in the soil available N and extractable P concentrations for maize. The P fertilizer was more pivotal in long-term wheat and maize growth and soil fertility conservation in NCP, although the N fertilizer input was important for both crops growth. The crop response to K fertilizers was much lower than that to N or P fertilizers, but for maize, the cumulative omission of K fertilizer decreased the yield by 26% and increased the yield gap at a rate of 322.7 kg ha^{-1} yr^{-1}. The soil indigenous K supply was not sufficiently high to meet maize K requirement over a long period. The proper application of K fertilizers is necessary for maize production in the region. Thus, the appropriate application of N and P fertilizers for the growth of both crops, while regularly combining K fertilizers for maize growth, is absolutely necessary for sustainable crop production in the NCP.

Editor: Carl J. Bernacchi, University of Illinois, United States of America

Funding: This study was financially supported by the Strategic Science Plan of Institute of Geographic Sciences and Natural Resources Research (2012ZD004)and National Science and Technology Support Projects of Ministry of Science and Technology of China (2013BAD05B03). The funders had no role in study design, data collection and analysis, decision to publish, or preparation of the manuscript.

Competing Interests: The authors have declared that no competing interests exist.

* E-mail: daixq@igsnrr.ac.cn (XD); liys@igsnrr.ac.cn (YL)

Introduction

The North China Plain (NCP) is the largest and the most important agricultural production region in China. It covers approximately 18.3% of total national farm lands (18 million hectares) and produces 21.6% of the total grain yield of edible crops in the country [1]. The main crops grown are wheat and maize, periodically rotated. The NCP now supplies more than 50% of the nation's wheat and 33% of its maize [2]. Grain production and the maintenance of soil fertility in the NCP are very important for the food security and agricultural sustainability of China.

Nutrient availability is the most yield-limiting factor. To produce higher yields, the over-application of chemical fertilizers has been a common practice in wheat-maize rotation systems and has led to severe environmental problems [3]. Improved nutrient management practices are urgently needed to maximize crop yields and maintain soil fertility while minimizing environmental impacts. The ability to better identify crop response to the application of fertilizers, soil indigenous nutrient supply capability, and the maintenance of soil fertility over time are crucial to the development of improved nutrient management practices. Various long-term experiments have been conducted to test the effects of fertilization on yield or soil fertility throughout the world [4–14], but over long time scales, crop response to the application of N, P and K fertilizers and soil indigenous nutrient supply capability has seldom been clearly understood.

The yield and soil fertility gap between a full NPK fertilizer plot and a fertilizer omission plot was used as a good diagnostic tool to assess the extent of macronutrient limitations. Mussgnug et al (2006) analyzed the yield gap resulting from nutrient limitation on a degraded soil in the Red River Delta (RRD) in Northern Vietnam. They found that in the absence of K application, the yield gap for rice and maize respectively averaged 1.7 Mg ha^{-1} and 3.4 Mg ha^{-1}, while when N or P was omitted the yield gap was less. Potassium was the most yield-limiting macronutrient in RRD of Vietnam [15]. The factors that primarily limit increasing crop yields varied by crop and region [16]. The objectives of the study were to examine the effects of the continuous use of inorganic fertilizers on wheat and maize yield and soil fertility gap to elucidate crop response to fertilizer inputs and the long-term N, P and K supply capability in the NCP.

Materials and Methods

Experimental Site

A long-term field experiment was conducted from 1990 to 2006 at the Yucheng Comprehensive Experimental Station (36°57′ N,

Table 1. Physical and chemical properties of soil at different layers at Yucheng, Shandong Province of China in 1990.

Soil parameters	0–20 cm	20–40 cm
Soil texture	Silt loam	Sandy loam
Clay (%)	17.3	13.7
Silt (%)	61.4	11.4
Sand (%)	21.3	74.9
Bulk density (g cm^{-3})	1.31	1.48
Field capacity (%)	23.3	22.7
Organic C (g kg^{-1})	5.4	4.8
Total N (g kg^{-1})	0.41	0.36
Total P (g kg^{-1})	1.11	1.13
Total K (g kg^{-1})	10.7	14.0
Available N (mg kg^{-1})	46.1	41.4
Available P (mg kg^{-1})	8.3	5.4
Available K (mg kg^{-1})	149.0	125.4

116°36′ E, 28 m a.m.s.l.) of the Chinese Academy of Sciences (CAS) located in Shandong Province, which lies in the NCP. This area represents the moderate- to high-yielding region of the NCP, where the dominant cropping system is double-cropped wheat-maize. The annual mean precipitation is 575 mm, with approximately 70% of the total precipitation falling between June and September. The soil is classified as fluvo-aquic loam soil. Before the initiation of the experiment, soil samples were collected from depths of 0–20 cm and 20–40 cm by stepwise soil augers in the experimental field in 1990. All samples from the same soil layers were mixed thoroughly to create one homogenous sample, and a representative sample was drawn to determine the soil texture, bulk density, organic carbon, total N, P and K, and available N, P and K. The physical and chemical properties of the soil in 1990 are listed in Table 1.

Experimental Design and Treatments

The study consisted of five treatments: no fertilization (control); N, P, and K applied (NPK); N and P applied (NP; K omission); N and K applied (NK; P omission); P and K applied (PK; N omission), with four replications performed for each condition in a randomized complete block design (Table 2). The application rates

Table 2. Treatment and fertilizer nutrient rates (kg ha^{-1}) applied to winter wheat and maize at Yucheng, Shandong Province of China.

Treatments	Winter wheat			Maize		
	N	P_2O_5	K_2O	N	P_2O_5	K_2O
NPK	253	90	171	262	52	298
PK	0	90	171	0	52	298
NK	253	0	171	262	0	298
NP	253	90	0	262	52	0
Control	0	0	0	0	0	0

of N, P and K, if they were applied, were the same in all of the treatments: 253 kg N ha^{-1} as urea (46% N), 90 kg P_2O_5 ha^{-1} as single superphosphate (16% P_2O_5), and 171 kg K_2O ha^{-1} as potassium sulfate (50% K_2O) for winter wheat and 262 kg N ha^{-1} as urea (46% N), 52 kg P_2O_5 ha^{-1} as single superphosphate (16% P_2O_5), and 298 kg K_2O ha^{-1} as potassium sulfate (50% K_2O) for maize (Table 2). Phosphate was applied before sowing and incorporated during land preparation. Winter wheat received 40% of the total N as a basal dressing before sowing, whereas the remainder was top-dressed in two splits (40% at green-up and 20% at flowering). Approximately 50% of the K was applied before sowing and 50% top-dressed at green-up (the first leaf growing in the early of spring growed to 1–2 cm from soil surface and the phenomena was seen on a half of wheat seedling in the field). During the maize season, all of the P and K fertilizers and 35% of the N fertilizer were applied as a basal fertilizer, and 65% of the N was applied at the elongation stage. Each plot area measured 6 m×5 m, and all plots were isolated from one another by a concrete wall to a soil depth of 1 m.

Winter wheat (*Triticum aestivum*) was double-cropped with maize (*Zea mays*), and the crops were grown along the border of the four blocks. Winter wheat was sown in mid-October and harvested in early June of the next year. Maize was sown in mid-June and harvested in early October of the same year. The winter wheat was seeded at a rate of 225 kg ha^{-1}, and the maize was planted at a rate of 75000 seeds ha^{-1}. Surface irrigations were conventionally conducted according to the conditions of crop growth and soil moisture. Wheat and maize were harvested to the level of the soil surface; thus, the stubble left in the field was negligible. However, the roots were left in the soil. All straws were removed from the field. All plots were kept free of weeds by the application of an herbicide. Pesticides were applied in accordance with good practices for crop protection.

Sampling and Chemical Analyses

At harvestable maturity, the grain yields were determined over the whole plot area. Soils samples were collected at a depth of 20 cm at six randomly selected points in every plot after wheat and maize were harvested each year. The soil samples were mixed thoroughly from all cores to obtain a representative soil sample for each plot. The soil samples were air-dried and sieved through a 1-mm mesh to perform available N, Olsen-P and available K measurements and through a 0.149-mm mesh to estimate the organic carbon and total N content. Soil available N was determined using the alkali-hydrolytic diffusion method [17]. Olsen-P was measured by ascorbic acid-molybdate blue colorimetry [18]. Available K was extracted with an ammonium acetate solution (NH_4OAc, 1 mol/L) and then determined with a flame photometer [17]. Soil organic carbon was determined using the wet oxidation method of Walkley and Black, and total N was measured following Kjeldahl digestion and distillation [17].

Data Analyses

The results of yield and soil fertility in the first three years of the experiment were not used because in the autumn of 1993 soybean was sown. Therefore, the results reported are from 1994 to 2006. Crop yield losses caused by N, P, K or NPK omission were calculated from the differences between an NPK treatment and the PK, NK, NP or control plot during the same years. The yield differences were called the "yield gap" induced by N (GYG_N), P (GYG_P), K (GYG_K) or NPK (GYG_{NPK}) fertilizer omission to assess the extent of macronutrient limitations in the study. The differences in soil available N, Olsen-P and available K between the NPK treatment and the PK, NK, NP or control plots were

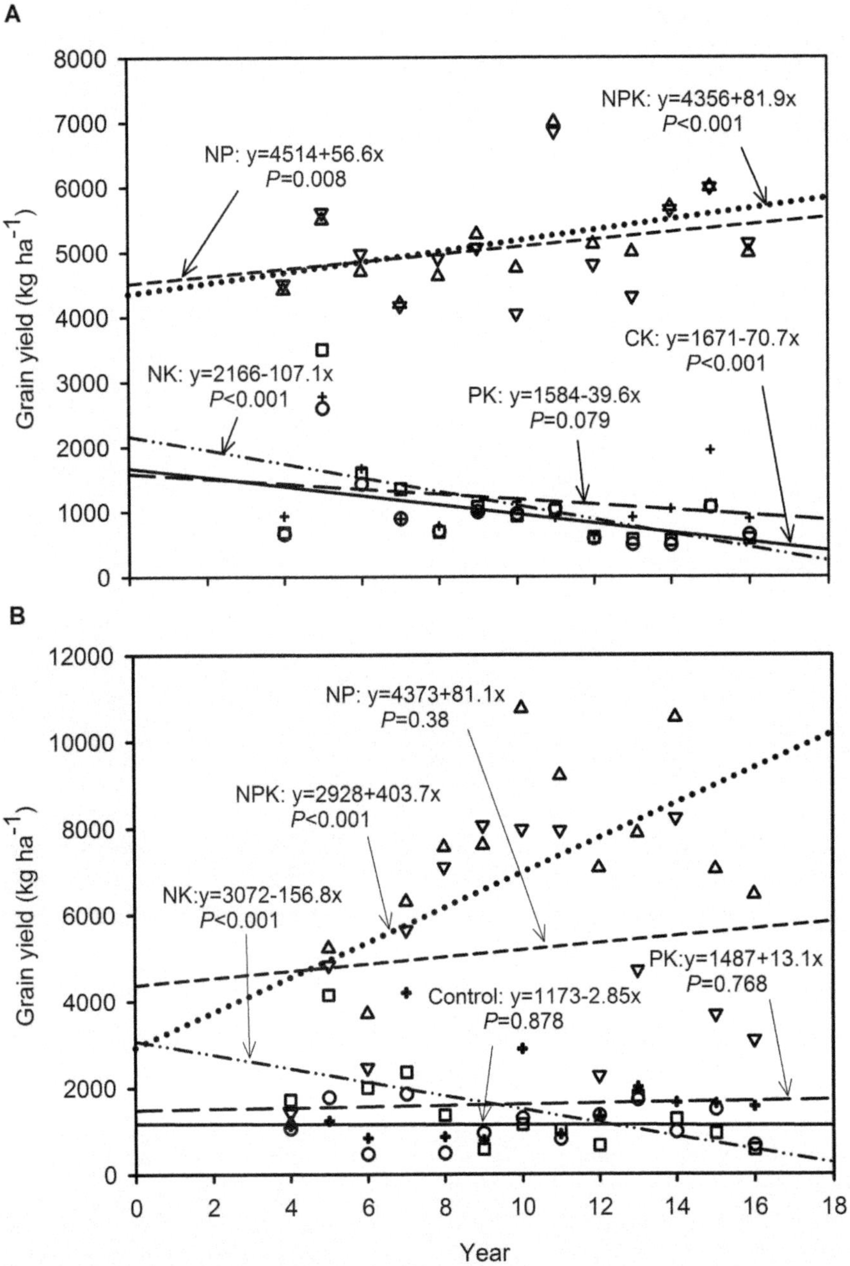

Figure 1. Grain yield of wheat (a) and maize (b) of different fertilizer treatments (Control, circle; NPK, triangle up; NP, triangle down; NK, square; PK, plus) under wheat-maize rotation system from 1994 to 2006 at Yucheng, Shandong Province of China.

calculated to evaluate the variation in soil fertility when nutrients were limited. The soil nutrient differences were called the "soil nutrient gap (SNG)" induced by N (SNG_N), P (SNG_P), K (SNG_P), or NPK (SNG_{NPK}) fertilizer omission.

All statistical analyses including the analysis of variance and regression were conducted using the SPSS package (version 16.0). Considering the data over time was lack of independence, the repeated measures ANOVA between differences in mean through years among treatments for various parameters (grain yield, soil organic C, total N, available N, Olsen-P, available K) were analyzed using a Fisher's protected least significant difference (LSD) test at $P = 0.05$. A mixed models with random block and block×treatment were done to assess trends (slopes) of grain yield, yield gap and various soil nutrients gap over the years. The *P*-values of the slopes were used to test whether the observed changes were significantly different from 0.

Results

Grain Yield and Yield Gap

The fertilizer treatments significantly affected the grain yields (Table 3). The wheat and maize yields were highest in the NPK treatment, followed by those in the NP treatment. In the PK or NK treatments, the yields of wheat and maize were significantly

Table 3. Multi-year average grain yield of wheat and maize, as affected by various treatments with fertilizer in a wheat-maize rotation system at Yucheng, Shandong Province of China (mean±standard error).

Treatment	Wheat (kg ha^{-1})	Maize (kg ha^{-1})
Control	963±155 b	1145±133 c
NPK	5174±201 a	6965±724 a
NP	5080±216 a	5184±666 b
NK	1094±218 b	1504±270 c
PK	1188±165 b	1618±269 c

Different letters within a column indicate a significant difference at $P<0.05$ between fertilizer treatments.

lower than those in the NP and NPK treatments. Compared to those in the NPK treatment, the yields of wheat and maize decreased, respectively, by 77% with the N omission, 79% and 78% with the P omission, 2% and 26% with the K omission, and 81% and 84% with no fertilizer. The grain yield losses were much greater in either N or P omission than in plots fertilized combined N and P. Over time wheat yield significantly increased at a rate of 81.9 kg ha^{-1} yr^{-1} in NPK plots ($P=0.008$), and maize yield also significantly increased at a rate of 403.7 kg ha^{-1} yr^{-1} in NPK plots ($P<0.001$, Fig. 1). The yields of wheat and maize in NK treatment decreased significantly at a rate of 107.1 kg ha^{-1} yr^{-1} ($P<0.001$) and 156.8 kg ha^{-1} yr^{-1} ($P<0.001$) over time during the study period, respectively (Fig. 1). The long-term omission of K fertilizer also significantly reduced the maize yield compared to the maize yield of the NPK-treated plot (Table 3). But both crops yields had no significant upward or downward trends when K fertilizer was omitted (Fig. 1).

The variation in the yield gap of wheat and maize between the NPK and PK, NK, NP and control plots is illustrated in Figure 2. For wheat, the yield gap significantly increased at a rate of 152.6 kg ha^{-1} yr^{-1} for the control plots ($P<0.001$), a rate of 189.1 kg ha^{-1} yr^{-1} for the NK plots ($P<0.001$) and a rate of 121.5 kg ha^{-1} yr^{-1} for the PK plots ($P<0.001$). In the NP plots, the yield gap also gradually increased with the cumulative effect of K omission, but the relationship was not significant ($P=0.246$). However, for maize, when the N, P K or NPK fertilizer was omitted, the yield gap significantly increased with the continuous omission of nutrients. The significant increase in the yield gap was 390.6 kg ha^{-1} yr^{-1} for the PK treatments ($P<0.001$), 560.6 kg ha^{-1} yr^{-1} for the NK treatments ($P<0.001$), 322.7 kg ha^{-1} yr^{-1} for the NP treatments ($P<0.001$), and 406.6 kg ha^{-1} yr^{-1} for the control treatments ($P<0.001$). For both crops, the increasing rate of the yield gap was highest in the P omission plots, which indicated that the cumulative effect of P omission was the most significant for wheat and maize in the region. The slopes of the yield gap were the second highest in the N omission plots for both crops, which indicated that N fertilizer inputs were important for wheat and maize growth, especially the latter. It is worth noting that the slopes of the maize yield gap increased significantly with continuous K omission. This result indicates that the soil indigenous K supply was not sufficiently high to meet the maize K requirement over a long period.

Soil Nutrient Status

Long-term continuous treatment with different fertilizers significantly affected the soil organic carbon, available N, Olsen-P and available K concentrations (Table 4). The multi-year average soil organic carbon content was significantly lower in the control, NK and PK plots than in the NPK and NP plots for wheat and maize. The decrease was 6.7% to 15.6% for wheat and 7.4% to 15.8% for maize. The different treatments except control had no significant influences on soil total N. The available N, Olsen-P and available K concentrations were higher in the plots treated with the corresponding N, P or K fertilizer inputs (Table 4). For wheat, the available N concentration increased significantly in the acquired N fertilizer inputs plots compared to that in the PK plots. The Olsen-P concentration in plots with P fertilizer inputs increased significantly by 7.2 to 12.3 times the concentration in the control plots. Meanwhile, the Olsen-P concentration also increased by 9% and 62% in the NP and PK plots, respectively, compared with that in the NPK plots, indicating that the P uptake of the crops was lower in the NP and PK plots. Similar to the Olsen-P concentration, the available K concentration increased significantly in the plots with K fertilizer inputs, and the increase was as high as 1.2 to 2.3 times the K concentration in the control plots. In the NK and PK plots, the available K concentration increased by 48% and 55%, respectively, compared to that in the NPK plots.

The differences in soil nutrients between the NPK and PK, NK, NP or NPK omission plots are shown in Figure 3&4. During the wheat growth season, the continuous omission of nutrients had no significant effects on the difference in the soil available N concentration (Fig. 3a). The continuous omission of P fertilizer significantly enhanced the differences in the soil Olsen-P concentration in the NPK plots ($P<0.001$), indicating that the long-term absence of P fertilizer resulted in a great deficit in the soil available P concentration. Similarly to the P omission plots, the control plots also possessed a significantly enhanced soil Olsen-P gap ($P<0.001$). However, a surplus concentration of Olsen-P was observed when the continuous N was omitted ($P=0.006$, Fig. 3b). Also a surplus concentration of available K was observed when the continuous P and N inputs were omitted ($P<0.001$) due to a large amount of K fertilizer inputs and lower crop uptake. When the K fertilizer was omitted, the soil available K concentration gap significantly increased ($P=0.001$ for NP and $P=0.002$ for Control), indicating a deficit in soil available K in the NP and Control plots.

During the maize growth season, the soil available N concentration gradually became deficient when the nutrients were omitted, but only when P fertilizer was omitted, the effects were significant (Fig. 4a). Like wheat, P fertilizer omission induced a severe deficit in soil Olsen-P ($P<0.001$). Although the PK plots feature a larger amount of P fertilizer inputs and lower crop uptake, no significant surplus in the soil Olsen-P concentration was observed (Fig. 4b). In the plots fed K fertilizer inputs, a surplus in the soil available K concentration was observed, but the effects were not significant (Fig. 4c). It is worthy to note that the NP plots induced a deficit in soil available K (Table 4). Although the effects were not significant, maize growth was significantly affected due to the shortage of soil available K.

Discussion

The response of wheat and maize growth of combined N and P fertilization (i.e. NP) was much greater than that of K fertilizer application combined with N or P (i.e. NK and PK treatments) (Table 3). The continuous 16-yr omission of N or P fertilizer significantly reduced the wheat and maize yield (Table 3). The grain yield in the N omission plots exhibited no significant decreasing trends over time (Fig. 1a). On the one hand, the

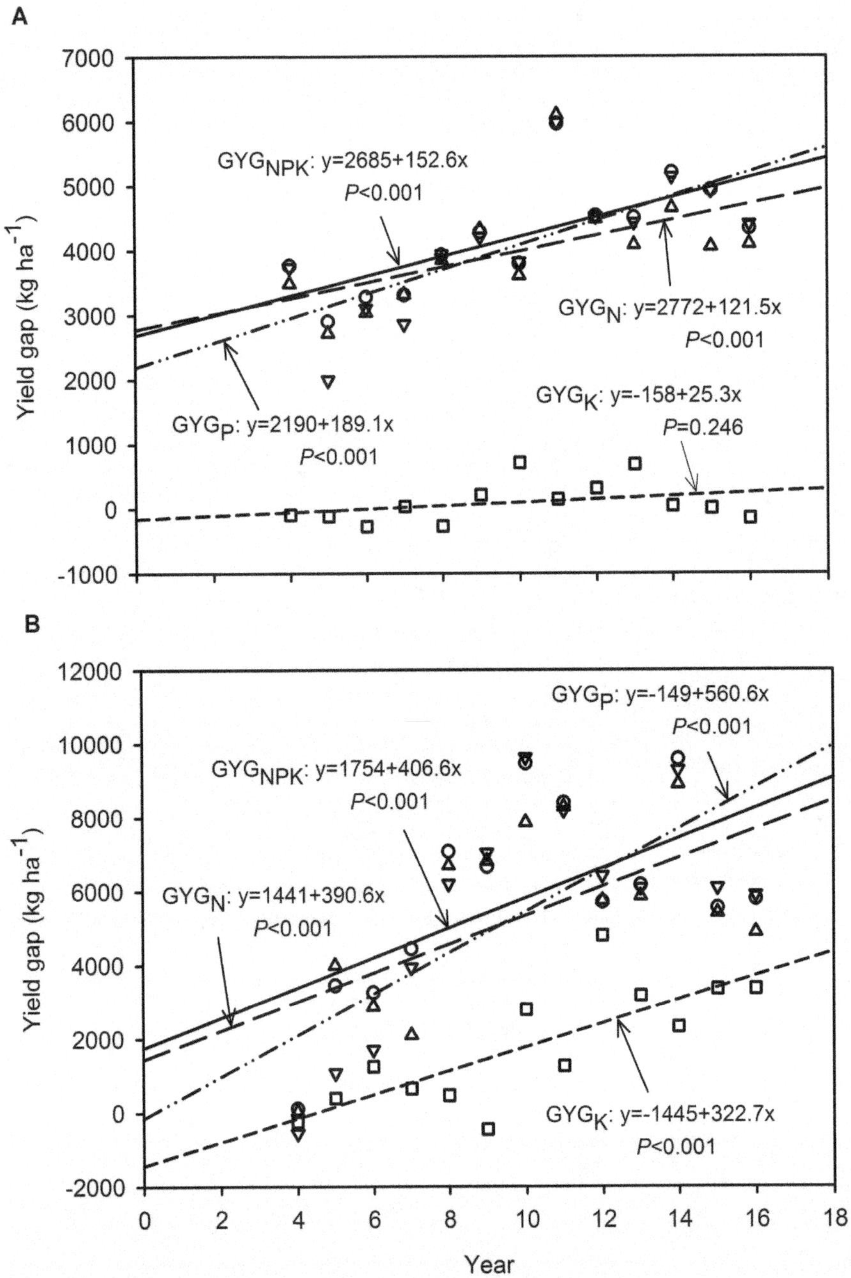

Figure 2. Yield gap variation of wheat (a) and maize (b) by N (GYG_N, the yield difference between NPK and PK plots in each year, triangle up), P (GYG_P, the difference between NPK and NK plots, triangle down), K (GYG_K, the difference between NPK and NP plots, square) or NPK (GYG_{NPK}, the difference between NPK and Control plots, circle) fertilizer omission from 1994 to 2006 at Yucheng, Shandong Province of China.

continuous N omission significantly reduced the soil available N concentration, but no significant effects were observed on the total N concentration (Table 4). The differences in soil available N between the NPK and PK plots also exhibited no significant increase over time (Fig. 3a&4a). It might be possible that atmospheric nitrogen deposition in the NCP soil, N mineralization and lower nitrogen depletion may have attributed to the trend in available N over the years in the nitrogen omitted plots. Liu et al. (2006) and Zhang et al. (2008) reported that the bulk deposition of inorganic N in the NCP is approximately 30 kg/ha per year, which has a significant effect on agricultural systems [19,20]. But yield gap significantly increased over time when the N was omitted (Fig. 2), indicating that the soil indigenous N supply combined with the deposition of N could not meet the N requirement for wheat and maize growth in the NCP.

Both of the wheat and maize yields had significant downward trends in P omission treatment (Fig. 1) and the yield gap of both crops significantly increased with the continuous omission of P fertilizer (Fig. 2), indicating that the cumulative absence of P fertilizers significantly inhibited wheat and maize growth. Tang

Table 4. Multi-year average soil nutrient status at wheat and maize harvest, as affected by various treatments with fertilizer in a wheat-maize rotation system at Yucheng, Shandong Province of China (mean±standard error).

Crops	Treatments	Organic C (g kg^{-1})	Total N (g kg^{-1})	Available N (mg kg^{-1})	Olsen-P (mg kg^{-1})	Available K (mg kg^{-1})
Wheat	Control	5.5±0.19 c	0.56±0.04b	52.5±5.15b	3.1±0.28c	104.5±5.50c
	NPK	6.5±0.21a	0.66±0.05a	63.3±4.35a	25.2±1.95b	226.1±16.0b
	NP	6.3±0.15ab	0.67±0.05a	62.7±4.07a	27.5±2.04b	91.6±8.72c
	NK	5.9±0.17abc	0.64±0.05a	62.5±4.17a	3.2±0.31c	333.4±40.3a
	PK	5.8±0.19bc	0.56±0.04ab	49.1±4.20b	40.8±3.98a	349.3±35.3a
Maize	Control	5.7±0.13c	0.66±0.02ab	56.9±6.11b	2.7±0.32c	106.5±5.22c
	NPK	6.7±0.15a	0.71±0.05a	65.8±6.45a	22.1±2.47b	283.7±24.6b
	NP	6.5±0.15ab	0.69±0.04ab	66.6±6.05a	21.7±1.78b	90.9±4.46c
	NK	6.0±0.13bc	0.66±0.04ab	68.4±6.76a	3.6±0.54c	423.8±43.7a
	PK	5.8±0.14bc	0.62±0.03b	52.1±5.90b	33.5±2.32a	438.8±39.8a

For each crop, different letters within a column indicate a significant difference at $P<0.05$ between fertilizer treatments.

et al. (2008) also reported that wheat and maize yields without P fertilization significantly decreased over time in several other sites in China [21]. The results are different from those reported under European conditions, most likely due to differences in the P supplying abilities of the different soils [22]. It is noted that the continuous omission of P fertilizer significantly reduced the soil Olsen-P concentration for both wheat and maize (Table 4). Meanwhile, the differences in the Olsen-P concentration between the NPK and NK plots significantly increased with continuous P omission (Fig. 3b&4b). The results indicate that the long-term absence of P fertilizer resulted in a great deficit in soil available P. It is interesting that the continuous omission of P fertilizer increased the available N concentration gap between the NPK and NK plots, and the effects were significant for maize (Fig. 4a), which indicates that a large amount of available N was depleted or lost, although the plots were fed N fertilizer inputs. The cumulative absence of P fertilizer intensified the shortage of N for crop growth. Because the treatments fertilized without P had a lower the number of cultivable microorganisms, microbial biomass and community functional diversity than in the treatments with P fertilization [23], which possibly induced the lower N and P mineralization in treatments fertilized without P.

The yield response to K fertilizer was much lower than that to N or P fertilizer (Table 3), most likely due to the high inherent soil K levels, which were in excess of the crop K demands. Most soils of the alluvial floodplain in Asia are high in K, and K is a rare limiting factor [24]. Shen et al. (2004) conducted a 14-yr field trial in Hebei Province of northern China and indicated that the application of N and P enhanced rice yields, while K had no yield-increasing effect due to the large resource of available soil K [25]. In our study, wheat and maize yield in the plots of combined NP and K fertilizer significantly improved, and the yield gap of maize significantly increased with continuous K omission; however, that of wheat yield did not (Fig. 1&2). Compared to that in the NPK plots, the soil available K concentration significantly decreased in the NP plots (Table 4). The differences in soil available K between NPK and NP also improved with the continuous omission of K, and the effects were significant during the wheat season (Fig. 3c). Although soil available K was gradually depleted, the soil indigenous K supply was sufficiently high to meet the requirement for normal wheat growth over the 16-yr period of the study. If the K fertilizer was continuously omitted for a longer period, the soil available K deficit would possibly inhibit wheat growth, which must be further verified. In addition, the soil available K gap between the NPK and NP plots during maize growth season showed an increasing trend with the omission of K, but the differences were not significant (Fig. 4c). This behavior is because the long-term omission of K fertilizer significantly reduced the maize yield, which resulted in lower K uptake. The results show that long-term maize production is much more sensitive to the absence of K than wheat production in the NCP. The results are supported by those of Tan et al. (2007), who concluded that the effect of K fertilizer on maize was higher than that on wheat under the wheat-maize rotation system of Hebei [26].

In our study, the combination of N and P fertilizers mostly sustained soil organic carbon, total N and available N, P and K levels over time (Table 4). This result is similar to that of Shen et al. (2004), who reported that the soil organic carbon and total N concentrations remained stable over time [25]. Manna et al. (2005) also showed that the recommended NPK plots are adequate for maintaining a constant SOC content under the sub-humid and semi-arid tropical conditions of India over a long period [27]. When N fertilizer was omitted, soil available P and K significantly improved in the plots with P and K fertilizer (PK). Meanwhile, when P fertilizer was omitted, soil available K was significantly accumulated in the plots with N and K fertilizer (NK, Table 4). Because the deficiency of other nutrients inhibited crop production, both crops exhibited lower P or K uptake. The application of inorganic fertilizer allowed the crops to exceed production needs and thus resulted in a substantial build-up of available P or available K, which inevitably induced resource waste and ecological pollution.

Conclusions

The P fertilizer was more pivotal in the long-term growth of wheat and maize and the conservation of soil fertility in the NCP, although the N fertilizer input was important for the growth of both crops as well. Although the crop yield response to K fertilizer was much lower than to N or P fertilizer, the proper application of K fertilizer is also necessary, especially for maize production in the

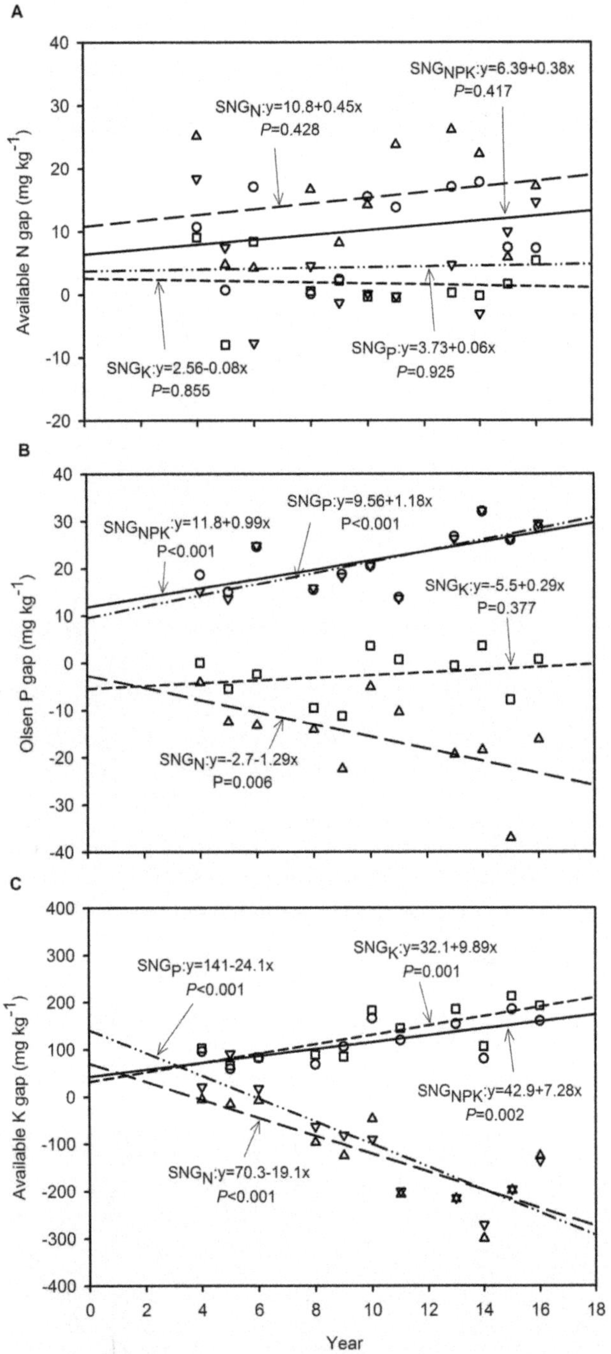

Figure 3. Soil available N (a), Olsen-P (b) and available K (c) gap variation by N (SNG_N, soil available nutrient difference between NPK and PK plots, triangle up), P (SNG_P, the difference between NPK and NK plots, triangle down), K (SNG_K, the difference between NPK and NP plots, square) or NPK (SNG_{NPK}, the difference between NPK and Control plots, circle) fertilizer omission from 1994 to 2006 during wheat season at Yucheng, Shandong Province of China.

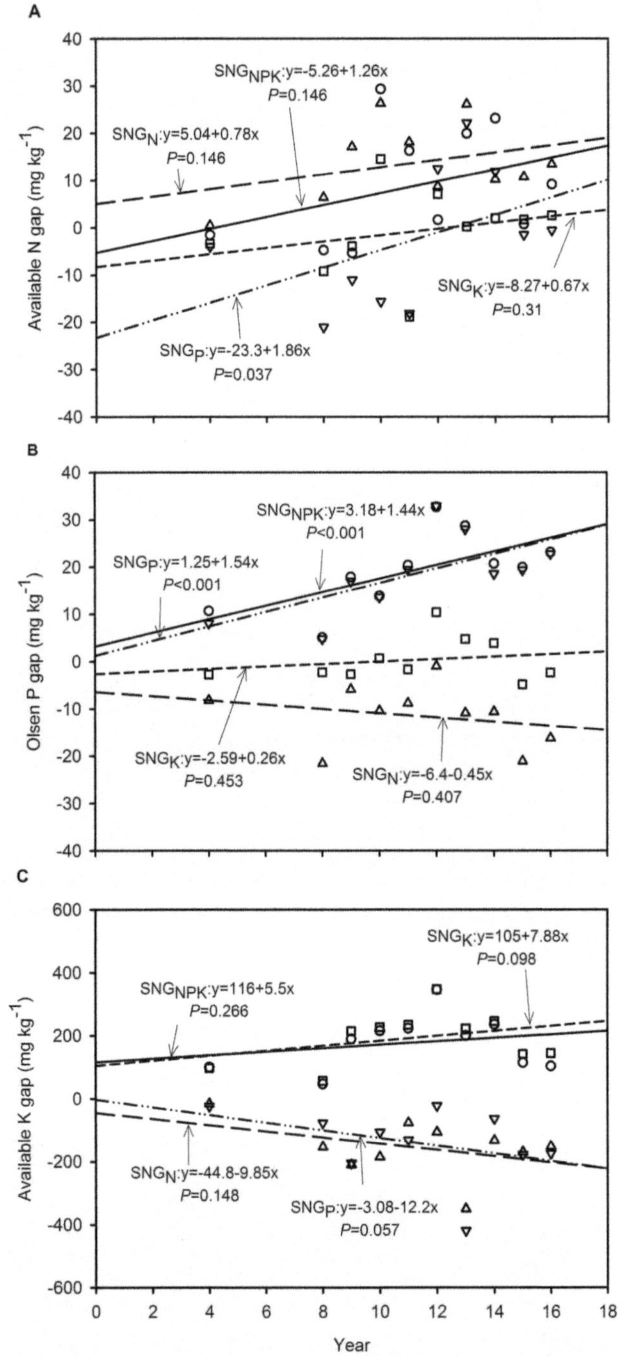

Figure 4. Soil available N (a), Olsen-P (b) and available K (c) gap variation by N (SNG_N, soil available nutrient difference between NPK and PK plots, triangle up), P (SNG_P, the difference between NPK and NK plots, triangle down), K (SNG_K, the difference between NPK and NP plots, square) or NPK (SNG_{NPK}, the difference between NPK and Control plots, circle) fertilizer omission from 1994 to 2006 during maize season at Yucheng, Shandong Province of China.

region. Thus, the appropriate application of N and P fertilizers for both crops, in combination with regular K fertilizers for maize, is absolutely necessary in terms of sustainable crop production in the NCP. However, a longer-range study is required to verify whether the soil indigenous K supply could continue to meet the requirement for wheat growth over many years.

Acknowledgments

The authors are grateful to Zhenrong Tian in the Yucheng Comprehensive Experimental Station of the Chinese Academy of Sciences (CAS) for many sampling work and Lynn M. Johnson in Cornell University for statistical analysis. Authors also thank the academic editor and anonymous reviewers for their constructive comments, which helped in improving the manuscript.

Author Contributions

Conceived and designed the experiments: ZO YL. Performed the experiments: XD YL. Analyzed the data: XD HW. Wrote the paper: XD.

References

1. Yu Q, Saseendran SA, Ma L, Flerchinger GN, Green TR, et al. (2006) Modeling a wheat-maize double cropping system in China using two plant growth modules in RZWQM. Agr Syst 89: 457–477.
2. China Statistics Bureau (2001) China Statistics Bureau, Annual Agricultural Statistics of Henan Province. Beijing, China Statistics Press. 112–115.
3. Vitousek PM, Naylor R, Crews T, David MB, Drinkwater LE, et al. (2009) Nutrient imbalances in agricultural development. Science 324: 1519–1520.
4. Berzsenyi Z, Győrffy B, Lap D (2000) Effect of crop rotation and fertilization on maize and wheat yields and yield stability in a long-term experiment. Eur J Agron 13: 225–244.
5. Bi LD, Zhang B, Liu GR, Li ZZ, Liu YR, et al. (2009) Long-term effects of organic amendments on the rice yields for double rice cropping systems in subtropical China. Agr Ecosyst Environ 129: 534–541.
6. Cai ZC, Qin SW (2006) Dynamics of crop yields and soil organic carbon in a long-term fertilization experiment in the Huang-Huai-Hai Plain of China. Geoderma 136: 708–715.
7. Fan T, Stewart BA, Wang Y, Luo J, Zhou G.(2005) Long-term fertilization effects on grain yield, water-use efficiency and soil fertility in the dryland of Loess Plateau in China. Agr Ecosyst Environ 106: 313–329.
8. Glendining MJ, Powlson DS, Poulton PR, Bradbury NJ, Palazzo D, et al. (1996) The effects of long-term applications of inorganic nitrogen fertilizer on soil nitrogen in the Broadbalk wheat experiment. J Agr Sci 127: 347–363.
9. Gong W, Yan XY, Wang JY, Hu TX, Gong YB (2009) Long-term manuring and fertilization effects on soil organic carbon pools under a wheat-maize cropping system in North China Plain. Plant Soil 314: 67–76.
10. Jiang D, Hengsdijk H, Dai T, de Boer W, Jing Q, et al. (2006) Long-term effects of manure and inorganic fertilizers on yield and soil fertility for a winter wheat-maize system in Jiangsu, China. Pedosphere16: 25–32.
11. Lu R, Shi Z (1998) Effect of long-term fertilization on soil properties. In: Lu R, Xie J, Cai G, Zhu Q (2010) editors. Soil-plant nutrients principles and fertilizer. Beijing: Chemical Industry Press. 102–110.
12. Wang YC, Wang E, Wang DL, Huang SM, Ma YB, et al. (2010) Crop productivity and nutrient use efficiency as affected by long-term fertilization in North China Plain. Nutr Cycl Agroecosys 86: 105–119.
13. Yan XY, Gong W (2010) The role of chemical and organic fertilizers on yield, yield variability and carbon sequestration – results of a 19-year experiment. Plant Soil 331: 471–480.
14. Zhang H, Xu M, Zhang F (2009) Long-term effects of manure application on grain yield under different cropping systems and ecological conditions in China. J Agr Sci 147: 31–42.
15. Mussgnug F, Becker M, Son TT, Buresh RJ, Vlek PLG (2006) Yield gaps and nutrient balances in intensive, rice-based cropping systems on degraded soils in the Red River Delta of Vietnam. Field Crop Res 98: 127–140.
16. Mueller ND, Gerber JS, Johnston M, Ray DK, Ramankutty N, et al. (2012) Closing yield gaps through nutrient and water management. Nature 490: 254–257.
17. Page AL, Miller RH, Keeney DR (1982) Methods of soil analysis. Part 2. Madison, WI: American Society of Agronomy. 539–871.
18. Olsen SR, Cole CV, Watanabe FS, Dean LA (1954) Estimation of available phosphorus in soils by extraction with sodium bicarbonate. USDA Circular 939.
19. Liu X, Ju X, Zhang Y, He C, Kopsch J, et al. (2006) Nitrogen deposition in agroecosystems in the Beijing area. Agr Ecosyst Environ 113: 370–377.
20. Zhang Y, Liu XJ, Fangmeier A, Goulding KTW, Zhang FS (2008) Nitrogen inputs and isotopes in precipitation in the North China Plain. Atmos Environ 42: 1436–1448.
21. Tang X, Li J, Ma Y, Hao X, Li X (2008) Phosphorus efficiency in long-term (15-years) wheat-maize cropping systems with various soil and climate conditions. Field Crop Res 108: 231–237.
22. Blake L, Mercik S, Koerschens M, Moskal S, Poulton PR, et al. (2000) Phosphorus content in soil, uptake by plants and balance in three European long-term field experiments. Nutr Cycl Agroecosys 56: 263–275.
23. Zhong W, Cai Z (2007) Long-term effects of inorganic fertilizers on microbial biomass and community functional diversity in a paddy soil derived from quaternary red clay. Appl Soil Ecol 36: 84–91.
24. Bajwa MI (1994) Soil K status, K fertilizer usage and recommendation in Pakistan. Potash Review No. 3/1994. Subject 1, 20th Suite. Basel: International Potash Institute. 67.
25. Shen J, Li R, Zhang F, Fan J, Tang C, et al. (2004) Crop yields, soil fertility and phosphorus fractions in response to long-term fertilization under the rice monoculture system on a calcareous soil. Field Crop Res 86: 225–238.
26. Tan D, Jin J, Huang S, Li S, He P (2007) Effect of long-term application of K fertilizer and wheat straw to soil on crop yield and soil K under different planting systems. Agr Sci China 6: 200–207.
27. Manna MC, Swarup A, Wanjari RH, Ravankar HN, Mishra B, et al. (2005) Long-term effect of fertilizer and manure application on soil organic carbon storage, soil quality and yield sustainability under sub-humid and semi-arid tropical India. Field Crop Res 93: 264–280.

21

Density and Stability of Soil Organic Carbon beneath Impervious Surfaces in Urban Areas

Zongqiang Wei[1,2], Shaohua Wu[1]*, Xiao Yan[2], Shenglu Zhou[1]

1 School of Geographic and Oceanographic Science, Nanjing University, Nanjing, China, **2** School of Environmental and Land Resource Management, Jiangxi Agricultural University, Nanchang, China

Abstract

Installation of impervious surfaces in urban areas has attracted increasing attention due to its potential hazard to urban ecosystems. Urban soils are suggested to have robust carbon (C) sequestration capacity; however, the C stocks and dynamics in the soils covered by impervious surfaces that dominate urban areas are still not well characterized. We compared soil organic C (SOC) densities and their stabilities under impervious surface, determined by a 28-d incubation experiment, with those in open areas in Yixing City, China. The SOC density (0–20 cm) under impervious surfaces was, on average, 68% lower than that in open areas. Furthermore, there was a significantly ($P<0.05$) positive correlation between the densities of SOC and total nitrogen (N) in the open soils, whereas the correlation was not apparent for the impervious-covered soils, suggesting that the artificial soil sealing in urban areas decoupled the cycle of C and N. Cumulative CO_2-C evolved during the 28-d incubation was lower from the impervious-covered soils than from the open soils, and agreed well with a first-order decay model ($C_t = C_1 + C_0(1-e^{-kt})$). The model results indicated that the SOC underlying capped surfaces had weaker decomposability and lower turnover rate. Our results confirm the unique character of urban SOC, especially that beneath impervious surface, and suggest that scientific and management views on regional SOC assessment may need to consider the role of urban carbon stocks.

Editor: Dafeng Hui, Tennessee State University, United States of America

Funding: This work was funded by the National Natural Science Foundation of China (41001047). The funders had no role in study design, data collection and analysis, decision to publish, or preparation of the manuscript.

Competing Interests: The authors have declared that no competing interests exist.

* Email: shaohuawu@126.com

Introduction

At present more than half of the world's population resides in cities and towns, and the percentage of urban population is projected to increase to 70% by 2050 [1]. As a result, urban areas are increasing in extent at a greater pace than any other land use type [2]. The rapid expanding of urban areas have caused large areas of agricultural, pasture, or forest soils to be changed to urban soils [3,4]. To date, urban land is estimated to cover 9% of the continent [5], in which the impervious surface (e.g., buildings, roads, and other pavements) is estimated to cover nearly 580 000 km^2 globally, an area larger than France [6]. Presently, the resulting impacts of the impervious surfaces on urban soil function, including soil organic carbon (SOC) storage and dynamics, remain largely unknown.

Yet a growing body of literature suggest that urban soils still have robust C storage capacity [7–10], especially in the areas covered by green vegetation (e.g., meadow, forest) which could provide fairly C sequestration [11,12]. Generally, the SOC storage capacity of urban open soils (without impervious surfaces) is comparable to that of adjacent agricultural soils and varies highly amongst different cities/regions, which is might be controlled by several factors, such as urbanization histories, land use types, soil parent materials, topography, and climate (Table 1). However, the storage and turnover of SOC beneath impervious surfaces are still poorly characterized due to their inaccessibility. The impervious surface in urban area is still rapidly expanding due to urbanization, thus it is critical to investigate the SOC stocks and dynamics beneath impervious, to provide accurate inventories in estimates of the entire SOC storage in urban areas, and to promote our understanding of the net impact of urbanization on terrestrial C pools. The relatively limited previous studies suggest that, in urban areas, the SOC density, an important parameter to calculate SOC storage, is significantly lower under impervious surfaces than in open sites [8,15]. These studies mainly focus on the amount or stocks of SOC in urban areas, whereas few attempts are made to study the SOC dynamics in the impervious-covered soils.

Here we collected some urban soil samples from typical impervious-covered and open areas in Yixing City, China, an area experienced rapid urbanization these decades. The main objectives of this study were to investigate the SOC density underlying impervious surfaces in urban areas, and to further study the stability or decomposability of SOC in the sealing environment.

Materials and Methods

Ethics Statement

This study was conducted in Yixing, China. The impervious-covered and open soils were collected from part of Yixing urban area (31°20′–31°25′N, 119°45′–119°50′E). The selected sampling

Table 1. Soil organic carbon densities (kg m^{-2} at a depth where indicated) for urban soils located in different cities.

City	Latitude/longitude	Urban SOC density	Suburban SOC density	Adjacent agricultural SOC density	Literature cited
		kg m^{-2}			
Liverpool, UK	53.4° N/2.98° W	4[a]	–	–	[13]
Boston, U.S.A.	42.35° N/71.06° W	4.02–4.24[b]	3.33–3.99	2.83–4.29	[14]
New York, U.S.A.	42.34° N/75.19° W	5.67[a]	–	–	[15]
New York, U.S.A.	42.34° N/75.19° W	5.1[c]	3.5	3.4	[7]
Baltimore, U.S.A.	39.28° N/76.62° W	4–7[a]	–	–	[7]
Chuncheon, South Korea	37.87° N/127.73° E	2.48[d]	–	3.16	[16]
Phoenix, U.S.A.	33.45° N/112.06° W	0.5–1.1[b]	–	0.75	[17]
Nanjing, China	32.05° N/118.77° E	4.52[e]	–	–	[18]
Chongqing, China	29.56° N/106.57° E	2–3.6[e]	–	–	[19]

a, 0–15 cm; b, 0–10 cm; c, 0–100 cm; d, 0–60 cm; e, 0–20 cm; –, not determined.
Cities were ranked by latitude.

areas did not involve endangered or protected species, and no specific permissions were required for the soil sampling.

Study area

The soils in this study were sampled in January 2008 from Yixing City (31°07′–31°37′N, 119°31′–120°03′E), which has an urban area of 13.4 km^2 in 2000 [3]. Yixing which locates on the plain of the lower reaches of the Yangtze River has had rapid urbanization in recent decades mainly at the expense of agricultural land. Average annual rainfall in Yixing is 1177 mm, and average annual temperature is 15.7°C. The soils in the studied area were formed on the alluvium of the Yangtze River, and the dominant soil type in the agricultural lands around Yixing City is Hydragric Anthrosols [20].

Soil sampling and analysis

Seven sites were selected for impervious-covered soils, and six sites with similar soil parent materials were selected for urban open soils (Figure 1). Soil in our study area had SOC content of 1.48 (±0.19 SE, $n = 9$) g kg^{-1}, TN content of 1.38 (±0.11 SE, $n = 9$) g kg^{-1}, and bulk density of 1.3 (±0.13 SE, $n = 61$) g cm^{-3} in 1982, which these parameters had relatively small variation as their CVs ranged from 8% to 13% (Table 2) [21]. The impervious sites consisted of road pavements, and paved residential squares and alleys, whereas the open area reference sites mainly comprised residential and commercial lawns and gardens, and public greenspaces. Outline description (0–20 cm) for the impervious-covered and open soils selected in our study is shown in Table 3. The soils sampled in our study area primarily had a slit loam texture but varied in the contents of soil admixtures. Compared with the soils in the open areas, the impervious-covered soils had little root penetration and greater amounts of artifacts, including asphalt, cement, brick, tiles, and gypsum. Generally, soil physico-chemical properties were more variable in impervious areas than in the grass areas, and more severely affected by human activities. The mean soil pH and particle fraction larger than 2 mm beneath impervious surfaces (0–20 cm) were 7.72 (range, 7.08–8.33) and 4.28% (range, 6.4%–7.74%), respectively (Table 3).

Each sample, consisting of 3 separate soil cores (5 cm in diameter), was taken from 0–20 cm depth, as SOC storage was supposed to be mainly allocated in the upper soil layers. To make the impervious-covered and open soils more comparable, in the impervious areas, only soils having roots penetrations and little mixture of artifacts were selected. We assumed that the selected soil horizon having these properties is of significance in sequestrating SOC. Specific soil horizon from which the soil was sampled was not determined, but the sampled soils had similar soil texture and chemical properties (Table 3). At the impervious sites, we punched holes on the hard ground near the central position of the selected sealing area (2 m×2 m) and removed the padding to make the covered soil accessible for sampling. After collection, soil samples were transported to the laboratory and air dried at room temperature. Then, the air-dried soils were grounded and sieved through a 2-mm nylon mesh; stones, artifacts, and coarse roots greater than 2 mm in size were weighed.

Soil pH was measured in water (1 soil: 2.5 water, w/v) using a glass electrode, and soil particle size analysis was determined using the hydrometer method [22]. Soil organic C concentrations in the samples were determined using the potassium dichromate sulfuric acid oxidation method [23]. Total nitrogen (N) concentrations were measured by Kjeldahl digestion [24]. Soil bulk density was measured by automated three-dimensional laser scanning [25] (NextEngine Desktop 3D Scanner, NextEngine, Inc., US). The densities of SOC and total N in a horizon of unit area (1 m^2) in each site were calculated as:

$$c = c_c \times BD \times (1 - \delta_{2mm}) \times H \quad (1)$$

where c (kg m^{-2}) represents SOC density or total N density, c_c (g kg^{-1}) represents SOC concentration or total N concentration, BD (g cm^{-3}) represents soil bulk density, δ_{2mm} (g g^{-1}) represents the fraction of material lager than 2 mm diameter, and H (m) represents the soil sampling depth. In our study, the densities of SOC and total N in each site were calculated based on 1 m^2 square to a 0.2 m depth.

To investigate the stability of SOC in urban areas, a thermostatic soil incubation experiment was conducted. Before the incubation, we first determined soil full water holding capacity. The sieved soil (<2 mm) was packed to the same bulk density (~1.5 g cm^3) in cutting ring (100 cm^3), saturated with water, weighed and then dried at 105°C for 48 hours to determine soil water content. Soil full water holding capacity was the water

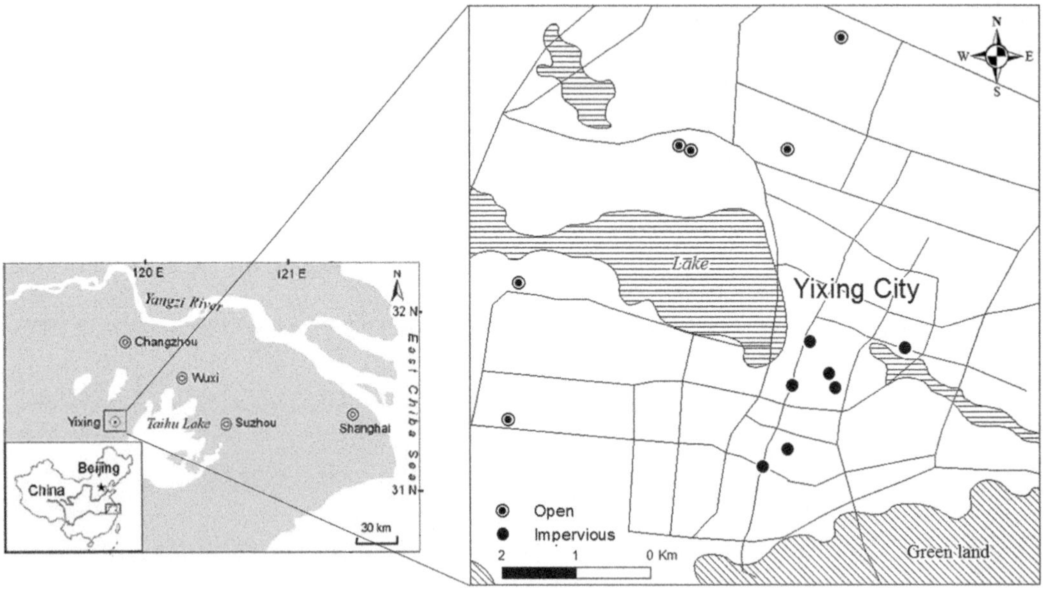

Figure 1. Soil sampling points in Yixing city. Seven sites were selected for impervious-covered soils, and six sites with similar soil parent materials were selected for open soils. Soil samples were collected at 0–20 cm depth.

content of the saturated soil. Fifty grams of each urban soil (< 2 mm) were placed into 500-mL capacity plastic bottles and incubated simultaneously in one incubator (SP-300B, Hengyu, China) for 28 days. The soil water content of each sample in the bottle was then adjusted to 60% of full water holding capacity, which the actual amount of water needed was calculated based on the preliminary test. Each plastic bottle was sealed airtight and incubated in the dark at 25°C. Each sample was pre-incubated without CO_2 absorption for 5 days [18]. Then, a beaker with 5 mL of 0.6 mol L^{-1} NaOH was placed in each jar to absorb the evolved CO_2 during the incubation. Control jars (without soil) were used to measure the background CO_2 concentration. The NaOH was renewed after 1, 3, 5, 7, 14, 21, and 28 days. Any loss of water from the cylinders (checked by weighing) was corrected by a mist sprayer. The CO_2 trapped in the NaOH was determined by back-titration of excess NaOH with 1.5 mol L^{-1} H_2SO_4 after precipitation with 1 mol L^{-1} $BaCl_2$ [26].

A first-order decay equation was used to describe the organic C mineralization in the samples studied [27]:

$$C_t = C_1 + C_0(1 - e^{-kt}) \qquad (2)$$

where C_t (mg C g^{-1} C) is the cumulative amount of SOC mineralized during time t (day), C_1 (mg C g^{-1} C) is the rapidly mineralizable SOC pool, C_0 (mg C g^{-1} C) is the potentially mineralizable SOC pool, and k (mg C g^{-1} C d^{-1}) is the corresponding mineralization rate constant.

The SOC mineralization half-time (i.e., time required to mineralize half of the potentially mineralizable SOC) was calculated as follows [28]:

Table 2. Initial properties of the urban soil in Yixing obtained from the second Chinese soil survey [a] ($n = 9$ except where noted).

Soil property	0–14 cm		14–25 cm	
	Mean ±SE	**CV, %**	**Mean ±SE**	**CV, %**
Bulk density [b], g cm^{-3}	1.3±0.13	10	1.44±0.15	10.4
pH	6.6±0.3	4.5	6.9±0.6	8.7
TOC [c], %	1.48±0.19	13.3	1.14±0.30	26.4
N, g kg^{-1}	1.38±0.11	8.0	1.13±0.28	24.8
C/N	10.6±1.2	11.3	9.9±1.5	15.2
P, %	0.095±0.01	10.5	0.09±0.125	16.7
K, %	1.67±0.09	5.4	1.68±0.11	6.7
Olsen P, mg kg^{-1}	5.8±2.8	48.3	5±3.8	76.0
Available K [d], mg kg^{-1}	85±30	35.3	96±30	31.2
CEC, cmol kg^{-1}	19.51±2.36	12.1	20.11±2.02	10.2

a, Office of Soil Survey in Yixing County, conducted in 1979–1982; b, $n = 61$; c, determined using the dichromate oxidation method; d, CH_3COONH_4 extractable K.

Table 3. Outline description for the urban soils (0–20 cm) selected in our study.

Soil sample	Texture	pH	artifacts, rock fragments, roots	> 2 mm fraction
				%
Impervious soils				
1	Silt loam	7.08	coarse angular weathered brick, roots	1.60
2	Silt	7.42	coarse angular freshly weathered rock fragment, roots	9.01
3	Silt loam	7.9	plastic, roots	5.19
4	Silt loam	8.15	coarse angular highly weathered brick and tiles	2.67
5	Silt loam	7.56		8.53
6	Silt	7.58	gypsum, roots	1.44
7	Silt loam	8.33		1.51
Open soils				
8	Silt loam	7.21	heavily roots	5.61
9	Silt loam	6.4		0.72
10	Silt loam	7.12	coarse angular weathered rock fragment, roots	4.25
11	Silt loam	7.58	heavily roots	2.78
12	Silt loam	7.74	heavily roots	2.33
13	Silt loam	6.94	heavily roots	2.35

$$t_{1/2} = \ln 2/k \quad (3)$$

Statistical analysis

Model fit was conducted using the Global curve fit wizard in SigmaPlot 12.0 software package (Systat Software, Inc., Chicago, IL, USA). Linear regression analysis and the comparisons in the densities of SOC and total N and the model parameters between the impervious and open land uses (group t-test) were conducted using SAS 8.2 software (SAS Institute, Cary, NC).

Results and Discussion

SOC density under impervious surfaces

The impervious-covered sites had lower SOC and TN concentrations (0–20 cm) compared with the open sites and those in 1982 (Figure 2a, b). The SOC and TN concentrations of the open soils were comparable with those in 1982. The mean SOC density beneath impervious surfaces was 68% (±7.7% SE, $P<0.05$) lower than that in open areas (2.46 versus 7.59 kg m^{-2}, respectively, Figure 2c). Similar results were also reported by Pouyat *et al.* (2006) and Raciti *et al.* (2012) that SOC densities under impermeable surfaces were lower than that of open soils [8,15]. It is interesting to note that the SOC densities observed at the impervious sites were comparable to those in New York City (Raciti *et al.* 2012) although the samples were distributed at different cities [15], suggesting that there might be an equilibrium value for the depletion of SOC densities under impervious surfaces. The total N densities were 0.25 kg m^{-2} and 0.32 kg m^{-2}, respectively, for the impervious-covered and open soils in Yixing City, but no significant difference was found between them (Figure 2d), probably due to the relatively small sample size in our study.

The exact mechanisms of SOC loss from the impervious areas are unknown, but likely possibilities include gaseous losses, aqueous losses as dissolved organic and inorganic C, and physical removal of topsoil during the construction process. The removal of surface soil may readily result in a significant depletion of organic C stock due to enhanced mineralization and re-use. The situation could be worse when topsoil is not re-used and is left slowly to decompose. If the potential loss of SOC inferred by this study hold true for other impervious-covered soils, this would suggest that roughly 0.45 Pg (1 Pg = 10^{15} g) of SOC would be potentially lost in China associated with 87128 km^2 of impervious surface [6]. The potential loss of SOC stock could be more significant in magnitude when taking into account the large vegetable C losses in response to initial land use change to urban.

Although the SOC densities beneath impermeable surfaces were relatively lower, it is not adequate to exclude these C pools for estimates of ecosystem C stocks in national scale as required by Kyoto Protocol signatories [10]. In fact, investigation of SOC stocks and dynamics in urban impervious areas are getting more important to the C study, since there is ongoing significant expanding of constructed impervious surfaces around the world [5,6]. In our study, soil samples were collected at 0–20 cm soil layer and therefore SOC densities were calculated to 0.2 m depth. Urban subsoil horizons may also contain considerable amounts of SOC since physical disturbance (e.g., mixing, burying) can result in translocation of topsoil that is rich in OC to deeper soil layer [29]. Thus, the SOC storage underlying capped surfaces in urban areas could be larger than we estimated when it was calculated to 1 m soil depth as usually did. The SOC inventory involving subsoil horizons beneath impervious surfaces in urban areas will strengthen our understanding on the impact of urbanization on the ecosystem. Furthermore, the SOC inventory in urban area should consider the definition and built-up age of the urban used by the investigators. Urban areas, regardless of definition, are rapidly expanding at unprecedented rates; inconsistent definitions of 'urban' will result in different conclusions about the size of urban C stocks [30]. Moreover, it was suggested that the built-up age could modify urban C stocks [31].

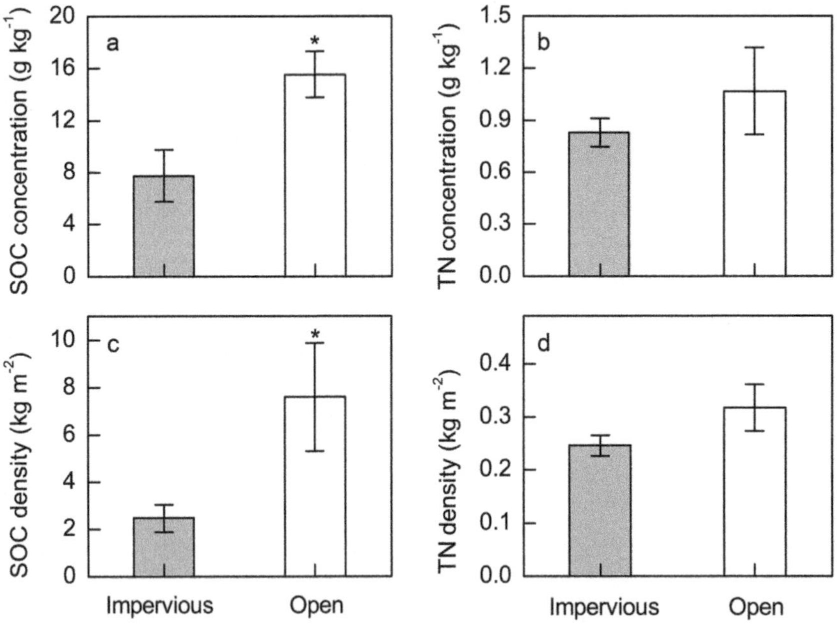

Figure 2. The concentrations and densities of SOC and TN for the impervious-covered and open soils in Yixing city. Values are means ± SE, SOC represents soil organic carbon, and TN represents total nitrogen, * $P<0.05$ ($n=7$ for urban impervious-covered soils, and $n=6$ for urban open soils).

In most soils, there is a tight coupling between the stocks and fluxes of C and N. The mean C/N ratio in the soils covered by impervious surfaces was significantly lower than the ratio in open soils (10.8 versus 22.1, respectively, $P<0.05$, data not shown), indicating that the microorganisms had a relative C deficit state in the impervious-covered soils. The higher C/N ratios in the open soils were probably due to the abundant input of organic materials (e.g., leaves, branches) from plant as these materials always had relatively higher C/N ratios [32]. Regression of SOC density as a function of total N density in the open soils revealed a strong linear relationship between the two variables at the 0–20 cm depth ($r^2=0.73$, $P<0.05$, Figure 3). In contrast, there was no clear relationship between SOC and total N for the soils beneath impervious surfaces, suggesting that the paving in urban areas decoupled the cycle of C and N in the soils [15].

Installation of impervious surfaces in urban areas had negative impact on urban ecosystem, which was indicated by large amount of SOC loss that can offset the C stored in trees or other green vegetation in urban areas, and perturbed soil C and N cycling in impervious area. More greenspaces or semi-pervious pavement systems were, therefore, recommended in future urban construction to mitigate the negative consequences of urban artificial soil sealing [5]. In addition, soil management (e.g., fertilization, tillage, irrigation) for the open sites could be optimized to enhance the SOC sequestration in urban soils.

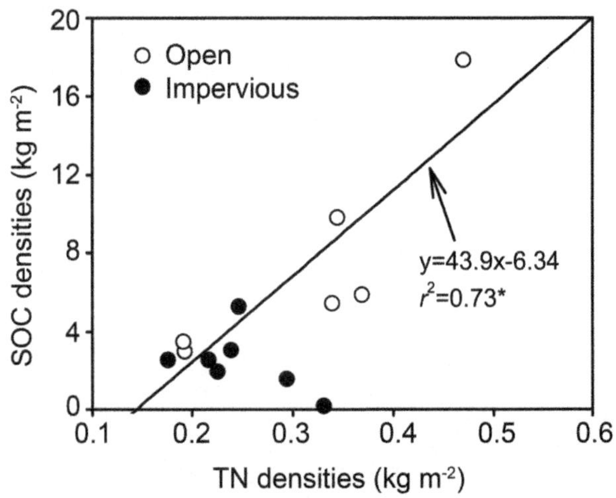

Figure 3. The correlations between the densities of SOC and TN for urban soils in Yixing city. SOC represents soil organic carbon, and TN represents total nitrogen, * $P<0.05$ ($n=7$ for urban impervious-covered soils, and $n=6$ for urban open soils).

SOC stability under impervious surfaces

The laboratory incubation can provide an insight of SOC dynamics (e.g., [33]). The amount of SOC mineralized (mg C g^{-1} C) over the 28-d incubation was calculated based on the total (whole soil) SOC content in each sample to enable us directly compare the potential C loss (i.e., CO_2 emissions) from the impervious-covered and open soils, because the portion of SOC is comparable amongst urban soils although the initial SOC content for the urban soils were unequal (Figure 4). It was found that less CO_2 was emitted from the soils underlying capped surfaces than from the open soils during the incubation, indicating that the SOC was more stable beneath impervious surfaces than in open areas. The lower transformation of organic C in the covered soils can be partially explained by a low microbial activity in these soils [34].

The first-order decay model ($C_t=C_1+C_0(1-e^{-kt})$) used in this study fitted the data well (r^2 were 0.97 and 0.96, Table 4). The easily decomposable SOC pool (C_1) was small relative to the potentially decomposable SOC pool (C_0) in all samples. The C_1 and C_0 of SOC under impervious surfaces were 47% and 27%, respectively, lower than those in open areas, indicating that there is a more severe depletion of readily decomposable SOC pool after

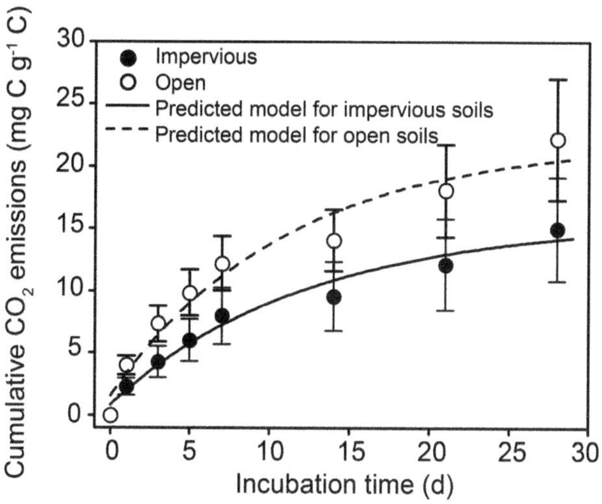

Figure 4. Cumulative carbon release (as CO_2) from the impervious-covered and open soils during the 28-d incubation. Data were fitted by the first-order decay model. The bars indicate standard errors ($n=7$ for urban impervious-covered soils, and $n=6$ for urban open soils).

paving. The impervious-covered and open soils had similar mineralization rate constants (*k*) of the potentially decomposable organic C. However, the lower value of C_0k, a parameter that could be comparable with the initial potential rate of SOC mineralization [35], together with the longer mineralization half-life time ($t_{1/2}$) of SOC under impervious surfaces revealed these soils had weaker organic C decomposability and lower turnover rate.

Edmondson *et al.* (2012) suggested that the turnover of SOC beneath impervious surface likely depend upon the type of capping and the extent of impervious surface [10]. In our study, the covered soils were collected conformably from areas with more than 95% impervious surface, in which the exchanges of gas and water between the soil and atmosphere were supposed to be rare. In fact, there are some urban impervious-covered soils (e.g., patio, garden path) which were distributed in areas dominated by vegetation. The soils underlying capped surfaces in these areas could be colonized by the root systems of lawn grasses and garden trees and shrubs. Thus, it is likely that below these smaller patches of impervious surfaces the soil remains active potentially accumulating SOC and has more robust organic C transformation. Accordingly, more systematic research and soil samples including different land use types and sealing degrees in urban areas are required to better understand the SOC turnover under impervious surfaces.

Additionally, some impervious surfaces in urban area may be removed due to urban land use change. The differences in soil biochemical properties between the impervious and open sites are supposed to be minimized, since the removal of impervious surface. Investigation on the temporal dynamic of SOC after removal of sealing will be beneficial for comprehensively understanding the ecological effects of urbanization. Therefore, management and time, which were needed to rebuild organic C sequestration capacity for the soil after removal of sealing, need to be studied.

Conclusions

Our data demonstrated that the SOC density decreased at 0–20 cm depth in urban areas (regardless of the definition) after the installation of impervious surfaces, although the precise mechanism of organic C loss from these soils was uncertain. The artificial soil sealing in urban areas also decoupled the cycle of C and N, since SOC density correlated positively with total N density in the open soils ($P<0.05$), but they did not exhibit an apparent relation in the impervious-covered soils. The SOC underlying capped surfaces had weaker decomposability and lower turnover rate compared with that in open areas, which was indicated by a smaller readily decomposable SOC pool, a longer half-time, and a smaller amount of CO_2-C emission during the 28-d incubation. More greenspaces or semi-pervious pavement systems in future urban construction will mitigate the negative consequences of urban artificial soil sealing.

Acknowledgments

We are grateful to the Center of Modern Analysis Nanjing University for supporting soil analysis in our study. We thank the three anonymous reviewers for their very helpful comments and revisions of the manuscript.

Author Contributions

Conceived and designed the experiments: ZQW SHW SLZ. Performed the experiments: ZQW SHW XY. Analyzed the data: ZQW SHW. Contributed reagents/materials/analysis tools: SHW XY. Wrote the paper: ZQW SHW.

Table 4. The first-order decay model (Eq. 2) parameters and coefficients of determination (r^2) for carbon mineralization in urban impervious-covered and open soils.

Parameters	Impervious	Open	
C_1 (mg C g^{-1} C)	0.84	1.59	n.s.
C_0 (mg C g^{-1} C)	14.9	20.53	n.s.
k (mg C g^{-1} C d^{-1})	0.08	0.09	n.s.
C_0k	1.17	1.87	n.s.
$t_{1/2}$ (d)	8.84	7.62	n.s.
r^2	0.97	0.96	n.s.
P	0.01	0.01	n.s.

C_1, rapidly mineralizable SOC pool; C_0, potentially mineralizable SOC pool; *k*, SOC mineralization rate constant; C_0k, a parameter that could be comparable with the initial potential rate of SOC mineralization [35]; $t_{1/2}$, SOC mineralization half-time.
n.s., not significant.

References

1. United Nations (2008) World urbanization prospects: the 2007 revision.
2. Hansen AJ, Knight RL, Marzluff JM, Powell S, Brown K, et al. (2005) Effects of exurban development on biodiversity: patterns, mechanisms, and research needs. Ecol Appl 15: 1893–1905. doi:10.1890/05-5221.
3. Pan XZ, Zhao QG (2007) Measurement of urbanization process and the paddy soil loss in Yixing city, China between 1949 and 2000. Catena 69: 65–73. doi:10.1016/j.catena.2006.04.016.
4. Su SL, Zhang Q, Zhang ZH, Zhi JJ, Wu JP (2011) Rural settlement expansion and paddy soil loss across an ex-urbanizing watershed in eastern coastal China during market transition. Reg Environ Change 11: 651–662. doi:10.1007/s10113-010-0197-2.
5. Scalenghe R, Ajmone-Marsan F (2009) The anthropogenic sealing of soils in urban areas. Lanscape Urban Plan 90: 1–10. doi:10.1016/j.landurbplan.2008.10.011.
6. Elvidge CD, Tuttle BT, Sutton PC, Baugh PC, Howard AT, et al. (2007) Global distribution and density of constructed impervious surfaces. Sensors 7: 1962–1979. doi:10.3390/s7091962.
7. Pouyat R, Groffman P, Yesilonis I, Hernandez L (2002) Soil carbon pools and fluxes in urban ecosystems. Environ Pollut 116: S107–S118. doi:10.1016/S0269-7491(01)00263-9.
8. Pouyat RV, Yesilonis ID, Nowak DJ (2006) Carbon storage by urban soils in the United States. J Environ Qual 35: 1566–1575. doi:10.2134/jeq2005.0215.
9. Churkina G, Brown DG, Keoleian G (2010) Carbon stored in human settlement: the conterminous United States. Global Change Biol 16: 135–143. doi:10.1111/j.1365-2486.2009.02002.x.
10. Edmondson JL, Davies ZG, McHugh N, Gaston KJ, Leake JR (2012) Organic carbon hidden in urban ecosystems. Sci Rep 2: 963. doi:10.1038/srep00963.
11. Townsend-Small A, Czimczik C (2010) Carbon sequestration and greenhouse gas emissions in urban turf. Geophys Res Lett 37: L02707. doi:10.1029/2009GL041675.
12. Raciti SM, Groffman PM, Jenkins JC, Pouyat RV, Fahey TJ, et al. (2011) Accumulation of carbon and nitrogen in residential soils with different land-use histories. Ecosystems 14: 287–297. doi:10.1007/s10021-010-9409-3.
13. Beesley L (2012) Carbon storage and fluxes in existing and newly created urban soils. J Environ Manage 104: 158–165. doi:10.1016/j.jenvman.
14. Rao P, Hutyra LR, Raciti SM, Finzi AC (2013) Field and remotely sensed measures of soil and vegetation carbon and nitrogen across an urbanization gradient in the Boston metropolitan area. Urban Ecosystems 16: 593–616. doi:10.1007/s11252-013-0291-6.
15. Raciti SM, Hutyra LR, Finzi AC (2012) Depleted soil carbon and nitrogen pools beneath impervious surfaces. Environ Pollut 164: 248–251. doi:10.1016/j.envpol.2012.01.046.
16. Jo HK (2002) Impacts of urban greenspace on offsetting carbon emissions for Middle Korea. J Environ Manage 64: 115–126. doi:10.1006/jema.2001.0491.
17. Kaye JP, Majumdar A, Gries C, Buyantuyev A, Grimm NB, et al. (2008) Hierarchical Bayesian scaling of soil properties across urban, agricultural, and desert ecosystems. Ecol Appl 18: 132–145. doi:10.1890/06-1952.1.
18. Wei ZQ, Wu SH, Zhou SL, Li JT, Zhao QG (2014) Soil organic carbon transformation and related properties in urban soil under impervious surfaces. Pedosphere 24: 56–64. doi:10.1016/S1002-0160(13)60080-6.
19. Liu Y, Wang C, Yue W, Hu Y (2013) Storage and density of soil organic carbon in urban topsoil of hilly cities: a case study of Chongqing Municipality of China. Chin Geogra Sci 23: 26–34. doi:10.1007/s11769-013-0585-x.
20. IUSS Working Group WRB (2007) World reference base for soil resources 2006, first update 2007. World Soil Resources Reports No. 103. FAO, Rome.
21. Office of Soil Survey in Yixing County (1988) The soils of Yixing County. Office of Soil Survey in Jiangsu Province, Nanjing. (in Chinese).
22. Gee GW, Or D (2002) Particle-size analysis. In: Dame JH, Topp GC, editors. Methods of Soil Analysis, Part 4. Physical Methods.pp.255–293.
23. Nelson DE, Sommers LE (1982) Total carbon, organic carbon, and organic matter. In: Page AL, Miller RH, Keeney DR, editors. Methods of soil analysis, Part 2. Chemical and microbiological properties. pp.539–580.
24. Bremner JM, Mulvaney CS (1982) Nitrogen total. In: Page AL, Miller RH, Keeney DR, editors. Methods of soil analysis. Part 2. Chemical and microbiological properties. pp.595–624.
25. Rossi AM, Hirmas DR, Graham RC, Sternberg PD (2008) Bulk density determination by automated three-dimensional laser scanning. Soil Sci Sco Am J 72: 1591–1593. doi:10.2136/sssaj2008.0072N.
26. Goyal S, Chander K, Mundra MC, Kapoor KK (1999) Influence of inorganic fertilizers and organic amendments on soil organic matter and soil microbial properties under tropical conditions. Biol Fertil Soils 29: 196–200. doi:10.1007/s003740050544.
27. Jones CA (1984) Estimation of an active fraction of soil nitrogen. Commun Soil Sci Plant Anal 15: 23–32. doi:10.1080/00103628409367451.
28. Zhang ZD, Yang XM, Drury CF, Reynolds WD, Zhao LP (2010) Mineralization of active soil organic carbon in particle size fractions of a Brookston clay soil under no-tillage and mouldboard plough tillage. Can J Soil Sci 90: 551–557. doi:10.4141/cjss09081.
29. Lorenz K, Kandeler E (2005) Biochemical characterization of urban soil profiles from Stuttgart, Germany. Soil Biol Biochem 37: 1373–1385. doi:10.1016/j.soilbio.2004.12.009.
30. Raciti SM, Hutyra LR, Rao P, Finzi AC (2012) Inconsistent definitions of "urban" result in different conclusions about the size of urban carbon and nitrogen stocks. Ecol Appl 22: 1015–1035. doi:10.1890/11-1250.1.
31. Scalenghe R, Malucelli F, Ungaro F, Perazzone L, Filippi N, et al. (2011) Influence of 150 years of land use on anthropogenic and natural carbon stocks in Emilia-Romagna region (Italy). Environ Sci Technol 45: 5112–5117. doi:10.1021/es1039437.
32. McGroddy ME, Daufresne T, Hedin LO (2004) Scaling of C:N:P stoichiometry in forests worldwide: implications of terrestrial Redfield-type ratios. Ecology 85: 2390–2401. doi:10.1890/03-0351.
33. Haile-Mariam S, Collins HP, Wright S, Paul EA (2008) Fractionation and long-term laboratory incubation to measure soil organic matter dynamics. Soil Sci Soc Am J 72: 370–378. doi:10.2136/sssaj2007.0126.
34. Wei ZQ, Wu SH, Zhou SL, Lin C (2013) Installation of impervious surfaces in urban areas affects microbial biomass, activity (potential C mineralization), and functional diversity of the fine earth. Soil Res 51: 59–67. doi:10.1071/SR12089
35. Fernández JM, Plaza C, Hernández D, Polo A (2007) Carbon mineralization in an arid soil amended with thermally-dried and composted sewage sludges. Geoderma 137: 497–503. doi:10.1016/j.geoderma.2006.10.013.

22

Interacting Microbe and Litter Quality Controls on Litter Decomposition

Daryl Moorhead[1,2,3*¶], Gwenaëlle Lashermes[2,3¶], Sylvie Recous[2,3], Isabelle Bertrand[2,3]

1 Department of Environmental Sciences, University of Toledo, Toledo, Ohio, United States of America, **2** INRA, UMR614 Fractionnement des AgroRessources et Environnement, Reims, France, **3** Université Reims-Champagne Ardenne, UMR614 Fractionnement des AgroRessources et Environnement, Reims, France

Abstract

The decomposition of plant litter in soil is a dynamic process during which substrate chemistry and microbial controls interact. We more clearly quantify these controls with a revised version of the Guild-based Decomposition Model (GDM) in which we used a reverse Michaelis-Menten approach to simulate short-term (112 days) decomposition of roots from four genotypes of *Zea mays* that differed primarily in lignin chemistry. A co-metabolic relationship between the degradation of lignin and holocellulose (cellulose+hemicellulose) fractions of litter showed that the reduction in decay rate with increasing lignin concentration (LCI) was related to the level of arabinan substitutions in arabinoxylan chains (i.e., arabinan to xylan or A:X ratio) and the extent to which hemicellulose chains are cross-linked with lignin in plant cell walls. This pattern was consistent between genotypes and during progressive decomposition within each genotype. Moreover, decay rates were controlled by these cross-linkages from the start of decomposition. We also discovered it necessary to divide the Van Soest soluble (labile) fraction of litter C into two pools: one that rapidly decomposed and a second that was more persistent. Simulated microbial production was consistent with recent studies suggesting that more rapidly decomposing materials can generate greater amounts of potentially recalcitrant microbial products despite the rapid loss of litter mass. Sensitivity analyses failed to identify any model parameter that consistently explained a large proportion of model variation, suggesting that feedback controls between litter quality and microbial activity in the reverse Michaelis-Menten approach resulted in stable model behavior. Model extrapolations to an independent set of data, derived from the decomposition of 12 different genotypes of maize roots, averaged within <3% of observed respiration rates and total CO_2 efflux over 112 days.

Editor: Andrew C. Singer, NERC Centre for Ecology & Hydrology, United Kingdom

Funding: This study was supported by United States National Science Foundation grants DEB-0946257 and DEB-0918718, and Institut National de la Recherche Agronomique. The funders had no role in study design, data collection and analysis, decision to publish, or preparation of the manuscript.

Competing Interests: The authors have declared that no competing interests exist.

* Email: daryl.moorhead@utoledo.edu

¶ These authors are co-first authors on this work.

Introduction

Recent studies are challenging the ways that we have traditionally perceived and modeled decomposition. Decomposers were once thought to rapidly degrade labile fractions of plant litter, such as carbohydrates and proteins, leaving more recalcitrant compounds, like lignin, to become the foundation of stabilized soil organic matter [1,2,3]. However, microbial products rather than plant lignin actually may represent the larger fraction of SOM [4,5], and lignin *per se* may not persist throughout decay [6]. Unfortunately, many mathematical models are based on the traditional view of decomposition. In part this is because the chemical composition of litter has often been evaluated by proximate carbon analysis, typically yielding three, qualitatively different pools of compounds: polar and nonpolar extractives, acid hydrolysable materials, and acid non-hydrolysable materials. These pools provide a convenient structural framework for modeling, but lack the resolution to address finer scale biochemical transformations revealed by more contemporary studies [4,5,6]. Mathematical models must change to reflect these observations.

One of the changes needed in traditional decomposition models is to explicitly simulate the activities of microorganisms. If microbial contributions to SOM are more substantial than lignin, then the litter compounds fueling microbial activity may be more important to C stabilization than the plant lignin pool. This point was suggested by Smith et al. [7] who noted that the proportion of initial litter C remaining over time (125 days) was directly related to the initial rate of decomposition prior to lignin decay. In other words, the incorporation of C into microbial biomass increased its overall persistence despite an initially higher fractional loss through respiration [2,8]. A simple explanation is that the higher carbon use efficiency (CUE) of readily decayed materials, like sugars and proteins, transfers a larger fraction of substrate C to microbial biomass than lignin, and that these microbial products are more persistent [4,5]. These results suggest that the early stage of litter decay when microbial activities are greatest should be an integral component of mathematical models.

A second change needed in decomposition models is how they respond to differences and interactions between the chemical constituents of decaying litter. Despite the questionable role of plant lignin in decomposition, both the initial lignin content and the lignocellulose index (LCI) of litter are often the best predictors of decay rate [1,3]. The lignocellulose index is normally calculated as the ratio of the acid non-hydrolysable/(non-hydrolysable+ hydrolysable) products of proximate C analysis (see above), and assumed to represent the lignin (non-hydrolysable) and holocellulose (hydrolysable) fractions of litter. Although this assumption may be nearly true for fresh litter, both microbial and lignin degradation products increasingly contribute to proximate C fractions during decay [5,6]. Thus LCI and proximate C fractions are ambiguous metrics of litter chemistry and models using them to regulate decomposition conflate litter quality controls with the decomposition process. In reality, plant cell walls are usually the largest component of plant litter and have a specific biophysical structure of interconnected saccharide and phenolic polymers [9]. This arrangement partly explains the relatively consistent relationship between LCI and decay, but raises questions about the precise relationships between lignin and other litter constituents [10,11,12] needed to more accurately model substrate dynamics during decomposition.

Few empirical studies have examined litter chemistry during decay with sufficient resolution to improve models beyond the lignin or LCI controls normally calculated. However, Machinet et al. [11,13,14] examined the detailed litter chemistry of maize (*Zea mays* L.) roots for naturally occurring genotypes decomposing in laboratory incubations. These genotypes varied primarily in their lignin content. Long-term C losses (>200 days) were most often correlated with compounds associated with lignin and cross-linked between lignin and polysaccharides, whereas short-term (< 200 days) losses were more closely related to soluble compounds and cell wall sugars [11]. Different loss rates of different sugars (e.g., glucans, xylans and arabinans) suggested that the relationship between lignin and decomposition was partly mediated by the specific composition of the polysaccharide fraction of the litter, probably because arabinan substitution in xylan chains interferes with the degradation of hemicellulose [9]. In other words, the sugar composition of cell walls appears to define the transition from short- to long-term patterns of decay, providing a mechanistic explanations for patterns in microbial activity, as well as the empirical relationships between LCI and decay most often used in decomposition models [15].

The overall goal of this study was to simulate the relationships between litter decay, microbial production and litter chemistry at the early stage of decomposition, using data collected by Machinet et al. [11,13] to test and refine the Guild Decomposition Model (GDM) developed by Moorhead and Sinsabaugh [16]. Our two specific objectives were to simulate patterns of microbial productivity during decomposition as litter chemistry changed, and to simulate the relationship between measured changes in cell wall chemistry (sugars and lignin) and decay rate. We chose GDM because it calculates the decay of specific substrate pools as a combined function of both microbial activity and substrate characteristics, using the Michaelis-Menten equation of substrate saturation. The maximum velocity of decomposition (V_{max}) is thus a function of microbial biomass [16], so that estimated decay rates reflect both the changing chemical composition of the decomposing litter and microbial productivity.

Methods

Our general approach was to revise GDM to use the empirical data collected by Machinet et al. [11,13,14] to derive parameters needed to simulate decomposition. Then, we analyzed the data from a 112-day laboratory study of four maize genotypes that differed from one another in initial litter chemistry [13,14], to evaluate interactions between litter qualities likely controlling the decomposition process, and to obtain estimates for parameter values used to describe these processes. Next, we selected two key parameters for GDM, which preliminary analyses indicated varied with initial litter chemistry and had large effects on model behavior (below). We then optimized these parameters to produce the best possible fit between simulated and observed patterns of decomposition, with respect to CO_2 efflux and chemical transformations in decaying litter. Best-fit parameter values were in turn compared to initial litter chemical characteristics to determine possible relationships. These relationships were then used to derive parameter values needed to simulate CO_2 efflux during decomposition of maize roots for 12 additional genotypes [11].

Experimental data

The data used to drive the revised GDM model were obtained from Machinet et al. [13]. Briefly fine roots (diameter 2–3 mm) of four natural genotypes of maize (F2, F2bm1, F292 and F292bm3), which differed in their chemical composition (Table 1), were cut into 5 mm lengths, added to soil in laboratory microcosms and incubated at 15°C. Potassium-nitrate fertilizer was added to microcosms to insure no nitrogen limitation [17]. They monitored CO_2-C efflux on days 3, 7, 10, 14, 21, 29, 36, 42, 51, 57, 70, 80, 87, 95 and 112. Chemical characteristics of litter were determined on days 0 (initial), 14, 36, 57 and 112, on roots manually removed from soils. Machinet et al. [11] used the same experimental methods for 12 additional maize genotypes, but examined only the initial litter chemistry and measured CO_2 efflux during decomposition.

The suite of litter chemical characteristics reported by Machinet et al. [11,13] in root residues included C and N content, Van Soest soluble C (C-SOL) and N (N-SOL) contents, and cell wall polysaccharides: glucan (Glu), arabinan (Ara) and xylan (Xyl), which are the major polymer carbohydrates in graminaea, as well as galactan (Gal) and uronic acids contents (galacturonic (Ac Gal) and glucuronic (Ac Glu)). They also determined Klason lignin (KL) and the lignin monomers, guaiacyl (G) and syringyl (S), as well as ester-linked *p*-coumaric (pCA) and ferulic acids (FA_{ester}), and ether-linked ferulic acids (FA_{ether}). From these data we calculated the sum of cell wall sugars ($\sum$Sug), arabinan:xylan ratio (A:X), carbon:nitrogen ratio (C:N), lignocellulose index ($LCI = C_3/[C_2+C_3]$) with C_1 (=C-SOL), C_2 (=$\sum$Sug) and C_3 (=KL), and non-detergent fiber (NDF) content.

Machinet et al. [13] estimated mass loss of litter based on cumulative CO_2 carbon efflux. Concentrations of litter chemical fractions at each date were multiplied by the estimated mass of remaining litter to estimate pool sizes during decomposition. Decay rate coefficients (k_i) were calculated for chemical constituents C_1, C_2 and C_3 for all 4 litter types over all 4 periods of observation (days 0–14, 14–36, 36–57 and 57–112), as the difference in the natural log of pool size between observation dates, divided by the time period.

Model revisions

This model was programmed in MATLAB (The MathWorks Inc., Natick, USA). We revised GDM to use Reverse Michaelis-Menten (RMM) functions to calculate decay rates as functions of

Table 1. Model parameters and state variables.

Factor	Description	Values[a] (genotype)	Unit
k_{1max}	Decay rate coefficient of substrate C_1	0.1	day^{-1}
k_{2max}	Maximum decay rate coefficient of substrate C_2	0.047[b]	day^{-1}
k_{3max}	Maximum decay rate coefficient of substrate C_3	0.001	day^{-1}
e_1	Substrate C_1 use efficiency	0.4	unitless
e_2	Substrate C_2 use efficiency	0.3	unitless
g	Coefficient for basal respiration	0.001	day^{-1}
BC_{max}	Maximum Biomass: total C ratio	0.05	unitless
LCI_T	LCI value at which C_3 decay starts	0.42 (F2)	unitless
		0.36 (F2bm1)	
		0.46 (F292)	
		0.45 (F292bm3)	
LCI_{max}	LCI value at which decay stops	0.7	unitless
K_{Bj1}	Half-saturation coefficient of all Guilds for substrate C_1	10	$mg\ C{\cdot}kg^{-1}$ soil
K_{B22}	Half-saturation coefficient of Guild 2 for	29 (F2)	$mg\ C{\cdot}kg^{-1}$
	substrate C_2	33 (F2bm1)	soil
		23 (F292)	
		19 (F292bm3)	
K_{B32}	Half-saturation coefficient of Guild 3 for	300	$mg\ C{\cdot}kg^{-1}$
	substrate C_2		soil
K_{B33}	Half-saturation coefficient of Guild 3 for substrate C_3	500	$mg\ C{\cdot}kg^{-1}$ soil
C_{1T}	Non-decomposable fraction of C_1	0.61[b] (F2)	unitless
		0.60 (F2bm1)	
		0.71 (F292)	
		0.90 (F292bm3)	
C_1	Litter C in C_1 pool (Van Soest soluble)	364 (F2)	$mg\ C{\cdot}kg^{-1}$
		419 (F2bm1)	soil
		361 (F292)	
		288 (F292bm3)	
C_2	Litter C in C_2 pool (acid hydrolysable fraction)	1238 (F2)	$mg\ C\ kg^{-1}$
		1166 (F2bm1)	soil
		1295 (F292)	
		1387 (F292bm3)	
C_3	Litter C in C_3 pool (acid non-hydrolysable	397 (F2)	$mg\ C{\cdot}kg^{-1}$
	fraction)	415 (F2bm1)	soil
		344 (F292)	
		325 (F292bm3)	
B_1	C in Guild 1	10	$mg\ C{\cdot}kg^{-1}$
			soil
B_2	C in Guild 2	10	$mg\ C{\cdot}kg^{-1}$
			soil
B_3	C in Guild 3	5	$mg\ C{\cdot}kg^{-1}$
			soil

[a]Initial values for variables.
[b]Empirically determined from experimental data.

microbial activity [18]: $dC_i/dt = k_i{\cdot}C_i{\cdot}B_j/(K_{Bji}+B_j)$, where C_i is the amount of substrate i, B_j is the amount of microbial biomass in guild j, and K_{Bji} is the half-saturation coefficient of guild j for substrate i. Note that we altered the more common RMM approach by using biomass (B) in place of an enzyme (E) pool, which assumes a constant ratio of E:B resulting from constitutive enzyme production [16,18]. We limited access to substrate pools by guilds: guild 1 was assumed to access only soluble resources (C_1); guild 2 was assumed to access both C_1 and C_2; guild 3 could access C_1, C_2 and C_3 (Figure 1).

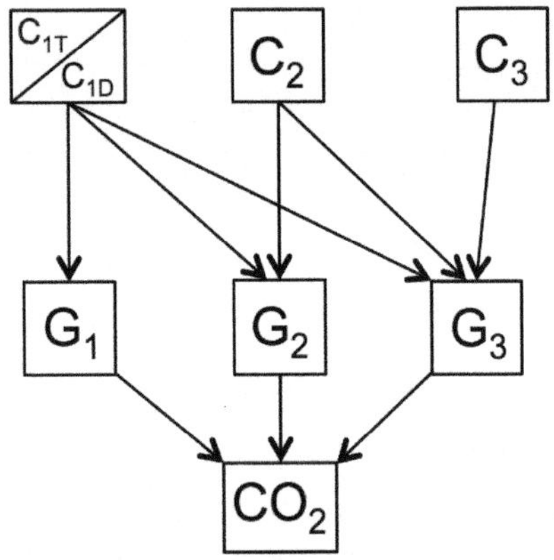

Figure 1. Carbon flow diagram for revised Guild Decomposition Model (GDM) simulating *Zea mays* root decomposition. Litter pools are two Van Soest soluble fractions, a decomposable fraction (C_{1D}) and a resistant fraction (C_{1T}), acid hydrolysable (C_2) and acid non-hydrolysable (C_3) fractions, and microbial guilds of opportunists (G_1), cellulolytic decomposers (G_2) and lignolytic decomposers (G_3).

A final revision to GDM was necessary to capture the dynamics of the soluble pool (C_1), a fraction of which persisted throughout the study by Machinet et al. [13]. This persistent fraction (C_{1T}) varied between genotypes and was considered to be non-decomposable during the time frame of the study (112 days).

Parameters and state variables

All parameters and state variables are reported in Table 1. Values of C_{1T}, the persistent fraction of the initial C_1 pool, were estimated from the average values of C_1 over time for each genotype (Table 1). We selected a decay rate coefficient (k_1) of 0.1 for the decomposable fraction of the C_1 pool (C_{1D}), due to its rapid loss.

The RMM approach made it possible to simplify the balance of C_2 and C_3 decay as functions of lignocellulose index (LCI) according to Moorhead et al. [19] (Text S1). In brief, the decay rate coefficients (k_2 and k_3) were described as linear functions of LCI: $k_i = m_i \cdot LCI + k_{imax}$, given empirically observed maximum values (k_{imax}) and slopes (m_i). This approach assumes that $k_3 = 0$ at a threshold level of $LCI = LCI_T$, wherein $LCI_T = 0.4$ [20], which defines the point at which LCI changes from being solely determined by C_2 decay ($LCI \leq LCI_T$) to also being determined by C_3 decay ($LCI > LCI_T$). The value of k_{2max} (maximum coefficient of C_2 decay) was estimated as the intercept of the linear regression of the observed decay rate coefficients for C_2 (k_2) against litter LCI for the four maize genotypes (Figure 2a; Table 1) over days 14–112, excluding values estimated over days 0–14 when we assumed that microorganisms had not begun to fully utilize pool C_2. Values of k_2 used during simulations were then estimated according to LCI of remaining litter. This revision to GDM provided a closer fit to observed patterns of C_2 decay (Figure 2a; $N = 12$, $R^2 = 0.97$, $P < 0.01$) than Moorhead and Sinsabaugh [16]. The value of k_{3max} was set to 0.001 because there was little evidence of C_3 decomposition for any litter type during incubations [13].

We selected identical values of K_{B11}, K_{B12} and K_{B13}, the half-saturation coefficients for the utilization of pool C_1 by all three guilds, assuming all microorganisms have similar affinities for soluble substrates (Table 1). We initially set K_{B22} and K_{B32} at the same values, again assuming that organisms capable of using the resource would have similar half-saturation coefficients, but higher than those for C_1, assuming that C_2 was generally less decomposable than C_1.

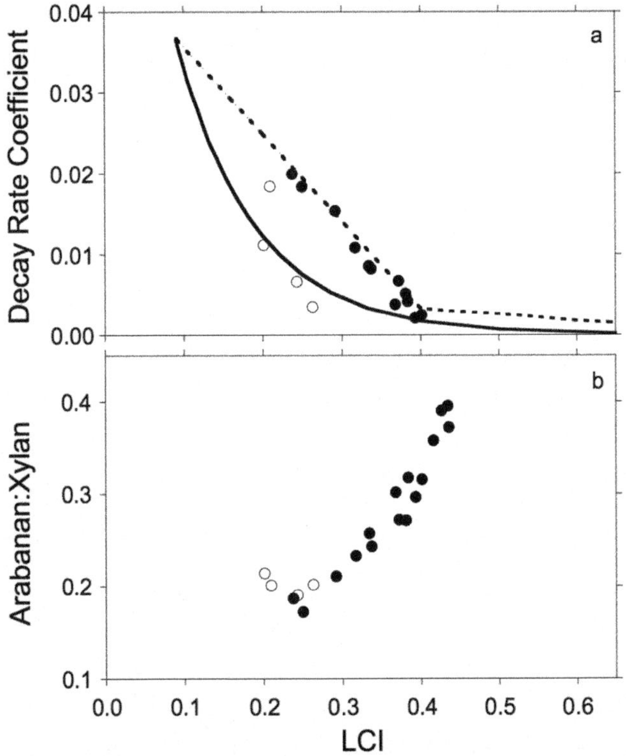

Figure 2. Observed and simulated decay rate coefficients for holocellulose and araban:xylan composition of remaining litter at residual values of lignocellulose index (LCI). Relationships between observed and simulated: a. decay rate coefficients (k_2) for litter pool C_2 and lignocellulose index (LCI) of litter (solid line is simulated according to Moorhead and Sinsabaugh [16], dashed line is simulated according to Moorhead et al. [19]), b. relationship between arabinan:-xylan (A:X) and LCI contents of decaying litter. Open circles are observations based on initial litter chemistry (day 0) and solid circles are from observations over time.

We then selected an even higher value for K_{B33} assuming that C_3 was even less decomposable than C_2. We are unaware of any published values for these parameters, and so followed the rationale of Moorhead and Sinsabaugh [16] in choosing values reflecting relative access to substrates of different qualities.

GDM requires an estimate of initial decomposer biomass, which it divides among three distinct microbial pools: early opportunists (guild 1), subsequent decomposition specialists (guild 2) and a final group of lignin degraders (guild 3) (Figure 1). Machinet et al. [13] estimated that microorganisms colonizing maize roots contributed 0–11% of the initial litter carbon content, and other studies suggest that microbial biomass rarely exceeds 2–3% of soil organic matter [21,22]. We set the initial microbial biomass pool at 1.25% (25 $mgC \cdot kg^{-1}$ soil) of total litter mass, assuming 10 $mgC \cdot kg^{-1}$ soil for guilds 1 and 2, and 5 $mgC\ kg^{-1}$ soil for guild 3 [16].

We estimated microbial production as the difference between the quantity of carbon released from decaying substrates and the amount mineralized through both growth- and maintenance-associated respiration. We assumed that this difference was immobilized in microbial biomass. GDM also calculates microbial turnover necessary to keep total microbial C less than 5% of the total system's organic C (microorganisms+substrates; Table 1). We estimated carbon use efficiency (CUE) as the difference between the amounts of C released from decaying substrates and mineralized through respiration, divided by the amount of C released by decomposition [23].

Optimizing parameter values

The values of two key model parameters that showed correlations with initial chemistry, LCI_T, and K_{B22}, were optimized for the four maize genotypes [13] using the *fmincon* function of the MATLAB Optimization Toolbox (The MathWorks, Inc., Natick, USA), which applies a sequential quadratic programming algorithm. The objective function was the sum of the root mean square errors calculated between the experimental and simulated CO_2 efflux rates, cumulative amount of carbon mineralized over 112-days and chemical evolution in the C_1 and C_2 pools during decomposition, normalized by the means of the observations.

We determined the best-fit estimates of parameters K_{B22} and LCI_T that most closely matched simulations to observed patterns of CO_2 efflux and both C_1 and C_2 pools over time (Table 1). We selected these parameters because the dynamics of pool C_2 were most closely related to cumulative CO_2 efflux and thus, estimated mass loss.

Sensitivity analysis

To evaluate the sensitivity of the model to parameter estimates (Table 1), we randomly varied model parameters e_1, e_2, k_{1max}, BC_{max} and K_{B11} within ±10% of their initial values (Table 1), 100 times for each litter type. We then calculated the differences between simulated and observed values of C_1, C_2, cumulative CO_2, and respiration rates on each date of observation [13] as a relative value = (observation-simulation)/observation. We summed these relative differences for each type of observation over all days of observation (i.e., through day 112), which produced a composite measure of the relative differences between observations and model output for each maize genotype. ANCOVA evaluated the contributions of variations in parameter values to variations in the relative differences between model output and observations by litter type. The type II sums of squares from ANCOVA were interpreted to represent the relative contribution of each parameter to model behavior.

Model extrapolation

The last set of simulations estimated decomposition for the 12 additional maize genotypes examined by Machinet et al. [13]. Relationships between values of C_{1T} and best-fit values of LCI_T and K_{B22} (Table 1) and the initial characteristics of litter chemistry were estimated for the four litter types described by Machinet et al. [13]: $LCI_T = -0.36 \cdot KL/AX + 0.70$ ($N = 4$; $R^2 = 0.923$; $P \leq 0.05$); $C_{1T} = 0.02 \cdot \sum Sug - 0.64$ ($N = 4$; $R^2 = 0.998$; $P \leq 0.05$); and $K_{B22} = 53.22 \cdot KL/AX - 13.65$ ($N = 4$; $R^2 = 0.975$; $P \leq 0.05$). These relationships were then used to estimate parameter values of C_{1T}, LCI_T and K_{B22} for simulations with each of the 12 additional genotypes reported by Machinet et al. [11], based on their initial chemistry. Principal components analyses evaluated relationships between the relative differences in observed and simulated values of respiration rates and cumulative CO_2 efflux, and initial litter chemistry characteristics for all 12 litters to determine if more detailed litter chemical characteristics than currently used in the model could provide additional insights to litter quality controls on decomposition.

Results

Decomposition dynamics

We discovered a close correspondence between the arabinan:-xylan ratio (A:X) and lignocellulose index (LCI) of the residue (Figure 2b) in the empirical data [11]. Between days 14 and 112, a linear regression of AX over LCI yielded an $R^2 = 0.91$ ($N = 12$, $P \leq 0.01$); we omitted days 0–14 because we expected decomposition to be limited by microbial activity rather than substrate [16,18]. Perhaps this relationship also explains why both AX and KL were related to best-fit model parameters LCI_T and K_{B22}.

Rates of CO_2 efflux for all four litter types rapidly increased to peak values within 14–21 days followed by gradual declines (Figure 3); cumulative CO_2 efflux rose rapidly during the first 36 days and then more slowly until day 112. For 3 of the 4 genotypes, the C_1 pool declined rapidly from day 1 to day 14, and then remained relatively constant during the rest of the incubation (Figure 4). For genotype F292bm3, the pool of C_1 remained essentially unchanged throughout the study. The C_2 pools of all four litters declined throughout the incubations, accounting for most of the litter mass loss over time (Figure 4). The C_3 fraction of the remaining litter showed a slight increase over the first 14 days of incubation (ca. 5–10%) for both F2 genotypes (not shown), but remained roughly constant for the two F292 genotypes [13].

Microbial production

Microbial biomass rapidly increased to peak values within 20–40 days, varying among genotypes (Figure 5a), declining most rapidly for those genotypes supporting the most rapid initial growth with the highest decay rates (F292 and F292bm3). Microbial turnover rates (Figure 5b), which should correlate with the generation of microbial products such as cell walls, were similar in shape but had lower peak values and lagged behind the patterns of rising and falling respiration rates (Figure 3). Genotypes with higher turnover rates also showed the greatest declines in biomass by day 112. Litter LCI increased over time in all litter types, coincident with declining biomass and turnover rates (Figure 5c), increasing most rapidly for litters decaying most rapidly (Figure 3). CUE declined over time for all litter types (Figure 5d), coincident with increasing LCI (Figure 5c), but started at higher values in litter types with larger pools of labile C_1 (i.e., C_{1D}), which had a higher C-assimilation efficiency (e_i) than C_2 (Table 1).

The total microbial production (C-immobilized into biomass including microbial turnover) by day 112 varied between litter types: 248, 195, 295 and 295 mgC for genotypes F2, F2bm1, F292 and F292bm3, respectively. These values were negatively related to initial litter KL and KL/AX contents, as well as final biomass and best-fit values of K_{B22}. They were positively related to total cumulative CO_2 efflux by day 112 and best-fit values of LCI_T (all $N = 4$, $P \leq 0.05$).

Model test

Our initial model parameter set, including best-fit estimates of K_{B22} and LCI_T, simulated rates of CO_2 efflux closely matching observations (all $N = 15$; omitting day 0), with R^2 values ranging from 0.87 for genotype F2bm3 to 0.98 for genotype F2bm1 (Fig. 3). Simulated rates peaked before observations for all litter types and at slightly lower values. However, simulated peak rates were within 10% of observed peaks for both F2 genotypes and 20% for both F292 genotypes. Simulated values of cumulative CO_2 efflux (to day 112) were even closer to observations, with R^2 values in excess of 0.99 for all genotypes (Figure 3). All simulations were within 5% of observed cumulative CO_2 efflux values. Simulated values of C_2 seldom differed by more than 5% of the observed values, with all $R^2 \geq 0.99$ (Figure 3). Finally, our model did not estimate any loss in pool C_3 during simulations and has no mechanism to generate an increase in this pool's size (Figure 1).

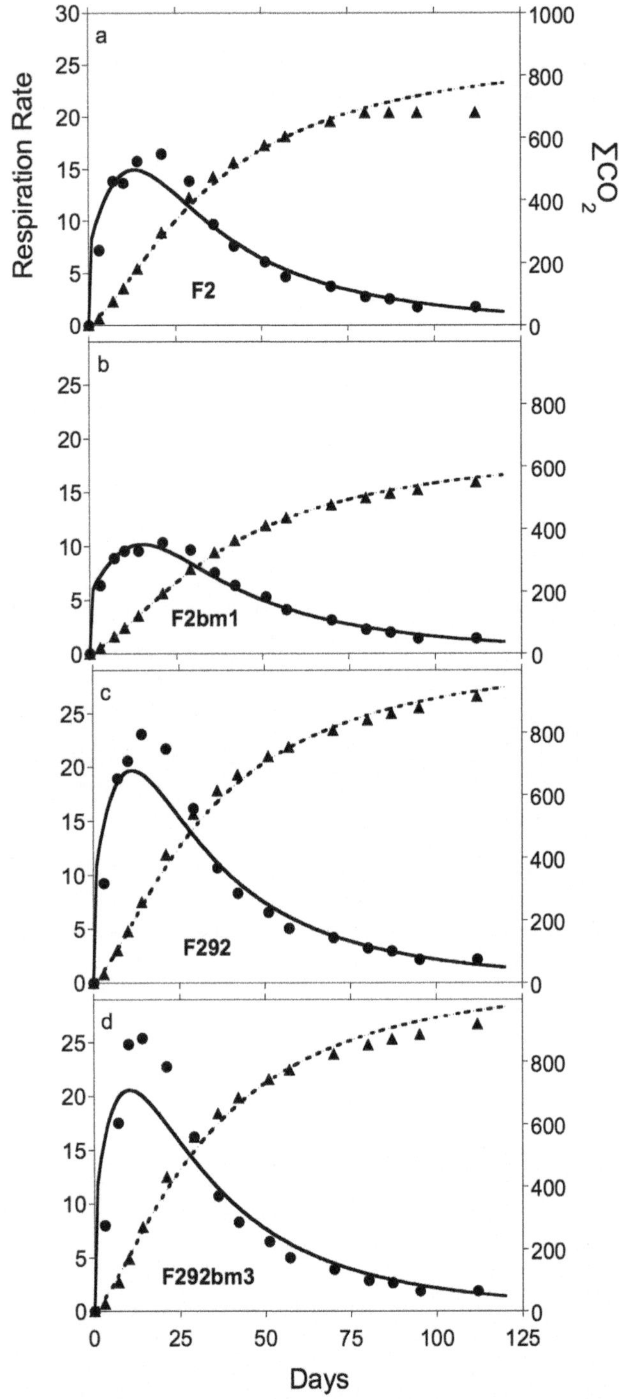

Figure 3. Observed and simulated respiration rates and cumulative CO_2 efflux during decomposition of *Zea mays* root litter. Observed (symbols) and simulated (lines) patterns of respiration rates (solid lines and circles, mgC·kg $soil^{-1}$·d^{-1}) and cumulative CO_2 efflux (dashed lines and triangles, mgC·kg $soil^{-1}$) for maize mutants, a. F2, b. F2bm1, c. F292 and d. F292bm3.

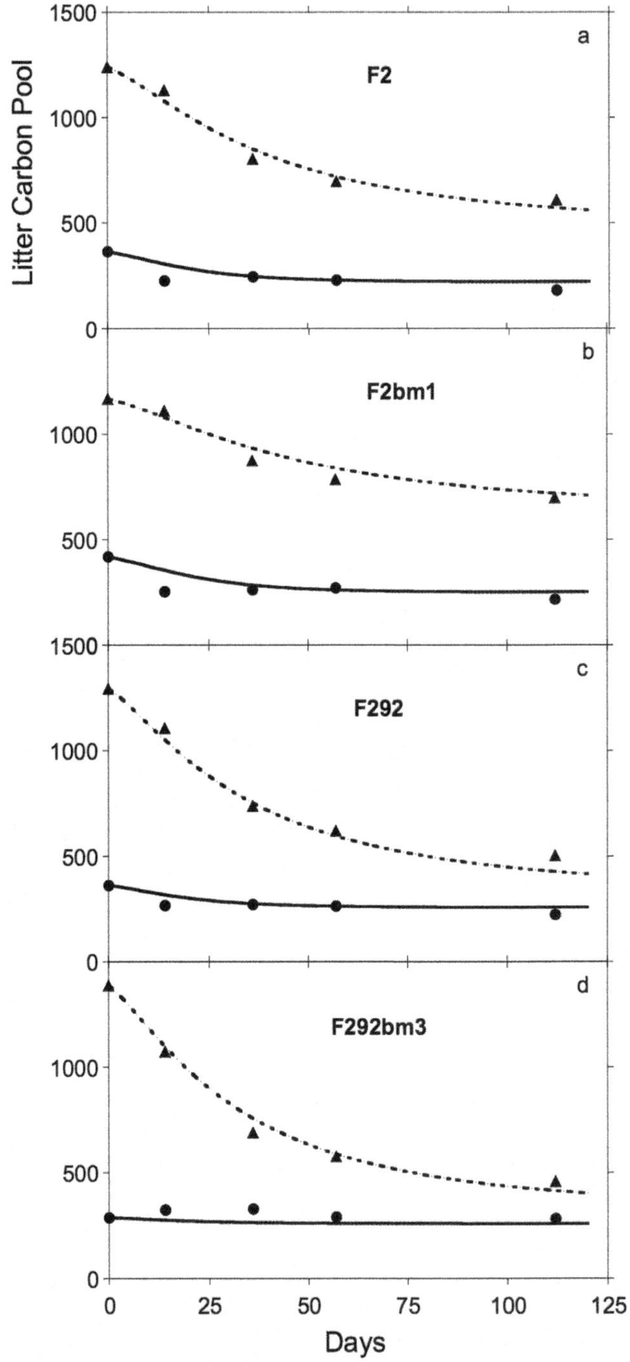

Figure 4. Observed and simulated mass remaining for soluble (C_1) and acid hydrolysable (C_2) chemical fractions of litter during decomposition of *Zea mays* roots. Observed (symbols) and simulated (lines) patterns of mass remaining for soluble C_1 pool (solid lines and circles, mgC·kg $soil^{-1}$) and acid hydrolysable C_2 pool (dashed lines and triangles, mgC·kg $soil^{-1}$) during decomposition of litter from maize mutants, a. F2, b. F2bm1, c. F292 and d. F292bm3.

Sensitivity analysis

The relative contributions of individual parameters to explaining variations in model behaviors differed by litter type and observation (Table 2). For example, the efficiency of C assimilation from substrate C_1 (e_1) rarely explained more than 1% of the variation in model behavior for any litter type. Moreover, no single parameter made a substantial contribution (≥10%) to explaining the variation in any model behavior when all four genotypes were pooled (not shown). When litter types were examined separately, variations in k_{1max} and K_{B11} often made their largest contributions to the same output variables for the

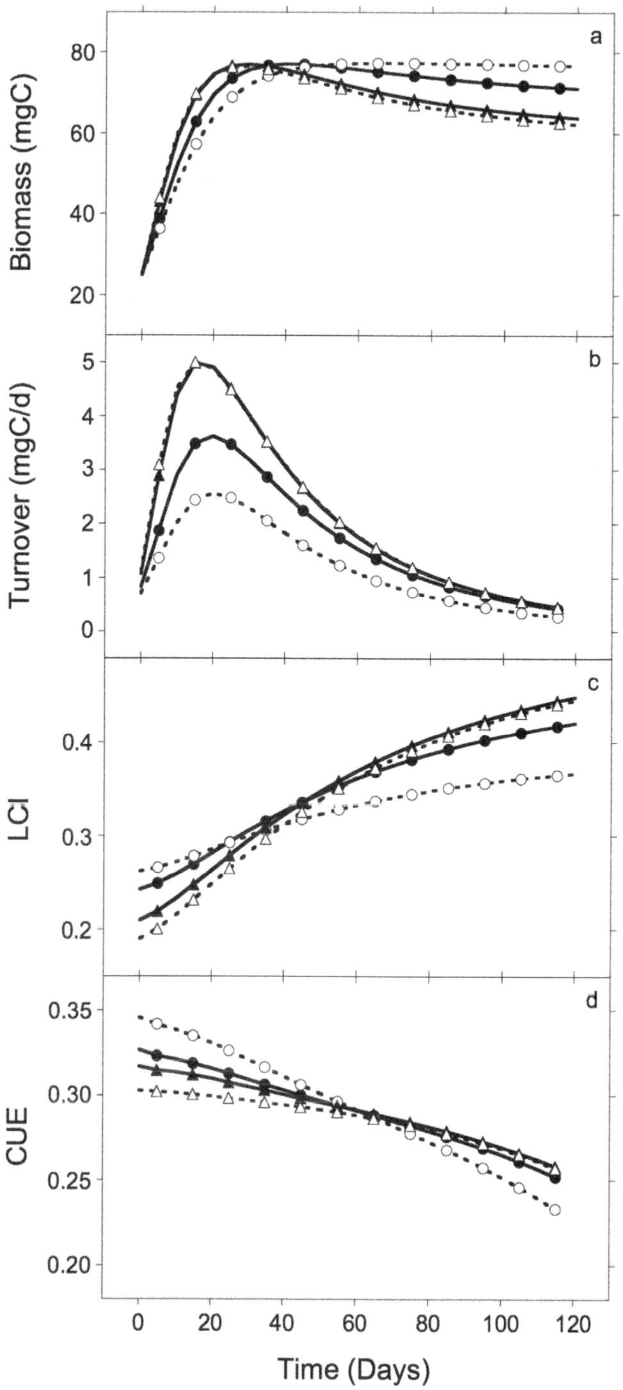

Figure 5. Simulated values of microbial biomass and turnover rate, remaining litter lignocellulose index (LCI), and carbon use efficiency (CUE), during decomposition of *Zea mays* root litter. Simulated values of a. total microbial biomass, b. daily microbial turnover rate, c. remaining litter lignocellulose index (LCI), and d. realized carbon use efficiency (CUE), over time, for litter genotypes F2 (filled circles), F2bm1 (open circles), F292 (filled triangles) and F292bm3 (open triangles).

same litter types. For example both parameters made substantial contributions (≥10%) to variations in C_1 for litter types F2, F2bm1 and F292bm3; cumulative CO_2 efflux for litter types F2, F2bm1 and F292; respiration rates for litter types F2bm1 and F292; and overall model fit for litter type F2bm1. Neither parameter made a substantial contribution to variation in C_2 for any litter type. In contrast, parameters e_2 and BC_{max} often made their largest contributions to model behaviors for litter types when k_{1max} and K_{B11} did not. For example, e_2 and BC_{max} made substantial contributions to variations in C_2 for litter types F2, F2bm1 and F292bm3; cumulative CO_2 efflux for litters F292 and F292bm1; respiration rates for F2, F292 and F292bm3; and overall model fit for F2.

Model extrapolation

When we simulated decomposition of the 12 different maize genotypes [11], we found that overall simulated rates of respiration were strongly related to observations ($N = 180$, $R^2 = 0.69$, $P \leq 0.01$), and that simulated rates of respiration averaged 1±17% lower than observed rates (data not shown). However, the relative differences ([observations-simulations]/observations) varied over time (Text S2). The first two axes of the principal components analysis explained 54% of the variation in the relative differences between observed and simulated respiration and litter chemical characteristics (Figure 6a). Differences in rates between days 10–21 were more strongly related to axis 2, along with litter LCI, KL/AX and ∑Sug. In contrast, differences in respiration for most days ≥50 were more strongly associated with axis 1, along with several chemical characteristics, the strongest being arabinans, AX, galactose, NDF and pCA (Figure 6a).

In contrast to the daily respiration rates, there was no significant relationship between simulated and observed peak respiration rates (means = 17.5±1.8 and 21.4±9.9 $mgC \cdot kg\ soil^{-1} \cdot d^{-1}$, respectively; not shown). The PCA showed that relative differences between peak rates (PR) were more closely related to axis 2 and opposite those of rates between days 10–21 (Figure 6a), with significant positive correlations with KL and KL/AX.

Simulated values of cumulative CO_2 efflux were also strongly related to observations ($N = 180$, $R^2 = 0.923$, $P < 0.01$), and averaged only 3±18% greater than observations. The first two axes of the PCA explained 69% of the variation in the relative differences between observed and simulated cumulative CO_2 efflux and initial litter chemistry characteristics (Figure 6b). All values for CO_2 efflux were tightly clustered and closely associated with the first axis. Litter chemical characteristics, NDF, arabinans, galactose and AX, were also closely related to the first axis. The relative differences between observations and simulations at day 36 (d36) showed a significant, positive correlation with the initial soluble content of litter (C_1 = C-SOL) and negative correlation with galactan (GAL) content. Differences by day 59 were also related to C_1 and galactan. Although there were no significant differences between simulations and observations at day 112, these variations were significantly related to several aspects of initial litter chemistry, including C_1.

Estimates of microbial production by day 112 for these 12 genotypes ranged 203–303 mgC (mean = 270±25 mgC), and were negatively correlated with KL/AX and final microbial biomass, as well as initial LCI. Production was positively related to total cumulative CO_2 efflux by day 112 and best-fit values of LCI_T, as well as initial litter concentrations of Glu, ∑Sug, and C_2 (all $N = 12$, $P \leq 0.05$). A stepwise regression explained nearly all variation in microbial production as a function of total CO_2 efflux and initial concentrations of xylans (Xyl) and guaiacyl (G) in litter ($N = 12$, $R^2 = 0.998$, $P \leq 0.01$).

Table 2. Results of sensitivity analysis quantifying the relative contributions (%) of random variations in model parameters (column headings) to resulting variations in model behaviors (row labels), based on sums of squares from ANOVA relating model output to parameters.

Source	Genotype	e_1	e_2	k_{1max}	BC_{max}	K_{B11}
C_1-Carbon	F2	0.2*	1.3*	47.5**	3.9*	30.6**
C_1-Carbon	F2bm1	2.0*	0.5*	63.7**	0.7*	38.9**
C_1-Carbon	F292	0.1*	1.7	3.9*	42.5**	1.2
C_1-Carbon	F292bm3	0.0	2.3*	65.6**	7.0*	22.1**
C_1-Carbon	All Litters	0.2	0.0	2.1*	0.1	6.3*
C_2-Carbon	F2	1.1*	38.4**	0.2*	53.9**	0.1
C_2-Carbon	F2bm1	1.2	13.2**	0.1*	71.4**	0.0
C_2-Carbon	F292	0.0	3.5*	0.8*	73.0**	0.1
C_2-Carbon	F292bm3	0.0	40.8**	0.0	40.9**	0.0
C_2-Carbon	All Litters	1.0*	1.6*	0.2	3.3*	0.4
Cumulative CO_2	F2	0.2*	1.2*	48.1**	3.8*	30.1**
Cumulative CO_2	F2bm1	0.4*	0.1	64.4**	0.2*	40.3**
Cumulative CO_2	F292	0.7*	11.7**	35.7**	12.3**	38.1**
Cumulative CO_2	F292bm3	0.5*	48.8**	6.6*	21.2**	4.5*
Cumulative CO_2	All Litters	0.2	0.0	0.6	0.3	1.6*
Respiration Rate	F2	1.0	28.7**	9.6*	17.8**	7.6*
Respiration Rate	F2bm1	0.3	0.1	48.5**	0.3	30.2**
Respiration Rate	F292	2.1*	40.8**	20.6**	11.4**	14.0**
Respiration Rate	F292bm3	0.2*	45.3**	1.8*	34.7**	0.5*
Respiration Rate	All Litters	0.0	0.2	0.0	0.4	0.1
Overall Fit	F2	3.0*	34.8**	4.0*	31.9**	2.1*
Overall Fit	F2bm1	0.6*	0.2	57.9**	0.6*	36.3**
Overall Fit	F292	2.6*	46.5**	4.6*	16.8*	7.9*
Overall Fit	F292bm3	0.3*	49.2**	0.9*	31.3*	0.7*
Overall Fit	All Litters	0.0	0.2	0.6	0.4	0.6
Maximum		30.4	49.2	65.6	73.0	40.3
Count>10%		0	11	9	12	9

*significant parameter contributions ($P \leq 0.05$) to variation in model behavior.
**contributions of parameters representing at least 10% of the explained variation in model behavior ($P \geq 0.05$).

Discussion

Lignin-cellulose interaction

Simulations provided a close match to observed patterns of holocellulose (C_2) decay (Figure 4). This pool represented the largest fraction of litter (Table 1), also explaining why simulations closely fit patterns of CO_2 efflux (Figure 3). Long-term decomposition (months to years) has often been negatively correlated to initial lignin content of litter [3], partly because lignin decays slowly and partly because some products of decomposition increase the size of the non-hydrolysable pool often interpreted as lignin [6]. However, Machinet et al. [11] also demonstrated the effects of biochemical connections between cell wall polysaccharides and lignin on the pattern of cumulative CO_2 efflux from decomposing maize roots. The fit between k_2 and LCI (Figure 2a) reveals an interaction between C_2 and C_3 in early stages of litter decay [19], well before C_3 begins to decline [13,20]. In other words, there is unlikely to be qualitatively separate pools of lignin-shielded and unshielded holocellulose, as is sometimes implied [1,3]. Instead, biochemical linkages between cellulose, hemicellulose and lignin components of cell walls influence patterns of decomposition throughout the process.

A brief explanation of these relationships is that the structural composition of hemicellulose is a primary chain consisting mainly of xylans with branching arabinan side chains that interact with other cell wall polymers. The level of arabinoxylan substitution (represented by A:X) increases with the progressive enzymatic degradation of plant material, during both digestion in the rumen [9,24] and decomposition in soils [12,13], as the more exposed elements of xylan chains are more readily hydrolyzed than those near arabinan branches. Gunnarsson et al. [25] found that initial xylan and arabinan concentrations in litter were as important as the total amount of hemicellulose in describing C mineralization over the first 9 days of laboratory incubations, and that arabinan was the single most important factor. These results were consistent with those of Machinet et al. [11], who identified initial arabinan content as an early predictor of C mineralization (days 3–7), followed by AX (days 7–14). In addition, hydroxycinnamic acids (ferulic and *p*-coumaric acids) play a key role in cross-linking arabinoxylans with lignin and this cohesive network also hampers decomposition [10,11,12]. These cross-linkages explain the

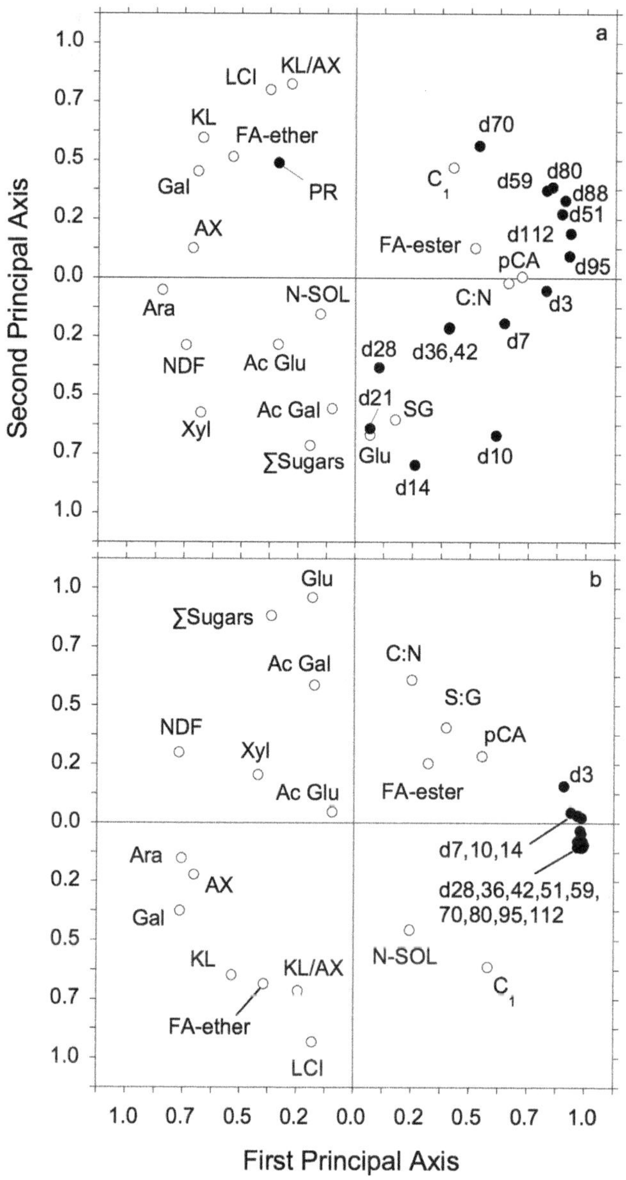

Figure 6. Principal components analysis of variations in initial litter chemistry characteristics and relative differences between observed and simulated rates of microbial respiration and cumulative CO_2 efflux during decomposition of *Zea mays* roots. Results of principal components analysis of variations in initial litter chemistry characteristics and relative differences = (observations − simulations)/observations from simulated decomposition of 12 novel maize mutants [11] for: a. rates of respiration on days 3–112 (e.g., d3 = rate on day 3) and peak respiration (PR), and b. cumulative CO_2 efflux by days 3–112 (parameter definitions given in text).

negative effect of lignin on holocellulose decay long reported in the literature [1,3] and also why k_2 decreased as AX increased early in decomposition (Figure 2).

Our model revision approximated this complex control with a straightforward relationship between LCI and k_2 (Figure 2) that included interactions between C_2 and C_3 (based on empirical sugar and lignin determinations). Moreover, key model parameters LCI_T and K_{B22} were most closely related to initial KL/AX ratios of litter, emphasizing the interaction between holocellulose and lignin throughout the decomposition process.

Soluble dynamics

Data from Machinet et al. [13] also revealed that the soluble C_1 pool included a fraction (C_{1T}) that persisted throughout the study (Figure 4). Although the soluble fraction of litter is usually considered to be highly decomposable, the persistence of a relatively large soluble pool is common in both soils and decaying litter [26,27]. The soluble pool usually contains a hydrophilic fraction including sugars and amino acids that are readily used by microorganisms [28,29], and a more hydrophobic portion, including soluble polyphenols (e.g., tannins) that are less rapidly utilized [26,30]. This differential use of compounds in the C_1 pool by decomposer microorganisms changes the pool's overall quality and decomposability with time [1]. Although the soluble pool is replenished with degradation products from non-soluble substrates [1,31], products of C_2 hydrolysis would enter the more labile fraction of the soluble pool, which cycles much more rapidly than the more persistent fraction [28,29]. In addition, Machinet et al. [13] found no decrease in the C_3 pool over time (not shown), so that it's degradation products could not have increased the more persistent fraction of the C_1 pool (C_{1T}). Simulating the persistence of a sizeable pool of C_1 with GDM required dividing the pool into labile and persistent fractions (Figure 1), with the labile fraction (C_{1D}) rapidly declining during decay and the persistent C_{1T} pool remaining intact (Figure 4). Moorhead and Sinsabaugh [16] found that simply routing the products of C_2 degradation through the whole C_1 pool GDM could not explain the size of this composite pool.

Respiration patterns

Simulations tended to overestimate rates of CO_2 efflux between days 0–14, and underestimate rates between days 14–36 (Figure 3). A positive relationship between the most labile components of litter and early respiration (hours to weeks) or decay rate is commonly reported [32,33], and most models, including GDM, assume a greater decay rate for soluble substrates [15,16]. In terms of substrate dynamics, GDM tended to underestimate early losses (day 14) of C_1 and overestimate C_2 losses (Figure 4), suggesting that values of k_{1max} should be slightly increased and K_{B1i} ($i = 1,2,3$) reduced (Table 1), to stimulate decay of C_{1D} and possibly initial CO_2 efflux. However, GDM underestimated respiration to a greater extent for litter types with lower amounts of C_{1D}, i.e., with higher C_{1T} (F292 and F292bm3). We found that C_{1T} was most closely related to the total sugar content of cell walls (∑Sug; $N = 4$; $R^2 = 0.998$; $P \leq 0.05$) and that xylan, arabinan and glucan concentrations were highly correlated with each other (not shown). Further resolution of the chemical composition of the soluble fraction (C_{1D} and C_{1T}) and its dynamics during decomposition are needed to improve the mathematical descriptions of these relationships.

Microbial production

Moorhead and Sinsabaugh [16] argued that litter decay is initially limited by microbial action, because there is a time lag in the colonization of fresh litter by decomposer microorganisms (Figure 2). Whether this lag was a numerical (biomass) or functional (physiological) response in the study by Machinet et al. [13] is unknown because they did not monitor microbial biomass. Therefore, the most speculative part of this study was our simulation of microbial dynamics. Nonetheless, the patterns and magnitudes of simulated microbial biomass and turnover rates were consistent with observed patterns of respiration (Figures 3,

5b), reported limits to microbial biomass concentration in soils [21,22], and changes in litter chemistry likely controlling microbial activities (Figures 2, 4 and 5c; [1]) as well as CUE (Figure 5d) [23]. Parameter estimates for C-assimilation efficiencies (e_i) for various substrate pools, coefficient of microbial basal respiration rate (g), and maximum biomass:total system C ratio (BC_{max}), most directly affected the relationships between substrate decomposition and biomass dynamics (Table 1). However, variations in any or all of these parameter values generate similar patterns with respect to different litter chemistry, although amplitudes and temporal regimes vary [16].

Our primary reason for simulating biomass dynamics was to determine if litter quality affected microbial production consistent with the idea that more decomposable chemical fractions of litter not only decay more rapidly but also generate more microbial products likely to enter stable soil organic matter pools [4,5]. In fact, simulated microbial production was significantly and negatively correlated to KL and KL/AX, which are inversely related to decomposability, but production showed no significant positive relationship to any simple measure of litter quality. However, it was positively related to the sum of the two most decomposable carbon pools, i.e., C_2+C_{1D}, (N = 4, rho = 0.982, P≤ 0.05). The contributions of the relatively larger pools of C_{1D} in litter types F2 and F2bm1 to simulated production were small compared to the larger pools of C_2 in the other litters, despite the lower C assimilation efficiency of C_2 versus C_1 (Table 1). Thus our results were consistent with the observations of Smith et al. [7] and others who found that higher initial rates of C incorporation into biomass coincided with higher losses of litter through respiration [2,8].

Sensitivity analysis

The most interesting result of our sensitivity analysis was the lack of any simple, overall interpretation. No single parameter consistently explained>10% of the observed variability in any model behavior (Table 2). Usually a parameter important to explaining one model behavior, such as the contributions of k_{1max} to simulated C_1 dynamics or cumulative CO_2 efflux (Table 2), made little contribution to other model outputs. The largest discrepancies between simulations and observations were in early respiration and C_1 loss (discussed above). Of the tested parameters, e_2, the C assimilation efficiency for C_2, appeared to be most important to respiration rates, which seems reasonable because C_2 was the largest substrate pool and provided most of the C respired (Table 1, Figure 4). As for the dynamics of the C_1 pool, parameter k_{1max} appeared to be most important, followed by K_{B11}; thus parameters controlling the degradation rate of this pool explained differences between simulations and observations (Figure 4). These results are consistent with our earlier conclusion that greater resolution of the C_1 pool composition and dynamics might provide a better understanding of decomposition.

Extrapolations

We simulated the decomposition of 12 additional maize genotypes in the second set of incubations conducted by Machinet et al. [11] based on the assumption that the relationships between key parameters, LCI_T, K_{B22} and C_{1T}, and litter chemical characteristics were consistent with those for the four genotypes examined by Machinet et al. [13]. The assumption seemed reasonable because all 16 genotypes were naturally occurring varieties of *Zea mays*, and likely to be more similar in chemistry and tissue architecture than unrelated species more commonly used in comparative decomposition experiments [6].

In general, simulations were within a few percent of observed values of respiration rates and cumulative CO_2 efflux over the entire period of incubation. Analyses of these differences between simulations and observations provided relatively little additional insight to patterns of cumulative CO_2 efflux (Figure 6b). The importance of the initial C_1 pool (as both C_{1T} and C_{1D} estimates) to cumulative CO_2 efflux through time again emphasized the importance of initial decay rate to longer-term patterns [7,8]. The relationships between CO_2 efflux on day 112 and initial arabinan and *p*-coumaric acid concentrations and AX suggest that the cross-linkages among hemicellulose and lignin became increasingly important with progressive decay. The tight cluster of cumulative CO_2 efflux along the first axis of our PCA also underscored the importance of cross-linkages among cell wall constituents (e.g., arabinan, AX, NDF and *p*-coumaric acid) to litter decay (Figure 6b), consistent with Machinet et al. [11].

The differences between simulated and observed respiration rates were more variable, suggesting temporally shifting controls on decomposition. Peak rates and those on days 14–28 were related to initial KL and KL/AX, as well as glucan and S:G (Figure 6a). In contrast, rates on days>42 were more closely associated with arabinan, AX, *p*-coumaric acid and ester-linked ferrulic acids, perhaps because polysaccharide-ester linked ferulic acids can form ether-links with lignin [34,35] and the syringyl units of lignin can be esterified by *p*-coumaric acids, which is typical of grass cell walls [11]. However, the biggest differences between simulated and observed rates were on days 36–42, which were most closely related to KL, ether-linked ferrulic acids and galactan (Figure 6b). The frequent importance of KL and AX (or their chemical constituents) to these patterns was surprising, because KL (in LCI) was used to estimate k_2, and KL/AX to estimate LCI_T and K_{B22} (previously discussed) used in simulations. Clearly, simple linear relationships were insufficient to capture the subtleties of these controls. The relationships between respiration and galactan and C:N ratio (Figure 6b) hint at a microbial control [14], in part because galactan is sometimes used as an index to microbial contributions [36], but it is also a hemicellulosic sugar, along with arabinan, rhamnan, and xylan [37].

Conclusions

We found that the level of arabinan substitution in xylan chains (AX) was an important control in early stages of decomposition, and was also linearly related to LCI calculated on the basis of fine scale cell wall chemistry. These relationships provide a plausible, mechanistic explanation for earlier, empirical descriptions of LCI effects on decomposition as a result of biochemical cross-linkages between polysaccharides and lignin. Thus lignin and LCI serve as convenient, negative proxies for the decomposability of litter even at the start of decay. However, additional research is needed to determine the chemical composition and dynamics of the non-hydrolysable product of proximate C analysis that is typically termed "lignin" if we are to discover the mechanistic relationships between LCI and latter stages of decomposition.

We also found that dividing the soluble pool of litter (C_1) into separate persistent (C_{1T}) and labile (C_{1D}) pools, was necessary to accurately simulate the dynamics of the composite soluble pool during early decomposition because not all soluble compounds are equally decomposable. The finer scale chemical composition and dynamics of the soluble component of litter is needed to determine the possible sources and fates of these compounds.

These relationships between simulated patterns of litter decay, litter chemistry (LCI, AX, C_{1T} and C_{1D}) and microbial productivity were consistent with the notion that the more rapid utilization of substrates with high carbon use efficiency generates greater amounts of microbial products that can contribute to stable soil organic matter pools than more persistent substrates with lower C-assimilation, like lignin. Our assumption that lignin decay provides little to no net C-acquisition by microorganisms is also consistent with recent observations that little lignin C enters stable SOC pools.

Finally, our extrapolations with litter types that differed in initial chemistry demonstrated that relationships we found between key model parameters, LCI_T, K_{B22} and C_{1T}, and decomposition were robust across a large range of maize litter types, highlighting the importance of access (K_{B22}) to the largest, rapidly decaying pool of substrate (C_2) by microorganisms able to use this resource (G_2), as well as negative controls imposed by the less-accessible substrate pools (KL, C_{1T}). Moreover, differences between simulations and observations indicated that temporal controls on decay rates shifted from relative substrate pool sizes (both accessible and persistent) dominating at the start to factors related to cross-linkages between structural polysaccharides and lignin with progressive decomposition.

In closing, our results suggest that interactions between decomposer microorganisms and litter quality characteristics at the earlier stages of decomposition may provide more insights to soil organic C stabilization than later stages dominated by the more persistent chemical characteristics of the cell wall, if indeed microbial products comprise a large fraction of stable soil C pools.

Supporting Information

Data S1 Data used to test and refine model. Observed respiration rates and chemical composition of residues of maize roots during decomposition [13].

Figure S1 Differences over time between observed and simulated respiration rates and cumulative CO_2 efflux from the decomposition of *Zea mays* roots. Relative differences between observations and simulations of litter decomposition from 12 novel maize genotypes [11] for: a. respiration rates (mgC·kg soil^{-1}·d^{-1}) over time (means±95% confidence intervals, all $N = 12$), b. cumulative CO_2 efflux (mgC·kg soil^{-1}) over time (means±95% confidence intervals, all $N = 12$).

Text S1 Lignocellulose controls. Relationships between decay rate coefficients for polysaccharides (C_2) and polyphenolics (C_3) as functions of lignocellulose index ($LCI = C_3/[C_2+C_3]$).

Text S2 Sensitivity analysis. Patterns of differences between observed and modeled patterns of respiration associated with decomposition of decaying maize roots.

Author Contributions

Conceived and designed the experiments: DM GL SR IB. Performed the experiments: DM GL. Analyzed the data: DM GL SR IB. Contributed reagents/materials/analysis tools: DM GL SR IB. Wrote the paper: DM GL SR IB.

References

1. Berg B, McClaugherty C (2008) Plant litter: Decomposition, humus formation, carbon sequestration, Second edition. New York: Springer. 338 p.
2. McMahon SK, Williams MA, Bottomley PJ, Myrold DD (2005) Dynamics of microbial communities during decomposition of carbon 13 labeled ryegrass fractions in soil. Soil Sci Soc Am J 69: 1238–1247.
3. Melillo JM, Aber JD, Muratore JF (1982) Nitrogen and lignin control of hardwood leaf litter decomposition dynamics. Ecology 63: 621–626.
4. Wickings K, Grandy AS, Reed SC, Cleveland CC (2012) The origin of litter chemical complexity during decomposition. Ecol Lett 15: 1180–1188.
5. Cotrufo MF, Wallenstein MD, Boot CM, Denef K, Paul E (2013) The microbial efficiency-matrix stabilization (MEMS) framework integrates plant litter decomposition with soil organic matter stabilization: do labile plant inputs form stable soil organic matter? Glob Change Biol 19: 988–995.
6. Preston CM, Nault JR, Trofymow JA (2009) Chemical changes during 6 years of decomposition of 11 litters in some Canadian forest sites. Part 2. 13C abundance, solid-state 13C NMR spectroscopy and the meaning of "Lignin". Ecosystems 12: 1078–1102.
7. Smith JL, Bell JM, Bolton H Jr, Bailey VL (2007) The initial rate of C substrate utilization and longer-term soil C storage. Biol Fertil Soils 44: 315–320.
8. Blagodatskaya EV, Blagodatsky SA, Anderson T -H, Kuzyakov Y (2009) Contrasting effects of glucose, living roots and maize straw on microbial growth kinetics and substrate availability in soil. Eur J Soil Sci 60: 186–197.
9. Chesson A (1988) Ligninopolysaccharide complexes of the plant cell wall and their effect on microbial degradation in the rumen. Anim Feed Sci Tech 21: 219–228.
10. Grabber JH, Mertens DR, Kim H, Funk C, Lu FC, et al. (2008) Cell wall fermentation kinetics are impacted more by lignin content and ferulate cross-linking than by lignin composition. J Sci Food Agr 89: 122–129.
11. Machinet GE, Bertrand I, Barriere Y, B Chabbert B, Recous S (2011) Impact of plant cell wall network on biodegradation in soil: Role of lignin composition and phenolic acids in roots from 16 maize genotypes. Soil Biol Biochem 43: 1544–1552.
12. Amin BAZ, Chabbert B, Moorhead D, Bertrand I (2014) Impact of fine litter chemistry on lignocellulolytic enzyme efficiency during decomposition of maize leaf and root in soil. Biogeochemistry 117: 169–183.
13. Machinet GE, Bertrand I, Chabbert B, Recous S (2009) Decomposition in soil and chemical changes of maize roots with genetic variations affecting cell wall quality. Eur J Soil Sci 60: 176–185.
14. Machinet GE, Bertrand I, Chabbert B, Watteau F, Villemin G, et al. (2009) Soil biodegradation of maize root residues: Interaction between chemical characteristics and the presence of colonizing micro-organisms. Soil Biol Biochem 41: 1253–1261.
15. Manzoni S, Porporato A (2009) Soil carbon and nitrogen mineralization: theory and models across scales. Soil Biol Biochem 41: 1355–1379.
16. Moorhead DL, Sinsabaugh RL (2006) A theoretical model of litter decay and microbial interaction. Ecol Monogr 76: 151–174.
17. Recous S, Robin D, Darwis D, Mary B (1995) Soil inorganic N availability: Effect on maize residue decomposition. Soil Biol Biochem 27: 1529–1538.
18. Schimel JP, Weintraub MN (2003) The implications of exoenzyme activity on microbial carbon and nitrogen limitation in soil: A theoretical model. Soil Biol Biochem 35: 549–563.
19. Moorhead DL, Lashermes G, Sinsabaugh RL, Weintraub MN (2013) Calculating co-metabolic costs of lignin decay and their impacts on carbon use efficiency. Soil Biol Biochem 66: 17–19.
20. Herman J, Moorhead D, Berg B (2008) The relationship between rates of lignin and cellulose decay in aboveground forest litter. Soil Biol Biochem 40: 2620–2626.
21. Anderson T-H, Domsch KH (1989) Ratios of microbial biomass carbon to total organic carbon in arable soils. Soil Biol Biochem 21: 471–479.
22. Wardle DA (1998) Controls of temporal variability of the soil microbial biomass: A global-scale synthesis. Soil Biol Biochem 30: 1627–1637.
23. Sinsabaugh RL, Manzoni S, Moorhead DL, Richter A (2013) Carbon use efficiency of microbial communities: stoichiometry, methodology and modeling. Ecol Lett 16: 930–939.
24. Vailhé MAB, Provan GJ, Scobbie L, Chesson A, Maillot MP, et al. (2000) Effect of phenolic structures on the degradability of cell walls isolated from newly extended apical internode of tall fescue (Festuca arundinacea Schreb.). J Agr Food Chem 48: 618–623.
25. Gunnarsson S, Marstorp H, Dahlin AS, Witter E (2008) Influence of non-cellulose structural carbohydrate composition on plant material decomposition in soil. Biol Fertil Soils 45: 27–36.
26. Cleveland CC, Neff JC, Townsend AR, Hood E (2004) Composition, dynamics, and fate of leached dissolved organic matter in terrestrial ecosystems: Results from a decomposition experiment. Ecosystems 7: 275–285.
27. Bengtson P, Bengtsson G (2007) Rapid turnover of DOC in temperate forests accounts for increased CO2 production at elevated temperatures. Ecol Lett 10: 783–790.
28. Kalbitz K, Schmerwits J, Schwesig D, Matzner E (2003) Biodegradation of soil-derived dissolved organic matter as related to its properties. Geoderma 113: 273–291.

29. Bertrand I, Prevot M, Chabbert B (2009) Soil decomposition of wheat internodes of different maturity stages: Relative impact of the soluble and structural fractions. Bioresource Technol 100: 155–163.
30. Nishimura S, Maie N, Baba M, Sudo T, Sugiura T, et al. (2012) Changes in the quality of chromophoric dissolved organic matter leached from senescent leaf litter during the early decomposition. J Environ Qual 41: 823–833.
31. Sinsabaugh RL (1994) Enzymatic analysis of microbial pattern and process. Biol Fertil Soils 17: 69–74.
32. Lagomarsino A, Moscatelli MC, De Angelis P, Grego S (2006) Labile substrates quality as the main driving force of microbial mineralization activity in a poplar plantation soil under elevated CO2 and nitrogen fertilization. Sci Total Environ 372: 256–265.
33. Hernandez DL, Hobbie SE (2010) The effects of substrate composition, quantity, and diversity on microbial activity. Plant Soil 335: 397–411.
34. Jung HG, Deetz DA (1993) Cell wall lignification and degradability. In: Jung HG, Buxton DR, Hatfield RD, Ralph J, editors. Forage cell wall structure and digestibility. Madison: ASA-CSSA-SSSA. Pp. 315–346.
35. Jacquet G, Pollet B, Lapierre C (1995) New ether-linked ferulic acid-coniferyl alcohol dimers identified in grass straws. J Agr Food Chem 43: 2746–2751.
36. Oades JM (1984) Soil organic matter and structural stability: mechanisms and implications for management. Plant Soil 76: 319–337.
37. Merila P, Malmivaara-Lamsa M, Spetz P, Stark S, Vierikko K, et al. (2010) Soil organic matter quality as a link between microbial community structure and vegetation composition along a successional gradient in a boreal forest. Appl Soil Ecol 46: 259–267.

23

Salinity and Bacterial Diversity: To What Extent Does the Concentration of Salt Affect the Bacterial Community in a Saline Soil?

Loredana Canfora[1]*, Giovanni Bacci[1,3], Flavia Pinzari[1], Giuseppe Lo Papa[2], Carmelo Dazzi[2], Anna Benedetti[1]

1 Consiglio per la Ricerca e la Sperimentazione in Agricoltura, Centro di Ricerca per lo studio delle relazioni tra Pianta e Suolo, Roma, Italy, **2** Dipartimento di Scienze Agrarie e Forestali, Università degli Studi di Palermo, Palermo, Italy, **3** Department of Biology, University of Florence, Florence, Italy

Abstract

In this study, the evaluation of soil characteristics was coupled with a pyrosequencing analysis of the V2-V3 16S rRNA gene region in order to investigate the bacterial community structure and diversity in the A horizon of a natural saline soil located in Sicily (Italy). The main aim of the research was to assess the organisation and diversity of microbial taxa using a spatial scale that revealed physical and chemical heterogeneity of the habitat under investigation. The results provided information on the type of distribution of different bacterial groups as a function of spatial gradients of soil salinity and pH. The analysis of bacterial 16S rRNA showed differences in bacterial composition and diversity due to a variable salt concentration in the soil. The bacterial community showed a statistically significant spatial variability. Some bacterial phyla appeared spread in the whole area, whatever the salinity gradient. It emerged therefore that a patchy saline soil can not contain just a single microbial community selected to withstand extreme osmotic phenomena, but many communities that can be variously correlated to one or more environmental parameters. Sequences have been deposited to the SRA database and can be accessed on ID Project PRJNA241061.

Editor: A. Mark Ibekwe, U. S. Salinity Lab, United States of America

Funding: The authors have no funding or support to report.

Competing Interests: The authors have declared that no competing interests exist.

* Email: loredana.canfora@entecra.it

Introduction

Saline soils are environments characterised by high concentrations of salts and by an uneven temporal and spatial water distribution. According to an early classification, a soil is considered to be saline when the Electrical Conductivity (ECe) of a saturated paste is greater than 4 dS m^{-1} [1]. More recently, the two international soil classification systems introduced higher minimum thresholds of ECe to classify a soil horizon as saline. In fact, the WRB (World Soil Resources Reports) [2] considers the reference value 15 dS m^{-1} of ECe in defining the saline horizon, while the Soil Taxonomy [3] fixes the threshold at 30 dS m^{-1}. A high concentration of salt in soil changes the availability of water and nutrients for both plants and microorganisms, and it has direct and indirect influences on soil organic matter, content, and stability [4]. Salinity, in fact, has been found to influence the size and the activity of soil microbial biomass [5,6], which in turns plays a key role in biogeochemical cycles.

A basic distinction must be made between primary and secondary salinisation processes. Primary salinisation consists of salt accumulation through natural processes, such as a high salt content of the parent material or in groundwater. Secondary salinisation is usually caused by human interventions such as inappropriate irrigation practices, i.e. after the use of salt-rich irrigation water, and/or insufficient drainage. A natural secondary soil salinisation mechanism is represented by the long-term effects of tsunami waves, which can deposit salty seawater on large flooded areas with dramatic consequences for agriculture. Depending on the climatic conditions, secondary soil salinisation can also be temporary, and the soils may recover by washing out the infiltrated salt deposits through rainfall; but this is not the case for Mediterranean and arid environments, which can hardly spontaneously recover from a secondary salinisation event. Microorganisms that occur in naturally saline habitats are supposed to share a strategy for resisting high salt concentrations, and to have developed multiple adaptations for maintaining their population active while coping with such extreme environmental conditions. From the genetic point of view, these species display an under- or over-expression of peculiar genes and metabolites, which confer them the capability of coping with an osmotic stress [7].

A naturally saline soil is also a mutable environment where rain and water movements can strongly change the distribution of salts, and create an evolving patchy landscape. The researcher's perception of environmental variability, and, consequently, the scale at which specific properties (like salinity) are measured, can misrepresent the spatial scale at which microbial groups differentiate their structure. Measurements of microbial community's structure and function are often based on broad-scale characterisations, and rarely consider the real spatial scale within which individuals and populations interact [8,9]. Space and scale in population, community, and ecosystem processes are increasingly recognised as fundamental factors in the study of microbial functions and activities in soil [8]. Understanding the spatial pattern in the abundance and structure of microbial communities occurring in saline soils represents a crucial target in ecology [9,10] as it sheds light on the selection mechanisms exerted by the environment on bacterial groups with specific functions and properties. The choice between a niche higher in salinity and instability, where instead of developing metabolic tools for resistance the bacterial communities wait for more favorable conditions, can represent the key mechanism that shapes microbial heterogeneity and taxa spatial composition in problematic soils.

In this framework, a series of surveys on "extreme" environments where salinity is the main determinant addressed the issue of some microbial taxa's specific ability to resist osmotic stress or other limiting factors associated with the presence of high salt concentrations. On the basis of these studies [11,12,13,14,15,16,17,18,19,20,21,22,23,24,25,26,27,28], we have attempted a rough discrimination between taxa frequently listed in these extreme environments (salinity related), and taxa less represented or even absent in salt affected sites (salinity unrelated). Clearly, this approach can lead to a dangerous generalisation, especially for the taxa that contain genera and species with strongly different physiologies and very broad geographical distributions. This is the case of the phylum *Proteobacteria*, which is probably the biggest group of bacteria associated to practically any environment.

A recent meta-analysis of soil microbial communities reported that the global microbial composition in a saline soil is influenced more by salinity than by any other extreme chemical factors such as temperature or pH [11,12]. Ma and Gong [12] recently examined the bacterial and archaeal diversity in saline soils using a meta-analysis approach indicating that approximately 50% of the archaeal diversity and less than 25% of the total bacterial diversity has been recorded from saline soil habitats. Ma and Gong [12] updated the available information on DNA sequences gained from a wide array of studies on soil microorganisms in saline environments, but also showed that there is a significant gap in the published information on the relevant soil properties where microbial communities had been sampled. Despite the useful and unique information collected, the authors could not address the beta diversity of the microbial species according to the different salt concentrations [12], thus failing to relate bacterial taxa composition with such drastic soil physical-chemical factors.

In the present study, a thorough evaluation of soil characteristics was supported by a pyrosequencing of the V2–V3 16S rRNA gene region in order to investigate the bacterial structure and diversity in the A horizon of a natural saline soil habitat located in Sicily (Italy). The application of metagenomic strategies has been recognised by several authors as a valid instrument to exploit the microbial biodiversity within soil habitats [29,30,31], but up to now saline habitats have still been largely unexplored by the metagenomic approach. The main aim of the research is to evaluate the organisation and diversity of microbial taxa by means of a spatial scale that reveals heterogeneity in the distribution of physical and chemical properties of the environment under investigation. As salinity is one of the most widespread soil degradation processes on earth affecting an estimated 1 million hectares in the European Union, mainly in the Mediterranean countries, and as it is a major cause of desertification, the understanding of soil microbial resilience and resistance in primary salinisation processes represents a cognitive platform for any application of bioremediation in secondary salinisation events.

Materials and Methods

Site description

The study was performed in Sicily, Italy, in an abandoned natural area, *Piana del Signore* -a semiarid Mediterranean environment- characterised by an alluvial flat land where the geomorphology has been modeled by the river *Gela* (Fig. S1). The area falls within the Natura 2000 network as a site of Community Importance (SCI), with the number SIC ITA050012- Gela. In the basin of the river *Gela* the prevalent lithology is made up by Messinian evaporites belonging to the *Gessoso-Solfifera* geological formation, among which many types of saline rocks crop out: Gypsum, Carbonates and Marls (with frequent chloride and sulfide rock inclusions). The area we surveyed is 12.3 hectares wide, and lies about 1 km far from the coastline and 1.2 km from the river estuary. The vegetation is a patchy mosaic plant association defined *Junco subulati-Salicornietum fruticosae* belonging to the *Thero-Salicornietea* class, in which the most plants are salt pioneer swards typical of salt marshes. The vegetation pattern consists of *Salicornia fruticosa* (L.), *Suaeda fruticosa* (L.), *Juncus subulatus* (Forsskal), *Juncus bufonius* (L.), *Phragmites australis* (Cav.), *Aster squamatus* (Spreng.), *Polypogon monspeliensis* (L.), *Hainardia cylindrica* (Wild.). The vegetation distribution is discontinuous: plant covered zones alternate with bare zones with visible salt crusts deposited above the soil surface. The whole area is temporarily flooded in the autumn and winter seasons, with a long permanence of water in those zones where salt crusts have formed on the soil surface after the water has dried out. Mean daily air temperature ranges from a maximum of 26.6°C in August to a minimum of 4.9°C in January. The average annual rainfall is 383 mm according to the nearest meteorological station (37°4′48″N, 14°13′12″E).

Soil Survey and sampling

Nine soil sites from A horizons (with a mean depth of 0–10 cm) were collected following a random simple sampling scheme (Fig. S2a) in Summer 2011. No specific permission was required for this location, as no endangered or protected species live in the area (Table S1). Sampling sites were positioned with a minimum distance between points of 50 m. Sites were accurately recorded with a GPS. In each site, three soil samples were collected in the vertices of 1-meter side equilateral triangle (Fig. S2b) and mixed together in a unique representative analytical sample [32]. Vegetation, salt crusts, and other features of the soil surface were described and recorded for each site. Soil samples from each site were subdivided in two representative subsamples: the first one was air dried, 2 mm sieved, then chemically and physically analysed, while the second one was stored at −80°C and later processed for the 454 pyrosequencing analysis reported below.

Soil physical and chemical analysis

Air-dried and sieved soil subsamples were analysed for the following physical and chemical properties: texture; reaction (pH), electrical conductivity (EC), and organic carbon (C_{org}). Texture was determined by the pipette method, without carbonate and organic matter removal, and after complete removal of soluble salts by using distilled water [33]; pH was measured on 1:2.5 (w/v) soil to water mixtures; $EC_{1:5}$ was measured on 1:5 (w/v) soil to water mixtures at 25°C; C_{org} was obtained using the Walkley and Black method; $EC_{1:5}$ was converted in electrical conductivity of the saturation paste extract (EC_e) using the correlation model proposed by Khorsandi and Yazdi [34] for arid and semiarid environments.

DNA extraction

DNA was extracted from the fresh soil sub-sample with the MoBio Power soil DNA extraction kit following the manufacturer's instructions. The samples were then purified from excess impurities with GeneReleaser [35]. DNA crude extract concentrations were measured using Qubit 2.0 Fluorometer by means of the kit Quanti HS assay Invitrogen, following the manufacturer's instructions. DNA was extracted in duplicates from each soil site, and then pooled and used for the following analytical steps.

Amplification of 16S rRNA genes and pyrosequencing

The V2–V3 region of the 16S rRNA bacterial gene was amplified by PCR. The PCR reaction mixture (50 µl) contained 10 µl 10-fold reaction buffer (Fusion GC buffer, FINNZYMES, Espoo, Finland), 800 mM of each of the four deoxynucleoside triphosphates, 3% DMSO, 1.2 mM of each of the primers, V2 For and V3 Rev, 0.5 U of Phusion hot start high- fidelity DNA Polymerase (FINNZYMES), and 20 ng of isolated DNA as template. The V2–V3 region was amplified with the following set of primers containing the Roche 454 pyrosequencing adaptors (underlined): V2for 5′-CGTATC**GCCTCCCTCGCGCCA**<u>TCAG</u>**ACGACTGCGTAGTGGCGGACGGGTGAGTAA**-3′ and V3rev 5′- CTATGC**GCCTTGCCAGCCCGC**<u>TCAG</u>**AGACGCACTCATTACCGCGGCTG**C-3′ ([31], modified following the instructions of BMR Genomics to better adapt the primers to the 454 project and overcome possible PCR biases. (In bold, the reader can find the starting bp number for each primer modified, adding the tag CGTATC and CTATGC at 5′ and a barcode in the middle of each primer TCAG). The following thermal cycling scheme was used: initial denaturation at 98°C for 5 min, 25 cycles of denaturation at 98°C for 45 s, annealing at 68°C for 45 s, and extension at 72°C for 25 s, followed by a final extension period at 72°C for 5 min. All samples were amplified in two series of triplicates, pooled in equal amounts, and then purified using the Qiaquick PCR purification kit (Qiagen Inc. Chatsworth, CA, USA). Six independent PCR products for each site were combined to minimise the impact of PCR errors. Quantification of the PCR products was performed using the Quant-iT dsDNA BR assay kit (Invitrogen GmbH, Karlsruhe, Germany) and a Qubit fluorometer (Invitrogen GmbH, Karlsruhe, Germany) as recommended by the manufacturer. The samples were stored at −20°C and sent to BMR Genomics s.r.l. (Padova, Italy) for pyrosequencing by means of a Genome Sequencer FLX System platform (454 Life Science Branford, CT, USA).

Analysis of pyrosequencing data: dataset clean up and taxonomic assignments

The obtained sequences were assigned to each site using a custom script developed by BMR Genomics. 9 different sequence files, with an average sequence number of 6108, and an average length of 492 nucleotides per sequence were obtained (Table S2, Table S3). In order to ensure a correct nucleotide assignment in the raw sequence files, 2 control steps were performed. First, the nucleotide distribution along each sequence was analysed. This analysis showed that each file had an unbalanced nucleotide distribution in the first 5–10 bases of each sequence (Fig. S4). Next, the quality distribution along each sequence was analysed in order to identify possible low quality segments. As a result, low quality segments of 50–100 bases were identified in the terminal region of each sequence. Finally, a trimming step was performed using StreamingTrim software [36]. An offset of 10 nucleotides was set in order to remove the first 10 bases of each sequence. After the trimming step, 6062 sequences were collected with an average length of 425 bases.

The last control step performed consisted in the identification of chimeric sequences in the dataset [37]. In order to detect all possible chimeric sequences, a dataset was constructed that contained all the 16s rRNA available genes in the NCBI (National center for Biotechnology Information). This database was used as reference for the UCHIME algorithm [38]. After UCHIME analysis 1936 chimeric sequences were detected and removed from the dataset. As a result, 9 different sequence files (one for each site), containing an average sequence number of 5847, were recovered (for additional details, see Fig. S3).

In order to construct a community data matrix, the cleaned files were analysed using the standalone version of RDP multiclassifier [37]. As the average length of the sequences to be analysed was bigger than 250 nucleotides, an assignment's confidence cutoff of 0.8 was set to perform a much stringent analysis (according to the RDP multiclassifier pipeline: http://rdp.cme.msu.edu/tutorials/classifier/RDPtutorial_MULTICLASSIFIER.html).

Statistical analysis of community data

The community data matrix obtained was used in a series of statistical and ecological evaluations carried out using the R software (http://cran.r-project.org/) [39] and the *vegan* package (http://cran.r-project.org/web/packages/vegan/index.html) [40]. A Rarefaction Analysis [41], based on *genus*-level data (all taxonomic assignments that reached the *genus* level), was performed in order to inspect the different grade of diversity explained in each site. *Richness* [42], *inverse Simpson* [43] and *evenness* [44] indexes were calculated on the same data set. *Richness* index was calculated on the basis of the number of the taxonomic assignments at the *genus* level gained for each site, while *evenness* and *inverse Simpson* were calculated as follows:

$$Evenness = \frac{-\sum_{i=1}^{S} p_i \log p_i}{\log S}$$

$$Inverse\ Simpson = \frac{1}{\sum_{i=1}^{S} p_i^2}$$

Where, S is the site *richness* (defined as the number of genera in each site) and p_i is the proportion of genera in the site [42]. In order to inspect the putative number of unseen *genera* present in the sites, Chao [45] has calculated the index using the following equation:

$$Chao = S + \frac{f_1^2}{2f_2}$$

Where, S is the site *richness* (defined above), f_1 and f_2 are the numbers of genera observed once and twice respectively. The community data matrix has then been transformed into a relative abundance matrix in order to highlight the differences in the community composition correlated to the salinity and pH levels of each site. The relative abundance values have been calculated dividing the number of reads assigned to each *genus* by the total number of reads present in each sequence file, in order to compare these values with each others. Moreover, a Canonical Correlation Analysis (CCA) was run with soil salinity and pH fitted onto the ordination analysis obtained, using the *envfit* function of the *vegan* R package, with a number of permutations of 10000. To emphasise the differences in the composition of the different *taxa* inside the community data matrix, a relative abundance matrix was generated, as described above, considering just the assignments at the Phylum level. This matrix was used to generate a heatmap plot of the ordered sites, following the soil degree of salinity (from the low salinity level, at the bottom, to the high salinity level, at the top). Above the generated heatmap, a coloured bar was placed to indicate the phyla correlated to a high level of salinity and the phyla not strictly correlated to high salinity levels. In addition, a dendrogram showing a cluster distribution was plotted and added to the map, using the Bray-Curtis distance and the UPGMA algorithm.

Furthermore, data on soil properties and on the abundance of microbial phyla were combined as variables in a principal component analysis (PCA), used as exploratory analysis, assuming that microbial groups (absolute abundance values) have a linear response to environmental gradients [46]. The PCA was performed on autoscaled data, and based on Spearman's rank correlation matrix; XLStat 7.0 (Addinsoft, Paris) statistical software was used for the purpose. Spearman's rank correlation test was used to define the degree of dependence among the variables. The advantage of using Spearman's rank correlation coefficient is the independence of the population's distribution, so that the data can be collected over regular spaced intervals. The Spearman's correlation coefficient (rho) was used as a measure of the correlation (dependence) between the variables, giving a value between +1 and −1, where 1 stands for the total positive correlation, 0 stands for no correlation, and −1 stands for the total negative correlation. The rho coefficient is based on the ranks of the observations; the Spearman's rank correlation coefficient does not assume that the relationship among the variables is linear [47].

Results

The 9 soil sites represented nine plots distributed in an area characterised by a great spatial variability, which encompassed different levels of salinity, and a significant variation in soil pH, organic carbon, vegetation type and cover percentage, saline crust percentage, and texture class (Table 1). Electrical conductivity ranged between 169.96 dS m^{-1} to 5.37 dS m^{-1} with values of pH between 6.4 and 8. Total organic Carbon ranged between 4.26 g kg^{-1} and 0.38 g kg^{-1}.

Microbial DNA extraction yielded between 4.5 ng/µl to 10.2 ng/µl, showing variability in DNA recovery. Given the very standardised methodology applied to all the samples during the extraction step, the reason for the yield variability can be searched in the different soil characteristics among the nine sites. A different salt concentration and the presence of a variable amount of organic carbon can modify the degree of interaction between microbial DNA and the used extracts [48]. In order to check whether possible interferences occur between the DNA extraction procedure here adopted and some of the main soil properties, namely Organic Carbon content, pH and Salt concentration, an evaluation of the degree of correlation (Spearman's correlation) between DNA recovery (ng/µl) and measured soil parameters has been carried out. The results (Table S5) excluded a statistically significant positive or negative role of soil organic carbon, pH and salt concentration in DNA recovery during extraction.

Table 1. Main physical and chemical properties of the soil sites and percentage of vegetation and salt crust cover of the sampling sites estimated in the field.

Plot	pH	EC_e	C_{org}	Clay	Silt	Sand	Texture class[1]	Vegetation cover	Salt crust cover
		dS m^{-1}	g kg^{-1}					%	
1	7.9	50.45	2.27	189	177	634	loam	90	70
2	7.0	15.06	1.32	80	98	822	loamy sand	85	5
3	7.2	92.73	0.67	76	74	850	loamy sand	40	100
4	7.4	28.00	4.26	245	294	461	silty loam	100	0
5	7.8	5.82	1.4	83	145	772	loamy sand	100	0
6	6.9	63.50	1.46	369	263	368	silty clay loam	90	0
7	7.2	169.96	2.35	188	217	595	loam	3	100
8	8.0	5.37	0.38	54	38	908	sand	70	0
9	7.5	54.88	1.27	107	105	788	sandy loam	90	100

[1]Classification according to the international scheme (Leeper and Uren, 1993).

General analyses of the pyrosequencing- derived dataset

The average length of the sequences analysed was bigger than 250 nucleotides, more than 75% of the reads analysed was assigned at the Phylum level, and just 25–30% of the reads was assigned to the *genus* level. The description of the assignments' number at each taxonomic level and for each site is reported in Figure 1. A total of 16342 sequences for the nine soil sites was used for bacterial diversity analysis. Figure S4 displays the nucleotide's relative frequency distribution along the sequences, where the first 10 bases of each sequence file have an unbalanced nucleotide distribution, suggesting an uneven distribution of sequences among the sites in the soil. The number of sequences per site ranged from 1968 to 15614, except for site 1. Figure 1 displays the number of sequences assigned by RDP classifier for each site above and below the domain of Bacteria 52617 sequences were assigned to the domain of Bacteria and 42459 of these sequences were classified below the domain level.

Bacterial diversity and richness

Figure 2 shows the rarefaction curves obtained for each soil site. Rarefaction curves were created by randomly re-sampling the pool of N samples multiple times, and then plotting the average number of species found in each site. This method generated the expected number of species in a small collection of n samples drawn at random from the large pool of N samples. Rarefaction curves grow rapidly at first, as the most common species are found, then the curves reach a plateau when the rare species remain to be sampled. In sites 1, 7 and 8 the curves didn't reach saturation, suggesting that taxonomic diversity was not fully exploited. In contrast, the remaining six sites showed rarefaction curves that reached a plateau. In particular, site 9 reached the point of saturation faster than the other sites.

A comparison of the rarefaction analyses with the Chao1 (Fig. 3) index revealed that a substantial fraction of soil sites (6 out of 9) showed a relevant number of putative "unseen genera", in agreement with the rarefaction analysis.

In table 2, the values of biodiversity indexes calculated on the nine soil sites were compared; site 9, despite showing a rarefaction curve that reached the plateau, was characterised by low values of the *Invsimpson* and *evenness* indexes, when compared with the other soil sites. The values of the *Invsimpson* index (Table 2) ranged between 1.6 and 22.7 while the *richness* index ranged between 34 and 148. The *evenness* index ranged from 0.2 to 0.8, indicating an uneven distribution of bacterial genera between the sites.

The comparison of the soil sites based on the values of the *richness* index showed the greatest bacterial richness in site 3, followed by sites 4, 6, and 2. Both *Invsimpson* and *richness* indexes (Table 2) discriminated between the different soil sites, but didn't succeed in emphasising peculiar behaviours of some sites. In the case of site 9, for example, which showed a curve of rarefaction very different from the other sites, the *evenness* index was the only indicator that showed sensitivity to this particular behaviour.

Bacterial phyla distribution along a salinity gradient

The 52623 sequences classified below the domain level were affiliated to 15 bacterial phyla. The dominant phyla across all sites were: *Acidobacteria*, *Actinobacteria*, *Bacteroidetes*, *BRC1*, *Chlor-*

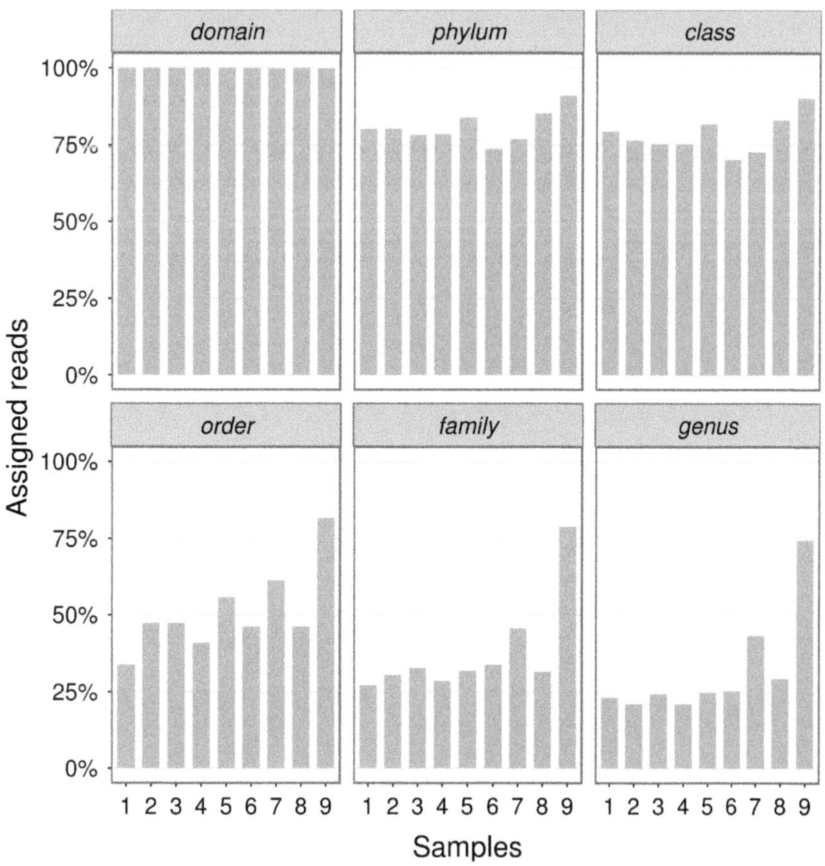

Figure 1. Number of sequences assigned by RDP multiclassifier for each site at each taxonomic level.

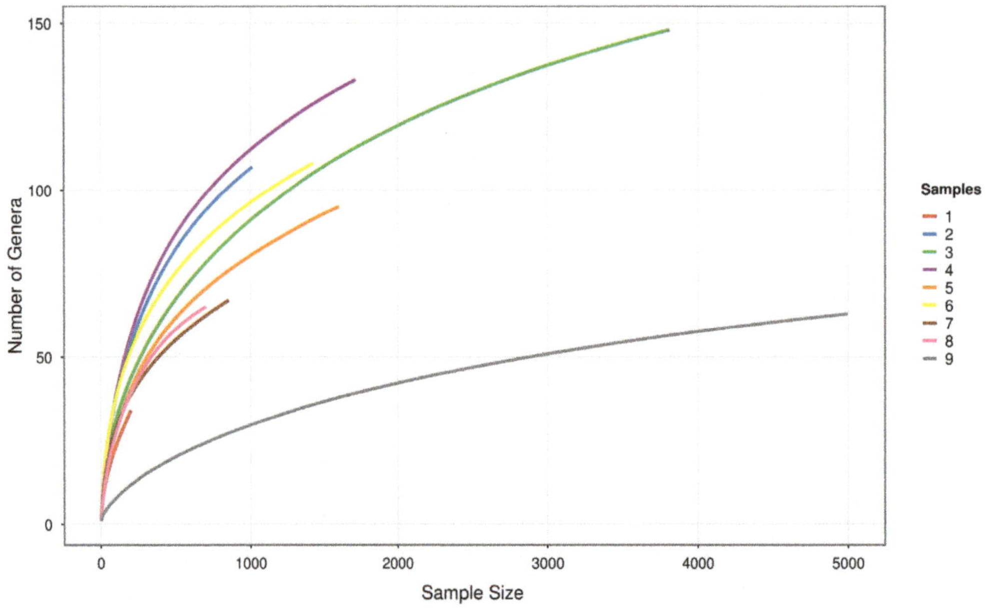

Figure 2. Rarefaction curves of the sites. These curves were obtained considering only the taxonomic assignments that reached the *genus* level in the RDP multiclassifier analysis.

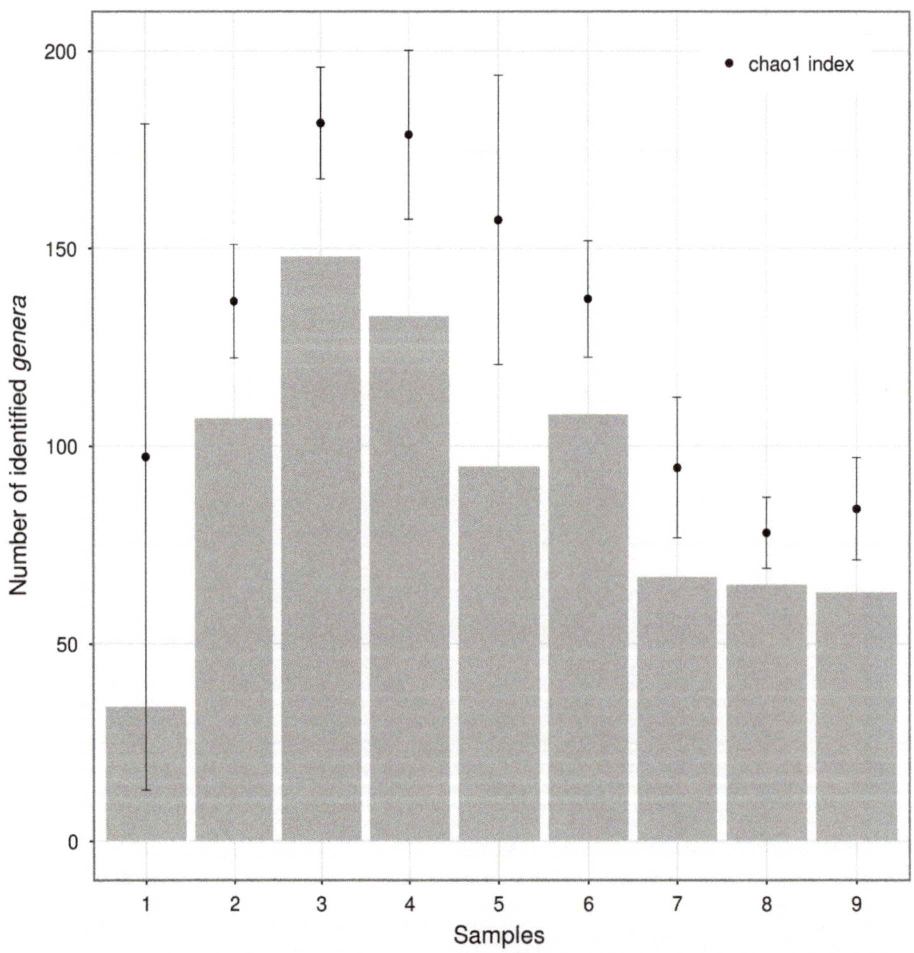

Figure 3. Richness values compared to Chao1 indexes. As shown in this plot almost all sites have a relevant number of putative "unseen *genera*" according to the Rarefaction analysis (Fig. 3).

Table 2. Diversity indexes. Each index has been calculated as reported in the section: "Statistical analysis of community data" of the "Material and Methods" chapter.

	value of biodiversity indices		
Plot	***Invsimpson***	***Richness***	***Evennes***
1	6.67	34	0.70
2	14.78	107	0.74
3	13.72	148	0.66
4	11.04	133	0.70
5	5.17	95	0.58
6	22.71	108	0.77
7	14.22	67	0.74
8	6.41	65	0.65
9	1.57	63	0.21

obi, *Chloroflexi*, *Cyanobacteria*, *Deferribacteres*, *Firmicutes*, *Gemmatimonadates*, *Nitrospira*, *Planctomycetes*, *Proteobacteria*, *Spirochaetes*, *Tenericutes*, *Verrucomicrobia*, *WS3*. The dominant taxa in the analysed soil sites were: *Proteobacteria* (95.95%), *Actinobacteria* (85.39%), *Acidobacteria* (72.12%), *Verrucomicrobia* (70.60%), *Firmicutes* (64.14%), followed by a second group showing a lower but still important percentage of distribution across all the soil sites (*Bacteroidetes*, *Chloroflexi*, *Chlorobi*, *Gemmatomonadates*, with a percentage between 70 and 50%). The taxa with a lower relative distribution were: *Planctomycetes*, *Tenericutes*, *Deferribacteres*, *Cyanobacteria*, *Spirochaetes*, *Nitrospira*, and the uncultured candidate bacterium divisions *WS3* and *BRC1*.

The relative distribution of the groups of bacteria varied among soil sites and along a salinity gradient, as showed in the heatmap of Figure 4, where the relative distribution of *phyla* assignments is clustered and plotted with respect to the different values of salinity of soil sites. Each row (sites) of the heatmap shows the phyla's relative abundance in a soil site. The rows of the heatmap are ordered according to the degree of salinity in the soil sites (from the lowest values, at the *bottom*, to the higher, at the *top*). The heatmap of Figure 4 reports the following annotations: "salinity related" and " salinity unrelated", referring to taxa frequently listed in saline environments and taxa less represented or even absent in salt affected sites, as roughly determined on the basis of what reported in previous studies [12,13,14,15,16,17,18,19,20,21,22,23,24,25,26,27,28]. Moreover, the cluster structure shows five main groups of phyla which share a peculiar composition and abundance among the sites. For example, site 3 presented a very peculiar abundance of three phyla (*Cyanobacteria*, *Deferribacteres*, *Nitrospira*), which resulted unrelated to salinity, while a large group of phyla, that appeared grouped together and uniformly distributed across the other sites, resulted strongly correlated to salinity, although showing a low abundance. In fact, the phyla *Acidobacteria*, *Actinobacteria*, *Firmicutes*, *Proteobacteria*, and *Bacteriodetes* resulted related to all the 9 sites, with almost the same abundance degree. *Spirochaetes* and *Tenericutes* resulted related just to site 7, and the BRC1 candidate Phylum was only found in site 5.

Salinity and pH values of each soil site were used as variables in the CCA ordination analysis, run in order to highlight the influence of these two soil properties on the bacterial community structure, and to highlight the phyla's relative abundance among sites (Fig. 5). The Canonical Correlation Analysis (CCA) (Fig. 5) summarises the joint variation of the two sets of variables, namely soil pH and salinity, in relation to the bacterial phyla ordination (obtained using the *envfit* function of the *vegan* R package, as described in the Materials and Methods Section). The CCA method combines a correspondence analysis with a multivariate regression analysis, taking into account the underlying model that assumes chi-squared dissimilarities among the sites. The plot obtained by CCA allowed us to extract synthetic environmental gradients from the metagenomic data-sets. The gradients are fundamental for succinctly describing and visualising the different habitat preferences (niches) of the phyla; in order to do so we used an ordination diagram that is shown in Figure 5. The graph obtained on the basis of permutations gave a clear picture of the variability expressed by the different sites, showing different ecological niches for both salinity and pH, but also with respect to the inhabiting microbial communities. In particular, Figure 5 showed a different spatial position within the plot of the sites 7, 9 and 1, due to a high salinity level in the first one, and due to a significantly lower pH in the other two sites. It also emerged that the bacterial communities of sites 5 and 8 were characterised by a tendency to low values of salinity, and that there was a core of sites with closer bacterial assemblages and soil characteristics (sites 2, 3, 4, 5, 6 and 8).

Spearman's rank correlation results (Table 3) show the dependence between a relative abundance of bacterial phyla and the soil properties. *Proteobacteria*, *Spirochaetes*, *Tenericutes*, *WS3 Plantomycetes* and *Bacteroidetes* showed a significant positive correlation with salinity. The *Acidobacteria* showed a statistically significant negative correlation with salinity (-0.717 for $p<0.01$). The only bacterial group that showed a relative abundance significantly correlated with soil pH was the *Plantomycetes*, which resulted distributed among sites with a negative correlation to the variable's values (a higher abundance of bacteria was related to lower pH values). Two phyla showed a statistically significant dependence on the organic carbon content (Corg) of soil sites: *Verrucomicrobia* and *Chlorobi*. Interestingly, the Spearman's rank correlation between the degree of salt crust coverage of the sites and the composition of bacterial phyla did not follow the trend showed by the variable ECe (salinity) and, while not showing statistically significant values, it divided the groups of bacteria into two categories, positively dependent and negatively dependent on the presence of a salt crust in the site.

Figure 6 shows the scatter plots of sites obtained by plotting the variables "abundance" and "soil salinity (ECe) for each bacterial phyla. The plots allow us to check for linearity, that is, for a monotonic relationship between the two variables, and to check

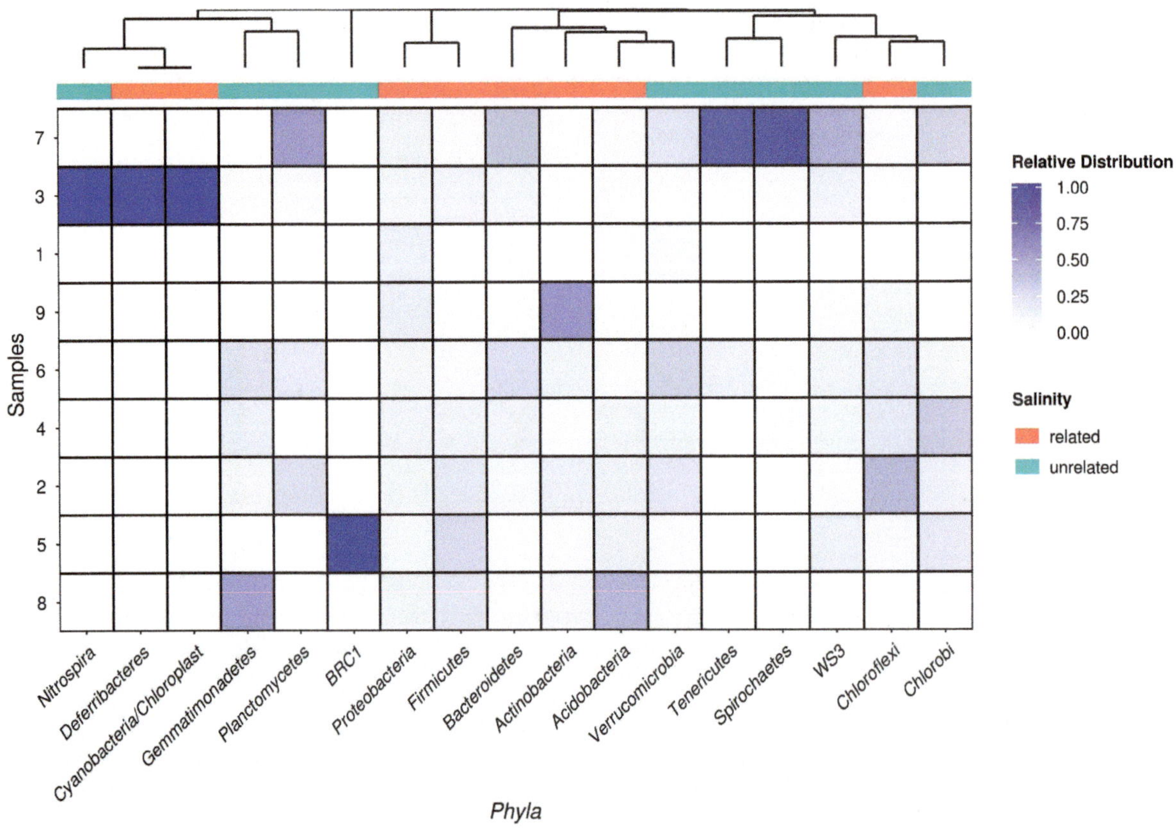

Figure 4. Heatmap of *Phyla* assignments. The heatmap reports the normalized values of the taxonomic assignments at *phylum* level. Each value has been normalized following this criterion:

$$X_{ij}^{norm} = \frac{X_{ij}}{\sum_{k=1}^{N} X_{ik}}$$

Where X_{ij} is the occurrence of the phyla 'j' in the site 'i' and N is the number of site in the dataset (in this case 9). Using this transformation each *phyla* assignment can be compared in all sites independently from its order of magnitude.

whether either the variables increase in value together, or, as one variable value increases, the other variable value decreases. The plots showed in Figure 6 relate to the bacterial phyla showing a significant rho value in the Spearman's rank correlation test. With regards to *Bacteroidetes*, *Spirochaetes* and *Tenericutes,* their relative abundances in the different sites increase positively with the salinity degree in the soil (from site 8, showing one of the lowest values of salinity, through to sites 5, 2, 4, 1, 9, until the sites 6,3, and 7 showing the highest values of salinity). *Proteobacteria* and *WS3* also showed a monotonic relationship between abundances and soil salinity, although some sites (i.e. site 1 for *Proteobacteria*) showed a non-linear behaviour. Furthermore, *Acidobacteria* showed an opposite trend, as relative abundances decrease with the increase of soil salinity, as also indicated by the Spearman's rank correlation test (Table 3).

A principal component analysis is reported as biplot in Figure 7. The bacterial phyla abundances in each sample were used as variables, together with soil chemical properties (C_{org}, pH, vegetation cover, salt crust, ECe), while the soil sites were showed as observations.

The biplot, obtained using the first two components, which together explained about 52.31% of the total variability of data, showed that salinity (ECe variable) is correlated with salt crust presence, and that both variables are positively correlated with the values of PC1 axes. Soil pH appeared correlated to the vegetation cover, and both variables are negatively correlated to the PC1 axes. Soil organic matter content (C_{org}) is positively correlated to PC2 axes. The loading of the sites on the principal components axes confirms the main patterns that were delineated by both Spearman's rank correlation analysis (Fig. 6; Table 3) and the cluster analysis (Fig. 4). The variability of soil properties between the 9 sites corresponded to a diverse relative abundance of bacterial phyla, which actually showed a different degree of correlation with soil characteristics. In some cases, there is an inverse relationship between bacterial phyla, thus showing a kind of vicariousness, suggesting that the presence of some groups of bacteria in a site could be the reason for the absence of others. Rather the opposite behaviour is found in some phyla that are strongly associated in the biplot, and seem to occur in the same sites and when the same micro-environmental conditions occur.

The percentages of the different bacterial phyla present in each soil site, and the percentages observed in the whole study site (values consisting of an average of what obtained in the nine sites) are reported in Figure 8, where, for each site, a pie-plot is showed, representing the bacterial community as emerged by the pyrosequencing analysis of the V2–V3 16S rRNA bacterial gene

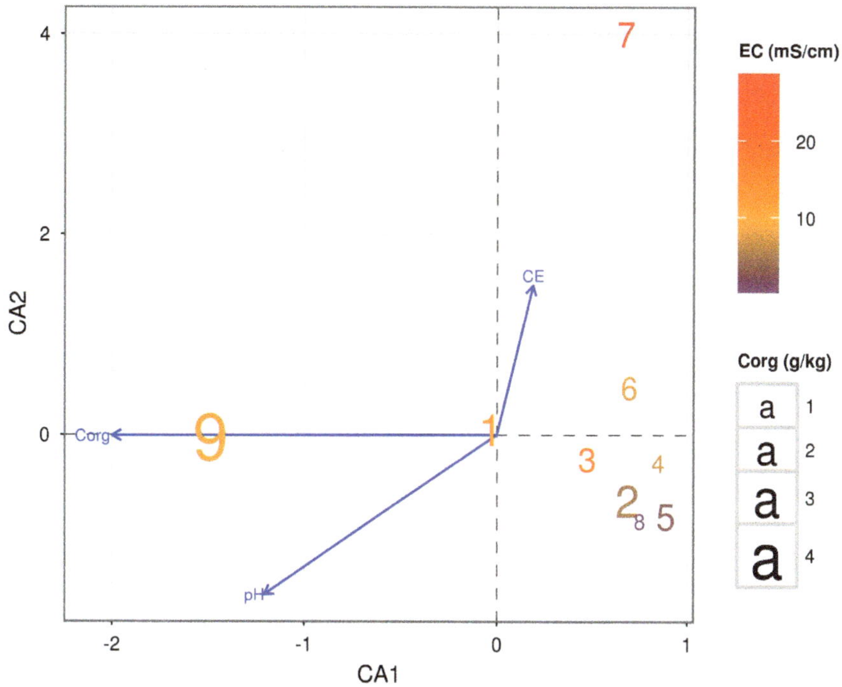

Figure 5. Canonical Correlation Analysis (CCA) based on community data matrix. The salinity levels and the pH levels of each site have been fitted onto the CCA ordination analysis in order to inspect the relevance of these two factors in relation to the bacterial communities distribution.

region; each slice of the pie represents a bacterial phyla. The large pie represents the average of the entire microbial community of the saline soil, and shows the labels for each slice/bacterial phyla that are also valid for the pies of each site. At a glance, if one compares the average composition of the saline soil under study and the composition of the bacterial community in each sampled

Table 3. Spearman's rank correlations between the relative abundances of the six most abundant bacterial phyla and the soil properties across the nine soil sites.

Taxonomic group	Correlation values				
	ECe dS m−1	pH	Corg	Salt Crust cover	Vegetation Cover
Proteobacteria	0.500	0,142	0.317	0.541	0.034
Actinobacteria	−0.317	−0.226	−0.467	−0.213	0.196
Acidobacteria	**−0.717****	0.084	−0.333	−0.638	0.009
Verrucomicrobia	0.217	−0.603	**0.667****	−0.301	−0.17
Firmicutes	−0.650	0.109	−0.367	−0.585	0.051
Bacteroidetes	0.533	−0.661	0.467	−0.027	−0.400
Chloroflexi	−0.033	−0.655	0.301	−0.236	0.393
Chlorobi	0	−0.345	**0.722***	−0.319	0.276
Gemmatimonadetes	−0.331	−0.179	−0.235	−0.625	−0.049
WS3	0.536	−0.445	0.301	0.183	−0.103
Plantomyccetes	0.566	**−0.816****	0.146	0.330	−0.634
Tenericutes	0.688*	−0.586	0.444	0.079	−0.231
Defferibacteres	0.411	−0.206	−0.411	0.437	−0.420
Cyanobacteria	0.411	−0.206	−0.411	0.437	−0.420
Spirochaetes	**0.712***	−0.468	0.390	0.271	−0.208
Nitrospira	0.411	−0.206	−0.411	0.437	−0.420
BRC1	−0.411	0.275	0	−0.364	0.490

In the table is reported the rho value. The significant correlation values are indicated as follows: * $P<0.05$; ** P 0.01.

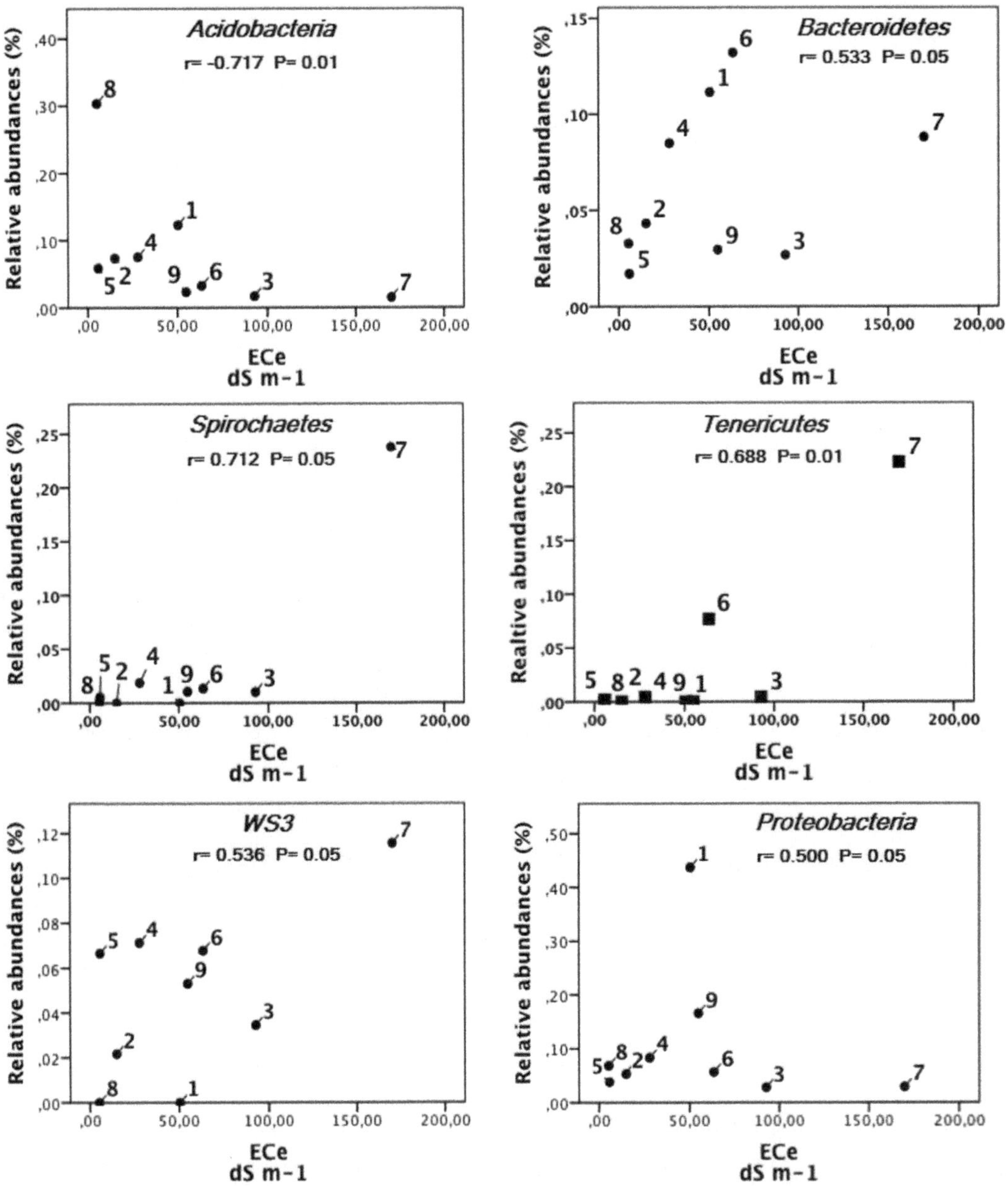

Figure 6. Correlations between relative abundances of different taxonomic groups and soil salinity. Circles represent the soil sites. Spearman rank correlation coefficient (r) with the related P values are shown for each taxonomic group.

site, a considerable spatial variability is noted. What is visually evident from Figure 8 is the way the microbial communities differ greatly both qualitatively and quantitatively from site to site, although the sites are only fifty meters away. Although some groups (like *Proteobacteria* and *Actinobacteria*) were dominant in almost all the sites, a very different composition of the microbial communities among the sites appeared. The graphs also show, from site to site, the relationships among the major microbial groups, and that those who dominate in one site can be absent or just visible in the next one. In each site, given the spatial proximity, it is likely that certain species of bacteria prevail better than others, but this likelihood is apparently expressed in a different way from point to point, probably depending on local environmental forces, and also on positive and/or negative interactions between the different microbial groups.

Discussion

Relative abundances of the phyla found in soils naturally affected by salt, and spatial variability

Some of the bacterial phyla found in the salt-affected soil under examination were also reported by Ma and Gong [12]. Using meta-analysis, these authors retrieved 10,082 sequences longer than 250 bp from the two databases GenBank and RDP, using the search terms "saline" OR "hyper saline", AND "soil" AND "16S". Ma and Gong [12] reported that 90% of the bacterial sequences they enumerated belonged to six phyla, namely *Proteobacteria*, *Actinobacteria*, *Firmicutes*, *Acidobacteria*, *Bacteroidetes*, and *Chloroflexi*, which were also detected in our study.

All the bacterial phyla reported as "salinity related" in previous studies were enumerated in the present survey, and, in addition,

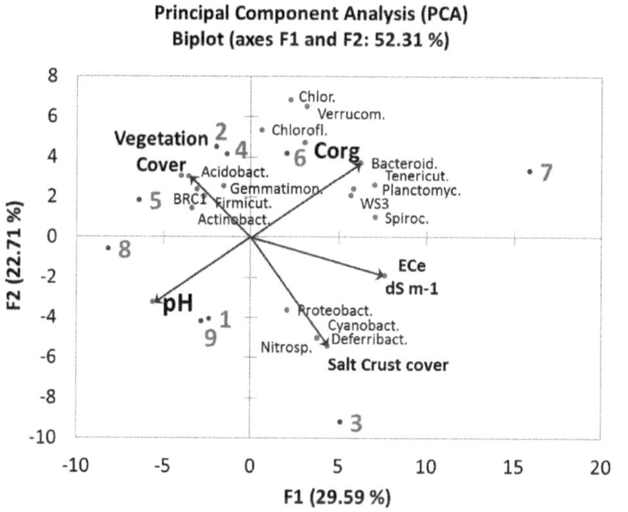

Figure 7. Principal component analysis of bacterial communities as affected by soil properties, based on the abundance of bacterial phyla. Every vector points to the direction of increase for a given variable so that soil sites with similar bacterial communities, are localized in similar positions in the diagram.

some phyla that, according to different authors [15,22], cannot be classified as "salinity related" were also found. This wide group contains the following phyla: *Nitrospira, Deferribacteres, Cyanobacteria/Chloroplast, Gemmatimonadetes, Planctomycetes, BRC1, Verrucomicrobia, Tenericutes, Spirochaetes, WS3* and *Chlorobi*. What is more, the following bacterial phyla were significantly related to soil salinity for the first time: *Nitrospira, Deferribacteres, Cyanobacteria/Chloroplast, Tenericutes* and *Spirochaete*. These phyla were abundant in the two sites showing the highest salinity grade (sites 7 and 3). On the other hand, in the present survey, some phyla classified as "salinity related" by some authors [12] showed an equal distribution all over the sites, apparently uninfluenced by the degree of salinity.

Proteobacteria seemed to be one of the most common bacterial taxon in saline soils [12,49]. In the present study, the occurrence of *Proteobacteria* is actually the highest, with 95.95% of frequencies in the sites, followed by the *Actinobacteria* that represented the second most spread taxon, as it was present in 83.39% of the sites. In contrast with what reported by other meta-analysis studies [12,50], the third largest phylum was *Acidobacteria* (72.12%), followed by *Verrucomicrobia* (70.60%), *Gemmatimonodates* (66.14%), *Firmicutes* (64.14%), *Chloroflexi* (62.69%), *Bacteroidetes* (56.62%), and *Chlorobi* (54.09%). In addition to the nine phyla reported above, 8 phyla characterised by a relative patchy abundance were observed. These taxa with a leopard-spot distribution among the sites had been related to hypersaline soil [12,13,17,23,24,25,27] by other authors. This is the case of groups like *Cyanobacteria* and *Deferribacteres* which characterised site 3, or the *BRC1* group which represented an isolated "spot" in the study area, being present just in site 5. These findings indicated that testing the variability of microbial community in a saline soil using a spatial scale was a successful operation, since it was effective in providing a picture of the subdivision of the microbial community according to a micro-environmental gradient. Although we examined bacterial communities at coarse levels of taxonomic resolution, we expect that a distribution of the bacterial groups based on a mosaic-like scheme would also apply at finer levels of taxonomic resolution.

We noticed that *Proteobacteria* and *Actinobacteria* were common denominators among the investigated sites, being dominant in soils with both high and low salinity. Both phyla contain representatives of the most abundant halophilic bacteria occurring in saline soils [12]. The *Chloroflexi* group, which was found by some authors in hypersaline wastewater [12,49], and is known to be a potential phototroph, was fairly well distributed between the different sites. *Firmicutes* can also be considered common denominators in all the nine sites, because they resulted homogeneously spread in the soil sites. This phylum was apparently absent in a number of hypersaline environments previously investigated [15,46], but it was found to be abundant in low salinity environments such as deep-sea sediments [21]. Among the dominant genera assigned to *Firmicutes*, the *Bacillus* outstands, as it proves to be an important resource for exploring halophilic enzymes and metabolic pathways for pollutant remediation in saline soil [22].

The phyla *Gemmatimonodates* is known for having members showing active roles in biogeochemical transformations, especially in hypersaline soils, where it was undeservingly described as a minor phylum [26]. The assignments attributed to this phylum showed a patchy distribution in the studied area, with a relatively high abundance in site 8 and in other sites characterised by a low salinity level. Another phylum showing a discontinuous distribution was *Planctomycetes,* found in previous studies as symbionts of marine algae and sponges [16,20]. *Nitrospira* dominated site 3 which was the site most affected by the salt crust coverage. Not surprisingly, *Nitrospira* phylum is represented by nitrite-oxidising bacterial species with a marked chemolithoautrophic nature. *Cyanobacteria* presented an identical distribution as that of *Nitrospira* and *Deferribactereres*, which were found as dominant groups in site 3. *Cyanobacteria* is a phylum represented by oxygen evolving and chlorophyll containing photosynthetic bacteria [17,24,25,46]. While *Nitrospira* and *Deferribactereres*, have an important role in biogeochemical cycles, being ammonia-oxidising bacteria (AOB) in saline soils [23] and sulphur oxidising, respectively. The phylum *Deferribacteres* comprehends also chemoorganotrophic heterotrophs that breath anaerobically with terminal electron acceptors including Fe(II), Mn(IV), SO, Co(III), and nitrate [23]. The phylum *Verrucomicrobia* showed, on the one hand, a uniform distribution among the considered sites, with no correlation to salinity, but also, on the other hand, a highly significant dependence on organic matter, with a Spearman's rho value for C_{org} of 0.667, for $p>0.01$. The bacteria belonging to this group is likely to overcome the selection imposed by the high salinity by means of an intimate association with the organic matter of the soil, and perhaps with a direct involvement in the carbon cycle. *Spirochaetes* showed a particular distribution, as they were present with low abundance in almost all sites, but also showing a particular association with the site 7, where the phylum exhibited a very high abundance and a correlation with the presence of another bacterial phylum, the *Tenericutes*. *Spirochetes* are widely distributed in nature; presumably they play an important role as free-living microbes in environments such as soil. The *Spirochaeta* phylum also contains moderately halophilic bacteria, such as species of the *Halomonas* and *Deleya* genera, being members of the gamma subclass of *Proteobacteria* [28,51]. Facultative aerobic halophilic *Spirochaeta* bacteria were isolated by various authors, close to salted lakes [18,28], although the phylum had never been recovered from saline soils, even according to meta-analysis based studies. *Tenericutes* is a phylum of bacteria that contains the Class *Mollicutes* and that, as reported above, presented almost the same distribution of *Spirochaeta*. *Tenericutes* comprehend denitrifying bacteria, and, due to the lack

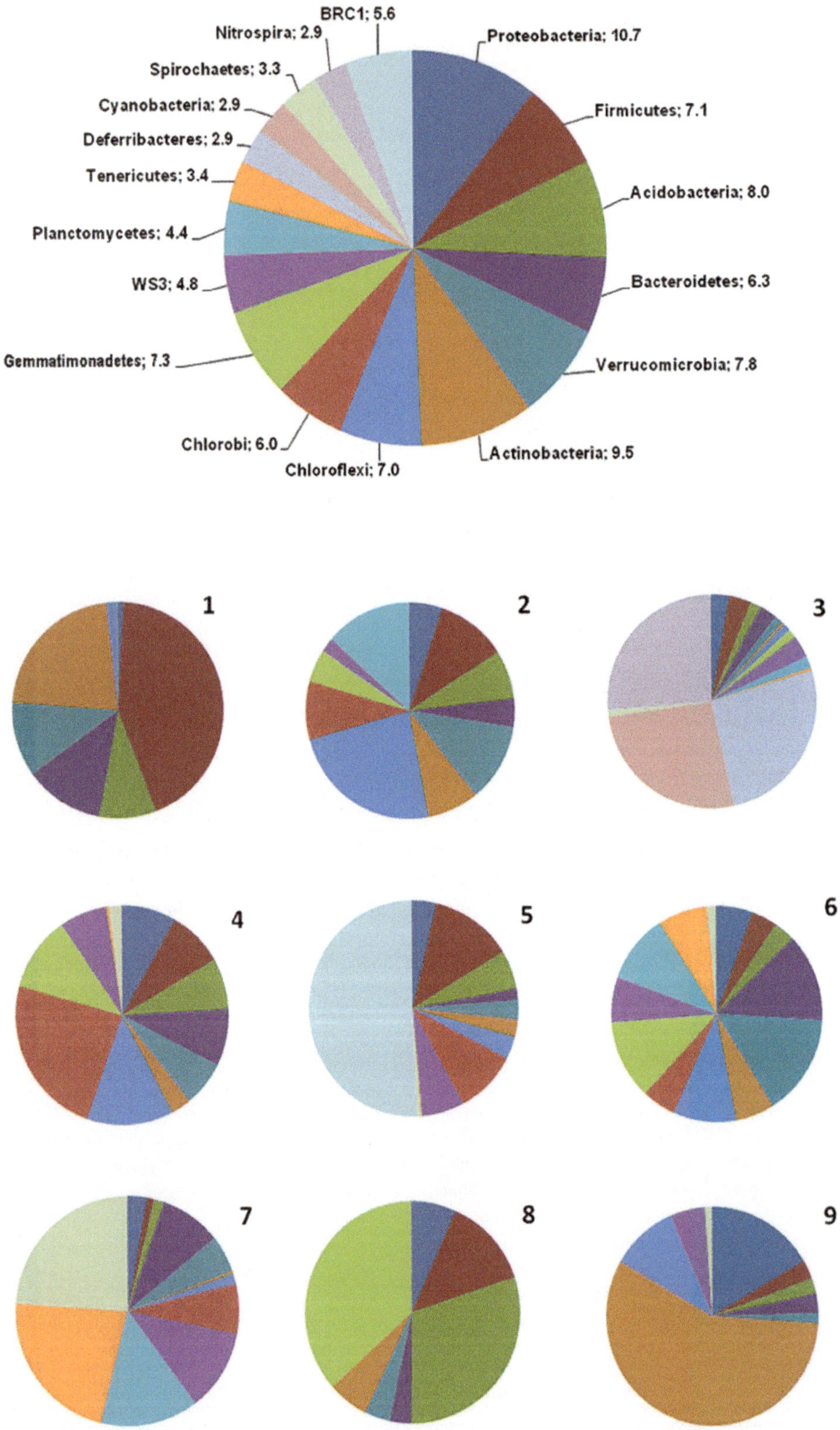

Figure 8. Piecharts based on the abundance (%) of bacterial phyla. Every pie shows the percent of the abundance (%) of the bacteria phyla.

of a cell wall, they are more sensitive to osmotic stress [28]. The correlation with *Spirochaetes* suggests a symbiotic or parasitic role of *Tenericutes* with respect to *Spirochaetes*.

A further interpretation key for understanding the relationship between the relative abundances of these bacterial phyla and salinity and the other soil properties, is given by the illuminating work done by Noah Fierer and colleagues [52] who suggested that certain bacterial phyla can be differentiated into copiotrophic and oligotrophic categories that correspond to the r- and K-selected categories used to describe the ecological attributes of plants and animals. By applying the copiotroph–oligotroph concept to saline soil phyla, we can further understand the structure and function of soil bacterial communities in extreme conditions, and in discontinuous/patchy environments. Copiotrophic bacteria that have higher growth rates, a greater degree of variability in population size, and lower substrate affinities than oligotrophic bacteria, should be dominating when there is abundance of nutritive substrates and, in general, a non-limiting situation. The oligotrophs should increase in relative abundance, as substrate quality and/or quantity declines over time and harsh environmental conditions prevail. Fierer et al. [52] found that bacteria belonging to the *Acidobacteria* phylum showed an oligotrophic behaviour while β-Proteobacteria, and *Bacteroidetes* exhibited copiotrophic attributes, changing their abundances in a predictable manner to changes in soil C availability. Moreover, Fierer et al. [52] found that, across 71 different samples, the *Acidobacteria* were less abundant, while the beta-Proteobacteria and the *Bacteroidetes* were more abundant. This study confirms the abundance of *Proteobacteria* all over the sites, and also shows a considerable presence of *Actinobacteria* as the second most spread taxon, and of *Acidobacteria* as the third largest phylum. Both *Acidobacteria* and *Actinobacteria* frequencies were not correlated to soil organic carbon contents (C_{org}), while a strong dependence to C_{org} is exhibited by *Verrucomicrobia* and *Chlorobi* phyla.

Conclusions

In the latest years, saline soils received a great attention because of the general shortage of arable land, and of the increasing demand for ecological restoration of areas affected by secondary salinisation processes. This is due to the fact that naturally salt-affected soils have a biotechnological potential in their microbial communities, which represent not only a gene reserve for future exploitation in biotechnological applications, assuming they could be used in some kind of restoration or conservation techniques of saline environments, but they can also serve as model systems for exploring the relationships between diversity and activity at the soil level in selective/limiting situations. As outlined in the introduction, very few studies succeeded in addressing the beta diversity of the microbial species in soils, according to the different salt concentrations and, at a different scale, to bacterial taxa distribution in relation to salinity gradients [12].

Although some of the enumerated phyla related to saline soils have already been found by other authors, this study complements the limited information available on these extreme habitats by providing specific information on the type of distribution of different bacterial groups as a function of spatial gradients in salinity and pH. The analysis of bacterial 16S rRNA-based datasets obtained from a naturally saline soil revealed significant differences in bacterial community composition and diversity, along an increasing salinity level, which underlies a multi-scale spatial variability. What is more, a spatial heterogeneity of microbial communities at a relatively small scale has emerged from this study, especially with respect to the macro-scale environmental scheme in terms of geography and soil. The soil of the study showed a patchy distribution of the vegetation structure and of chemical properties, which coincided with an heterogeneous distribution of many bacterial groups. Some bacterial phyla appeared, however, spread in the whole study area.

It is possible to make some assumptions that could be the basis for future in-depth studies on the association between groups of bacteria, or on their variance in certain extreme environments. The first assumption is that spatial autocorrelation in terms of microbial diversity can hardly be found at the soil scales used for physical-chemical studies. According to some authors [53], spatial autocorrelation in soil ranges from 30 cm to more than 6 m, depending on the sampling extent considered. In some locations, Franklin et al. [53] found up to four different correlation length scales. The presence of nested scales of variability suggests that the environmental factors regulating the development of the communities in the saline soil of the present study may have operated at different scales. The presence of spatial patterns in the distribution of bacteria was demonstrated at the microscale by Nunan et al. [54] who showed ranges of spatial autocorrelation of 1 mm and below. The second assumption is that an environment in which some limiting factors favour some microbial groups and not others is in fact compared to a set of islands that allow the formation of different communities, separated from each other by the discontinuity of the chemical-physical factors and by the availability of nutrients. One could imagine that in spite of the same element of "noise" (salinity), the spatial discontinuity allows the formation of more possible microbial assortments. Therefore, a patchy saline environment can contain not just a single microbial community selected to withstand extreme osmotic phenomena, but many different though efficient communities.The occurrence of a significant number of "salinity unrelated" phyla (e.g. *Nitrospira*, *Spirochaetes*) captures our interest, therefore we strongly believe a further analysis, and a further step in metatrascriptomic of functional genes, are needed.

Responding to the initial question on the role of salt concentration in defining the diversity of the bacterial community in a saline soil, we can say that salinity had the strongest effect on bacterial community structure, as revealed by the study of the correlation between soil properties and bacterial phyla occurrence. Soil pH and other chemical properties seemed to have a minor impact on bacterial group distribution when analysed at the considered spatial scale. The relative abundances of a number of taxonomic groups, as a matter of fact, changed significantly between soil sites according to differences in soil salt content. Nevertheless, the abundance of some other taxa resulted almost unaffected by the salinity level (e.g. *BRC1*, *Gemmatimonodates*). This may indicate, on the one hand, a high plasticity of bacterial phyla that evidently possess genera and species adaptable to different conditions, while on the other hand that the sensitivity to salinity of some groups is poor or, in any case, less dependent on other factors, such as the presence of organic matter, plant roots, etc.

Furthermore, it is not certain that bacterial phyla co-occurring at a given site occupy the same ecological niche; rather, the spatial variability can indicate the existence of different scales in the distribution of some major environmental factors, just as the salinity factor. In any case it is evident that the correlation of some groups (*Nitrospira*, *Deferribacteres*, *Spirochaetes*) to the degree of salinity seems to be a necessary condition for the proliferation of the species belonging to those particular groupings.

In conclusion, we feel the need to deepen the scale at which we analyse the bacterial communities in extreme environments. To go back to the more general discussion on saline system ecology, and to the measurement of the "extent of species replacement or biotic change along environmental gradients", which corresponds to the beta diversity *sensu* Whittaker [55], one should distinguish between two rather antithetical phenomena: nestedness and turnover. In the saline soil here studied, we have seen that nestedness occurred only for some taxa, when the biota of a site with a lower number of representatives was a subset of a biota with a greater number of elements of the same taxa (i.e. *Bacteroidetes*, *Chloroflexi*, *Chlorobi*, *Gemmatomonadates*). In this case, the dissimilarity between two sites is related to the difference in specific *richness*, and it occurs even in the absence of a real turnover of species. In contrast, the spatial turnover implies that he replacement of some species by others can easily occur in a mutable environment, where rain and water movements can strongly change the distribution of salts, although it requires a different experimental scheme, with time-related samplings.

It appears evident that the assortment and distribution of microorganisms in a heavily fragmented environment depend on very complex dynamics of colonisation and dispersion, and that the analysis of the correlation between the population of microorganisms and environmental parameters, such as the organic matter, pH, and salinity, adds important information that can help to unravel the mechanisms of formation and structure of the bacterial communities.

Supporting Information

Figure S1 Location of the study area.

Figure S2 Sampling scheme.

Figure S3 Sequences lost during "quality refinement" steps. The piecharts report the fraction of the sequences maintained and the fraction of sequences lost during quality refinement steps. The green portion of each piechart is the maintained portion of sequences (approximately more than 90% of the total sequences) while the other two portions (the red and the blue ones) are the portion of sequences lost during the trimming and the chimera check steps, respectively.

Figure S4 Nucleotide relative frequency distribution along the sequences. The first 10 bases of each sequences file showed an unbalanced nucleotide distribution.

Author Contributions

Conceived and designed the experiments: LC GB AB. Performed the experiments: LC GB GLP. Analyzed the data: LC GB FP GLP. Contributed reagents/materials/analysis tools: CD AB GLP. Contributed to the writing of the manuscript: LC FP GB.

References

1. Richards LA (1954) Diagnosis and improvement of saline and alkali soils. US Salinity Lab., US Department of Agriculture Handbook 60 California, USA.
2. IUSS Working Group WRB (2007) World Reference Base for Soil Resources 2006, first update. 2007. World Soil Resources Reports No. 103. FAO, Rome.
3. Soil Survey Staff (2010) Keys to Soil Taxonomy, 11th ed. USDA-Natural Resources Conservation Service, Washington, DC.
4. Mavi MS, Sandarman J, Chittleborough DJ, Cox JW, Marchner P (2012) Sorption of dissolved organic matter in salt-affected soils: Effect of salinity, sodicity and texture. Sci. Total Environ. 435–436.
5. Rietz DN, Haynes RJ (2003) Effects of irrigation induced salinity and sodicity on soil microbial activity. Soil Biol. Biochem. 35, 845–854.
6. Tripathi S, Kumari S, Chakraborty A, Gupta A, Chakrabarti K, et al. (2006) Microbial biomass and its activities in salt-affected coastal soils. Biol. Fert. Soils. 42: 273–277.
7. Dion P, Nautiyal CS (2008) Microbiology of Extreme Soils. Soil Biology 13 3© Springer-Verlag Berlin Heidelberg.
8. Ettema CH, Wardle DA (2002) Spatial soil ecology. Trends Ecol Evol 17: 177–183.
9. Parkin TB (1993) Spatial variability of microbial processes in soil- a review. J. Environ. Qual. 22: 409–417.
10. Pereira MC S, Cavalcante AF D, van Elsas JD, Salles JF (2012) Spatial and temporal variation of archaeal, bacterial and fungal communities in agricultural soils. Plos One, 7, 12.
11. Lozupone CA, Knight R (2007) Global patterns in bacterial diversity. P. Natl. Acad. Sci. USA. 104: 11436–11440.
12. Ma B, Gong J (2013) A meta- analysis of the publicly available bacterial and archaeal sequence diversity in saline soils. World J Microbiol Biotechnol. DOI 10.1007/s11274-013-1399-9.
13. Benlloch S, Lopez-Lopez A, Casamayor EO, Ovreas L, Goddard V, et al. (2002) Prokaryotic genetic diversity throughout the salinity gradient of a coastal solar saltern. Environ Microbiol 4(6), 349–360.
14. de los Rios A, Valea S, Ascaso C, Davila A, Kastovsky J, et al. (2010) Comparative analysis of the microbial communities inhabiting halite evaporates of the Atacama deserts. International Microbiol 13: 79–89.
15. Demergasso C, Casamayor EO, Chong G, Galleguillos P, Escudero L, et al. (2004) Distribution of prokaryotic genetic diversity in athalassohaline lakes of the Atacama Desert, Northern Chile. FEMS Micro- biol Ecol 48: 57–69.
16. Fukunaga Y, Kurahashi M, Sakiyama Y, Ohuchi M, Yokota A, et al. (2009) Phycisphaera mikurensis gen. nov., sp. nov., isolated from a marine alga, and proposal of Phycisphaeraceae fam. nov., Phycisphaerales ord. nov. and Phycisphaerae classis nov. in the phylum Planctomycetes. J Gen Appl Microbiol 55: 267–275.
17. Garcia-Pichel F, Cortes AL, Nubel U (2001) Phylogenetic and morphological diversity of cyanobacteria in soil desert crusts from the Colorado plateau. Appl. Environ. Microbiol. 67: 1902–1910.
18. Greenberg EP, Canale-Parola E (1976) *Spirochaeta halophila* sp. nov., a fermentative anaerobic bacterium from a high-salinity pond. Arch. Microbiol. 110: 185–194.
19. Hollister EB, Engledow AS, Hammett A, Provin JM, Wilkinson TL, et al. (2010) Shifts in microbial community structure along an ecological gradient of hypersaline soils and sediments. The ISME Journal 4, 829–838.
20. Jogler C, Glockner FO, Kolter R (2011) Characterization of *Planctomyces limnophilus* and development of genetic tools for its manipulation establish it as a model species for the phylum *Planctomycetes*. Appl Environ Microbiol 77: 5826–5829.
21. Li LN, Kato C, Horikoshi K (1999) Bacterial diversity in deep-sea sediments from different depths. Biodivers Conserv 8(5): 659–677
22. Liszka M, Clark M, Schneider E, Clark DS (2012) Nature versus nurture: developing enzymes that function under extreme conditions. Ann Rev Chem Biomol Eng 3: 77–102
23. Mohamed NM, Saito K, Tal Y, Hill RT (2010) Diversity of aerobic and anaerobic ammonia-oxidizing bacteria in marine sponges. ISME J 4: 38–48.
24. Nagy ML, Pérez A, Garcia- Pichel F (2005) The prokaryotic diversity of biological soil crusts in the Sonoran Desert (Organ pice cactus national monument, AZ). FEMS Microbiol Ecol 233–245.
25. Srivastava AK, Bhargava P, Kumar A, Rai Lal C, Neilan BA (2009) Molecular characterization and the effect of salinity on cyanobacterial diversity in the rice fields of Eastern Uttar India. Saline Systems 5: 4.
26. Zhang H, Sekiguchi Y, Hanada S, Hugenholtz P, Kim H, et al. (2003) *Gemmatimonas aurantiaca* gen. nov., sp nov., a gram-negative, aerobic, polyphosphate-accumulating micro-organism, the first cultured representative of the new bacterial phylum Gemmatimonadetes phyl. nov. Int J Syst Evol Microbiol 53: 1155–1163.
27. Yousuf B, Sanadhya P, Keshri J, Jha B (2012) Comparative molecular analysis of chemolithoautitrophic bacterial diversity and community structure from coastal saline soils, Gujarat, India. BMC Microbiol 12: 150.
28. Ventosa A, Nieto JJ, Oren A (1998) Biology of moderately halophilic aerobic bacteria. Microbiol. Mol. Biol. Rev. 62: 504–544.
29. Roesch LFW, Fulthorpe RR, Riva A, Casella G, Hadwin AKM, et al. (2007) Pyrosequencing enumerates and contrasts soil microbial diversity. ISME J 1: 283–290.
30. Acosta-Martinez V, Dowd S, Sun Y, Allen V (2008) Tag-encoded pyrosequencing analysis of bacterial diversity in a single soil type as affected by management and land use. Soil Biol Biochem 40: 2762–2770.
31. Nacke H, Thürmer A, Wollherr A, Will C, Hodac L, et al. (2011) Pyrosequencing- based assessment of bacterial community structure along a different management types in German forest and grassland soils. 6, 2 e17000.
32. FAO (2006) Guidelines for Soil Description (Fourth edition). Food and Agriculture Organisation of the United States, Rome.
33. U.S. Salinity Lab Staff (1954) Diagnosis and improvement of saline and alkali soils. Agr. Handbook 60, USDA, Washington, DC, 122–124.

34. Khorsandi F, Yazdi FA (2011) Estimation of Saturated Paste Extracts' Electrical Conductivity from 1: 5 Soil/Water Suspension and Gypsum. Communications in Soil Science and Plant Analysis 42(3), 315–321.
35. Zidan WMH, Radwan SMA, El-Khawas H, Zahra MK, Badr El-Din SM (2005) Extraction of Microbial Community DNA from Soil for Polymerase Chain Reaction. Egyptian Journal of Applied Science 20(10B): 430–441.
36. Bacci G, Bazzicalupo M, Benedetti A, Mengoni A (2013) StreamingTrim 1.0: a Java software for dynamic trimming of 16S rRNA sequence data from metagenetic studies. Molecular ecology resources.
37. Wang Q, Garrity GM, Tiedje JM, Cole JR (2007) Naive Bayesian classifier for rapid assignment of rRNA sequences into the new bacterial taxonomy. Applied and environmental microbiology, 73(16), 5261–5267.
38. Edgar RC, Haas BJ, Clemente JC, Quince C, Knight R (2001) UCHIME improves sensitivity and speed of chimera detection. Bioinformatics, 27(16), 2194–2200.
39. R Core Team (2013) R: A language and environment for statistical computing. R Foundation for Statistical Computing, Vienna, Austria. URL http://www.R-project.org/.
40. Oksanen J, Guillaume Blanchet F, Kindt R, Legendre P, Minchin PR, et al. (2013) vegan: Community Ecology Package. R package version 2.0–10. http://CRAN.R-project.org/package = vegan
41. Gotelli NJ, Colwell RK (2001) Quantifying biodiversity: procedures and pitfalls in the measurement and comparison of species richness. Ecology letters 4(4): 379–391.
42. Tuomisto H (2010) A diversity of beta diversities: straightening up a concept gone awry. Part 1. Defining beta diversity as a function of alpha and gamma diversity. Ecography, 33(1): 2–22.
43. Simpson EH (1949) Measurement of diversity. Nature.
44. Magurran AE (2004) Measuring biological diversity. Blackwell Science, Oxford.
45. Chao A, Gotelli NJ, Hsieh TC, Sander EL, Ma KH, et al. (2014) Rarefaction and extrapolation with Hill numbers: a framework for sampling and estimation in species diversity studies." Ecological Monographs 84(1): 45–67.
46. Ramette A (2007) Multivariate analyses in microbial ecology, Minireview. FEMS Microbiol Ecol 62: 142–16.
47. Gautier TD (2001) Detecting trends using Spearman's rank correlation coefficient. Environ Forensics 2: 359–362.
48. Purohit MK, Singh SP (2009) Assessment of various methods for extraction of metagenomic DNA from saline habitats of coastal Gujarat (India) to explore molecular diversity. Letters in Applied Microbiology 49: 338–34.
49. Valenzuela-Encinas C, Neria-Gonzalez I, Alcantara-Hernandez RJ, Estrada-Alvarado I, Zavala-Diaz de la Serna FJ, et al. (2009) Changes in the bacterial populations of the highly alkaline saline soil of the former lake Texcoco (Mexico) following flooding. Extremophiles 13: 609–621.
50. Siddikee MA, Tipayno SC, Kim K, Chung JB, Sa T (2011) Influence of varying degree of salinity-sodicity stress on enzyme activities and bacterial populations of coastal soils of Yellow Sea, South Korea. J Microbiol Biotechnol 21: 341–346.
51. DasSarma S, Arora P (2001) Halophiles. Encyclopedia of Life sciences.
52. Fierer N, Bradford MA, Jackson RB (2007) Toward and ecological classification of soil bacteria. Ecology 88 (6) 1354–1364.
53. Franklin RB, Mills AL (2003) Multi-scale variation in spatial heterogeneity for microbial community structure in an eastern Virginia agricultural field. FEMS Microbiology Ecology 44: 335–346.
54. Nunan N, Kejian Wu, Young IM, Crawford JW, Ritz K (2003) Spatial distribution of bacterial communities and their relationships with the micro-architecture of soil. FEMS Microbiology Ecology 44 203–215.
55. Whittaker RH (1972) Evolution and measurement of species diversity. Taxon, 21: 213–251.

Long-Term Fertilization Modifies the Structures of Soil Fulvic Acids and Their Binding Capability with Al

Jun Wu[1], Minjie Wu[2,3], Chunping Li[1], Guanghui Yu[1,2,3]*

1 Key Laboratory of Soil Environment and Pollution Remediation, Institute of Soil Science, Chinese Academy of Sciences, Nanjing, Jiangsu, PR China, **2** State Key Laboratory of Soil Sustainable Agriculture, Institute of Soil Science, Chinese Academy of Sciences, Nanjing, Jiangsu, PR China, **3** National Engineering Research Center for Organic-based Fertilizers, College of Resources and Environmental Sciences, Nanjing Agricultural University, Nanjing, PR China

Abstract

The binding characteristics of organic ligands and minerals in fulvic acids (FAs) with Al are essential for understanding soil C sequestration, remain poorly understood. In this study, Fourier transform infrared (FTIR) spectroscopy combined with two-dimensional correlation spectroscopy (2DCOS) analysis was applied for the first time to explore the binding of Al with organic ligands and minerals in soil FAs. For these analyses, two contrasting treatments were selected from a long-term (i.e., 22-year) fertilization experiment: chemical (NPK) fertilization and swine manure (SM) fertilization. The results showed that the long-term application of organic and inorganic fertilizers to soils had little effect on the compositions of the fluorescent substances and organic ligands in the soil FAs. However, long-term SM fertilization increased the weathered Al and Si concentrations in the soil FAs compared with long-term chemical fertilization. Furthermore, organic ligands in the soil FAs were mainly bound with Al in the NPK treatment, whereas both organic ligands and minerals (Al-O-Si, Si-O) were bound with Al under the M fertilization conditions. Both transmission electron microscopy (TEM) images and X-ray diffraction spectra demonstrated that amorphous and short-range-ordered nanominerals were abundant in the soil FAs from the SM plot in contrast to the soil FAs from the NPK plot. This result illustrates the role nanominerals play in the preservation of soil FAs by during long-term organic fertilization. In summary, the combination of FTIR and 2D correlation spectroscopy is a promising approach for the characterization of the binding capability between soil FAs and Al, and a better understanding FA-Al binding capability will greatly contribute to global C cycling.

Editor: Dafeng Hui, Tennessee State University, United States of America

Funding: This work was funded by the National Natural Science Foundation of China (41371248, 41301241), the National Basic Research Program of China (2011CB100503), the Natural Science Foundation of Jiangsu Province of China (BK20131321, BK2013171), and the Qing Lan Project. The funders had no role in study design, data collection and analysis, decision to publish, or preparation of the manuscript.

Competing Interests: The authors have declared that no competing interests exist.

* Email: yuguanghui@njau.edu.cn

Introduction

Humic substances (HSs), the dominant component of soil organic matter (SOM), are formed by the decomposition of plant, animal, and microbial materials. They are complex, heterogeneous, organic compounds that possess different structures and solubilities and a wide range of molecular weights [1–3]. HSs are sequestered in the soil systems and contribute to increases in soil fertility and reductions in greenhouse gas emissions following complexation with metals. HSs are categorized into the following three groups on the basis of their solubility: humic acids (HAs), fulvic acids (FAs), and humins. Among these groups, FAs have the lowest -molecular -weight, with molecular weights ranging from approximately 500 to 2000 Da, and they contain the highest oxygen content in their complex humic groups [4]. Additionally, FAs are the most mobile components among these three groups. Large specific surface areas and strong structural binding sites make FAs potentially important as carrying agents with regard to the binding and bioavailability of metals [5,6]. Recently, studies have demonstrated that nanominerals are abundant in soil FAs, which affect the nanominerals' binding behavior with metals [7,8]. However, the binding behavior between nanominerals in soil FAs and metals remains unclear.

Fourier transform infrared (FTIR) spectroscopy is a commonly used technique that can distinguish between the principal organic ligands and minerals that are found in soil FAs, such as carbohydrates, lignin, cellulose, lipids, proteins, and compounds containing Al-O bonds, through the vibration characteristics of their structural chemical bonds [8,9]. By determining the FTIR spectra of FAs after a series of metal titrations, it is possible to explore the complexes between the organic ligands in FAs and metals. However, individual FTIR spectral features often overlap [8,10], complicating investigations of binding between the organic ligands in FAs and metals. Recently, two-dimensional correlation spectroscopy (2DCOS) analysis, which can solve the overlapping peaks problem by distributing the spectral intensity trends within a data set over a second dimension [9–11], has been applied to characterize the binding of different organic ligands with metals [8,12]. It is expected that FTIR spectroscopy combined with 2DCOS analysis can facilitate investigations of the binding of organic ligands in soil FAs with Al at the molecular level.

Fertilization practices have been shown to affect the levels and the quality of SOM within cropping systems [8,13–15]. However, little is known about the binding of organic ligands and minerals in soil FAs with Al. By understanding the binding between organic ligands in soil FAs and Al, the stability and bioavailability of SOM under different fertilization conditions can be better understood. The objectives of this study were 1) to investigate the binding of organic ligands in soil FAs with Al and 2) to explore the mechanism by which SOM is sequestered under different fertilization practices. For these purposes, the following two contrasting treatments were selected from a long-term (i.e., 22-year) fertilization experiment: chemical (NPK) fertilization and swine manure (SM) fertilization.

Materials and Methods

Site and soil sample collection

Soils were collected from two contrasting treatments, NPK and SM, at a long-term fertilization experiment station. The long-term fertilization experiment was initiated in September 1990 in fields that were double cropped with wheat and corn at an experiment station of the Chinese Academy of Agricultural Sciences, Qiyang (26°45′ N, 111°52′ E, 120 m altitude), Hunan Province, Southern China. No specific field permits were required for this study. The land accessed is not privately owned or protected. No protected species were sampled. The red soil was classified as Ferralic Cambisol. Concentration of allophane in soils from NPK and SM treatments in 2012 was 0.25±0.05 mg/kg and 1.80±0.13 mg/kg, respectively. A detailed description of the long-term fertilization experiment site can be found elsewhere [8,16]. In brief, the experimental area receives 1,255 mm of annual precipitation, approximately 70–80% of which occurs from April to October. Mean annual temperature, annual evaporation, annual frost-free days, and sunshine hours are 18°C, 1,470 mm, 300 d, and 1,610 h, respectively. Prior to the experiment, the field had been under an annual wheat-corn rotation for 3 yrs to give soil uniform fertility. The chemical fertilizers used were urea for N, superphosphate for P, and potassium chloride (KCl) for K. The total N applied in the NPK and SM treatments per year is 300 kg/ha.

Soil samples at depths of 0–20 cm were collected in October 2012 using a 5-cm internal diameter auger. Each sample was a composite of ten random cores that were collected from two replicate plots. The fresh soil was mixed thoroughly, air-dried, roots removed, and then sieved through a 2.0-mm sieve for further analysis.

Extraction and purification of fulvic acids (FAs)

The FA fraction was extracted from the surface soil (0–20 cm) samples following the International Humic Substance Society (IHSS) procedure [17]. The sample was equilibrated to a pH value between 1 and 2 with 1 M HCl at room temperature. The solution volume was adjusted with 0.1 M HCl to provide a final concentration ratio of 10 mL liquid per g dry sample. The suspension was shaken for 1 h, and the supernatant was separated from the residue by low speed centrifugation (i.e., 2000 *g*). The residues were then neutralized with 1 M NaOH. A 0.1 M solution of NaOH was then added to the residues under nitrogen gas using an extractant to sample ratio of 10:1. The suspension was extracted under nitrogen gas with intermittent shaking for a minimum of 4 h. After centrifugation, the supernatant solution was collected and acidified to pH 1.0 with 6 M HCl and was allowed to settle for 12 h before centrifuging again [17]. The FA fraction was concentrated and purified by flowing the FAs through an XAD-8 adsorption resin column (AMBERLITE), and the 0.1 M NaOH eluate was passed through a H^+-saturated cation exchange resin column (CER, Dowex MAC-3, Sigma) (Hiradate, 2006). The purified FA samples were stored at 4°C until further analysis.

Fluorescence excitation-emission matrix (EEM) determination

The fluorescence EEMs were measured on a Varian Eclipse fluorescence spectrophotometer in scan mode. Scanning emission (Em) spectra from 250 to 600 nm were obtained in 2 nm increments by varying the excitation (Ex) wavelength from 200 to 500 nm in 10 nm increments [18]. The spectra were recorded at a scan rate of 1200 nm/min using excitation and emission slit band widths of 5 nm. Several preprocessing steps were used to minimize the influence of scatter lines and other attributes of the EEM landscape. The EEM of a control Milli-Q water was subtracted from each sample, and other Rayleigh and Raman scatters were removed by the protocol of Bahram et al. [19]. After that, the EEMs were normalized by dividing the spectra by the corresponding DOC concentrations.

Al titration test

Aliquots of 25 mL of the diluted solution of FAs were titrated into 40 mL brown, sealed vials containing 0.01 mol/L $AlCl_3$ using an automatic syringe. The Al concentrations in the final solutions ranged from 0 to 70 μmol/L. No more than 25 μL of the metal titrant was added during titration to maintain constant pH before and after titration. All titrated solutions were shaken for 24 h at 25°C to ensure complexation equilibrium [20–22]. Finally, after freeze-drying the solutions, they were analyzed by FTIR spectroscopy.

FTIR spectroscopy determination and analysis of two-dimensional correlation spectroscopy

The samples were prepared as a mixture of 1 mg of freeze-dried FA or of the FA-Al complex and 100 mg of potassium bromide (KBr, IR grade), and this mixture was then ground and homogenized [9,10]. A subsample was then compressed twice in

Table 1. Physiochemical characteristics of soil fulvic acids in the NPK and M treatments*.

Treatments	SOC	DOC (mg/L)	EC (m S/cm)	Al	Si
NPK	10.67±0.24c	570.6±11.8b	0.69±0.11b	3.3±0.2b	0.3±0.1b
SM	15.01±0.47a	893.3±23.5a	1.93±0.24a	6.8±0.5a	0.9±0.2a

*Note: Significance was determined using one-way ANOVA followed by Tukey's HSD post hoc tests, where conditions of normality and homogeneity of variance were met, means (n = 3) followed by different letters indicate significant differences between treatments at P<0.05. Abbreviations: NPK, chemical nitrogen, phosphorus and potassium; SM, swine manure; SOC, soil organic carbon; DOC, dissolved organic carbon; EC, electricity conductivity.

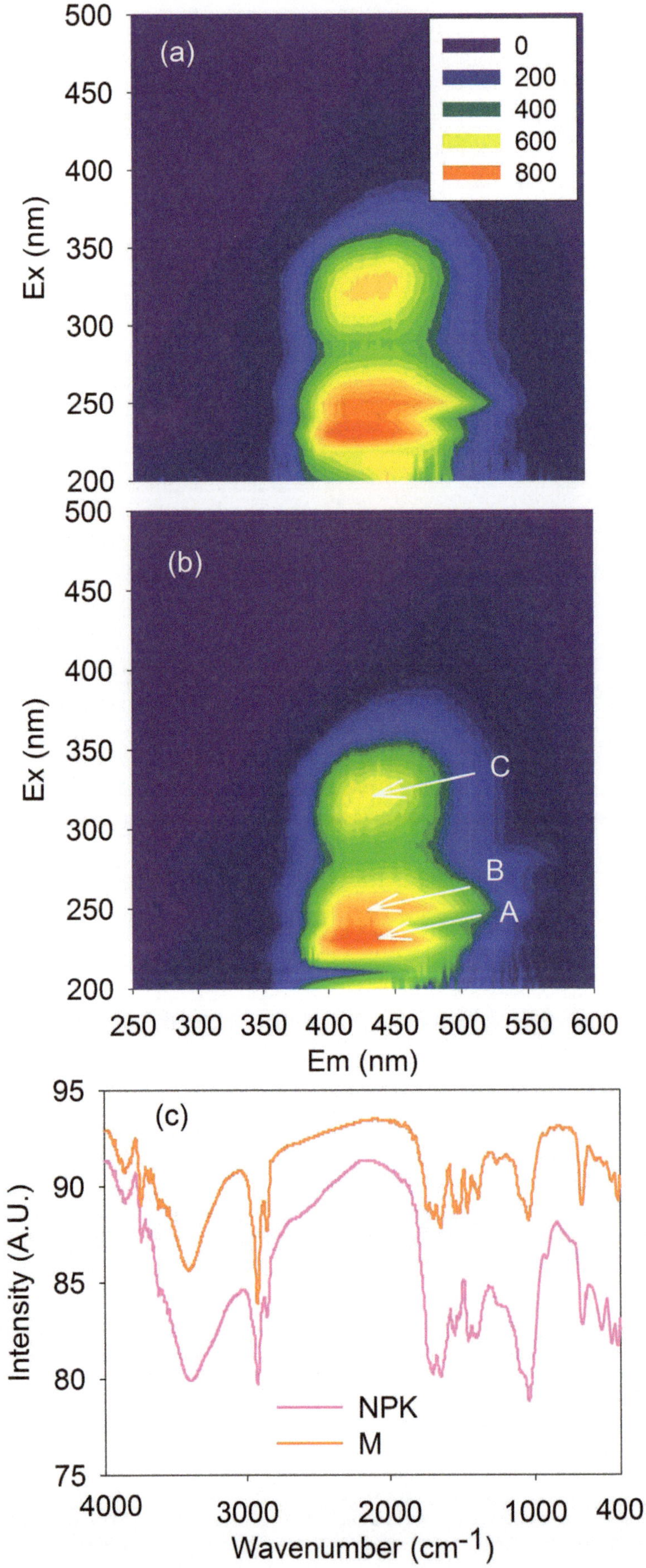
(a)
0
200
400
600
800
Ex (nm)
500
450
400
350
300
250
200
(b)
C
B
A
250 300 350 400 450 500 550 600
Em (nm)
(c)
Intensity (A.U.)
95
90
85
80
75
NPK
M
4000 3000 2000 1000 400
Wavenumber (cm-1)

Figure 1. Fluorescence excitation-emission matrix (EEM) spectra of soil fulvic acids from the NPK (a) and SM (b) plots and FTIR spectra of the fulvic acids (c) during a long-term (i.e., 22-year) fertilization experiment. Note that peaks A, B and C were located at Ex/Em of 230/410, 250/410, and 330/410, respectively. NPK, chemical nitrogen, phosphorus and potassium; SM, swine manure.

a hydraulic press at 20,000 psi between two clean, polished iron anvils to form a KBr window. The FTIR spectra were obtained by collecting 200 scans with a Nicolet 370 FTIR spectrometer.

The 2DCOS spectra were produced according to the method of Noda and Ozaki [11]. In this study, the Al concentration was applied as an external perturbation, and a set of concentration-dependent FTIR spectra was obtained. For example, in an analytical spectrum where $I(x, t)$, the variable x is the index variable representing the FTIR spectra induced by the perturbation variable t. The x was used instead of the general notation in conventional 2D correlation equations based on the spectral index v. The analytical spectrum $I(x, t)$ at m evenly spaced points in t (between $T\ min$ and $T\ max$) can be represented as

$$I_j(x) = I(x, t_j),\ j = 1, 2, \cdots, m. \tag{1}$$

A set of dynamic spectra is given by

$$\tilde{I}(x,t) = I(x,t_j) - \bar{I}(x) \tag{2}$$

where $\bar{I}(x)$ denotes the reference spectrum, which is typically the average spectrum that is expressed as $\bar{I}(x) = \frac{1}{m}\sum_{j=1}^{m} I(x, t_j)$. The synchronous correlation intensity can be directly calculated from the following dynamic spectra:

$$\varnothing(x_1, x_2) = \frac{1}{m-1}\sum_{j=1}^{m} \tilde{I}_j(x_1)\tilde{I}_j(x_2) \tag{3}$$

The intensity of a synchronous correlation spectrum ($\varnothing(x_1, x_2)$) represents the simultaneous changes in two spectral intensities measured at x_1 and x_2 during the interval between T_{min} and T_{max}.

Prior to 2DCOS analysis, the FTIR spectra were normalized by summing the absorbance from 4000-400 cm^{-1} and multiplying by 1000. Subsequently, the normalized FTIR spectra were analyzed using principal component analysis (PCA) to reduce the level of noise [23]. Finally, 2DCOS was conducted using 2D shige software (Kwansei-Gakuin University, Japan).

In this study, we focused on the FTIR regions from 3600 cm^{-1} to 3100 cm^{-1}, 1800 cm^{-1} to 900 cm^{-1}, and 900 cm^{-1} to 400 cm^{-1} because they contain the major excitation bands of OH bonds, amides, carboxylic acids, esters, carbohydrates, Al-O and Si-O, and aliphatic C-H stretching [24,25].

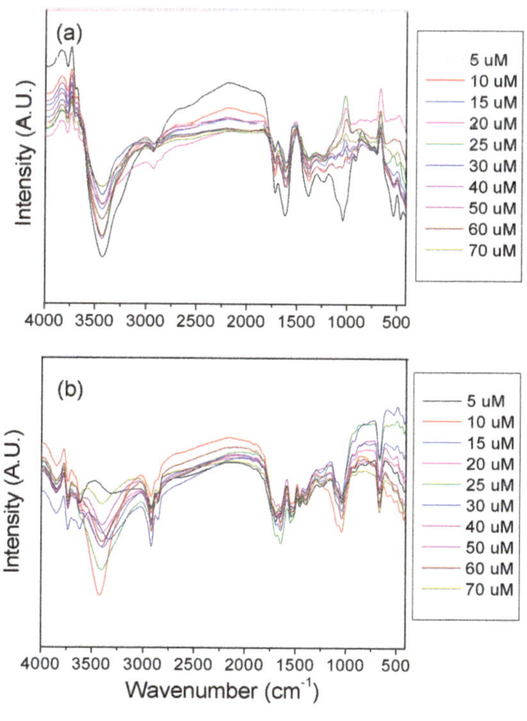

Figure 2. One-dimensional FTIR spectra of soil FAs from the NPK (a) and M (b) treatments during a long-term (i.e., 22-year) fertilization experiment. NPK, chemical nitrogen, phosphorus and potassium; SM, swine manure.

Chemical analysis

The carbon (C) contents in the soil were analyzed using a Perkin-Elmer 2400 CHN elemental analyzer. The filtered (0.45 μm PTFE filter) extract was measured for DOC using a TOC/TN analyzer (multi N/C 3000, Analytik Jena AG, Germany). The pH was determined in 1:2.5 (w/v) soil:water extracts using a pH electrode. Electrical conductivity (EC) was determined using a conductivity meter (LF91, German). The Al and Si concentrations in the soil FA extracts were measured using ICP-AES (Optima 7000, PerkinElmer, USA). Mineralogical identification was performed using X-ray powder diffraction (Nanjing Normal University). The diffraction patterns were recorded from 3 to 60° 2-theta using Ni-filtered Cu K-alpha radiation at a rate of 300 s per step on a PANalytical X-pert Pro diffractometer using an X-celerator position-sensitive detector.

Statistical analysis

The data (means ± SD, $n = 3$) were assessed via one-way analysis of variance (ANOVA) using SPSS software version 18.0 for Windows (SPSS, Chicago, IL). Significance was determined using one-way ANOVA followed by Tukey's HSD post hoc tests, where conditions of normality and homogeneity of variance were met, and means followed by different letters indicated significant differences between treatments at $P<0.05$.

Results and Discussion

Characterization of soil FAs in long-term fertilization treatments

As one of the main components of SOC, the soil FAs in the SM plot after 22 years of fertilization treatment contained a higher DOC concentration and EC than the FAs in the NPK plot ($p<0.01$, Table 1), which was consistent with the results that also showed that the SOC concentrations in the SM plot were significantly ($p<0.01$) greater (~54±5%) than those in the NPK plot (Table 1). Therefore, consistent with previous publications, organic fertilization could markedly increase SOC content and soil fertility [26,27]. Moreover, organic fertilization significantly

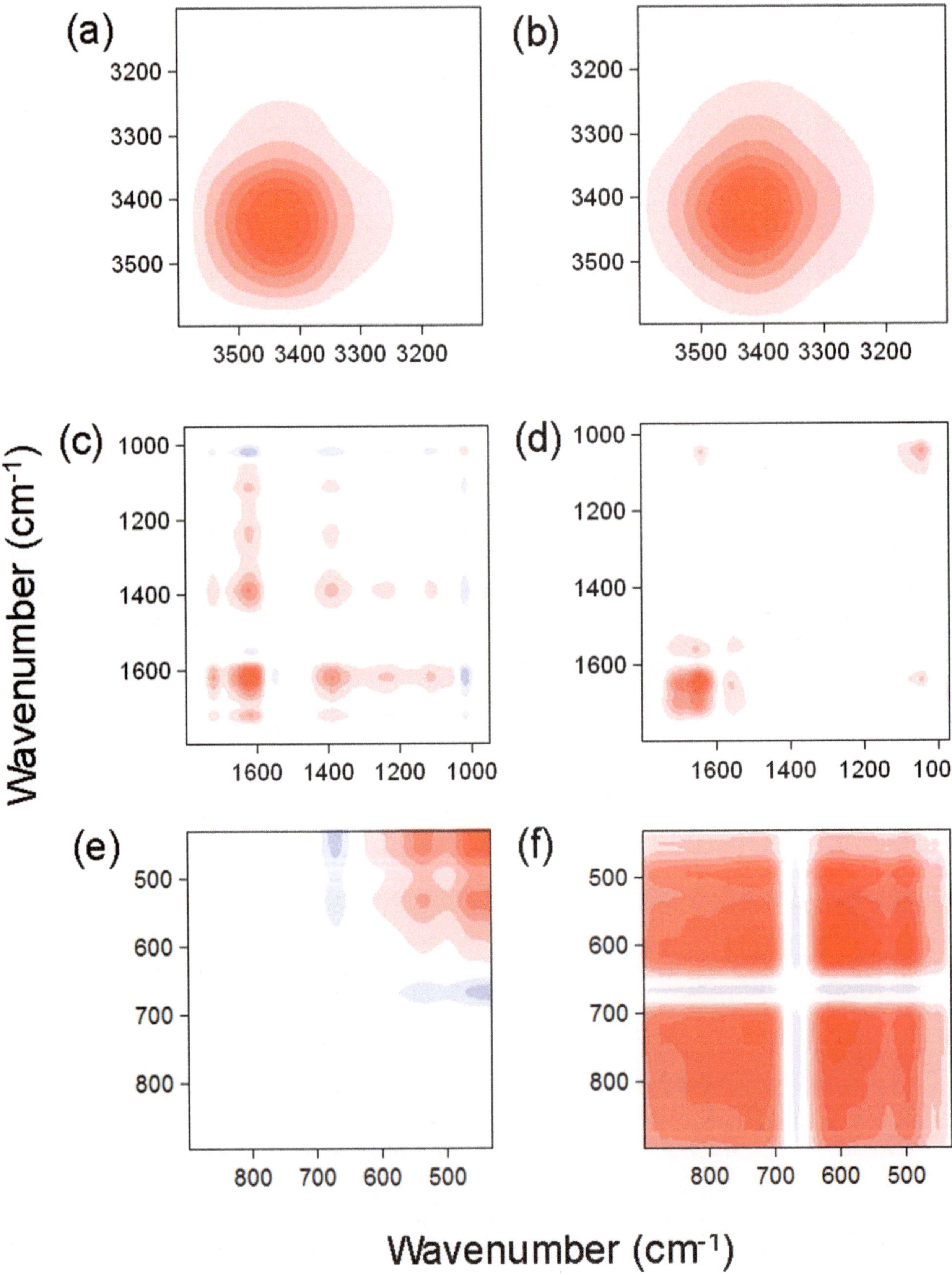

Figure 3. Synchronous 2D correlation maps generated from the FTIR spectra of soil FAs in the NPK (a, c and e) and SM (b, d and f) treatments with Al. Red represents positive correlation, and blue represents negative correlation; a higher color intensity indicates a stronger positive or negative correlation. NPK, chemical nitrogen, phosphorus and potassium; SM, swine manure.

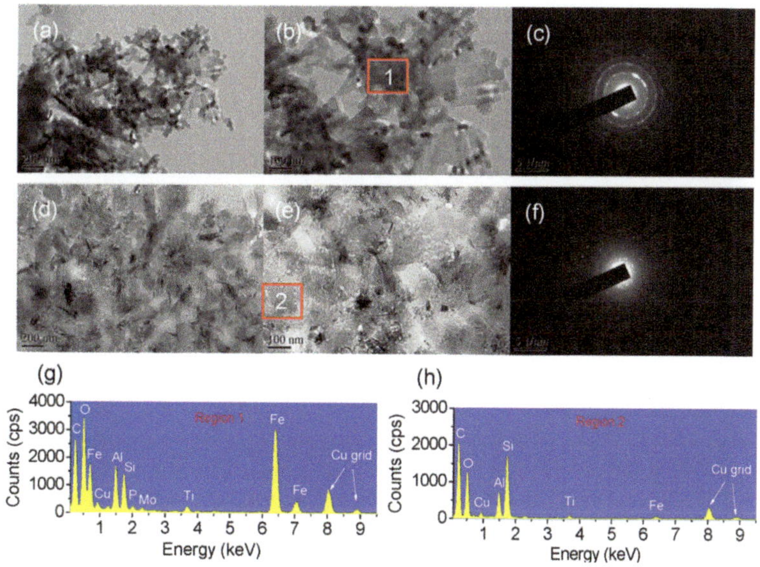

Figure 4. TEM images of the soil FAs from the NPK (a–b) and SM (d–e) treatments during a long-term (i.e., 22-year) fertilization experiment. Selected area electron diffraction (SAED) patterns of the two images indicated by the blue squares, demonstrate that the black regions are polycrystal (c) or short-ranged-order (SRO) (f) minerals. NPK, chemical nitrogen, phosphorus and potassium; SM, swine manure. (g) and (h) energy dispersive X-ray spectroscopy (EDS) of region 1 and 2 located at the red square frames.

increased ($p<0.01$) the concentration of weathered Al and Si and incorporated them into the soil FAs.

Fluorescence EEM spectra demonstrated that three fluorescence peaks, i.e., peaks A, B, and C, which were located at Ex/Em of 230/420, 250/420, and 330/420, respectively, were present in the soil FAs from both the NPK and SM plots (Figures 1-a and 1-b) and were attributed to fulvic-like (i.e., peaks A and B) and humic-like (i.e., peak C) substances, respectively [8,18,28]. Although the fulvic-like (i.e., peaks A and B) substances were dominant, a considerable fraction of humic-like (i.e., peak C) substances was also present in the extracted soil FAs, suggesting that the soil FAs extracted by the operationally defined method actually include both FAs and HAs as indicated by fluorescence spectroscopy.

The organic ligands of the soil FAs from the NPK and SM plots were quite similar in the FTIR functional region (i.e., 4000-1000 cm^{-1}), including OH from polysaccharides (3400 cm^{-1}), CH_2 from lipids (2920 cm^{-1}), C=O stretching of the amide groups in proteins (1650 cm^{-1}), C-N stretching and N-H bending vibrations of the amide groups in proteins (1540 cm^{-1}), and Si-O (1020 cm^{-1}) and Al-O-Si bands (610 cm^{-1} and 720 cm^{-1}) in allophanes. However, the lignads were distinctly different in the fingerprint region (1000-400 cm^{-1}) (Figure 1-c). In the fingerprint region, peaks at 720 cm^{-1} and 610 cm^{-1} were only present in the SM plot, both of them belonging to the Al–O–Si band in Al nanominerals [29,30]. Different from the fluorescence EEM spectra, which report information only about fluorescence substances (i.e., protein- and humic-like substances), FTIR spectroscopy provides information about non-fluorescent substances (e.g., polysaccharides and lipids) and nanominerals (e.g., allophane) present in the soil FAs.

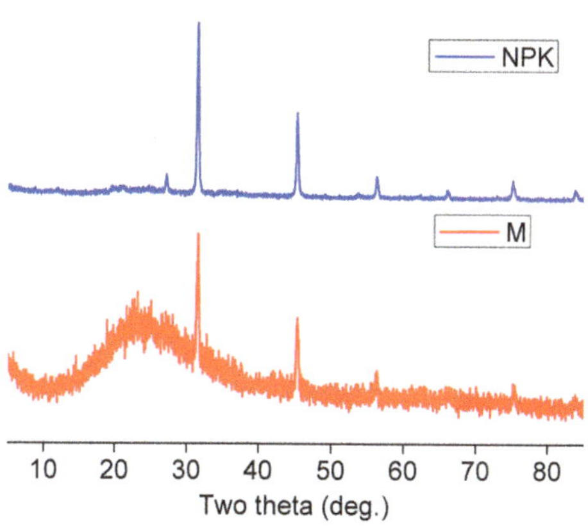

Figure 5. The X-ray diffraction pattern of soil FAs from the NPK and SM treatments during a long-term (i.e., 22-year) fertilization experiment. NPK, chemical nitrogen, phosphorus and potassium; SM, swine manure.

In summary, the long-term application of different fertilizers onto soils had little effect on the composition of fluorescent substances and organic ligands. The results were similar to those of a study by Randall et al. [31], in which mineral fertilizer and manure differed little in their effects on the chemical composition of SOM. Similar FTIR band characteristics were also previously reported between soil HAs in soils that were fertilized with manure and inorganic N fertilizer and an unfertilized control [32]. The minor effect of fertilization on the structures of the soil FAs may have been because the chemical composition of SOM is determined primarily by the interaction between the organisms responsible for decomposition and the mineral soil matrix rather than by the nature of the substrate input [31].

Binding of organic ligands and nanominerals with Al in soil FAs

A better understanding of the binding of organic ligands and minerals to cations is essential to the stability of the SOC. Following the addition of Al, the major IR peaks appeared deformed in the one-dimensional FTIR spectra (Figure 2). The peaks were significantly overlapped, and no further information could be extracted. Therefore, two-dimensional correlation spectroscopy analysis was applied in this study to further analyze the binding of the organic ligands and minerals in the soil FAs to Al.

Synchronous 2DCOS spectra of samples from the two plots are shown in Figure 3. In the 3600-3100 cm^{-1} region, only one auto-peak located at 3420 cm^{-1} (OH stretching in polysaccharides) was observed in both soil FAs from the two plots. In the 1800-900 cm^{-1} region, four and three major auto-peaks were present in the samples from the NPK and SM plots, respectively. The changes in band intensity decreased in the following orders: for the NPK plot, C = O stretching of amide I (1630 cm^{-1}) > CH deformations in aliphatic groups (1380 cm^{-1}) >> C = O stretching of COOH (1720 cm^{-1}) ≈Si-O asymmetric stretching (1020 cm^{-1}); for the SM plot, C = O stretching of amide I (1650 cm^{-1}) > Si-O asymmetric stretching (1020 cm^{-1}) >> C = N stretching of amide II (1540 cm^{-1}). In the fingerprint region (900-400 cm^{-1}), two and three auto-peaks were present in the NPK and SM plots, respectively. The band intensities decreased in the following orders: for the NPK plot, Si-O bending vibrations (430 cm^{-1}) > Al-O-Si deformation (540 cm^{-1}); for the SM plot, coupled Al-O and Si-O (out-of-plane) (610 cm^{-1}) > Si-O stretching (720 cm^{-1}) > Al-O-Si deformation (510 cm^{-1}).

Seven positive cross-peaks at Φ(1630, 1380), Φ(1630, 1220), Φ(1630, 1080), Φ(1720, 1630), Φ(1720, 1380), Φ(1380, 1220), and Φ(1380, 1080) were present in the NPK fertilization, indicating that the C = O stretching of amide I, the CH deformations in the aliphatic groups, and the OH deformation of COOH were co-binding to Al. Moreover, four negative cross-peaks at Φ(1630, 1540), Φ(1630, 1010), Φ(1380, 1010), and Φ(1080, 1010) were also identified in the NPK fertilization, suggesting that the C = O stretching was negatively correlated with N–H deformations in the plane of amide II and that the Si-O stretching of the nanominerals was negatively correlated with C = O stretching, OH deformation of COOH, and CH deformations. In contrast, only two positive cross-peaks at Φ(1650, 1540) and Φ(1650 1020) were present in the SM fertilization, revealing that the C = O stretching of the amide groups were co-bound to Al, with N-H deformations of amide II and Si-O stretching of the nanominerals. In the fingerprint region, the presence of positive cross-peaks at Φ(540, 430) in the NPK plot and both Φ(720, 610) and Φ(720, 510) in the SM plot indicated that Si-O and Al-O-Si were co-bound to Al.

Although the long-term application of different fertilizers to soils had little effect with regard to organic ligands, this study showed, for the first time, that the different fertilization treatments had a marked effect on the binding of FAs with Al. First, more organic ligands in the 1800-900 cm^{-1} region were bound to Al in the NPK plot than in SM (Figures 3-b and 3-d). In contrast, except for the binding of organic ligands in the soil FAs with Al, the Si-O content (i.e., 1020, 720, and 610 cm^{-1}) in minerals in the SM plot played a more important role in the binding to Al than in the NPK plot. This result was consistent with the finding of Yu et al. [8], in which a large amount of silicates was present in an NPK plus manure (NPKM) plot when compared to an NPK plot, suggesting that Al was bound to a silicate network in the SM plot. In addition to the binding of Al to the H-bond network (i.e., 3400 cm^{-1}) (Figures 3-a and 3-b), it is reasonable to surmise that a large amount of allophone $[Al_2O_3(SiO_2)_{1-2}(H_2O)_{2-4}]$ and/or imogolite $[(OH)_3Al_2O_3SiOH]$ was formed in the soil FAs in the M plot in contrast to the NPK plot.

Morphology, diffraction pattern, and X-ray diffraction spectra of soil FAs

To confirm whether nanominerals were formed in the soil FAs in the SM and the NPK plots, TEM images and XRD spectra were used to observe the morphology and mineralogy of the nanominerals. A large amount of nanominerals was present in the TEM images of the soil FAs from both NPK (Figures 4-a and 4-b) and SM plots (Figures 4-d and 4-e). The electron diffraction patterns indicated that, although the soil FAs from both NPK and SM plots had a determinate crystalline and amorphous pattern, the obtained nanominerals were more crystalline under NPK conditions (Figure 4-c) than under SM conditions (Figure 4-f). Furthermore, the XRD spectra of the soil FAs corroborated that crystalline minerals were dominant under NPK fertilization conditions, while amorphous minerals were predominant under SM fertilization conditions (Figure 5).

Collectively, the TEM images and XRD spectra showed that nanominerals, i.e., allophone $[Al_2O_3(SiO_2)_{1-2}(H_2O)_{2-4}]$ and/or imogolite $[(OH)_3Al_2O_3SiOH]$, were found abundantly in the soil FAs from the SM plot rather than from the NPK plot. In combination with the results of the 2DCOS spectra, the TEM images and XRD spectra showed that the nanominerals in soil FAs may play an important role in the binding of metals, which should benefit both soil remediation and soil C sequestration.

Implications to soil organic carbon sequestration

The results from the 2D FTIR correlation spectroscopy, TEM images and XRD spectra point to the formation of soil nanominerals in the SM rather than in the NPK treatments. The formation of soil nanominerals (i.e., allophone and imogolite) could greatly benefit the preservation of SOC [33]. Torn et al. [34] previously demonstrated that soil nanominerals controlled SOC storage and turnover. Under favorable conditions, the turnover of SOC in allophonic soils could persist in tephra beds for at least 250,000 years [7]. Globally, approximately 190–332 million tons of C can be sequestered by silicate minerals annually [35]. As one of the main components of SOC, the abundance of nanominerals in the soil FAs from the SM plot would definitely contribute to the stability of the SOC. The knowledge of the binding of organics and metals improves our understanding of the sequestration process of SOC and provides valuable information with regard to fertilization techniques and scientific research.

Conclusions

In this study, FTIR spectroscopy combined with 2DCOS analysis has been developed to investigate the binding capability between soil FA and mineral elements.

The results show that the organic ligands in the soil FAs were mainly bound to Al in the NPK treatment, whereas both organic ligands and minerals (Al–O–Si, Si–O) were bound to Al in the SM treatment. The abundance of amorphous and short-range-ordered (SRO) nanominerals in the soil FAs of the SM plot, rather than the NPK plot, illustrated the preservation role of the soil FAs by nanominerals during long-term organic fertilization. In summary, the combination of FTIR and 2D correlation spectroscopy is a promising approach for the characterization of the binding capability between soil FA and mineral elements. A better understanding of this binding capability will benefit the understanding of soil C sequestration and greatly contribute to global C cycling.

Acknowledgments

This work was funded by the National Natural Science Foundation of China (41371248, 41301241), the National Basic Research Program of China (2011CB100503), the Natural Science Foundation of Jiangsu Province of China (BK20131321, BK2013171), and the Qing Lan Project.

Author Contributions

Conceived and designed the experiments: GY. Performed the experiments: MW JW CL. Analyzed the data: GY JW MW. Contributed reagents/materials/analysis tools: GY JW MW. Contributed to the writing of the manuscript: GY MW JW CL.

References

1. Lead JR, Balnois E, Hosse M, Menghetti R, Wilkinson KJ (1999) Characterization of Norwegian natural organic matter: size, diffusion coefficients and electrophoretic mobilities. Environment International 25: 245–258.
2. Maurice PA, Namjesnik-Dejanovic K (1999) Aggregate structures of sorbed humic substances observed in aqueous solution. Environmental Science and Technology 33: 1538–1541.
3. Sutton R, Sposito G (2005) Molecular structure in soil humic substances: the new view. Environmental Science and Technology 39: 9009–9015.
4. Gaffney JS, Marley NA, Clark SB (1996) Humic and Fulvic Acids and the Organic Colloidal Materials in the Environment, Humic and Fulvic Acids: Isolation, Structure and Environmental Role. In: Gaffney JS, Marley NA, Clark SB (Eds.), American Chemical Society, Washington, DC, 2–16.
5. Dwane GC, Tipping E (1998) Testing a humic speciation model by titration of copper amended natural waters. Environment International 24: 609–616.
6. Tipping E (2002) Cation Binding by Humic Substances. Cambridge University Press., Cambridge.
7. Parfitt RL (2009) Allophane and imogolite: role in soil biogeochemical processes. Clay Minerals 44: 135–155.
8. Yu GH, Wu MJ, Wei GR, Luo YH, Ran W, et al. (2010) Binding of organic ligands with Al(III) in dissolved organic matter from soil: implications for soil organic carbon storage. Environmental Science and Technology 46: 6102–6109.
9. Wang LP, Shen QR, Yu GH, Ran W, Xu YC (2012) Fate of biopolymers during rapeseed meal and wheat bran composting as studied by two-dimensional correlation spectroscopy in combination with multiple fluorescence labeling techniques. Bioresource Technology 105: 88–94.
10. Yu GH, Tang Z, Xu YC, Shen QR (2011) Multiple fluorescence labeling and two dimensional FTIR-^{13}C NMR heterospectral correlation spectroscopy to characterize extracellular polymeric substances in biofilms produced during composting. Environmental Science and Technology 45: 9224–9231.
11. Noda I, Ozaki Y, Eds. (2004) Two-Dimensional Correlation Spectroscopy-Applications in Vibrational and Optical Spectroscopy. John Wiley & Sons., England.
12. Xu HC, Yu GH, Jiang HL (2013) Investigation on extracellular polymeric substances from mucilaginous cyanobacterial blooms in eutrophic freshwater lakes. Chemosphere 93: 75–81.
13. Paustian K, Parton WJ, Persson J (1992) Modeling soil organic matter in organic-amended and nitrogen-fertilized long-term plots. Soil Science Society of America Journal 56: 476–488.
14. Varvel GE (1994) Rotation and nitrogen fertilizer effects on changes in soil carbon and nitrogen. Agronomy Journal 86: 319–325.
15. Campbell CA, Biederbeck VO, McConkey BG, Curtin D, Zentner RP (1999) Soil quality effect of tillage and fallow frequency. Soil organic matter quality as influenced by tillage and fallow frequency in a silt loam in southwestern Saskatchewan. Soil Biology & Biochemistry 31: 1–7.
16. Zhang HM, Wang BR, Xu MG, Fan TL (2009) Crop yield and soil responses to long-term fertilization on a red soil in southern China. Pedosphere 19: 199–207.
17. Swift RS (1996) Methods of Soil Analysis: Part 3. Chemical Methods. Soil Science Society of America Book Series 5, 1018–1020.
18. Yu GH, Luo YH, Wu MJ, Tang Z, Liu DY, et al. (2010) PARAFAC modeling of fluorescence excitation–emission spectra for rapid assessment of compost maturity. Bioresource Technology 101: 8244–8251.
19. Bahram M, Bro R, Stedmon C, Afkhami A (2007) Handling of Rayleigh and Raman scatter for PARAFAC modeling of fluorescence data using interpolation. Journal of Chemometrics 20: 99–105.
20. Wu J, Zhang H, He PJ, Shao LM (2011) Insight into the heavy metal binding potential of dissolved organic matter in MSW leachate using EEM quenching combined with PARAFAC analysis. Water Research 45: 1711–1719.
21. Wu J, Zhang H, Shao LM, He PJ (2012) Fluorescent characteristics and metal binding properties of individual molecular weight fractions in municipal solid waste leachate. Environmental Pollution 162: 63–71.
22. Wen YL, Li H, Xiao J, Wang C, Shen QR, et al. (2014) Insights into complexation of dissolved organic matter and Al(III) and nanominerals formation in soils under contrasting fertilizations using two-dimensional correlation spectroscopy and high resolution- transmission electron microscopy techniques. Chemosphere 111: 441–449.
23. Abdulla HA, Minor EC, Dias RF, Hatcher PG (2010) Changes in the compound classes of dissolved organic matter along an estuarine transect: A study using FTIR and ^{13}C NMR. Geochim Cosmochim Acta 74: 3815–3838.
24. Hinterstoisser B, Salmen L (2000) Application of dynamic 2D-FTIR to cellulose. Vibrational Spectroscopy 22: 111–118.
25. Hiradate S, Yonezawa T, Takesako H (2006) Isolation and purification of hydrophilic fulvic acids by precipitation. Geoderma 132: 196–205.
26. Gattinger A, Muller A, Haeni M, Skinner C, Fliessbach A, et al. (2012) Enhanced top soil carbon stocks under organic farming. Proc Natl Acad Sci USA 109: 18226–18231.
27. Maillard E, Angers DA (2014) Animal manure application and soil organic carbon stocks: a meta-analysis. Global Change Biology 20: 666–679.
28. Chen W, Westerhoff P, Leenheer JA, Booksh K (2003) Fluorescence excitation–emission matrix regional integration to quantify spectra for dissolved organic matter. Environmental Science and Technology 37: 5701–5710.
29. Madejova J, Komadel P (2001) Baseline studies of the clay minerals society source clays: infrared methods. Clays and Clay Minerals 49: 410–432.
30. Liu C, Huang PM (2002) Role of hydroxy-aluminosilicate ions (proto- imogolite sol) in the formation of humic substances. Organic Geochemistry 33: 295–305.
31. Randall EW, Mahieu N, Powlson DS, Christensen BT (1995) Fertilization effects on organic matter in physically fractionated soils as studied by ^{13}C NMR: results from two long-term field experiments. European Journal of Soil Science 46: 557–565.
32. Mao JD, Olk DC, Fang XW (2008) Influence of animal manure application on the chemical structures of soil organic matter as investigated by advanced solid-state NMR and FT-IR spectroscopy. Geoderma 146: 353–362.
33. Aiken GR, Hsu-Kim H, Ryan JN (2011) Influence of dissolved organic matter on the environmental fate of metals, nanoparticles, and colloids. Environmental Science and Technology 45: 3196–3201.
34. Torn MS, Trumbore SE, Chadwich OA, Vitousek PM, Hendricks DM (1997) Mineral control of soil organic carbon storage and turnover. Nature 389: 170–173.
35. Renforth P, Washbourne CL, Taylder J, Manning DAC (2011) Silicate production and availability for mineral carbonation. Environmental Science and Technology 45: 2035–2041.

Spatial and Temporal Variations of Crop Fertilization and Soil Fertility in the Loess Plateau in China from the 1970s to the 2000s

Xiaoying Wang[1,2], Yanan Tong[1,2]*, Yimin Gao[1], Pengcheng Gao[1], Fen Liu[1], Zuoping Zhao[1], Yan Pang[1]

1 College of Natural Resources and Environment, Northwest A&F University, Yangling, China, **2** Key Laboratory of Plant Nutrition and the Agri-environment in Northwest China, Ministry of Agriculture, Yangling, China

Abstract

Increased fertilizer input in agricultural systems during the last few decades has resulted in large yield increases, but also in environmental problems. We used data from published papers and a soil testing and fertilization project in Shaanxi province during the years 2005 to 2009 to analyze chemical fertilizer inputs and yields of wheat (*Triticum aestivum* L.) and maize (*Zea mays* L.) on the farmers' level, and soil fertility change from the 1970s to the 2000s in the Loess Plateau in China. The results showed that in different regions of the province, chemical fertilizer NPK inputs and yields of wheat and maize increased. With regard to soil nutrient balance, N and P gradually changed from deficit to surplus levels, while K deficiency became more severe. In addition, soil organic matter, total nitrogen, alkali-hydrolysis nitrogen, available phosphorus and available potassium increased during the same period. The PFP of N, NP and NPK on wheat and maize all decreased from the 1970s to the 2000s as a whole. With the increase in N fertilizer inputs, both soil total nitrogen and alkali-hydrolysis nitrogen increased; P fertilizer increased soil available phosphorus and K fertilizer increased soil available potassium. At the same time, soil organic matter, total nitrogen, alkali-hydrolysis nitrogen, available phosphorus and available potassium all had positive impacts on crop yields. In order to promote food safety and environmental protection, fertilizer requirements should be assessed at the farmers' level. In many cases, farmers should be encouraged to reduce nitrogen and phosphate fertilizer inputs significantly, but increase potassium fertilizer and organic manure on cereal crops as a whole.

Editor: Cheng-Sen Li, Institute of Botany, China

Funding: The authors thank the Special Fund for Agro-scientific Research in the Public Interest of China (201103003) and the Soil Quality Foundation of China (2012BAD05B03) for their financial support. The funders had no role in study design, data collection and analysis, decision to publish, or preparation of the manuscript.

Competing Interests: The authors have declared that no competing interests exist.

* Email: tongyanan@nwsuaf.edu.cn

Introduction

China has only 9% of the world's arable land and feeds nearly 22% of the world population [1–2]. This depends heavily on increasing grain production with the use of chemical fertilizers. Before the 1970s, farmers maintained the original agricultural practices, such as crop rotation, diversified plantation, manure application and legume crop integration, for soil fertility maintenance and pest and disease control. Since the late 1980s, the practice of applying organic manure in arable cropping systems has nearly come to an end [2–6]. From then on, almost all available organic manure has been used on vegetables and fruit trees, while the nutrients for cereal crops have been mainly in the form of chemical fertilizers. From 1970 to 2010, total annual grain production in China increased from 240 to 546 million tons (a 128% increase). However, inorganic fertilizer application increased from 3.51 to 55.62 million tons (a 1485% increase) over the same period [7].

Soil quality indicators are measurable soil properties that benefit food production or other specific functions, including physical, chemical and biological characteristics [8]. The increase or decrease in single soil index values, such as soil organic matter, total nitrogen and available nutrients, amplitude of variation and variation in time, can be used as a monitoring index for agricultural land management [9–11]. Given the spatial and temporal variation in characteristics of soil quality, it is necessary to compare or analyze two or more phase changes to understand the nature and mechanisms of soil quality [12].

Farmland fertilization is one of the most effective ways to maintain soil fertility and increase crop yields [13–15]. For this reason, information on household fertilization levels is of great value. In addition, wheat and maize are two of the most important food crops throughout the world, and they account for 51.7% of the total area for food crops and 53.5% of the total food production in 2010 in China [7]. Chemical fertilizer consumption data from official Chinese statistics do not contain information on usage for each kind of crop. It is imprecise to analyze and evaluate fertilizer efficiency using total amounts, because the distribution and application of fertilizer on specific crops are ambiguous [16].

Thus, the objectives of this study were to: (1) reveal the spatial and temporal variations of chemical fertilization and yields of wheat and maize at the farmers' level from the 1970s to the 2000s in the Loess Plateau in China; (2) reveal the spatial and temporal variations of soil fertility over the same period; and (3) reveal the relationships among fertilizer inputs, crop yields and soil fertility.

Materials and Methods

Ethics Statement

This study has been approved by the Agricultural Technology Extension Center of Shaanxi province, which is responsible for fertilization and soil fertility in Shaanxi province. All data in this study can be published and shared.

Study area

Shaanxi province (Figure 1) is located in the middle reaches of the Yellow River and the upper reaches of the Yangtze River of the eastern part of northwest China, and it falls between latitudes 31°42′ and 39°35′N, and longitudes 105°29′ and 111°15′E. The area is 2.058×10^5 km^2, extending about 880 km from north to south and 160 to 490 km from east to west. The whole province from north to south can be divided into four agro-ecological zones, which include the Loess Plateau area of northern Shaanxi, the Weibei dry plateau, the Guanzhong irrigated area and the Qin-Ba mountain area of southern Shaanxi; the previous three regions belong to the Loess Plateau and in this study they are abbreviated as North, Weibei, and Guanzhong, respectively. The Loess Plateau region in China, covers five provinces (including Shaanxi province), stretches over an area of 0.62 million km^2, and consists of typical semiarid and arid areas with rainfed farming [17–18]. Winter wheat is planted in the regions of Weibei and Guanzhong, while summer maize is planted in the Guanzhong region and spring maize in the North and Weibei regions. Main soil types and climatic conditions in the different regions are shown in Table 1.

Data sources

The data from the 1970s to the 1990s was extracted from 380 published papers reporting household fertilization and soil fertility in the study area; the screening process and results are shown in Figure 2. Data from the 2000s was collected from the project "soil testing and formulated fertilization in Shaanxi province during the years 2005 to 2009."

Statistics

The data were analyzed by EXCEL software. In this study, we used the following equations to analyze the soil nutrient balance and partial factor productivity (PFP) of fertilizer:

$$\textit{Soil nutrient balance} = \textit{nutrient input rate} - \textit{nutrient output rate} \quad \text{(Eq 1)}$$

where the nutrient input rate represents chemical fertilizer input, and the nutrient output rate represents amounts extracted in crop products and above ground biomass;

$$PFP = Y/F \quad \text{(Eq 2)}$$

where Y represents crop yields, and F represents chemical fertilizer input.

Results

Spatial and temporal variations of chemical fertilization and yields of wheat and maize at the farmers' level in different regions of Shaanxi province

The average chemical fertilizer NPK inputs for both wheat and maize at the farmers' level increased for decades in the different regions (Figure 3). In the Weibei and Guanzhong regions, chemical fertilizer N inputs for wheat in the 1970s were 45 kg ha^{-1} and 52 kg ha^{-1}, respectively, and in the 2000s they increased to 185 kg ha^{-1} and 195 kg ha^{-1}, respectively. In these two regions, chemical fertilizer P_2O_5 inputs were 45 kg ha^{-1} and 46 kg ha^{-1} in the 1970s and they increased to 112 kg ha^{-1} and 115 kg ha^{-1} in the 2000s. In the 1980s, farmers started to use the chemical fertilizer K_2O for wheat, which was increased from 0.5 kg ha^{-1} and 2.3 kg ha^{-1} to 22.8 kg ha^{-1} and 22.5 kg ha^{-1}, respectively, during the 1980s to the 2000s in the two regions. For maize in the North, Weibei and Guanzhong regions, chemical fertilizer N inputs were 48 kg ha^{-1}, 89 kg ha^{-1} and 36 kg ha^{-1} and they increased to 237 kg ha^{-1}, 223 kg ha^{-1} and 244 kg ha^{-1}, respectively, from the 1970s to the 2000s. Unlike wheat, from the 1980s onward farmers were awarded for using the chemical fertilizers P_2O_5 and K_2O for maize, and their use has increased greatly.

In accordance with increased chemical fertilizer NPK inputs (Figure 3), the average yields of wheat and maize showed increasing trends in the different regions over the four decades (Figure 4). In the Weibei and Guanzhong regions, from the 1970s to the 2000s, yields of wheat changed from 1883 kg ha^{-1} and 3377 kg ha^{-1} to 4269 kg ha^{-1} and 6437 kg ha^{-1}, with increase rates of 127% and 91%, respectively. In the North, Weibei and Guanzhong regions, yields of maize changed from 3636 kg ha^{-1}, 2519 kg ha^{-1} and 4232 kg ha^{-1} to 7867 kg ha^{-1}, 7077 kg ha^{-1} and 6886 kg ha^{-1}, with increase rates of 116%, 181% and 63%, respectively, for the same period.

Spatial and temporal variations of soil nutrient balance from the inputs and uptake on wheat and maize plots in different regions of Shaanxi province

Because the farmers tended not to use organic manure for cereal crops, especially from the 1980s onward, the soil nutrient inputs only include chemical fertilizers, and the nutrient uptakes include those extracted in crop products and above ground biomass. The nutrient balance was calculated as the difference between the average input and uptake (Eq. 1). Other losses, from leakage and gaseous loss, were not included in these calculations. In the 1970s, N was deficient on wheat and maize plots in the different regions (except for maize plots in the Weibei region). Then from the 1980s N was consistently at surplus levels, and it displayed an upward trend with time. In the 2000s, N surpluses on wheat plots were 74 kg ha^{-1} and 29 kg ha^{-1} in the Weibei and Guanzhong regions, respectively; meanwhile N surpluses on maize plots were 64 kg ha^{-1}, 67 kg ha^{-1} and 93 kg ha^{-1} in the North, Weibei and Guanzhong regions, respectively (Figure 5).

In the Weibei and Guanzhong regions, the amount of surplus P_2O_5 on wheat plots increased each year from the 1970s to the 2000s, and surplus amounts increased from 24 kg ha^{-1} and 9 kg ha^{-1} to 65 kg ha^{-1} and 44 kg ha^{-1}, respectively. In the North, Weibei and Guanzhong regions, P_2O_5 was deficient on maize plots in the 1980s; then it gradually reached surplus levels until the 2000s with the increased application of chemical fertilizer phosphorus. The balance of P_2O_5 on maize plots increased from −34 kg ha^{-1}, −7 kg ha^{-1} and −35 kg ha^{-1} to 29 kg ha^{-1}, 28 kg ha^{-1} and −11 kg ha^{-1}, respectively, in the three regions from the 1980s to the 2000s. It is worth noting, that winter wheat and summer maize were in a rotation system in the Guanzhong region, so total P_2O_5 was in surplus in this region in the 2000s and the amount was 33 kg ha^{-1} (Figure 5).

Although farmers have been awarded for using K_2O chemical fertilizer in recent years, the amount used was still small (Figure 3), and it was usually from compound fertilizers. So K_2O deficiency has become more serious (Figure 5). In the 2000s, K_2O deficiency

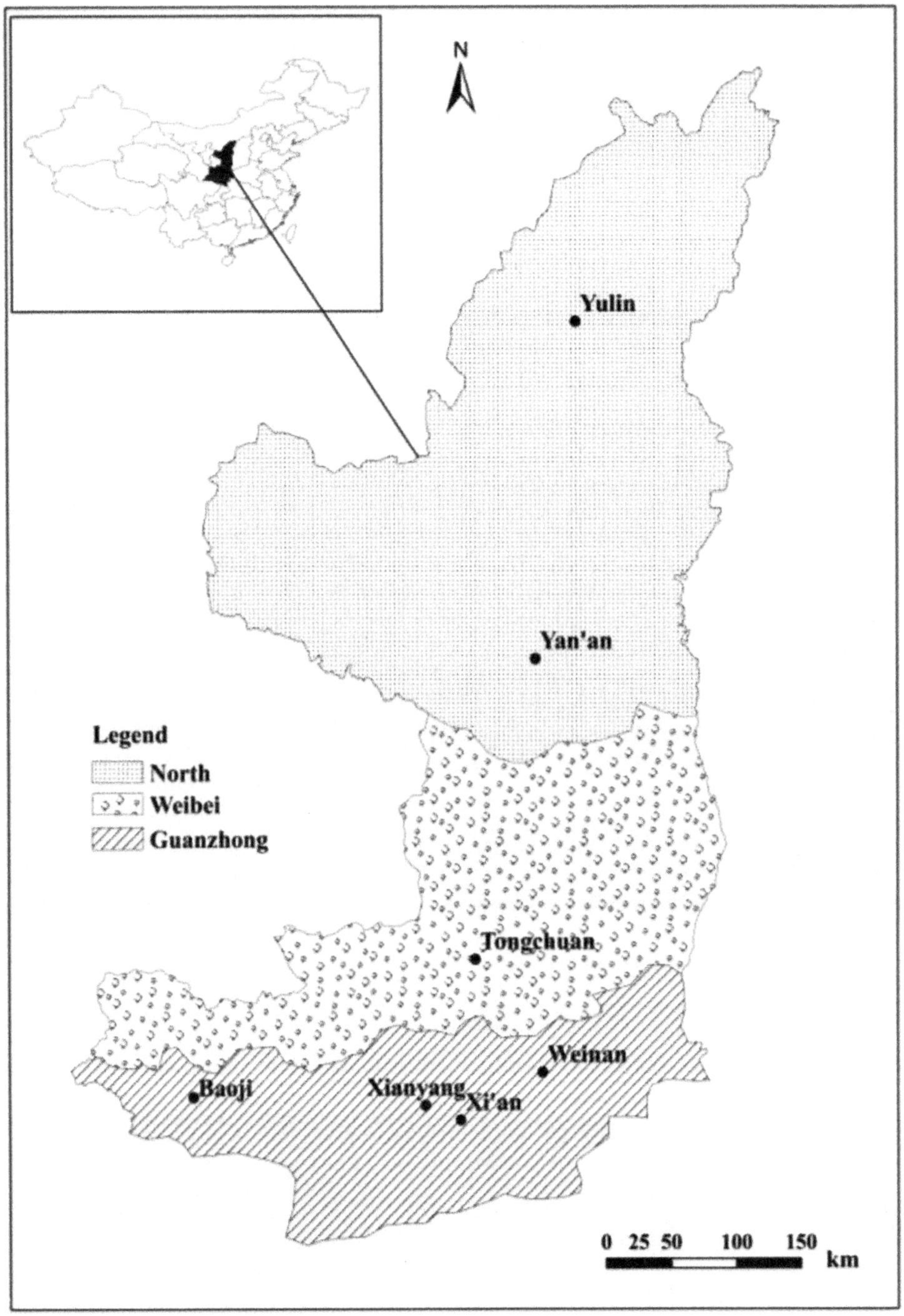

Figure 1. Map of the study area.

Table 1. Main soil types and climatic conditions in the different regions.

Region	Main soil types	Annual mean temperature (°C)	Annual precipitation (mm)
North	Castanozems, Sierozems, Loess soils	8~11	275~590
Weibei	Black loess soils, Loess soils	9~13	530~630
Guanzhong	Cinnamon soils	10~14	600~720

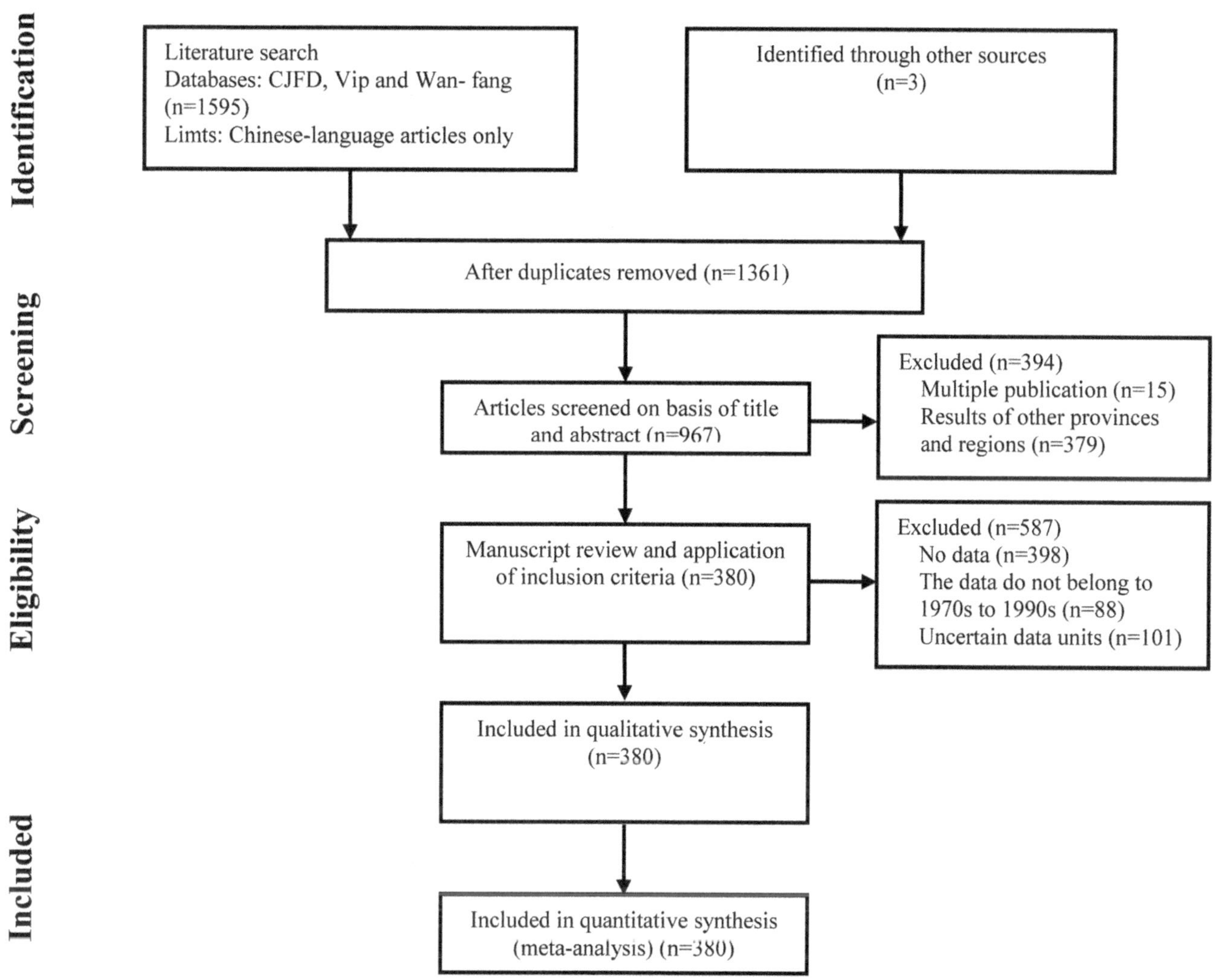

Figure 2. The screening process and results for literature from the 1970s to the 1990s.

levels on wheat plots were −102 kg ha^{-1} and −165 kg ha^{-1} in the Weibei and Guanzhong regions, respectively; meanwhile K_2O deficiency levels on maize plots were −179 kg ha^{-1}, −137 kg ha^{-1} and −147 kg ha^{-1} in the North, Weibei and Guanzhong regions, respectively.

Spatial and temporal variations of soil fertility in different regions of Shaanxi province

In the different regions of Shaanxi province, soil fertility indexes, including organic matter, total nitrogen, alkali-hydrolysis nitrogen, available phosphorus and available potassium, all increased from the 1970s to the 2000s. Simultaneously, each of these five indicators increased from the north to the south during the same period (North<Weibei<Guanzhong) (Figure 6). In the North, Weibei and Guanzhong regions from the 1970s to the 2000s, organic matter varied from 0.57%, 1.01% and 1.12% to 0.83%, 1.26% and 1.50%, with increase rates of 46%, 26% and 43%, respectively; total nitrogen varied from 0.04%, 0.07% and 0.07% to 0.05%, 0.08% and 0.09%, with increase rates of 42%, 9% and 14%, respectively; alkali-hydrolysis nitrogen varied from 29.95 mg kg^{-1}, 20.43 mg kg^{-1} and 30.81 mg kg^{-1} to 35.20 mg kg^{-1}, 58.70 mg kg^{-1} and 68.40 mg kg^{-1}, with increase rates of 18%, 187% and 122%, respectively; available phosphorus varied from 4.98 mg kg^{-1}, 7.13 mg kg^{-1} and 9.90 mg kg^{-1} to 8.10 mg kg^{-1}, 14.60 mg kg^{-1} and 26.40 mg kg^{-1}, with increase rates of 63%, 105% and 167%, respectively; available potassium varied from 85.60 mg kg^{-1}, 56.78 mg kg^{-1} and 111.75 mg kg^{-1} to 99.60 mg kg^{-1}, 160.70 mg kg^{-1} and 170.40 mg kg^{-1}, with increase rates of 16%, 183% and 52%, respectively.

Relationships among fertilizer inputs, crop yields and soil fertility in different regions of Shaanxi province

Because farmers used little P and K fertilizers in the 1970s and 1980s (Figure 3), only PFP of N, NP and NPK were calculated in the study (Eq. 2). The PFP of N, NP and NPK on wheat and maize decreased from the 1970s to the 2000s as a whole in the different regions (Table 2). The PFP of N on wheat in the Weibei and Guanzhong regions were 42 kg kg^{-1} and 65 kg kg^{-1}, respectively, in the 1970s, which decreased to 23 kg kg^{-1} and 33 kg kg^{-1}, respectively, in the 2000s. Meanwhile the PFP of N on maize in the North and Guanzhong regions were 76 kg kg^{-1} and 118 kg kg^{-1}, respectively, and they decreased to 33 kg kg^{-1} and 28 kg kg^{-1} from the 1970s to the 2000s. In the Weibei region, the PFP of N on maize changed slightly from 28 kg kg^{-1} to 32 kg kg^{-1}, which resulted from the use of high N inputs relative to the other two regions (up to 89 kg ha^{-1}) in the 1970s (Figure 3). This led to

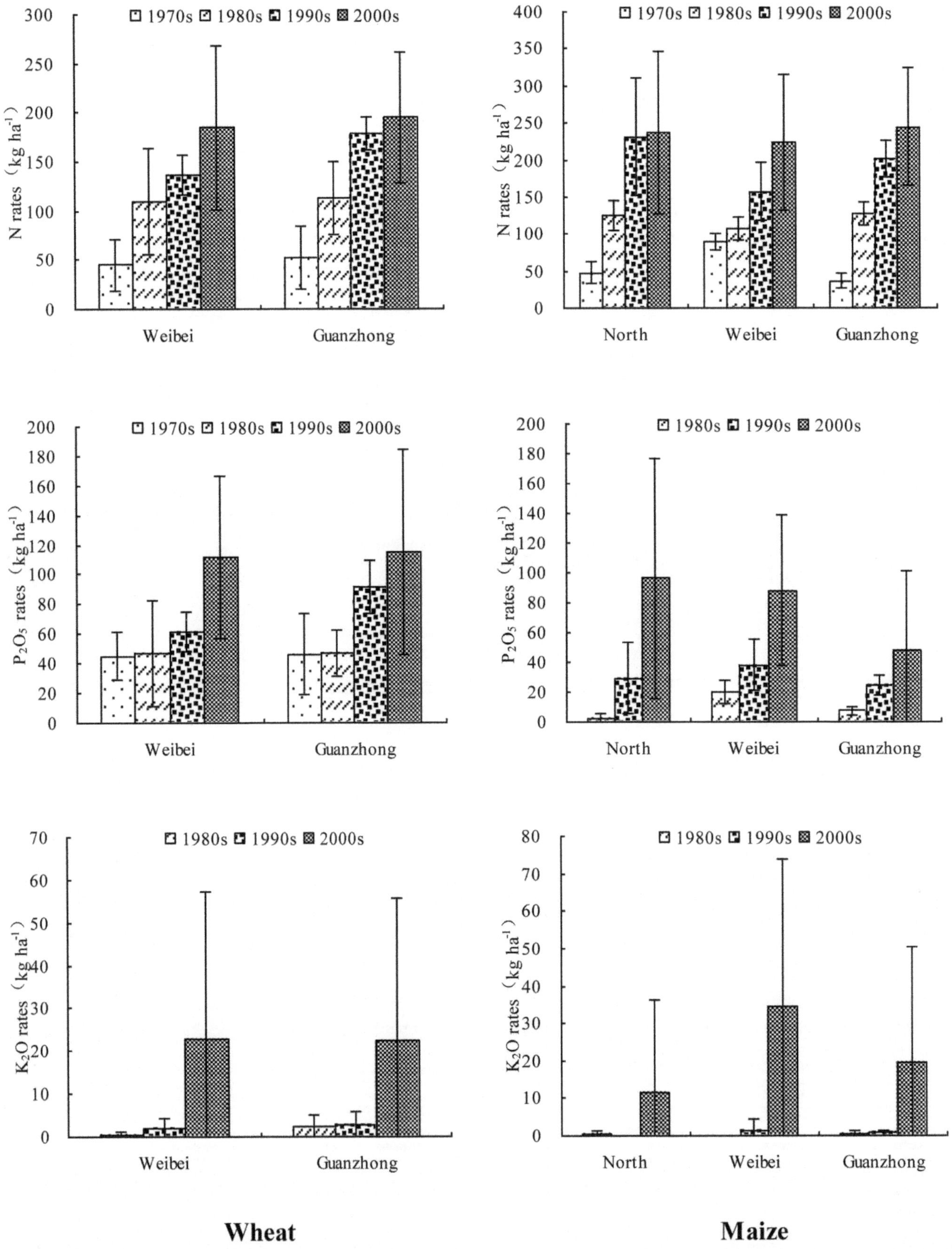

Figure 3. Variations of chemical fertilization for wheat and maize at the farmers' level in different regions of Shaanxi province (error bars show standard deviations).

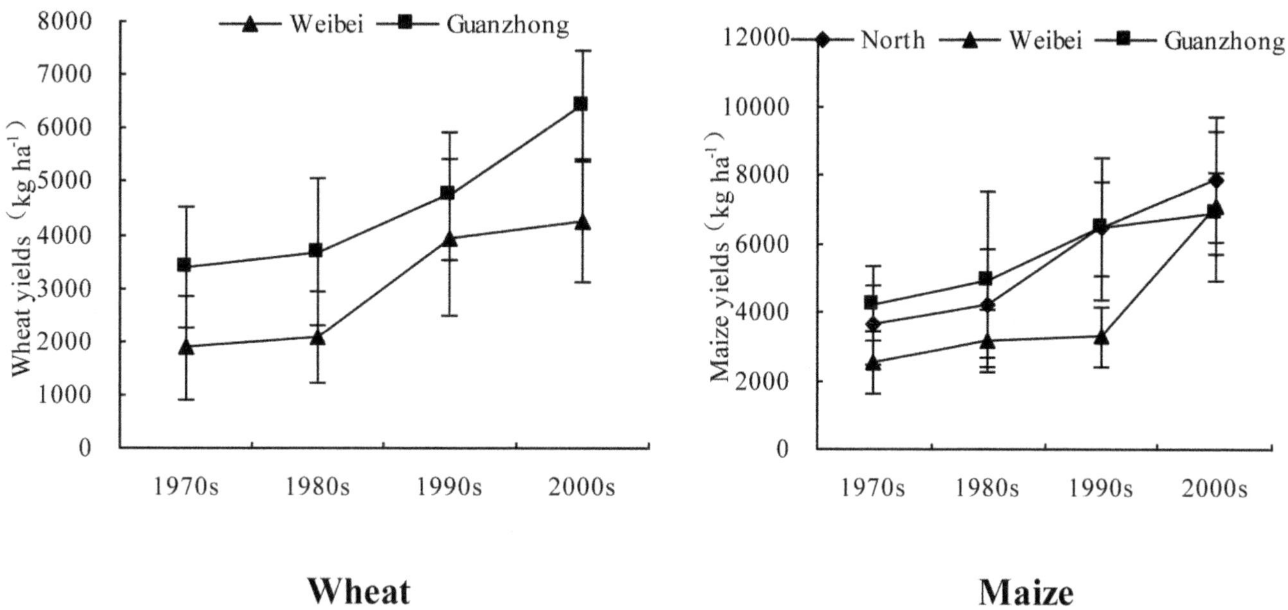

Figure 4. Variations of yields for wheat and maize at the farmers' level in different regions of Shaanxi province (error bars show standard deviations).

a low PFP of N in that period. The PFP of NP on wheat decreased to 14 kg kg^{-1} and 21 kg kg^{-1} in the Weibei and Guanzhong regions, respectively; in maize it decreased to 24 kg kg^{-1}, 23 kg kg^{-1} and 24 kg kg^{-1} in the North, Weibei and Guanzhong regions, respectively. Similar to N and NP, the PFP of NPK on wheat decreased to 13 kg kg^{-1} and 19 kg kg^{-1} in the Weibei and Guanzhong regions, respectively; in maize it decreased to 23 kg kg^{-1}, 20 kg kg^{-1} and 22 kg kg^{-1} in the North, Weibei and Guanzhong regions, respectively (Table 2).

In order to find relationships among soil fertility, crop yields and fertilizer rates, we used the Weibei region as an example. The values of fertilization, crop yields and soil fertility did not have one to one correspondence from the 1970s to the 1990s, so their mean value from each period was examined (Figures 7 and 8). Although the sample size was small and some relationships did not reach significant levels, with the increase in N fertilizer inputs, soil total nitrogen and alkali-hydrolysis nitrogen both increased. P fertilizer increased soil available phosphorus and K fertilizer increased soil available potassium significantly (Figure 7). At the same time, soil organic matter, total nitrogen, alkali-hydrolysis nitrogen, available phosphorus and available potassium all had positive impacts on wheat yields (Figure 8).

Discussion

Fertilizer use efficiency of both wheat and maize decreased from the 1970s to the 2000s as a whole in the Loess Plateau of Shaanxi (Table 2), which was consistent with national trends. Nitrogen fertilizer, phosphorus fertilizer and potassium fertilizer use efficiencies were 30–35%, 15–20% and 35–50%, respectively, from 1981 to 1983, and the average values decreased to 28%, 12% and 32% on cereal crops by 2001 to 2005 in China [19]. This suggested that the effect of chemical fertilizers on increasing grain production had diminished. The PFP of N on wheat in the Weibei and Guanzhong regions decreased to 23 kg kg^{-1} and 33 kg kg^{-1}, respectively, and the PFP of N on maize in the North, Weibei and Guanzhong regions were 33 kg kg^{-1}, 32 kg kg^{-1} and 28 kg kg^{-1}, respectively, in the 2000s (Table 2). Zhang et al. [19] reported average PFP values of N for wheat and maize of 43 kg kg^{-1} and 52 kg kg^{-1}, respectively, in China. Dobermann and Cassman [20] reported a global average PFP of N for cereals of 44 kg kg^{-1}. This indicated that nitrogen use efficiency on wheat and maize in the Loess Plateau of Shaanxi was much lower than the current national and global levels. Excessive fertilization has been the main reason for low fertilizer use efficiency in China [19]. In addition, Liu et al. [21] reported that in agro-ecosystems, surplus N increased from 1978 to 2005 throughout the country, and our findings on the Loess Plateau were consistent with this trend. For example, in the 2000s, chemical fertilizer N inputs on maize were 237 kg ha^{-1}, 223 kg ha^{-1} and 244 kg ha^{-1} in the North, Weibei and Guanzhong regions, respectively (Figure 3); meanwhile N surpluses on maize plots were 64 kg ha^{-1}, 67 kg ha^{-1} and 93 kg ha^{-1}, respectively, in the three regions (Figure 5). This indicated that excessive N fertilization was a serious problem in the Loess Plateau, and the same phenomenon has been reported many times in China, for example, in Beijing [16,22], Shandong [1,23–25], and Jiangsu [26–27]. Excessive N fertilization not only wastes resources, but also leads to many serious environmental problems [28–31] including nitrate pollution of groundwater [32–37], eutrophication of surface water [38–39], greenhouse gas emissions and other forms of air pollution [40–42], acid rain [43–46], soil acidification [36,47–50] and so on. On the other hand, a lower fertilization rate does not necessarily reduce crop yields [51]. Many studies have shown that reducing the current N application rates by 30 to 60% could increase N fertilizer efficiency, while still maintaining crop yields and substantially reducing N losses to the environment [31,52–53].

Like nitrogen, phosphate fertilizer inputs (Figure 3), P surpluses (Figure 5) and soil available phosphorus levels (Figure 6) all increased in the last 40 years on the Loess Plateau in Shaanxi. Similar results have been noted in north China and all over the country [25,54]. Yang et al. [55] reported that maintaining soil available phosphorus at a relatively high level requires a P application rate of about 80 kg ha^{-1} yr^{-1} in winter wheat/summer maize rotation systems in the Guanzhong region. Our

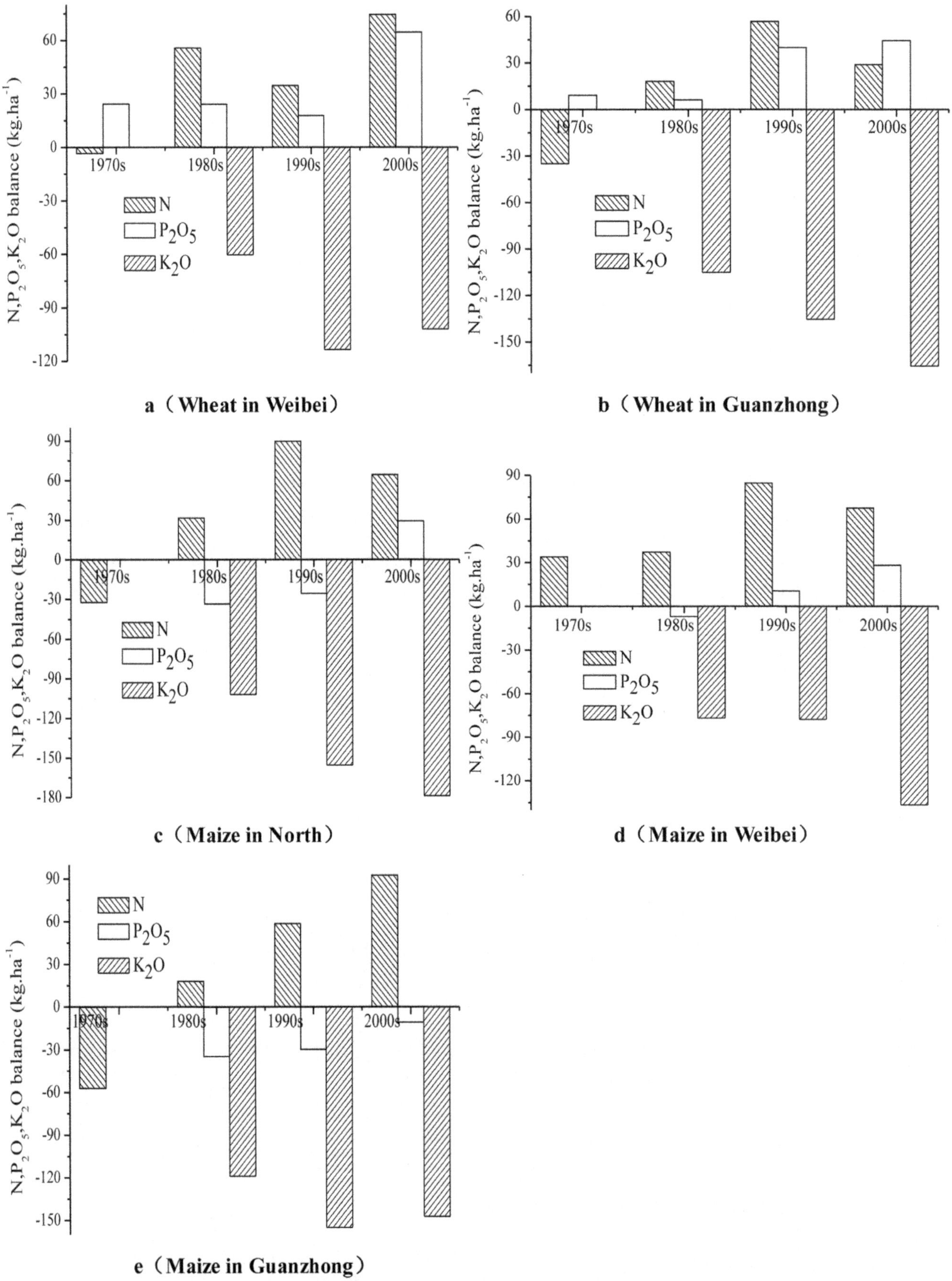

Figure 5. Variations of soil nutrient balance on wheat and maize plots in different regions of Shaanxi province.

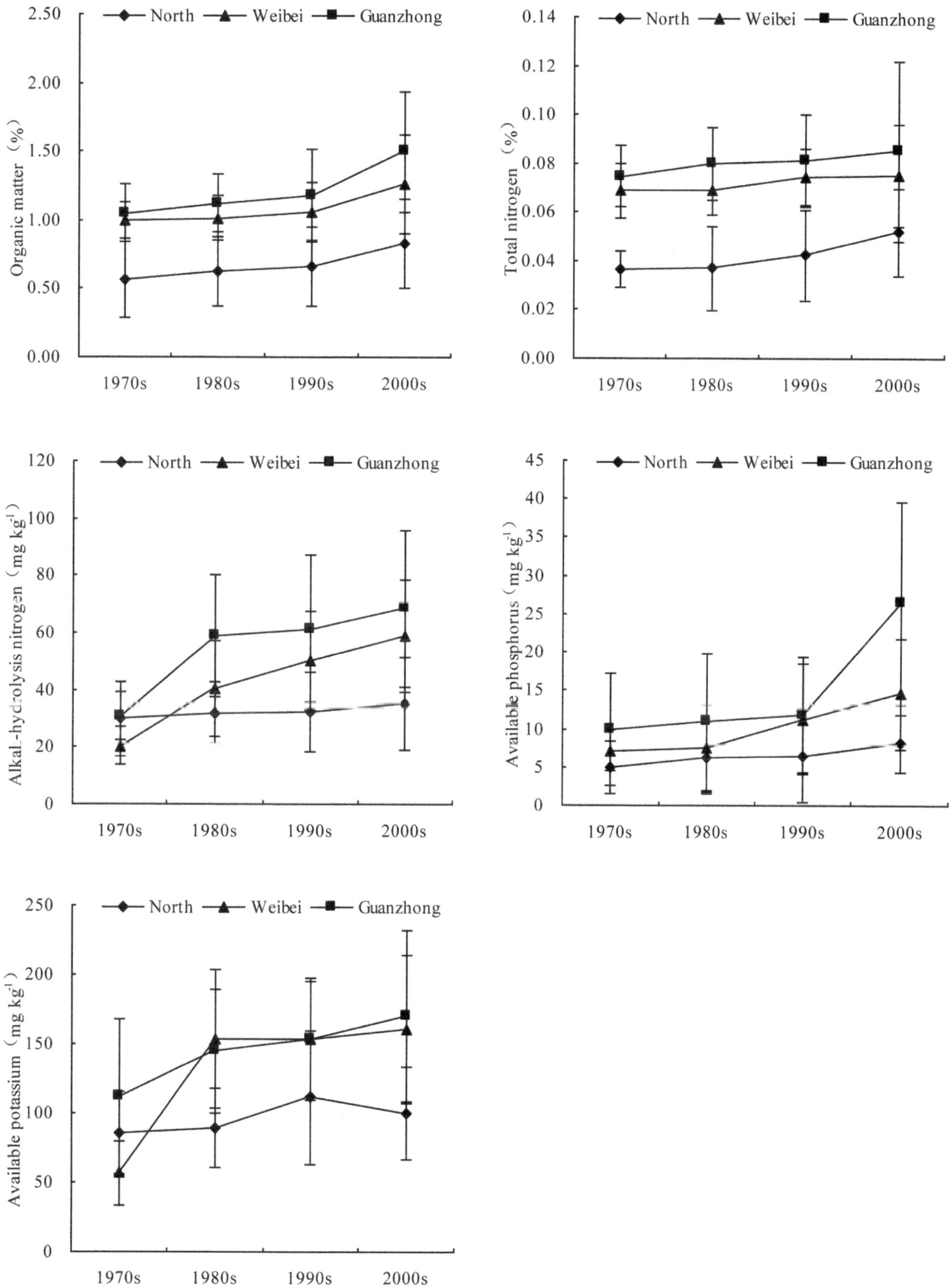
North
Weibei
Guanzhong
Organic matter (%)
2.50
2.00
1.50
1.00
0.50
0.00
1970s
1980s
1990s
2000s
Total nitrogen (%)
0.14
0.12
0.10
0.08
0.06
0.04
0.02
0.00
Alkal-hydrolysis nitrogen (mg kg^{-1})
120
100
80
60
40
20
0
Available phosphorus (mg kg^{-1})
45
40
35
30
25
20
15
10
5
0
Available potassium (mg kg^{-1})
250
200
150
100
50
0

Figure 6. Variations of soil organic matter, total nitrogen, alkali-hydrolysis nitrogen, available phosphorus and available potassium in different regions of Shaanxi province (error bars show standard deviations).

results showed phosphate fertilizer inputs of up to 163 kg ha^{-1} in winter wheat/summer maize rotation systems in this region in the 2000s (Figure 3). This indicated that P fertilization was also excessive, which not only wasted resources but also led to many serious environmental problems [28–31]. Phosphate fertilizer production consumes more than 80% of the phosphate rock resources [56], but phosphate rock resources are limited and high grade material is in short supply [57]. In addition, the phosphate fertilization utilization ratio of the main crops ranged from 7% to 20%. It averages 12% in China [19], which has led to phosphorus accumulation in the soil, increasing the risk of non-point source pollution from surface runoff [58]. Agricultural non-point source pollution has become an increasingly serious problem in China, primarily because it leads to eutrophication.

In spite of increased K fertilizer inputs on wheat and maize in recent years (Figure 3), the soil K balance has become increasingly negative (Figure 5) and soil available potassium has increased (Figure 6) in the last 40 years. This phenomenon was previously reported in northwest and north China [55,59–60]. Evidently, K fertilizer application was not the only source of K absorbed by crops. The primary sources of K for crops were weathering of parent materials [60–61], release of K into the soil from increased soil organic matter and changes in soil pH [61]. Yang et al. [55] found that soil organic matter content in all treatments (including those without fertilizer) significantly increased over time and soil pH dropped from the initial value of 8.65 to 8.58 from 1991 to 2010 during long-term field trials in the Guanzhong region. Our results showed that in the North, Weibei and Guanzhong regions soil organic matter increased from 0.57%, 1.01% and 1.12% to 0.83%, 1.26% and 1.50%, respectively, from the 1970s to the 2000s (Figure 6). The average soil pH has declined 0.5 units with the overuse of N fertilizer in the past two decades in China [62]. Li et al. [63] reported that the soil pH decreased from the initial value of 8.76 to 8.56 from 1992 to 2008 during long-term field trials in the North region. There may be other mechanisms involved, for example, crops might draw on K in the deeper soil layers or from the non-exchangeable pool. The contribution of K from the subsoil could be considerable [64]. Witter and Johansson [65] found that 41–47% of the K was from the subsoil for green manure crops. Many studies have shown that crops use non-exchangeable K [66–67]. Decreases in the abundance of non-exchangeable K with simultaneous increases in exchangeable and water-soluble K concentrations suggest that much of the K taken up by crops comes from non-exchangeable species via solution and exchangeable phases in a way that establishes and maintains the equilibrium between various forms of K in the soil [66].

Fertilizer rates had a large effect on soil fertility. With the increase in N fertilizer inputs, both soil total nitrogen and alkali-hydrolysis nitrogen increased; P fertilizer increased soil available phosphorus and K fertilizer increased soil available potassium significantly in the Weibei region (Figure 7). It has been reported that after 25 years of N fertilization, soil organic carbon and total nitrogen had increased by 18% and 26%, respectively, from 1984 to 2009 in the Weibei region [18]. Cai and Hao [68] also found that accumulation of soil nitrogen initially increased and then decreased with increasing nitrogen, and total nitrogen and alkali-hydrolysis nitrogen content reached the highest value or the second highest value of 135 kg ha^{-1} on wheat plots in the Weibei region, which was in accordance with findings in northwest and north China by Li et al. [63] and Lin et al. [69]. Through long-term field experimentation on the Loess Plateau in Shaanxi, Li et al. [63] and Hao et al. [70] found that with increases in P fertilizer inputs, soil available P increased significantly. Similar results have been obtained in northeast and northwest China by Geng et al. [71] and Zhao et al. [72], and also in America by Griffin et al. [73]. In addition, Li et al. [74] found that with increased K fertilizer inputs, soil available K increased significantly in a long-term field experiment on the Loess Plateau. Further-

Table 2. Variations of PFP of fertilizer on wheat and maize in the different regions (kg kg^{-1}).

Crop	Fertilizer type	Region	1970s	1980s	1990s	2000s
Wheat	N	Weibei	42	19	29	23
		Guanzhong	65	33	26	33
	$N+P_2O_5$	Weibei	21	13	20	14
		Guanzhong	34	23	17	21
	$N+P_2O_5+K_2O$	Weibei	21	13	20	13
		Guanzhong	34	23	17	19
Maize	N	North	76	34	28	33
		Weibei	28	30	21	32
		Guanzhong	118	39	32	28
	$N+P_2O_5$	North	76	33	25	24
		Weibei	28	25	17	23
		Guanzhong	118	37	29	24
	$N+P_2O_5+K_2O$	North	76	33	25	23
		Weibei	28	25	17	20
		Guanzhong	118	37	29	22

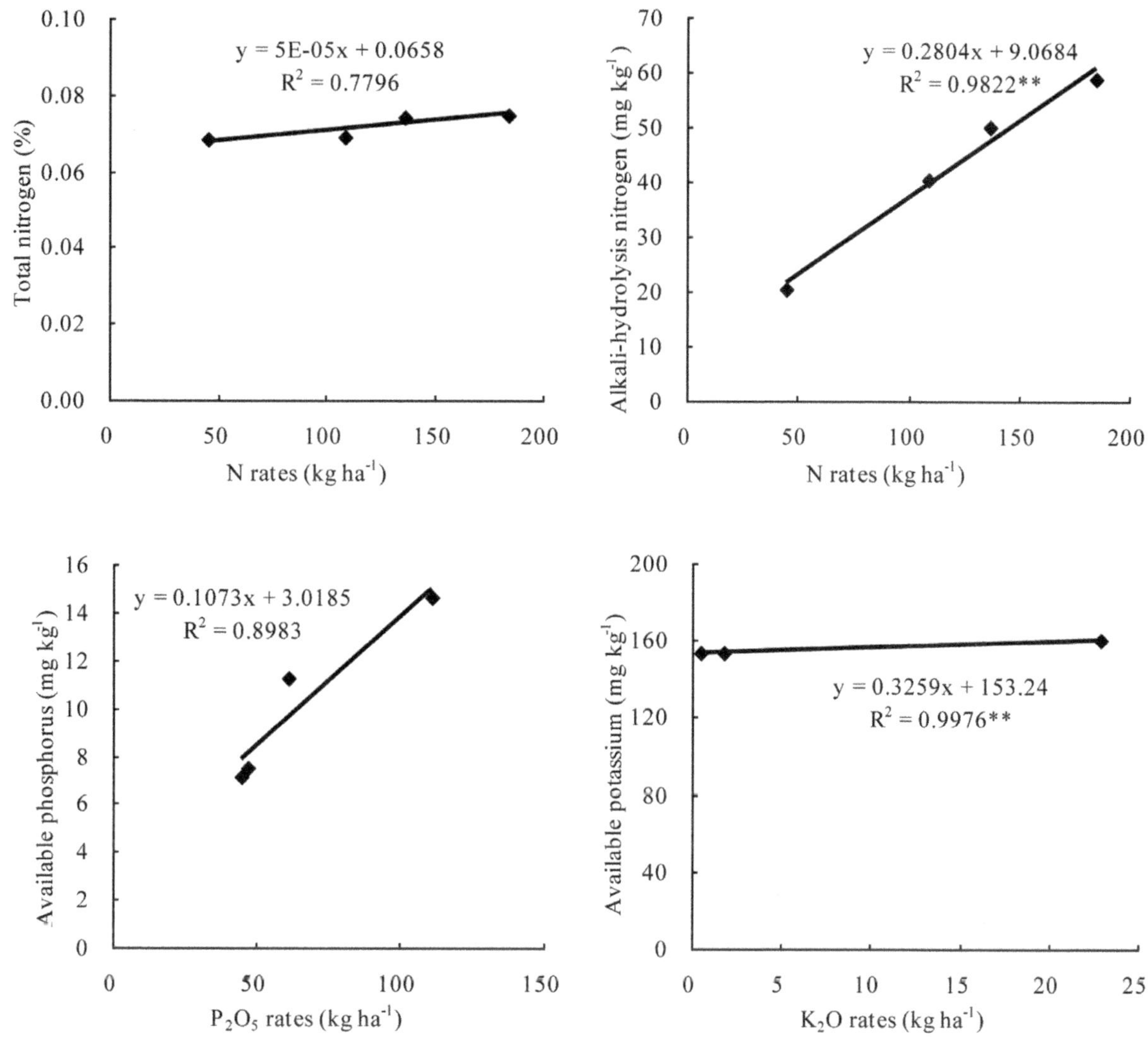

Figure 7. Relationships between N rates and total nitrogen, N rates and alkali-hydrolysis nitrogen, P_2O_5 rates and available phosphorus and K_2O rates and available potassium on wheat plots in the Weibei region of Shaanxi province. **Significance level: P< 0.01.

more, many studies in this area have shown that on the basis of N and P fertilizer application, long-term K fertilizer application can increase soil available K and grain yields [75–76].

Our research also found that soil fertility had a positive impact on crop yields (Figure 8). Zhou et al. [77] revealed that soil organic carbon and total nitrogen concentrations had a significant effect on crop yields in the semi-arid Loess Plateau by long-term experimentation. Higher yields without fertilizer were generally obtained in soils with higher average soil organic matter concentrations. For example, yields without fertilizer <4000 kg ha^{-1} were obtained with average soil organic matter concentrations of 1.41% for winter wheat and 1.46 for summer maize. In contrast, average soil organic matter concentrations were 1.69% for winter wheat and 1.61% for summer maize for plots with yields>6000 kg ha^{-1} without fertilizer in north China [78]. Gong et al. [79] also found that the contribution percentage of basic soil productivity to wheat yield was significantly correlated with soil organic carbon, total nitrogen, available nitrogen, available phosphorus and available potassium in long-term soil fertility experiments in north China. Similar results have been obtained in other parts of mainland China [80], indicating that inherent soil productivity contributed to the substantial increase in China's crop yields.

In addition, although the use of chemical fertilizers to supplement NPK nutrients in the soil is important, many researchers at home and abroad reported that the application of chemical fertilizer in combination with organic manure is helpful in maintaining soil fertility (especially soil organic carbon) and buffering capacity, and in reducing NO_3-N accumulation in the soil, while maintaining high soil productivity [4,81–87].

Conclusions

From the 1970s to the 2000s in the North, Weibei and Guanzhong regions of the Loess Plateau in Shaanxi province, chemical fertilizer NPK inputs and yields of wheat and maize increased at the farmers' level. In the 1970s, N was deficient on wheat and maize plots in the different regions; thereafter N was in surplus. In the same way, P gradually changed from deficit to surplus levels. In addition, soil organic matter, total nitrogen, alkali-hydrolysis nitrogen, available phosphorus and available

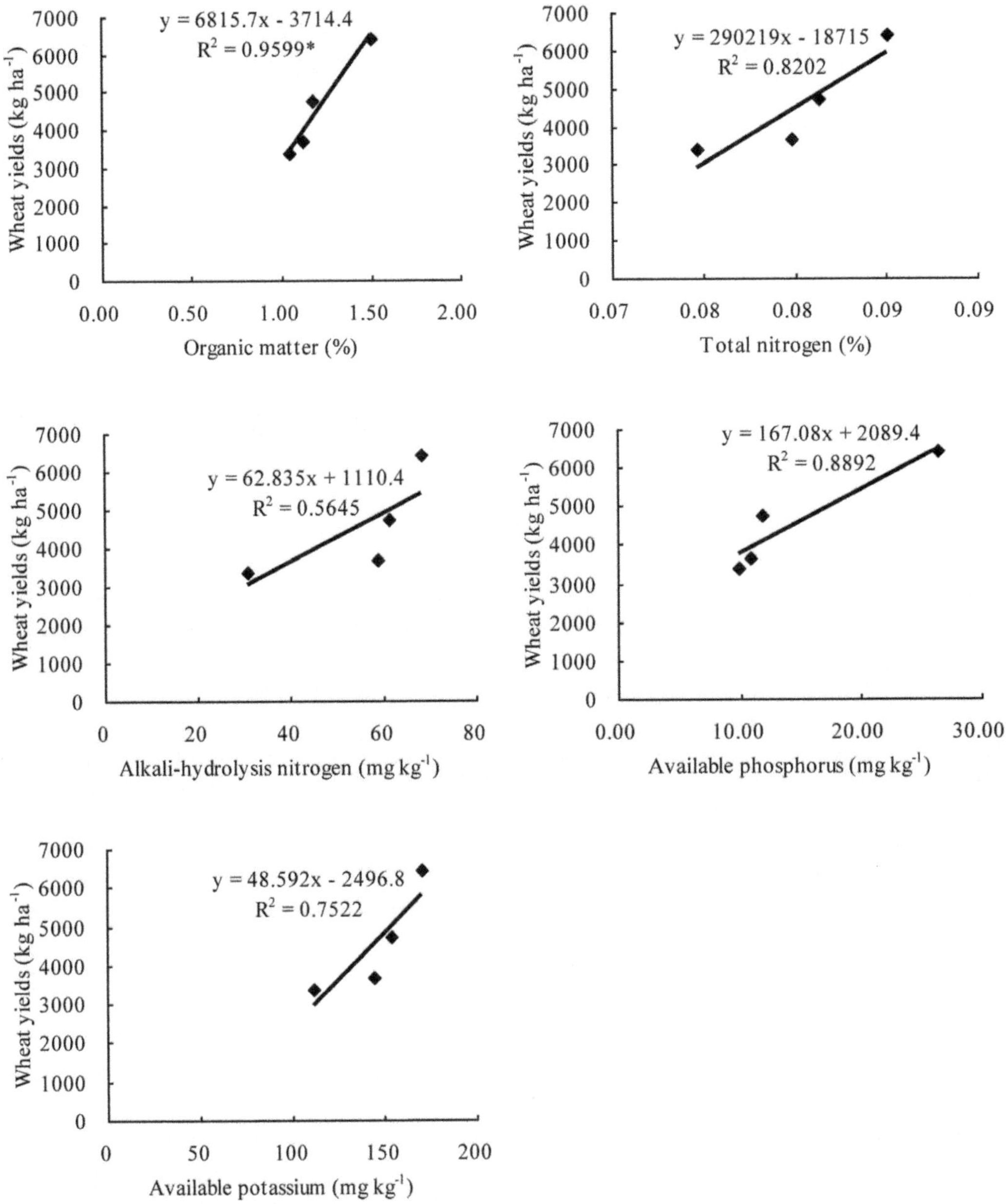

Figure 8. Relationships between wheat yield and soil organic matter, total nitrogen, alkali-hydrolysis nitrogen, available phosphorus and available potassium in the Weibei region of Shaanxi province. *Significance level: $P<0.05$.

potassium increased over the same period. However, K deficiencies became more and more severe. The PFP of N, NP and NPK on wheat and maize all decreased from the 1970s to the 2000s as a whole. With the increase in N fertilizer inputs, both soil total nitrogen and alkali-hydrolysis nitrogen increased; P fertilizer increased soil available phosphorus and K fertilizer increased soil available potassium significantly. At the same time, soil organic matter, total nitrogen, alkali-hydrolysis nitrogen, available phosphorus and available potassium all had positive impacts on crop yields. In order to promote food safety and environmental protection, farmers should be encouraged to assess their fertilizer needs carefully. Many can reduce nitrogen and phosphate fertilizer inputs significantly and increase potassium fertilizer and organic manure on cereal crops.

Acknowledgments

We are grateful to Harald Grip and Lars Lövdahl for their help in writing this paper. The authors would also like to thank the Agricultural Technology Extension Center of Shaanxi province for the help with data collection.

Author Contributions

Conceived and designed the experiments: YT PG. Analyzed the data: XW YG. Contributed reagents/materials/analysis tools: YT YG PG. Wrote the paper: XW. Collected the data: XW FL ZZ YP.

References

1. Cui ZL, Chen XP, Zhang FS (2010) Current nitrogen management status and measures to improve the intensive wheat–maize system in China. AMBIO 39: 376–384.
2. Gao C, Sun B, Zhang TL (2006) Sustainable nutrient management in Chinese agriculture: challenges and perspective. Pedosphere 16(2): 253–263.
3. Zhu ZL, Chen DL (2002) Nitrogen fertilizer use in China-Contributions to food production, impacts on the environment and best management strategies. Nutrient Cycling in Agroecosystems 63: 117–127.
4. Jiang D, Hengsdijk H, Dai TB, de Boer W, Qi J, et al. (2006) Long-term effects of manure and inorganic fertilizers on yield and soil fertility for a winter wheat-maize system in Jiangsu, China. Pedosphere 16(1): 25–32.
5. Luo SM (2007) To discover the secret of traditional agriculture and serve the modern ecoagriculture. Geographical Research 26(3): 609–615. (in Chinese).
6. Zhang F, Qiao Y, Wang F, Zhang W (2007) A perspective on organic agriculture in China: Opportunities and challenges. Proceedings of 9th German Scientific Conference on Organic Agriculture.
7. Department of Rural Surveys, National Bureau of Statistics (1971–2011) China Rural Statistical Yearbook. China Statistics Press. Beijing, China. (in Chinese).
8. Karlen DL, Mausbach MJ, Doran JW, Cline RG, Harris RF, et al. (1997) Soil quality: A concept, definition and framework for evaluation. Soil Science Society of America Journal 61: 4–10.
9. Wang XJ, Gong ZT (1998) Assessment and analysis of soil quality changes after eleven years of reclamation in subtropical China. Geoderma 81: 339–355.
10. Arshad MA, Martin S (2002) Identifying critical limits for soil quality indicators in agro-ecosystems. Agriculture, Ecosystems & Environment 88: 153–160.
11. Huang B, Sun WX, Zhao YC, Zhu J, Yang RQ, et al. (2007) Temporal and spatial variability of soil organic matter and total nitrogen in an agricultural ecosystem as affected by farming practices. Geoderma 139: 336–345.
12. Hoosbeek MR, Bryant RB (1992) Towards the quantitative modeling of pedogenesis: A review. Geoderma 55: 183–210.
13. Smil V (2001) Enriching the Earth: Fritz Haber, Carl Bosch, and the Transformation of World Food Production. MIT Press. Cambridge, UK.
14. Gong W, Yan XY, Wang JY (2011) Effect of long-term fertilization on soil fertility. Soils 43: 336–342. (in Chinese).
15. Bierman PM, Rosen CJ, Venterea RT, Lamb JA (2012) Survey of nitrogen fertilizer use on corn in Minnesota. Agricultural Systems 109: 43–52.
16. Wang SR (2002) Current status and evaluation of crop fertilization in Shaanxi province and Beijing city, Ph.D. thesis, China Agricultural University, Beijing, China. (in Chinese).
17. Liu GB (1999) Soil conservation and sustainable agriculture on the Loess Plateau: Challenges and prospects. AMBIO 28: 663–668.
18. Guo SL, Zhu HH, Dang TH, Wu JS, Liu WZ, et al. (2012) Winter wheat grain yield associated with precipitation distribution under long-term nitrogen fertilization in the semiarid Loess Plateau in China. Geoderma 189: 442–450.
19. Zhang FS, Wang JQ, Zhang WF, Cui ZL, Ma WQ, et al. (2008) Nutrient use efficiencies of major cereal crops in China and measures for improvement. Acta Pedologica Sinica 45: 915–924. (in Chinese).
20. Dobermann A, Cassman KG (2005) Cereal area and nitrogen use efficiency are drivers of future nitrogen fertilizer consumption. Science in China (Series C: Life Sciences) 48: 745–758.
21. Liu Z, Li BG, Fu J (2009) Nitrogen balance in agro-ecosystem in China from 1978 to 2005 based on DSS. Transactions of the CSAE 25(4): 168–175. (in Chinese).
22. Zhao JR, Guo Q, Guo JL, Wei DM, Wang CW, et al. (1997) The chemical fertilizer inputs and yields of grain fields in the suburbs of Beijing. Beijing Agricultural Sciences 15(2): 36–38. (in Chinese).
23. Ma WQ (1999) Current status and evaluation of crop fertilization in Shandong province, Ph.D. thesis, China Agricultural University, Beijing, China. (in Chinese).
24. Li JL, Cui DJ, Meng XX, Li XL, Zhang FS (2002) The study of fertilization condition and question in protectorate vegetable in Shouguang Shandong. Chinese Journal of Soil Science 33: 126–128. (in Chinese).
25. Zhen L, Zoebisch MA, Chen GB, Feng ZM (2006) Sustainability of farmers' soil fertility management practices: A case study in the North China Plain. Journal of Environmental Management 79: 409–419.
26. Richter J, Roelcke M (2000) The N-cycle as determined by intensive agriculture–examples from central Europe and China. Nutrient Cycling in Agroecosystems 57: 33–46.
27. Ma LH, Zhang Y, Sui B, Liu CL, Wang P, et al. (2011) The impact factors of excessive fertilization in Jiangsu province. Journal of Yangzhou University 32(2): 48–52, 80. (in Chinese).
28. Gao XZ, Ma WQ, Du S, Zhang FS, Mao DR (2001) Current status and problems of fertilization in China. Chinese Journal of Soil Science 32(6): 258–261. (in Chinese).
29. Cui ZL, Chen XP, Miao YX, Zhang FS, Sun QP, et al. (2008) On-farm evaluation of the improved soil N_{min}-based nitrogen management for summer maize in North China Plain. Agronomy Journal 100: 517–525.
30. Cui ZL, Zhang FS, Chen XP, Miao YX, Li JL, et al. (2008) On-farm evaluation of an in-season nitrogen management strategy based on soil N_{min} test. Field Crops Research 105: 48–55.
31. Ju XT, Xing GX, Chen XP, Zhang SL, Zhang LJ, et al. (2009) Reducing environmental risk by improving N management in intensive Chinese agricultural systems. Proceedings of the National Academy of Sciences 106: 3041–3046.
32. Tong YA, Emteryd O, Lu DQ, Grip H (1997) Effect of organic manure and chemical fertilizer on nitrogen uptake and nitrate leaching in a Eum-orthic anthrosols profile. Nutrient Cycling in Agroecosystems 48: 225–229.
33. Ju XT, Kou CL, Zhang FS, Christie P (2006) Nitrogen balance and groundwater nitrate contamination: Comparison among three intensive cropping systems on the North China Plain. Environmental Pollution 143: 117–125.
34. Ju XT, Liu XJ, Zhang FS, Roelcke M (2004) Nitrogen fertilization, soil nitrate accumulation, and policy recommendations in several agricultural regions of China. AMBIO 33: 300–305.
35. Yan X, Jin JY, He P, Liang MZ (2008) Recent advances on the technologies to increase fertilizer use efficiency. Agricultural Sciences in China 7(4): 469–479.
36. Guo SL, Wu JS, Dang TH, Liu WZ, Li Y, et al. (2010) Impacts of fertilizer practices on environmental risk of nitrate in semiarid farmlands in the Loess Plateau of China. Plant and Soil 330: 1–13.
37. Gao Y, Yu G, Luo C, Zhou P (2012) Groundwater nitrogen pollution and assessment of its health risks: A case study of a typical village in rural-urban continuum, China. PloS ONE 7(4): e33982.
38. Tilman D, Fargione J, Wolff B, D'Antonio C, Dobson A, et al. (2001) Forecasting agriculturally driven global environmental change. Science 292: 281–284.
39. Huang GQ, Wang XX, Qian HY, Zhang TL, Zhao QG (2004) Negative impact of inorganic fertilizer application on agricultural environment and its countermeasures. Ecology and Environment 13(4): 656–660. (in Chinese).
40. Mosier AR, Duxbury JM, Freney JR, Heinemeyer O, Minami K (1996) Nitrous oxide emissions from agricultural fields: Assessment, measurement and mitigation. Plant and Soil 181: 95–108.
41. Zhang JF, Han XG (2008) N_2O emission from the semi-arid ecosystem under mineral fertilizer (urea and superphosphate) and increased precipitation in northern China. Atmospheric Environment 42: 291–302.
42. Li H, Qiu JJ, Wang LG, Tang HJ, Li CS, et al. (2010) Modelling impacts of alternative farming management practices on greenhouse gas emissions from a winter wheat-maize rotation system in China. Agriculture, Ecosystems and Environment 135: 24–33.
43. Krusche AV, de Camargo PB, Cerri CE, Ballester MV, Lara LBLS, et al. (2003) Acid rain and nitrogen deposition in a sub-tropical watershed (Piracicaba): ecosystem consequences. Environmental Pollution 121: 389–399.
44. Menz FC, Seip HM (2004) Acid rain in Europe and the United States: an update. Environmental Science & Policy 7: 253–265.
45. Wu D, Wang SG, Shang KZ (2006) Progress in research of acid rain in China. Arid Meteorology 24(2): 70–77. (in Chinese).
46. Huang DY, Xu YG, Peng PA, Zhang HH, Lan JB (2009) Chemical composition and seasonal variation of acid deposition in Guangzhou, South China: Comparison with precipitation in other major Chinese cities. Environmental Pollution 157: 35–41.
47. Dai ZH, Liu YX, Wang XJ, Zhao DW (1998) Changes in pH, CEC, and exchangeable acidity of some forest soils in southern China during the last 32–35 years. Water, Air, and Soil Pollution 108: 377–390.
48. Zhang HM, Wang BR, Xu MG, Fan TL (2009) Crop yield and soil responses to long-term fertilization on a red soil in southern China. Pedosphere 19(2): 199–207.
49. Zhao X, Xing GX (2009) Variation in the relationship between nitrification and acidification of subtropical soils as affected by the addition of urea or ammonium sulfate. Soil Biology & Biochemistry 41: 2584–2587.
50. Huang S, Zhang WJ, Yu XC, Huang QR (2010) Effects of long-term fertilization on corn productivity and its sustainability in an Ultisol of southern China. Agriculture, Ecosystems and Environment 138: 44–50.
51. Ma WQ, István S (2008) Can sharp decrease of fertilizer input lead obvious reduction of crop yield?. Ecology and Environment 17: 1296–1301. (in Chinese).
52. Peng SB, Buresh RJ, Huang JL, Yang JC, Zou YB, et al. (2006) Strategies for overcoming low agronomic nitrogen use efficiency in irrigated rice systems in China. Field Crops Research 96: 37–47.
53. Yi Q, Zhang XZ, He P, Yang L, Xiong GY (2010) Effects of reducing N application on crop N uptake, utilization, and soil N balance in rice-wheat rotation system. Plant Nutrition and Fertilizer Science 16: 1069–1077. (in Chinese).
54. Cao N, Zhang YB, Chen XP (2009) Spatial-temporal change of phosphorus balance and the driving factors for agroecosystems in China. Chinese Agricultural Science Bulletin 25: 220–225. (in Chinese).
55. Yang XY, Sun BH, Zhang SL (2014) Trends of yield and soil fertility in a long-term wheat-maize system. Journal of Integrative Agriculture 13: 402–414.
56. Zhang WX (2011) Development and utilization trend of phosphate resources in China. Journal of Wuhan Institute of Technology 33: 1–5. (in Chinese).
57. Zhang WF, Ma WQ, Zhang FS, Ma J (2005) Comparative analysis of the superiority of China's phosphate rock and development strategies with that of the United States and Morocco. Journal of Natural Resources 20: 378–386. (in Chinese).

58. van Bochove E, Thériault G, Dechmi F, Leclerc ML, Goussard N (2007) Indicator of risk of water contamination by phosphorus: Temporal trends for the Province of Quebec from 1981 to 2001. Canadian Journal of Soil Science 87: 121–128.
59. Liu EK, Yan CR, Mei XR, He WQ, Bing SH, et al. (2010) Long-term effect of chemical fertilizer, straw, and manure on soil chemical and biological properties in northwest China. Geoderma 158: 173–180.
60. Tan DS, Jin J Y, Jiang LH, Huang SW, Liu ZH (2012) Potassium assessment of grain producing soils in North China. Agriculture, Ecosystems and Environment 148: 65–71.
61. Munson RD (1985) Potassium in Agriculture. Soil Science Society of America Madison, Wisconsin, USA.
62. Guo JH, Liu XJ, Zhang Y, Shen JL, Han WX, et al. (2010) Significant acidification in major Chinese croplands. Science 327: 1008–1010.
63. Li Q, Xu MX, Liu GB, ZhaoYG, Tuo DF (2013) Cumulative effects of a 17-year chemical fertilization on the soil quality of cropping system in the Loess Hilly Region, China. Journal of Plant Nutrition and Soil Science 176: 249–259.
64. Kautz T, Amelung W, Ewert F, Gaiser T, Horn R, et al. (2013) Nutrient acquisition from arable subsoils in temperate climates: A review. Soil Biology & Biochemistry 57: 1003–1022.
65. Witter E, Johansson G (2001) Potassium uptake from the subsoil by green manure crops. Biological Agriculture & Horticulture 19: 127–141.
66. Singh M, Singh VP, Reddy DD (2002) Potassium balance and release kinetics under continuous rice-wheat cropping system in Vertisol. Field Crops Research 77: 81–91.
67. Sharma A, Jalali VK, Arora S (2010) Non-exchangeable potassium release and its removal in foot-hill soils of North-west Himalayas. Catena 82: 112–117.
68. Cai Y, Hao MD (2013) Effects of long-term nitrogen fertilization on wheat in Loess Plateau. Journal of Triticeae Crops 33: 983–987. (in Chinese).
69. Lin ZA, Zhao BQ, Yuan L, Bing-So H (2009) Effects of organic manure and fertilizers long-term located application on soil fertility and crop yield. Scientia Agricultura Sinica 42: 2809–2819. (in Chinese).
70. Hao MD, Fan J, Wei XR, Pen LF, Lu L (2005) Effect of fertilization on soil fertility and wheat yield of dryland in the Loess Plateau. Pedosphere 15(2): 189–195.
71. Geng YH, Cao GJ, Ye Q, Qi QG, Wu P, et al. (2013) Effects of different phosphorus applications on soil available phosphorus, phosphorus absorption and yield of spring maize. Journal of South China Agricultural University 34: 470–474. (in Chinese).
72. Zhao J, Hou ZA, Li SX, Liu LP, Huang T, et al. (2014) Effects of P rate on soil available P, yield and nutrient uptake of maize. Journal of Maize Sciences 22: 123–128. (in Chinese).
73. Griffin TS, Honeycutt CW, He Z (2003) Changes in soil phosphorus from manure application. Soil Science Society of America Journal 67: 645–653.
74. Li LF, Hao MD, Li YM, Gao CQ (2009) Research on characteristics of spatial distribution and availability of soil potassium forms under long-term fertilization in the dryland of the Loess Plateau. Agricultural Research in the Arid Areas 27: 127–131, 142. (in Chinese).
75. Wang HT, Jin JY, Wang B, Zhao PP (2010) Effects of long-term potassium application and wheat straw return to cinnamon soil on wheat yields and soil potassium balance in Shanxi. Plant Nutrition and Fertilizer Science 16: 801–808. (in Chinese).
76. Zhang YL, Lu JL, Jin JY, Li ST, Chen ZQ, et al. (2012) Effects of chemical fertilizer and straw return on soil fertility and spring wheat quality. Plant Nutrition and Fertilizer Science 18: 307–314. (in Chinese).
77. Zhou ZC, Gan ZT, Shangguan ZP, Zhang FP (2013) Effects of long-term repeated mineral and organic fertilizer applications on soil organic carbon and total nitrogen in a semi-arid cropland. European Journal of Agronomy 45: 20–26.
78. Fan MS, Lai R, Cao J, Qiao L, Su YS, et al. (2013) Plant-based assessment of inherent soil productivity and contributions to China's cereal crop yield increase since 1980. PloS ONE, 8(9): e74617.
79. Gong FF, Zha Y, Wu XP, Huang SM, Xu MG, et al. (2013) Analysis on basic soil productivity change of winter wheat in fluvo-aquic soil under long-term fertilization. Transactions of the Chinese Society of Agricultural Engineering 29(12): 120–129. (in Chinese).
80. Tang YH, Huang Y (2009) Spatial distribution characteristics of the percentage of soil fertility contribution and its associated basic crop yield in mainland China. Journal of Agro-Environment Science 28: 1070–1078. (in Chinese).
81. Gami SK, Ladha JK, Pathak H, Shah MP, Pasuquin E, et al. (2001) Long-term changes in yield and soil fertility in a twenty-year rice-wheat experiment in Nepal. Biology and Fertility of Soils 34: 73–78.
82. Yang SM, Li FM, Malhi SS, Wang P, Suo DR, et al. (2004) Long-term fertilization effects on crop yield and nitrate nitrogen accumulation in soil in northwestern China. Agronomy Journal 96: 1039–1049.
83. Mando A, Ouattara B, Somado AE, Wopereis MCS, Stroosnijder L, et al. (2005) Long-term effects of fallow, tillage and manure application on soil organic matter and nitrogen fractions and on sorghum yield under Sudano-Sahelian conditions. Soil Use and Management 21: 25–31.
84. Li J, Zhao BQ, Li XY, Jiang RB, Bing SH (2008) Effects of long-term combined application of organic and mineral fertilizers on microbial biomass, soil enzyme activities and soil fertility. Agricultural Sciences in China 7(3): 336–343.
85. Banger K, Kukal SS, Toor G, Sudhir K, Hanumanthraju TH (2009) Impact of long-term additions of chemical fertilizers and farmyard manure on carbon and nitrogen sequestration under rice–cowpea cropping system in semi-arid tropics. Plant and Soil 318: 27–35.
86. Majumder B, Mandal B, Bandyopadhyay PK (2008) Soil organic carbon pools and productivity in relation to nutrient management in a 20-year-old rice-berseem agroecosystem. Biology and Fertility of Soils 44: 451–561.
87. Moharana PC, Sharma BM, Biswas DR, Dwivedi BS, Singh RV (2012) Long-term effect of nutrient management on soil fertility and soil organic carbon pools under a 6-year-old pearl millet–wheat cropping system in an Inceptisol of subtropical India. Field Crops Research 136: 32–41.

Permissions

All chapters in this book were first published in PLOS ONE, by The Public Library of Science; hereby published with permission under the Creative Commons Attribution License or equivalent. Every chapter published in this book has been scrutinized by our experts. Their significance has been extensively debated. The topics covered herein carry significant findings which will fuel the growth of the discipline. They may even be implemented as practical applications or may be referred to as a beginning point for another development.

The contributors of this book come from diverse backgrounds, making this book a truly international effort. This book will bring forth new frontiers with its revolutionizing research information and detailed analysis of the nascent developments around the world.

We would like to thank all the contributing authors for lending their expertise to make the book truly unique. They have played a crucial role in the development of this book. Without their invaluable contributions this book wouldn't have been possible. They have made vital efforts to compile up to date information on the varied aspects of this subject to make this book a valuable addition to the collection of many professionals and students.

This book was conceptualized with the vision of imparting up-to-date information and advanced data in this field. To ensure the same, a matchless editorial board was set up. Every individual on the board went through rigorous rounds of assessment to prove their worth. After which they invested a large part of their time researching and compiling the most relevant data for our readers.

The editorial board has been involved in producing this book since its inception. They have spent rigorous hours researching and exploring the diverse topics which have resulted in the successful publishing of this book. They have passed on their knowledge of decades through this book. To expedite this challenging task, the publisher supported the team at every step. A small team of assistant editors was also appointed to further simplify the editing procedure and attain best results for the readers.

Apart from the editorial board, the designing team has also invested a significant amount of their time in understanding the subject and creating the most relevant covers. They scrutinized every image to scout for the most suitable representation of the subject and create an appropriate cover for the book.

The publishing team has been an ardent support to the editorial, designing and production team. Their endless efforts to recruit the best for this project, has resulted in the accomplishment of this book. They are a veteran in the field of academics and their pool of knowledge is as vast as their experience in printing. Their expertise and guidance has proved useful at every step. Their uncompromising quality standards have made this book an exceptional effort. Their encouragement from time to time has been an inspiration for everyone.

The publisher and the editorial board hope that this book will prove to be a valuable piece of knowledge for researchers, students, practitioners and scholars across the globe.

List of Contributors

Jorge Omar López-Martínez, Lucía Sanaphre-Villanueva, Juan Manuel Dupuy and José Luis Hernández-Stefanoni
Centro de Investigación Científica de Yucatán, Mérida, México

Jorge Arturo Meave and José Alberto Gallardo-Cruz
Departamento de Ecología y Recursos Naturales, Facultad de Ciencias, Universidad Nacional Autónoma de México, México, México

Wei Huang and Honghua Ruan
Faculty of Forest Resources and Environmental Science, Nanjing Forestry University, Nanjing, Jiangsu, China

William H. McDowell
Department of Natural Resources and the Environment, University of New Hampshire, Durham, New Hampshire, United States of America

Xiaoming Zou
Faculty of Forest Resources and Environmental Science, Nanjing Forestry University, Nanjing, Jiangsu, China
Institute for Tropical Ecosystem Studies, University of Puerto Rico, San Juan, Puerto Rico, United States of America

Jiashe Wang
Administrative Bureau of Wuyishan National Nature Reserve, Wuyishan, Fujian, China

Liguang Li
Faculty of Forest Resources and Environmental Science, Nanjing Forestry University, Nanjing, Jiangsu, China
Soil and Water Science Department, University of Florida, Gainesville, Florida, United States of America

Letícia Bianca Pereira, Renato Vicentini and Laura M. M. Ottoboni
Center for Molecular Biology and Genetic Engineering (CBMEG), State University of Campinas - UNICAMP, Campinas, SP, Brazil

Wenming He
Key Laboratory of Aquatic Botany and Watershed Ecology, Wuhan Botanical Garden, Chinese Academy of Sciences, Moshan, Wuchang, Wuhan, Hubei Province, China
Graduate University of Chinese Academy of Sciences, Beijing, China

Fang Chen
Key Laboratory of Aquatic Botany and Watershed Ecology, Wuhan Botanical Garden, Chinese Academy of Sciences, Moshan, Wuchang, Wuhan, Hubei Province, China
International Plant Nutrition Institute, Wuhan, China

Rebecca T. Barnes, Morgan E. Gallagher, Caroline A. Masiello, Zuolin Liu and Brandon Dugan
Department of Earth Science, Rice University, Houston, Texas, United States of America

Sara Palacio
Instituto Pirenaico de Ecología (IPE-CSIC), Jaca, Huesca, Spain

Matt Aitkenhead and A. H. Jean Robertson
The James Hutton Institute, Aberdeen, Scotland, United Kingdom

Adrián Escudero
Biodiversity and Conservation Group, E.S.C.E.T., Universidad Rey Juan Carlos, Móstoles, Madrid, Spain

Gabriel Montserrat-Martí and Melchor Maestro
Instituto Pirenaico de Ecologı´a (IPE-CSIC), Zaragoza, Spain

Zhen Zhen
State Key Laboratory of Vegetation and Environmental Change, Institute of Botany, the Chinese Academy of Sciences, Beijing, China
Laboratory of Crop, Guangdong Ocean University, Zhanjiang, China
University of the Chinese Academy of Sciences, Beijing, China

Haitao Liu, Liyue Guo, Jie Meng, Guanglei Wu and Gaoming Jiang
State Key Laboratory of Vegetation and Environmental Change, Institute of Botany, the Chinese Academy of Sciences, Beijing, China

Na Wang
Development and Promotion Center, Shandong Small and Mediumsized Enterprises, Ji'nan, China

Na Ding
State Key Laboratory of Crop Biology, Shandong Agricultural University, Tai'an, China

Ana Luisa Caetano, Catarina R. Marques and Fernando Gonçalves
Department of Biology, University of Aveiro, Campus Universitário de Santiago, Aveiro, Portugal
CESAM, University of Aveiro, Campus Universita´rio de Santiago, Aveiro, Portugal

Ana Gavina
CESAM, University of Aveiro, Campus Universita´rio de Santiago, Aveiro, Portugal
Interdisciplinary Centre of Marine and Environmental Research (CIIMAR/CIMAR), University of Porto, Porto, Portugal

Fernando Carvalho
Nuclear and Technological Institute (ITN) Department of Radiological Protection and Nuclear Safety, Sacavém, Portugal

Eduardo Ferreira da Silva
Department of Geosciences, University of Aveiro, GeoBioTec Research Center, Campus Universitário de Santiago, Aveiro, Portugal

Ruth Pereira
Department of Biology, Faculty of Sciences of the University of Porto, Porto, Portugal
Interdisciplinary Centre of Marine and Environmental Research (CIIMAR/CIMAR), University of Porto, Porto, Portugal

Fan Ding, Guangfu Jiang and Yue Chen
State Key Laboratory of Vegetation and Environmental Change, Institute of Botany, Chinese Academy of Sciences, Beijing, China
University of Chinese Academy of Sciences, Beijing, China

Yao Huang and Wenjuan Sun
State Key Laboratory of Vegetation and Environmental Change, Institute of Botany, Chinese Academy of Sciences, Beijing, China
University of Chinese Academy of Sciences, Beijing, China

Antje Lauer, Joe Daryl Hugo Baal, Kayla Casimiro and Natasha Shroff
Department of Biology, California State University, Bakersfield, California, United States of America

Jorge Talamantes and Luis Jaime Medina
Department of Physics & Engineering, California State University, Bakersfield, California, United States of America

Laura Rosío Castañón Olivares
Laboratorio de Micología Médica, Facultad de Medicina, Universidad Nacional Autónoma de México, Mexico City, Mexico

Kirt W. Emery
County of Kern Public Health Services Department, Bakersfield, California, United States of America

Jean-Baptiste Ramond, Annelize Pienaar, Alacia Armstrong and Don A. Cowan
Center for Microbial Ecology and Genomics (CMEG), Genomic Research Institute, University of Pretoria, Pretoria, South Africa

Mary Seely
Gobabeb Research and Training Center (GRTC), Walvis Bay, Namibia

Feifei Zhu
Key Laboratory of Vegetation Restoration and Management of Degraded Ecosystems, South China Botanical Garden, Chinese Academy of Sciences, Guangzhou, China
United Graduate School of Agriculture Science, Tokyo University of Agriculture and Technology, Tokyo, Japan
University of Chinese Academy of Sciences, Beijing, China

Muneoki Yoh
Institute of Agriculture, Tokyo University of Agriculture and Technology, Tokyo, Japan

Frank S. Gilliam
Department of Biological Sciences, Marshall University, Huntington, West Virginia, United States of America

Xiankai Lu and Jiangming Mo
Key Laboratory of Vegetation Restoration and Management of Degraded Ecosystems, South China Botanical Garden, Chinese Academy of Sciences, Guangzhou, China

Andrew Siefert, Jason D. Fridley and Mark E. Ritchie
Department of Biology, Syracuse University, Syracuse, New York, United States of America

Aude Locatelli, Aymé Spor and Alain Hartmann
INRA, UMR1347 Agroécologie, Dijon, France

Claudy Jolivet
INRA, US-1106 InfoSol, BP20619, Orléans, France

Pascal Piveteau
INRA, UMR1347 Agroécologie, Dijon, France
INRA, US-1106 InfoSol, BP20619, Orléans, France
Université de Bourgogne, UMR1347 Agroécologie, Dijon, France

Hua Yang
Institute of Agricultural Resources and Regional Planning, Chinese Academy of Agricultural Sciences, Key Laboratory of Plant Nutrition and Fertilizer, Ministry of Agriculture, Beijing, China
Guizhou Key Laboratory of Mountain Environment, Guizhou Normal University, Guiyang, China

Zhaojun Li and Yongchao Liang
Institute of Agricultural Resources and Regional Planning, Chinese Academy of Agricultural Sciences, Key Laboratory of Plant Nutrition and Fertilizer, Ministry of Agriculture, Beijing, China

Lu Lu
Graduate School of the Chinese Academy of Agricultural Sciences, Beijing, China

Jian Long
Guizhou Key Laboratory of Mountain Environment, Guizhou Normal University, Guiyang, China

Zhiyong Li
College of Agriculture, Henan University of Science and Technology, Luoyang, China
Institute of Forest Ecology, Environment and Protection, Chinese Academy of Forestry, Beijing, China

Yanhui Wang and Zhen-Hua Li
Institute of Forest Ecology, Environment and Protection, Chinese Academy of Forestry, Beijing, China

Yuan Liu
University of Chinese Academy of Sciences, Beijing, China

Hao Guo
Institute of Desertification Studies, Chinese Academy of Forestry, Beijing, China

Guoan Shi
College of Agriculture, Henan University of Science and Technology, Luoyang, China

Tao Li
Hebei University, Baoding, China

Xiaoqin Dai, Zhu Ouyang, Yunsheng Li and Huimin Wang
Key Laboratory of Ecosystem Network Observation and Modeling, Institute of Geographic Sciences and Natural Resources Research, Chinese Academy of Sciences, Beijing, China

Zongqiang Wei
School of Geographic and Oceanographic Science, Nanjing University, Nanjing, China
School of Environmental and Land Resource Management, Jiangxi Agricultural University, Nanchang, China

Shaohua Wu and Shenglu Zhou
School of Geographic and Oceanographic Science, Nanjing University, Nanjing, China

Xiao Yan
School of Environmental and Land Resource Management, Jiangxi Agricultural University, Nanchang, China

Daryl Moorhead
Department of Environmental Sciences, University of Toledo, Toledo, Ohio, United States of America
INRA, UMR614 Fractionnement des AgroRessources et Environnement, Reims, France
Université Reims-Champagne Ardenne, UMR614 Fractionnement des AgroRessources et Environnement, Reims, France

Gwenaëlle Lashermes, Sylvie Recous and Isabelle Bertrand
INRA, UMR614 Fractionnement des AgroRessources et Environnement, Reims, France
Université Reims-Champagne Ardenne, UMR614 Fractionnement des AgroRessources et Environnement, Reims, France

Jun Wu and Chunping Li
Key Laboratory of Soil Environment and Pollution Remediation, Institute of Soil Science, Chinese Academy of Sciences, Nanjing, Jiangsu, PR China

Minjie Wu
State Key Laboratory of Soil Sustainable Agriculture, Institute of Soil Science, Chinese Academy of Sciences, Nanjing, Jiangsu, PR China
National Engineering Research Center for Organic-based Fertilizers, College of Resources and Environmental Sciences, Nanjing Agricultural University, Nanjing, PR China

Guanghui Yu
Key Laboratory of Soil Environment and Pollution Remediation, Institute of Soil Science, Chinese Academy of Sciences, Nanjing, Jiangsu, PR China
State Key Laboratory of Soil Sustainable Agriculture, Institute of Soil Science, Chinese Academy of Sciences, Nanjing, Jiangsu, PR China
National Engineering Research Center for Organic-based Fertilizers, College of Resources and Environmental Sciences, Nanjing Agricultural University, Nanjing, PR China

Xiaoying Wang and Yanan Tong
College of Natural Resources and Environment, Northwest A&F University, Yangling, China
Key Laboratory of Plant Nutrition and the Agri-environment in Northwest China, Ministry of Agriculture, Yangling, China

Yimin Gao, Pengcheng Gao, Fen Liu, Zuoping Zhao and Yan Pang
College of Natural Resources and Environment, Northwest A&F University, Yangling, China

Index